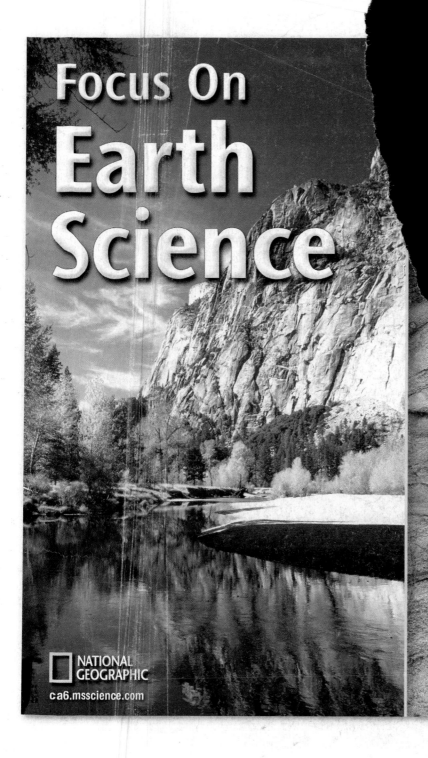

Focus On Earth Science

NATIONAL GEOGRAPHIC

ca6.msscience.com

Glencoe

New York, New York　Columbus, Ohio　Chicago, Illinois　Peoria, Illinois　Woodland Hills, California

Science Online ca6.msscience.com

Check out the following features on your
Online Learning Center:

Study Tools

- Concepts In Motion
  - Interactive Tables
  - Interactive Time Line
  - Animated Illustrations
- Lesson Self-Check Quizzes
- Chapter Test Practice
- Standardized Test Practice
- Vocabulary PuzzleMaker
- Interactive Tutor
- Multilingual Science Glossary
- Study to Go
- Online Student Edition
- BrainPop Movies

Extensions

- Virtual Labs
- Microscopy Links
- Periodic Table Links
- Career Links
- Prescreened Web Links
- WebQuest Project
- Science Fair Ideas
- Internet Labs

For Teachers

- Teacher Bulletin Board
- Teaching Today, and much more!

...ice

...ntains lakes,
...lleys, and polished
... are evidence of the
...d the Sierra Nevada.

Glencoe

The McGraw·Hill Companies

Send all inquiries to:
Glencoe/McGraw-Hill
8787 Orion Place
Columbus, OH 43240-4027

ISBN-13: 978-0-07-879428-5
ISBN-10: 0-07-879428-5

Printed in the United States of America.

3 4 5 6 7 8 9 10 079/043 11 10 09 08 07

Contents in Brief

Teacher Advisory Board

The California Science Teacher Advisory Board provided valuable input in the development of the 2007 edition of *Focus On Earth Science.* They helped create the scope and sequence of the Student Edition, provided content and pedagogical comments, and provided feedback for the Teacher Wraparound Edition.

Charles Beecroft
8th Grade Science
   Teacher
Columbia School District
Redding, CA

Douglas Fisher
Director of Professional
   Development
City Heights Educational
   Collaborative
San Diego, CA

Patricia Juárez
Coordinator III
Sacramento City Unified
   School District
Sacramento, CA

Tom Castro
Science Teacher
Martinez JHS/
   Martinez USD
Martinez, CA

Mindi Fisher
Leadership Team
   Administrator
Peninsula Union School
   District
Samoa, CA

Kathy Molnar
Professional
   Development Mentor
Etiwanda School District
Etiwanda, CA

Lisa L. Cordes
Science Department
   Chair
Rivera Middle School/
   El Rancho USD
Pico Rivera, CA

Frederick W. Freking
Faculty Advisor
University of California,
   Los Angeles
Los Angeles, CA

Carol Orton
Teacher
Bernardo Heights
   Middle School
San Diego, CA

Justin Cunningham EdD
Coordinator, Small
   School District Services
San Diego, County Office
   of Education
San Diego, CA

Nancy Frey
Associate Professor of
   Literacy
San Diego State
   University
San Diego, CA

Joycalyn Peoples
Science Specialist
Riverside Unified School
   District
Riverside, CA

Richard Filson
Science Department
   Chair
Edison High School,
   Stockton Unified
   School District
Stockton, CA

Maria C. Grant
Teacher
Hoover High School/
   San Diego City School
   and San Diego State
   University
San Diego, CA

Wendi L. Rodriguez
Teacher
Heritage/Snowline JUSD
Phelan, CA

Bruce Fisher
Distinguished Teacher
   in Residence
Humboldt State
   University
Arcata, CA

Patrick Horton
Science Teacher
Day Creek Intermediate
   School
Etiwanda, CA

Gladys Sorensen
Science Department
   Chair
Patrick Henry Middle
   School
Grenada Hills, CA

Patty Horton
Professional
   Development Provider
Etiwanda School District
Etiwanda, CA

Granger B. Ward
California
   Superintendent and
   Former Science Teacher
San Diego, CA

Acknowledgements

Authors

Science Online Learn more about the authors at ca6.msscience.com.

Juli Berwald, PhD
Science Writer
Austin, TX

Sergio A. Guazzotti, PhD
Teaching Faculty
Department of Chemistry and Biochemistry
UC, San Diego
La Jolla, CA

Douglas Fisher, PhD
Director of Professional Development and Professor, City Heights Educational Collaborative, San Diego State University
San Diego, CA

Joseph J. Kerski, PhD
Geographer
University of Denver
Denver, CO

Elizabeth A. Nagy-Shadman, PhD
Assistant Professor
Department of Geological Sciences
California State University, Northridge
Northridge, CA

Donna L. Ross, PhD
Associate Professor of Science Education
San Diego State University
San Diego, CA

Julie Meyer Sheets, PhD
Science Writer
Columbus, OH

Nancy Trautmann, PhD
Director of Environmental Inquiry Program
Cornell University
Ithaca, NY

Jan Vermilye, PhD
Assistant Professor
Whittier College
Whittier, CA

Dinah Zike, MEd
Educational Consultant
Dinah-Might Activities, Inc.
San Antonio, TX

NATIONAL GEOGRAPHIC

National Geographic
Education Division
Washington, D.C.

Series Consultants

Content consultants reviewed the chapters in their area of expertise and provided suggestions for improving the effectiveness of the science instruction.

Science Consultants

Richard Allen, PhD
University of California, Berkeley
Berkeley, CA

Karamjeet Arya, PhD
San Jose State University
San Jose, CA

Teaster Baird, PhD
San Francisco State University
San Francisco, CA

Natalie Batalha, PhD
San Jose State University
San Jose, CA

Robin Bennett, MS
University of Washington
Seattle, WA

William B. N. Berry, PhD
University of California, Berkeley
Berkeley, CA

Diane Clayton, PhD
NASA
Santa Barbara, CA

Susan Crawford, PhD
California State University
Sacramento, CA

Stephen F. Cunha, PhD
Humboldt State University
Arcata, CA

Jennifer A. Dever, PhD
University of San Francisco
San Francisco, CA

Alejandro Garcia, PhD
San Jose State University
San Jose, CA

Alan Gishlick, PhD
National Center for Science Education
Oakland, CA

Juno Hsu, PhD
University of California, Irvine
Irvine, CA

Martha Jagucki, MS
Geologist
Columbus, OH

Lee Kats, PhD
Pepperdine University
Malibu, CA

Christopher Kim, PhD
Chapman University
Orange, CA

Monika Kress, PhD
San Jose State University
San Jose, CA

Steve Lund, PhD
University of Southern California
Los Angeles, CA

Michael Manga, PhD
University of California, Berkeley
Berkeley, CA

Kate Schafer, PhD
Aquamarine Research
Mountain View, CA

Julio G. Soto, PhD
San Jose State University
San Jose, CA

Acknowledgements

Dr. Edward Walton
California Polytechnical
  Institute
Pomona, CA

VivianLee Ward
National Health Museum
Washington, DC

Math Consultant

Grant Fraser, PhD
California State
  University
Los Angeles, CA

Reading Consultant

ReLeah Cossett Lent
Author/Educational
  Consultant
Alford, FL

Safety Consultant

Jeff Vogt, MEd
Federal Hocking Middle
  School
Stewart, OH

Series Teacher Reviewers

Each Teacher Reviewer reviewed at least two chapters, providing feedback and suggestions for improving the effectiveness of the science instruction.

Joel Austin
Roosevelt Middle School
San Francisco, CA

Nicole Belong
Coronado Middle School
Coronado, CA

Patrick Brickey
Lakeview Junior High School
Santa Maria, CA

Mary Pilles Bryant
Henry J. Kaiser High School
Fontana, CA

Edward Case
Washington Academic Middle
  School
Sanger, CA

Monaliza Chian
E. O. Green Junior High School
Oxnard, CA

Valesca Lopez Dwyer
Park View Middle School
Yucaipa, CA

Kathryn Froman
North Davis Elementary School
Davis, CA

Brian Gary
Margaret Landell Elementary
Cypress, CA

Jeanette George-Becker
Roosevelt Elementary School
San Gabriel, CA

Bret Harrison
Frank Ledesma Elementary
Soledad, CA

Rick Hoffman
Kastner Intermediate School
Fresno, CA

Kimberly Klein
Barstow Intermediate School
Barstow, CA

David Kulka
South Peninsula Hebrew Day
  School
Sunnyvale, CA

Christina Lambie
Highland Elementary School
Richmond, CA

Kathleen Magnani
Center Junior High School
Antelope, CA

Tara McGuigan
Monroe Clark Middle School
San Diego, CA

Shelia Patterson
K–12 Alliance-California
Oceano, CA

Sharon Pendola
St. Albans Country Day School
Roseville, CA

Lori Poublon-Ramirez
Herman Intermediate School
San Jose, CA

Martha Romero
E. O. Green Junior High School
Oxnard, CA

Arlene Sackman
Earlimart Middle School
Earlimart, CA

Rex Scates
Herman Intermediate School
San Jose, CA

Robert Sherriff
Winston Churchill Middle School
Carmichael, CA

Maria Mendez Simpson
School Programs Coordinator/
  Birch Aquarium
La Jolla, CA

Lorre Stange
Laytonville Elementary School
Laytonville, CA

Louann Talbert
Laytonville Middle School
Laytonville, CA

Gina Marie Turcketta
St. Joan of Arc School
Los Angeles, CA

Table of Contents

A view of Earth from space

Table of Contents

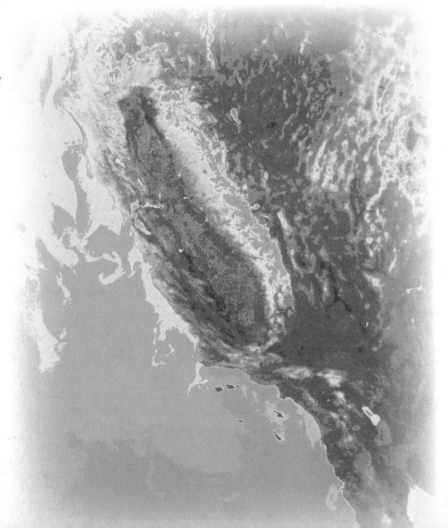

Satellite image colored to show temperatures of land and water

Table of Contents

Lassen Peak is an active volcano

Table of Contents

*Wind energy farm
Palm Springs,
California*

Table of Contents

Black-tailed prairie dogs

Table of Contents

Laguna Beach, California

BrainPOP Movies

To view BrainPOP Movies go to **ca6.msscience.com**. The features listed here correlate to their respective chapter's science content.

Brain POP Movies

Real World Science

The features listed here correlate to their respective chapter's science content.

Chapter/ Page	Science & Career	Science & Technology	Science & History	Science & Society
1 *66–67*	You can map Earth's surface.	Finding Your Exact Location	The Importance of Mapping in Early America	GIS as a Tool to Understanding Earth Changes
2 *112–113*	Studying the Earth's Magnetic Field	Ways of Measuring the Earth's Magnetic Field	The History of Geomagnetism	How the Changing Magnetic Field Affects Us
3 *154–155*	Heating and Cooling	Hybrid Cars	From the Caloric Theory to Kinetic Theory	Steam Engines in the Industrial Revolution
4 *198–199*	Paleontologists Validate Wegener in Dinosaur Cove	Defining Plate Boundaries	Behind a Revolutionary Theory	Hole 504B
5 *230–231*	Studying Earth's Plate Boundaries	Finding Faults	A View from Space	A Shaky History
6 *282–283*	Designing Safer Structures	Measuring Major Earthquakes	Earthquake of 1906	Preparing for the Next Tsunami
7 *322–323*	Predicting Volcanic Eruptions	From Blast to Blast	"The Year Without a Summer"	Making the Most of Volcanoes
8 *366–367*	You can build dams!	Preventing Beach Erosion	Fossils at Lyme Regis	Acid Rain
9 *412–413*	Managing Atmospheric Expeditions	Wind Machines	Sailing in the Jet Stream	The Santa Ana Winds
10 *456–457*	Shipboard Instructor	OTEC	Marie Tharp	Overfishing the Oceans
11 *500–501*	Mediterranean Climate	Super Scooper	Chaparral	Fire's Role
12 *540–541*	You can help save wildlife!	Tracking Animals	Prairie Dogs—Friend or Foe?	Recovering Threatened Species
13 *576–577*	You could be an oceanographer!	Tracing Mercury	California's First Oceanographers	The California Sea Otter
14 *618–619*	You could dig up coal, gold, or gravel!	Burning ice as fuel?	Burning Biomass for Fuel	Using Geothermal Energy

Labs

Labs

Launch Labs (continued)

MiniLabs

MiniLabs (continued)

DataLabs

 This lab might be performed at home.

Math & Language Arts

The California Science, Math and Language Arts correlations for these features can be found on the referenced page.

Get Ready to Read

Target Your Reading

Applying Math

A Guide to California Standards

For Students and Their Families

What is the purpose of the California Content Standards?

Content standards were designed to encourage the highest achievement of every student, by defining the knowledge, concepts, and skills that students should acquire at each grade level.

This Guide Contains:

California State Capitol Building, Sacramento

California Science Content Standards

Grade 6 Focus On Earth Science

The science curriculum in grade six emphasizes the study of earth sciences. Students at this age are increasing their awareness of the environment and are ready to learn more. The standards in grade six present many of the foundations of geology and geophysics, including plate tectonics and earth structure, topography, and energy. The material is linked to resource management and ecology, building on what students have learned in previous grades. Unless students take a high school earth science class, what they learn in grade six will be their foundation for earth science literacy. Items within the text that relate to a Science Content Standard will be represented like this: **1.g**

California Science Content Standards

Correlated to *Focus On Earth Science*

Science Content Standards	Page Numbers
Plate Tectonics and Earth's Structure	
1. Plate tectonics accounts for important features of Earth's surface and major geologic events. As a basis for understanding this concept:	
1.a Students know evidence of plate tectonics is derived from the fit of the continents; the location of earthquakes, volcanoes, and midocean ridges; and the distribution of fossils, rock types, and ancient climatic zones.	49–50, 163, **167, 168–171,** **172, 174–180,** 182, **185, 186,** 223, 228–232, 427–428
1.b Students know Earth is composed of several layers: a cold, brittle lithosphere; a hot, convecting mantle; and a dense, metallic core.	**102–105, 110–111, 188,** 228–232
1.c Students know lithospheric plates the size of continents and oceans move at rates of centimeters per year in response to movements in the mantle.	183, **186–187, 189, 191,** 196–197, **218–219**
1.d Students know that earthquakes are sudden motions along breaks in the crust called faults and that volcanoes and fissures are locations where magma reaches the surface.	82, 103, **110–111,** 207, **211,** 214, 243, **246–247,** 291, **295,** 300, **302–306, 308–310, 312,** 320–321

Bold page numbers indicate in-depth coverage of standard.

California Science Content Standards

Science Content Standards	Page Numbers
Plate Tectonics and Earth's Structure (continued)	
1.e Students know major geologic events, such as earthquakes, volcanic eruptions, and mountain building, result from plate motions.	49–50, **78–79, 81–82,** **218–219, 223–225,** 227, 228–229, **249–250,** **296–298, 320–321,** **446–447**
1.f Students know how to explain major features of California geology (including mountains, faults, volcanoes) in terms of plate tectonics.	79, **223–225, 307,** **355–361, 446–447**
1.g Students know how to determine the epicenter of an earthquake and know that the effects of an earthquake on any region vary, depending on the size of the earthquake, the distance of the region from the epicenter, the local geology, and the type of construction in the region.	49–50, **253–255, 256,** **261–266, 268–269, 271,** **276, 279, 280–281**
Shaping Earth's Surface	
2. Topography is reshaped by the weathering of rock and soil and by the transportation and deposition of sediment. As a basis for understanding this concept:	
2.a Students know water running downhill is the dominant process in shaping the landscape, including California's landscape.	**80–81, 83–84,** 99, **334–341**
2.b Students know rivers and streams are dynamic systems that erode, transport sediment, change course, and flood their banks in natural and recurring patterns.	83, 99, **342, 345–348,** **351–353, 356–357,** **359, 362**
2.c Students know beaches are dynamic systems in which the sand is supplied by rivers and moved along the coast by the action of waves.	83, **349–350,** 354, **360,** **438–444**
2.d Students know earthquakes, volcanic eruptions, landslides, and floods change human and wildlife habitats.	**271–272, 275–276,** **280–281, 313–315,** 318, **343–344, 348, 436, 447,** **479–481**

Bold page numbers indicate in-depth coverage of standard.

California Science Content Standards

Science Content Standards	Page Numbers

Heat (Thermal Energy) (Physical Sciences)

3. Heat moves in a predictable flow from warmer objects to cooler objects until all the objects are at the same temperature. As a basis for understanding this concept:

3.a	Students know energy can be carried from one place to another by heat flow or by waves, including water, light and sound waves, or by moving objects.	97–98, 121, **124–126**, 127, **129, 131–134, 142**
3.b	Students know that when fuel is consumed, most of the energy released becomes heat energy.	**135–136**, 137, 564
3.c	Students know heat flows in solids by conduction (which involves no flow of matter) and in fluids by conduction and by convection (which involves flow of matter).	104, 106–107, **145–149**, 152–153, **393–396**
3.d	Students know heat energy is also transferred between objects by radiation (radiation can travel through space).	129, 134, **150**, 151, **397–400**, 410–411

Energy in the Earth System

4. Many phenomena on Earth's surface are affected by the transfer of energy through radiation and convection currents. As a basis for understanding this concept:

4.a	Students know the sun is the major source of energy for phenomena on Earth's surface; it powers winds, ocean currents, and the water cycle.	334–336, **382–390**, **401–403, 431–432**, **472–474, 517–520**, 553–555, 557, 559–562
4.b	Students know solar energy reaches Earth through radiation, mostly in the form of visible light.	**384–387**, 398–399, 410–411
4.c	Students know heat from Earth's interior reaches the surface primarily through convection.	**106–108, 188**, 194
4.d	Students know convection currents distribute heat in the atmosphere and oceans.	**394–396, 402–409**, **430–437, 448–449**, **486–487, 491–495**
4.e	Students know differences in pressure, heat, air movement, and humidity result in changes of weather.	**401–409, 468–474**, **475–482, 483–488**, **490–496**

Bold page numbers indicate in-depth coverage of standard.

California Science Content Standards

Science Content Standards	Page Numbers
Ecology (Life Sciences)	
5. Organisms in ecosystems exchange energy and nutrients among themselves and with the environment. As a basis for understanding this concept:	
5.a Students know energy entering ecosystems as sunlight is transferred by producers into chemical energy through photosynthesis and then from organism to organism through food webs.	**553–567**
5.b Students know matter is transferred over time from one organism to others in the food web and between organisms and the physical environment.	520, **557–672, 574–575**
5.c Students know populations of organisms can be categorized by the functions they serve in an ecosystem.	**552–559**
5.d Students know different kinds of organisms may play similar ecological roles in similar biomes.	**536, 564**
5.e Students know the number and types of organisms an ecosystem can support depends on the resources available and on abiotic factors, such as quantities of light and water, a range of temperatures, and soil composition.	450, **516–522,** 534, **538–539,** 568–571
Resources	
6. Sources of energy and materials differ in amounts, distribution, usefulness, and the time required for their formation. As a basis for understanding this concept:	
6.a Students know the utility of energy sources is determined by factors that are involved in converting these sources to useful forms and the consequences of the conversion process.	389, 412, **595–617**
6.b Students know different natural energy and material resources, including air, soil, rocks, minerals, petroleum, fresh water, wildlife, and forests, and know how to classify them as renewable or nonrenewable.	**132–138, 141–147,** 389, 450–452, **588–517**
6.c Students know the natural origin of the materials used to make common objects.	140, **588–594, 607–608**

Bold page numbers indicate in-depth coverage of standard.

California Science Content Standards

Science Content Standards	Page Numbers
Investigation and Experimentation	
7. Scientific progress is made by asking meaningful questions and conducting careful investigations. As a basis for understanding this concept and addressing the content in the other three strands, students should develop their own questions and perform investigations. Students will:	
7.a Develop a hypothesis.	19, 29, 36, **152, 196–197, 228–229, 280–281,** 320–321, 331, **338, 362, 538–539, 557, 569, 574–575, 616–617**
7.b Select and use appropriate tools and technology (including calculators, computers, balances, spring scales, microscopes, and binoculars) to perform tests, collect data, and display data.	7–17, 31, 38, 64–65, **93, 103, 228–229, 268–269, 279, 300, 318, 364–365, 534, 564, 567, 569, 574–575, 606**
7.c Construct appropriate graphs from data and develop qualitative statements about the relationships between variables.	24–27, 32, **62, 85, 144, 410–411,** 428, 450, **530, 590**
7.d Communicate the steps and results from an investigation in written reports and oral presentations.	5, 22–23, 28–33, 37, **110–111, 280–281, 320–321, 362, 498–499, 538–539, 564, 574–575,** 614, **616–617**
7.e Recognize whether evidence is consistent with a proposed explanation.	5, 6, 19, 20, 21, 33, 36–41, 163, **194, 196–197,** 207, 227, **228–229,** 243, **250, 256,** 291, **362, 567, 606, 608**

Bold page numbers indicate in-depth coverage of standard.

California Science Content Standards

Bold page numbers indicate in-depth coverage of standard.

Joshua Tree National Park, California

California Math Content Standards

Items within the text that relate to a Math Content Standard will be represented like this: **MA6: NS 1.0**

Number Sense

MA6: NS 1.0 Students compare and order positive and negative fractions, decimals, and mixed numbers. Students solve problems involving fractions, ratios, proportions, and percentages:

MA6: NS 1.1 Compare and order positive and negative fractions, decimals, and mixed numbers and place them on a number line.

MA6: NS 1.2 Interpret and use ratios in different contexts (e.g., batting averages, miles per hour) to show the relative sizes of two quantities, using appropriate notations ($\frac{a}{b}$, a to b, a:b).

MA6: NS 1.3 Use proportions to solve problems (e.g., determine the value of N if $\frac{4}{7} = \frac{N}{21}$, find the length of a side of a polygon similar to a known polygon). Use cross-multiplication as a method for solving such problems, understanding it as the multiplication of both sides of an equation by a multiplicative inverse.

MA6: NS 1.4 Calculate given percentages of quantities and solve problems involving discounts at sales, interest earned, and tips.

MA6: NS 2.0 Students calculate and solve problems involving addition, subtraction, multiplication, and division:

MA6: NS 2.1 Solve problems involving addition, subtraction, multiplication, and division of positive fractions and explain why a particular operation was used for a given situation.

MA6: NS 2.2 Explain the meaning of multiplication and division of positive fractions and perform the calculations (e.g., $\frac{5}{8} \div \frac{15}{16} = \frac{5}{8} \times \frac{16}{15} = \frac{2}{3}$).

MA6: NS 2.3 Solve addition, subtraction, multiplication, and division problems, including those arising in concrete situations, that use positive and negative integers and combinations of these operations.

MA6: NS 2.4 Determine the least common multiple and the greatest common divisor of whole numbers; use them to solve problems with fractions (e.g., to find a common denominator to add two fractions or to find the reduced form for a fraction).

Algebra and Functions

MA6: AF 1.0 Students write verbal expressions and sentences as algebraic expressions and equations; they evaluate algebraic expressions, solve simple linear equations, and graph and interpret their results:

MA6: AF 1.1 Write and solve one-step linear equations in one variable.

MA6: AF 1.2 Write and evaluate an algebraic expression for a given situation, using up to three variables.

MA6: AF 1.3 Apply algebraic order of operations and the commutative, associative, and distributive properties to evaluate expressions; and justify each step in the process.

MA6: AF 1.4 Solve problems manually by using the correct order of operations or by using a scientific calculator.

MA6: AF 2.0 Students analyze and use tables, graphs, and rules to solve problems involving rates and proportions:

MA6: AF 2.1 Convert one unit of measurement to another (e.g., from feet to miles, from centimeters to inches).

MA6: AF 2.2 Demonstrate an understanding that *rate* is a measure of one quantity per unit value of another quantity.

MA6: AF 2.3 Solve problems involving rates, average speed, distance, and time.

MA6: AF 3.0 Students investigate geometric patterns and describe them algebraically:

MA6: AF 3.1 Use variables in expressions describing geometric quantities (e.g., $P = 2w + 2l$, $A = \frac{1}{2}bh$, $C = \pi d$ — the formulas for the perimeter of a rectangle, the area of a triangle, and the circumference of a circle, respectively).

MA6: AF 3.2 Express in symbolic form simple relationships arising from geometry.

Measurement and Geometry

MA6: MG 1.0 Students deepen their understanding of the measurement of plane and solid shapes and use this understanding to solve problems:

MA6: MG 1.1 Understand the concept of a constant such as π; know the formulas for the circumference and area of a circle.

MA6: MG 1.2 Know common estimates of π (3.14; $\frac{22}{7}$) and use these values to estimate and calculate the circumference and the area of circles; compare with actual measurements.

MA6: MG 1.3 Know and use the formulas for the volume of triangular prisms and cylinders (area of base $\times$ height); compare these formulas and explain the similarity between them and the formula for the volume of a rectangular solid.

MA6: MG 2.0 Students identify and describe the properties of two-dimensional figures:

MA6: MG 2.1 Identify angles as vertical, adjacent, complementary, or supplementary and provide descriptions of these terms.

MA6: MG 2.2 Use the properties of complementary and supplementary angles and the sum of the angles of a triangle to solve problems involving an unknown angle.

MA6: MG 2.3 Draw quadrilaterals and triangles from given information about them (e.g., a quadrilateral having equal sides but no right angles, a right isosceles triangle).

Statistics, Data Analysis, and Probability

MA6: SP 1.0 Students compute and analyze statistical measurements for data sets:

MA6: SP 1.1 Compute the range, mean, median, and mode of data sets.

MA6: SP 1.2 Understand how additional data added to data sets may affect these computations of measures of central tendency.

MA6: SP 1.3 Understand how the inclusion or exclusion of outliers affects measures of central tendency.

MA6: SP 1.4 Know why a specific measure of central tendency (mean, median, mode) provides the most useful information in a given context.

MA6: SP 2.0 Students use data samples of a population and describe the characteristics and limitations of the samples:

MA6: SP 2.1 Compare different samples of a population with the data from the entire population and identify a situation in which it makes sense to use a sample.

MA6: SP 2.2 Identify different ways of selecting a sample (e.g., convenience sampling, responses to a survey, random sampling) and which method makes a sample more representative for a population.

MA6: SP 2.3 Analyze data displays and explain why the way in which the question was asked might have influenced the results obtained and why the way in which the results were displayed might have influenced the conclusions reached.

MA6: SP 2.4 Identify data that represent sampling errors and explain why the sample (and the display) might be biased.

MA6: SP 2.5 Identify claims based on statistical data and, in simple cases, evaluate the validity of the claims.

MA6: SP 3.0 Students determine theoretical and experimental probabilities and use these to make predictions about events:

MA6: SP 3.1 Represent all possible outcomes for compound events in an organized way (e.g., tables, grids, tree diagrams) and express the theoretical probability of each outcome.

MA6: SP 3.2 Use data to estimate the probability of future events (e.g., batting averages or number of accidents per mile driven).

MA6: SP 3.3 Represent probabilities as ratios, proportions, decimals between 0 and 1, and percentages between 0 and 100 and verify that the probabilities computed are reasonable; know that if P is the probability of an event, $1 - P$ is the probability of an event not occurring.

MA6: SP 3.4 Understand that the probability of either of two disjoint events occurring is the sum of the two individual probabilities and that the probability of one event following another, in independent trials, is the product of the two probabilities.

MA6: SP 3.5 Understand the difference between independent and dependent events.

Mathematical Reasoning

MA6: MR 1.0 Students make decisions about how to approach problems:

MA6: MR 1.1 Analyze problems by identifying relationships, distinguishing relevant from irrelevant information, identifying missing information, sequencing and prioritizing information, and observing patterns.

MA6: MR 1.2 Formulate and justify mathematical conjectures based on a general description of the mathematical question or problem posed.

MA6: MR 1.3 Determine when and how to break a problem into simpler parts.

MA6: MR 2.0 Students use strategies, skills, and concepts in finding solutions:

MA6: MR 2.1 Use estimation to verify the reasonableness of calculated results.

MA6: MR 2.2 Apply strategies and results from simpler problems to more complex problems.

MA6: MR 2.3 Estimate unknown quantities graphically and solve for them by using logical reasoning and arithmetic and algebraic techniques.

MA6: MR 2.4 Use a variety of methods, such as words, numbers, symbols, charts, graphs, tables, diagrams, and models, to explain mathematical reasoning.

MA6: MR 2.5 Express the solution clearly and logically by using the appropriate mathematical notation and terms and clear language; support solutions with evidence in both verbal and symbolic work.

MA6: MR 2.6 Indicate the relative advantages of exact and approximate solutions to problems and give answers to a specified degree of accuracy.

MA6: MR 2.7 Make precise calculations and check the validity of the results from the context of the problem.

MA6: MR 3.0 Students move beyond a particular problem by generalizing to other situations:

MA6: MR 3.1 Evaluate the reasonableness of the solution in the context of the original situation.

MA6: MR 3.2 Note the method of deriving the solution and demonstrate a conceptual understanding of the derivation by solving similar problems.

MA6: MR 3.3 Develop generalizations of the results obtained and the strategies used and apply them in new problem situations.

California English-Language Arts Content Standards

Items within the text that relate to an English-Language Arts Content Standard will be represented like this: **ELA6: R 1.4**

Reading

ELA6: R 1.0   Word Analysis, Fluency, and Systematic Vocabulary Development

Word Recognition

ELA6: R 1.1   Read aloud narrative and expository text fluently and accurately and with appropriate pacing, intonation, and expression.

Vocabulary and Concept Development

ELA6: R 1.2   Identify and interpret figurative language and words with multiple meanings.

ELA6: R 1.3   Recognize the origins and meanings of frequently used foreign words in English and use these words accurately in speaking and writing.

ELA6: R 1.4   Monitor expository text for unknown words or words with novel meanings by using word, sentence, and paragraph clues to determine meaning.

ELA6: R 1.5   Understand and explain "shades of meaning" in related words (e.g., *softly* and *quietly*).

ELA6: R 2.0   Reading Comprehension (Focus on Informational Materials)

Structural Features of Informational Materials

ELA6: R 2.1   Identify the structural features of popular media (e.g., newspapers, magazines, online information) and use the features to obtain information.

ELA6: R 2.2   Analyze text that uses the compare-and-contrast organizational pattern.

Comprehension and Analysis of Grade-Level-Appropriate Text

ELA6: R 2.3   Connect and clarify main ideas by identifying their relationships to other sources and related topics.

ELA6: R 2.4   Clarify an understanding of texts by creating outlines, logical notes, summaries, or reports.

ELA6: R 2.5   Follow multiple-step instructions for preparing applications (e.g., for a public library card, bank savings account, sports club, league membership).

Expository Critique

ELA6: R 2.6   Determine the adequacy and appropriateness of the evidence for an author's conclusions.

ELA6: R 2.7   Make reasonable assertions about a text through accurate, supporting citations.

ELA6: R 2.8   Note instances of unsupported inferences, fallacious reasoning, persuasion, and propaganda in text.

ELA6: R 3.0   Literary Response and Analysis

Structural Features of Literature

ELA6: R 3.1   Identify the forms of fiction and describe the major characteristics of each form.

Narrative Analysis of Grade-Level-Appropriate Text

ELA6: R 3.2   Analyze the effect of the qualities of the character (e.g., courage or cowardice, ambition or laziness) on the plot and the resolution of the conflict.

ELA6: R 3.3   Analyze the influence of setting on the problem and its resolution.

ELA6: R 3.4   Define how tone or meaning is conveyed in poetry through word choice, figurative language, sentence structure, line length, punctuation, rhythm, repetition, and rhyme.

ELA6: R 3.5   Identify the speaker and recognize the difference between first- and third-person narration (e.g., autobiography compared with biography).

ELA6: R 3.6   Identify and analyze features of themes conveyed through characters, actions, and images.

ELA6: R 3.7   Explain the effects of common literary devices (e.g., symbolism, imagery, metaphor) in a variety of fictional and nonfictional texts.

Literary Criticism

ELA6: R 3.8   Critique the credibility of characterization and the degree to which a plot is contrived or realistic (e.g., compare use of fact and fantasy in historical fiction).

Writing

ELA6: W 1.0   Writing Strategies

Organization and Focus

ELA6: W 1.1   Choose the form of writing (e.g., personal letter, letter to the editor, review, poem, report, narrative) that best suits the intended purpose.

ELA6: W 1.2   Create multiple-paragraph expository compositions:
   a. Engage the interest of the reader and state a clear purpose.
   b. Develop the topic with supporting details and precise verbs, nouns, and adjectives to paint a visual image in the mind of the reader.
   c. Conclude with a detailed summary linked to the purpose of the composition.

ELA6: W 1.3   Use a variety of effective and coherent organizational patterns, including comparison and contrast; organization by categories; and arrangement by spatial order, order of importance, or climactic order.

Research and Technology

ELA6: W 1.4   Use organizational features of electronic text (e.g., bulletin boards, databases, keyword searches, e-mail addresses) to locate information.

ELA6: W 1.5   Compose documents with appropriate formatting by using word-processing skills and principles of design (e.g., margins, tabs, spacing, columns, page orientation).

Evaluation and Revision

ELA6: W 1.6   Revise writing to improve the organization and consistency of ideas within and between paragraphs.

ELA6: W 2.0   Writing Applications (Genres and Their Characteristics) Using the writing strategies of grade six outlined in Writing Standard 1.0, students:

ELA6: W 2.1   Write narratives:
   a. Establish and develop a plot and setting and present a point of view that is appropriate to the stories.
   b. Include sensory details and concrete language to develop plot and character.
   c. Use a range of narrative devices (e.g., dialogue, suspense).

California English-Language Arts Content Standards

ELA6: W 2.2 Write expository compositions (e.g., description, explanation, comparison and contrast, problem and solution):

a. State the thesis or purpose.
b. Explain the situation.
c. Follow an organizational pattern appropriate to the type of composition.
d. Offer persuasive evidence to validate arguments and conclusions as needed.

ELA6: W 2.3 Write research reports:

a. Pose relevant questions with a scope narrow enough to be thoroughly covered.
b. Support the main idea or ideas with facts, details, examples, and explanations from multiple authoritative sources (e.g., speakers, periodicals, online information searches).
c. Include a bibliography.

ELA6: W 2.4 Write responses to literature:

a. Develop an interpretation exhibiting careful reading, understanding, and insight.
b. Organize the interpretation around several clear ideas, premises, or images.
c. Develop and justify the interpretation through sustained use of examples and textual evidence.

ELA6: W 2.5 Write persuasive compositions:

a. State a clear position on a proposition or proposal.
b. Support the position with organized and relevant evidence.
c. Anticipate and address reader concerns and counterarguments.

Written and Oral English Language Conventions

ELA6: WO 1.0   Written and Oral English Language Conventions

Sentence Structure

ELA6: WO 1.1 Use simple, compound, and compound-complex sentences; use effective coordination and subordination of ideas to express complete thoughts.

Grammar

ELA6: WO 1.2 Identify and properly use indefinite pronouns and present perfect, past perfect, and future perfect verb tenses; ensure that verbs agree with compound subjects.

Punctuation

ELA6: WO 1.3 Use colons after the salutation in business letters, semicolons to connect independent clauses, and commas when linking two clauses with a conjunction in compound sentences.

Capitalization

ELA6: WO 1.4 Use correct capitalization.

Spelling

ELA6: WO 1.5 Spell frequently misspelled words correctly (e.g., *their, they're, there*).

Listening and Speaking

ELA6: LS 1.0   Listening and Speaking Strategies

Comprehension

ELA6: LS 1.1 Relate the speaker's verbal communication (e.g., word choice, pitch, feeling, tone) to the nonverbal message (e.g., posture, gesture).

ELA6: LS 1.2 Identify the tone, mood, and emotion conveyed in the oral communication.

ELA6: LS 1.3 Restate and execute multiple-step oral instructions and directions.

Organization and Delivery of Oral Communication

ELA6: LS 1.4 Select a focus, an organizational structure, and a point of view, matching the purpose, message, occasion, and vocal modulation to the audience.

ELA6: LS 1.5 Emphasize salient points to assist the listener in following the main ideas and concepts.

ELA6: LS 1.6 Support opinions with detailed evidence and with visual or media displays that use appropriate technology.

ELA6: LS 1.7 Use effective rate, volume, pitch, and tone and align nonverbal elements to sustain audience interest and attention.

Analysis and Evaluation of Oral and Media Communications

ELA6: LS 1.8 Analyze the use of rhetorical devices (e.g., cadence, repetitive patterns, use of onomatopoeia) for intent and effect.

ELA6: LS 1.9 Identify persuasive and propaganda techniques used in television and identify false and misleading information.

ELA6: LS 2.0   Speaking Applications (Genres and Their Characteristics) Using the speaking strategies of grade six outlined in Listening and Speaking Standard 1.0, students:

ELA6: LS 2.1 Deliver narrative presentations:

a. Establish a context, plot, and point of view.
b. Include sensory details and concrete language to develop the plot and character.
c. Use a range of narrative devices (e.g., dialogue, tension, or suspense).

ELA6: LS 2.2 Deliver informative presentations:

a. Pose relevant questions sufficiently limited in scope to be completely and thoroughly answered.
b. Develop the topic with facts, details, examples, and explanations from multiple authoritative sources (e.g., speakers, periodicals, online information).

ELA6: LS 2.3 Deliver oral responses to literature:

a. Develop an interpretation exhibiting careful reading, understanding, and insight.
b. Organize the selected interpretation around several clear ideas, premises, or images.
c. Develop and justify the selected interpretation through sustained use of examples and textual evidence.

ELA6: LS 2.4 Deliver persuasive presentations:

a. Provide a clear statement of the position.
b. Include relevant evidence.
c. Offer a logical sequence of information.
d. Engage the listener and foster acceptance of the proposition or proposal.

ELA6: LS 2.5 Deliver presentations on problems and solutions:

a. Theorize on the causes and effects of each problem and establish connections between the defined problem and at least one solution.
b. Offer persuasive evidence to validate the definition of the problem and the proposed solutions.

Reading for Information

When you read *Focus On Earth Science*, you are reading for information. Science is nonfiction writing—it describes real-life events, people, ideas, and technology. Here are some tools that *Focus On Earth Science* has to help you read.

Before You Read

By reading **(The BIG Idea)** and **(Main Idea)** prior to reading the chapter or lesson, you will get a preview of the coming material.

On the first page of each chapter you will find **(The BIG Idea)**. The Big Idea is a sentence that describes what you will learn about in the chapter.

CHAPTER 6

Earthquakes

(The BIG Idea)
Earthquakes cause seismic waves that can be devastating to humans and other organisms.

LESSON 1 1.d, 1.e, 7.c
Earthquakes and Plate Boundaries
(Main Idea) Most earthquakes occur at plate boundaries when rocks break and move along faults.

LESSON 2 1.g, 7.e
Earthquakes and Seismic Waves
(Main Idea) Earthquakes cause seismic waves that provide valuable data.

LESSON 3 1.g, 7.b, 7.g
Measuring Earthquakes
(Main Idea) Data from seismic waves are recorded and interpreted to determine the location and size of an earthquake.

LESSON 4 1.g, 2.d, 7.a, 7.b, 7.d
Earthquake Hazards and Safety
(Main Idea) Effects of an earthquake depend on its size and the types of structures and geology in a region.

Now how did *that* happen?   On January 17, 1995, at 5:46 A.M., the people of Kobe, Japan, awoke to a major earthquake that toppled buildings, highways, and homes. The Kobe earthquake, also known as the Great Hanshin earthquake, killed 6,433 people and injured 43,792.

Science Journal Have you ever experienced an earthquake? If so, write a paragraph about the event. If not, write how you imagine it would feel to experience an earthquake.

Source: Chapter 6, p. 242

LESSON 2

Science Content Standards
1.g Students know how to determine the epicenter of an earthquake and know that the effects of an earthquake on any region vary, depending on the size of the earthquake, the distance of the region from the epicenter, the local geology, and the type of construction in the region.
7.e Recognize whether evidence is consistent with a proposed explanation.

Reading Guide

What *You'll* Learn
▸ **Explain** how energy released during earthquakes travels in seismic waves.
▸ **Distinguish** among primary, secondary, and surface waves.
▸ **Describe** how seismic waves are used to investigate Earth's interior.

Why *It's* Important
Scientists can locate the epicenter of an earthquake by analyzing seismic waves.

Vocabulary
seismic wave
epicenter
primary wave
secondary wave

Review Vocabulary
wave: a disturbance in a material that transfers energy without transferring matter (p. 132)

Earthquakes and Seismic Waves

(Main Idea) Earthquakes cause seismic waves that provide valuable data.

Real-World Reading Connection If you throw or drop a rock into a pond, you might notice that ripples form in the water. Circles of waves move outward from the place where the rock entered the water. In a similar way, an earthquake generates complex waves that move outward through rock.

What are seismic waves?

During an earthquake, the ground moves forward and backward, heaves up and down, and shifts from side to side. Usually this motion is felt as vibrations, or shaking.

Large earthquakes can cause the ground surface to ripple like the waves shown in **Figure 7.** Imagine trying to stand on Earth's surface if it had waves traveling through it. This is what people and structures experience during a strong earthquake. These waves of energy, produced at the focus of an earthquake, are called **seismic** (SIZE mihk) **waves.**

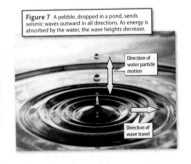

Figure 7 A pebble, dropped in a pond, sends seismic waves outward in all directions. As energy is absorbed by the water, the wave heights decrease.

Direction of water particle motion

Direction of wave travel

252 Chapter 6 • Earthquakes

Source: Chapter 6, Lesson 2, p. 252

(The BIG Idea) is divided into Main Ideas. Each lesson of the chapter has a **(Main Idea)** that describes the focus of the lesson.

Other Ways to Preview

- Read the chapter title to find out what area of science you will study.
- Skim the photo, illustrations, captions, graphs, and tables.
- Look for key terms that are boldfaced and . highlighted.

Reading for Information

The Get Ready to Read section allows you to learn, practice, and apply a reading skill before you start reading the chapter's first lesson. Target Your Reading will help you keep the main idea in focus as you read the chapter.

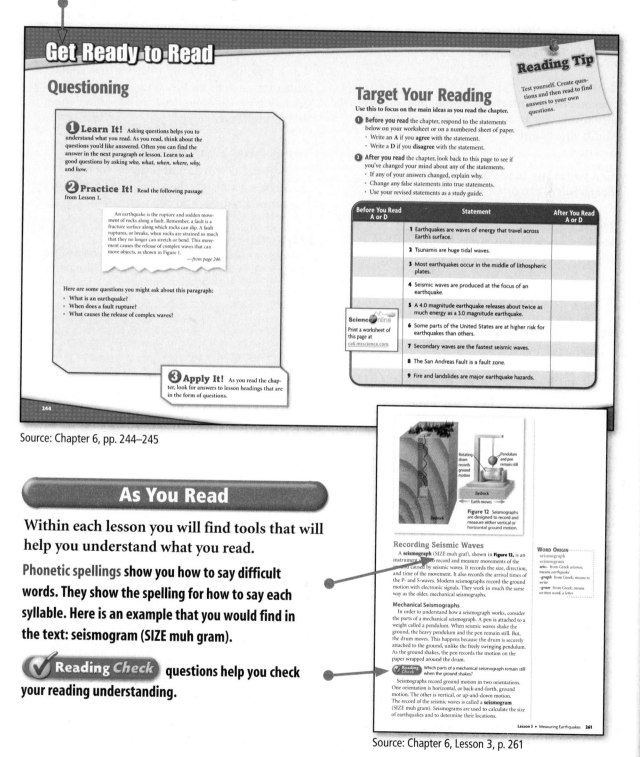

Get Ready to Read

Questioning

1 Learn It! Asking questions helps you to understand what you read. As you read, think about the questions you'd like answered. Often you can find the answer in the next paragraph or lesson. Learn to ask good questions by asking *who, what, when, where, why,* and *how.*

2 Practice It! Read the following passage from Lesson 1.

> An earthquake is the rupture and sudden movement of rocks along a fault. Remember, a fault is a fracture surface along which rocks can slip. A fault ruptures, or breaks, when rocks are strained so much that they no longer can stretch or bend. This movement causes the release of complex waves that can move objects, as shown in Figure 1.
>
> ---from page 246

Here are some questions you might ask about this paragraph:
- What is an earthquake?
- When does a fault rupture?
- What causes the release of complex waves?

3 Apply It! As you read the chapter, look for answers to lesson headings that are in the form of questions.

Target Your Reading
Use this to focus on the main ideas as you read the chapter.

1 Before you read the chapter, respond to the statements below on your worksheet or on a numbered sheet of paper.
- Write an A if you **agree** with the statement.
- Write a D if you **disagree** with the statement.

2 After you read the chapter, look back to this page to see if you've changed your mind about any of the statements.
- If any of your answers changed, explain why.
- Change any false statements into true statements.
- Use your revised statements as a study guide.

Science Online
Print a worksheet of this page at ca6.msscience.com.

Before You Read A or D	Statement	After You Read A or D
	1 Earthquakes are waves of energy that travel across Earth's surface.	
	2 Tsunamis are huge tidal waves.	
	3 Most earthquakes occur in the middle of lithospheric plates.	
	4 Seismic waves are produced at the focus of an earthquake.	
	5 A 4.0 magnitude earthquake releases about twice as much energy as a 3.0 magnitude earthquake.	
	6 Some parts of the United States are at higher risk for earthquakes than others.	
	7 Secondary waves are the fastest seismic waves.	
	8 The San Andreas Fault is a fault zone.	
	9 Fire and landslides are major earthquake hazards.	

Reading Tip
Test yourself. Create questions and then read to find answers to your own questions.

244

Source: Chapter 6, pp. 244–245

As You Read

Within each lesson you will find tools that will help you understand what you read.

Phonetic spellings show you how to say difficult words. They show the spelling for how to say each syllable. Here is an example that you would find in the text: seismogram (SIZE muh gram).

✔ Reading Check questions help you check your reading understanding.

Rotating drum records ground motion

Pendulum and pen remain still

Bedrock

Earth moves

Bedrock

Figure 12 Seismographs are designed to record and measure either vertical or horizontal ground motion.

Recording Seismic Waves

A **seismograph** (SIZE muh graf), shown in **Figure 12,** is an instrument used to record and measure movements of the ground caused by seismic waves. It records the size, direction, and time of the movement. It also records the arrival times of the P- and S-waves. Modern seismographs record the ground motion with electronic signals. They work in much the same way as the older, mechanical seismographs.

Mechanical Seismographs

In order to understand how a seismograph works, consider the parts of a mechanical seismograph. A pen is attached to a weight called a pendulum. When seismic waves shake the ground, the heavy pendulum and the pen remain still. But, the drum moves. This happens because the drum is securely attached to the ground, unlike the freely swinging pendulum. As the ground shakes, the pen records the motion on the paper wrapped around the drum.

✔ Reading Check Which parts of a mechanical seismograph remain still when the ground shakes?

Seismographs record ground motion in two orientations. One orientation is horizontal, or back-and-forth, ground motion. The other is vertical, or up-and-down motion. The record of the seismic waves is called a **seismogram** (SIZE muh gram). Seismograms are used to calculate the size of earthquakes and to determine their locations.

WORD ORIGIN
seismograph
seismogram
seis– from Greek *seismos*, means *earthquake*
-graph from Greek; means to *write*
-gram from Greek; means *written word, a letter*

Lesson 3 • Measuring Earthquakes **261**

Source: Chapter 6, Lesson 3, p. 261

Reading for Information

Other Skills to Exercise as You Read

Question
- What is the **Main Idea** ?
- What is **The BIG Idea** ?

Connect
- As you read, think about people, places, and situations you've encountered. Are there any similarities with those in *Focus On Earth Science*?
- Can you relate the information in *Focus On Earth Science* to other areas of your life?

Predict
- Predict events or outcomes by using clues and information you already know.
- Change your prediction as you read and gather new information.

Visualize
- Create a picture in your mind about what you are reading. Picture the setting—for example, a laboratory, a roller coaster, or a mountain.
- A mental image can help you remember what you read for a longer time.

Compare and Contrast Sentences
- Look for clue words and phrases that signal comparison, such as *similar to, just as, both, in common, also,* and *too.*
- Look for clue words and phrases that signal contrast, such as *on the other hand, in contrast to, however, different, instead of, rather than, but,* and *unlike.*

Cause-and-Effect Sentences
- Look for clue words and phrases such as *because, as a result, therefore, that is why, since, so, for this reason,* and *consequently.*

Sequential Sentences
- Look for clue words and phrases such as *after, before, first, next, last, during, finally, earlier, later, since,* and *then.*

After You Read

Follow up your reading with a summary and an assessment of the material to evaluate if you understood the text.

Summarize
- Describe **The BIG Idea** and how the details support it.
- Describe the **Main Idea** and how the details support it.
- Use your own words to explain what you read.
- Complete the Summary Activity at the end of the lesson.

Assess
- What was **The BIG Idea** ?
- What was the **Main Idea** ?
- Did you learn anything new from the material?
- Can you use this new information in other school subjects or at home?
- What other sources could you use to find out more information about the topic?

Previewing Your Textbook

Follow the tour through the next few pages to learn about using your textbook, *Focus On Earth Science*. This tour will help you understand what you will discover as you read *Focus On Earth Science*. Before you begin reading, take the tour so that you are familiar with how this textbook works.

Unit Preview

West-Coast Events Time Line See significant events that occurred on the West Coast of the United States and compare them to events that occurred around the world.

World Events Time Line See significant events that occurred around the world and compare them to events that occurred on the West Coast.

Science Online A visual reminder to explore online tools to learn more about a scientist's career.

Source: Unit 1, pp. 42–43

Unit Review

Reading on Your Own a listing of books recommended by the California State Board of Education

Unit Test multiple-choice questions and written-response questions that review the unit

Source: Unit 1, pp. 238–239

Previewing Your Textbook

Chapters

The BIG Idea The Big Idea is a sentence that describes what you will learn about in the chapter.

Main Idea The Main Ideas support the Big Idea. Each lesson of the chapter has a Main Idea that describes the focus of the lesson.

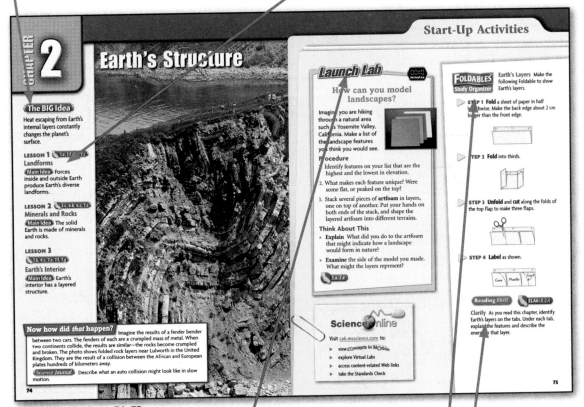

Source: Chapter 2, pp. 74–75

Launch Lab a short investigation that introduces the chapter's subject

Foldables™ Study Organizer an easy way to take notes as you read the chapter and a valuable tool for review

Reading Skill This is a reading skill that you will practice throughout the chapter.

Previewing Your Textbook

Lessons

Main Idea The Big Idea is supported by Main Ideas. Each lesson of the chapter has a Main Idea that describes the focus of the lesson.

Science Content Standards a listing of the California Science Content Standards that are covered within the lesson

Reading Check a question that tests your reading comprehension

Visual Check and Caption Questions questions found throughout the lesson about important graphs, photos, or illustrations

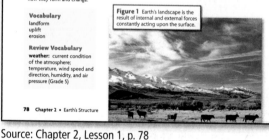

LESSON 1

Science Content Standards

1.e Students know major geologic events, such as earthquakes, volcanic eruptions, and mountain building, result from plate motions.

1.f Students know how to explain major features of California geology (including mountains, faults, volcanoes) in terms of plate tectonics.

2.a Students know water running downhill is the dominant process in shaping the landscape, including California's landscape.

7.c Construct appropriate graphs from data and develop qualitative statements about the relationships between variables.

Reading Guide

What You'll Learn
▶ **Classify** landforms.
▶ **Explain** how landforms are produced.
▶ **Relate** your knowledge of landforms to California landscapes.

Why It's Important
You'll appreciate landforms around you as you discover how they form and change.

Vocabulary
landform
uplift
erosion

Review Vocabulary
weather: current condition of the atmosphere; temperature, wind speed and direction, humidity, and air pressure (Grade 5)

Landforms

Main Idea Forces inside and outside Earth produce Earth's diverse landforms.

Real-World Reading Connection Imagine you're making a sculpture by piling up sand near the shore. Suddenly, a wave comes and washes away part of your new artwork. Through different and slower processes, landforms are constantly being built up and worn down on Earth's surface.

How do landscapes form?

You live on the surface of Earth. Look out the window at this surface, or look at a photograph or drawing of a landscape. **Figure 1** is an example. There are tall mountains, deep valleys, and flat plains. Why does the landscape have different shapes and forms?

An endless interaction of forces reshapes Earth's topography. The transfer of matter and energy from Earth's interior builds mountains. Forces on the surface continuously wear down the mountains. These forces are caused by uneven heating of the surface by the Sun. In turn, this energy is transferred to the atmosphere. This makes weather that constantly bombards surface material and erodes it away, especially in higher areas. Without these competing forces the planet's surface would be a flatter and less exciting place to live.

Reading Check What is the source of energy for Earth's weather?

Figure 1 Earth's landscape is the result of internal and external forces constantly acting upon the surface.

78 Chapter 2 • Earth's Structure

Source: Chapter 2, Lesson 1, p. 78

Lesson Review

Summarize Use this exercise to help you create your own summary of the lesson's content.

Self Check A series of questions to check your understanding of the lesson's content.

Changing Landforms

Although they might seem like permanent features, landforms in your surroundings change continuously. Heat energy from the Sun and from Earth's interior provides the energy to change these landscapes. The constant movement of energy from Earth's interior to the surface results in forces that uplift the land into mountains and plateaus. At the same time, thermal energy from the Sun provides the energy for weather that includes precipitation, which wears down the uplifted landforms. At times, these changes are abrupt and dramatic, as when volcanoes erupt. Most often though, the changes are slow and steady, but endlessly sculpt Earth's landforms.

LESSON 1 Review

Summarize

Create your own lesson summary as you design a **visual aid.**

1. **Write** the lesson title, number, and page numbers at the top of your poster.

2. **Scan** the lesson to find the red main headings. Organize these headings on your poster, leaving space between each.

3. **Design** an information box beneath each red heading. In the box, list 2–3 details, key terms, and definitions from each blue subheading.

4. **Illustrate** your poster with diagrams of important structures or processes next to each information box.

 ELA6: R 2.4

Using Vocabulary

1. A glacier scraping sediment and rock from the sides of a mountain is an example of _____. **2.a**

2. In your own words, write a definition for *landform*. **1.e**

Understanding Main Ideas

3. How did the landform shown above most likely form? **1.e**
   A. when a block of rock uplifted
   B. when sediment was piled up by a river
   C. when a volcano erupted
   D. when a glacier passed over a valley

4. **Identify** a landform you have seen that was made by erosion. **2.a**

Standards Check

5. **Compare and contrast** the ways that internal and external forces produce surface landforms. **1.f**

6. **Compare and contrast** the formation of Lassen Peak with the formation of the Sierra Nevada. **1.e**

Applying Science

7. **Predict** what would happen to Earth's surface if all of Earth's internal heat escaped. **1.e**

8. **Decide** if a constantly changing landscape is beneficial for people. **1.e**

Landscape	Benefit	Harm

Science Online

For more practice, visit Standards Check at ca6.msscience.com.

84 Chapter 2 • Earth's Structure

Source: Chapter 2, Lesson 1, p. 84

Previewing Your Textbook

Hands-On Science

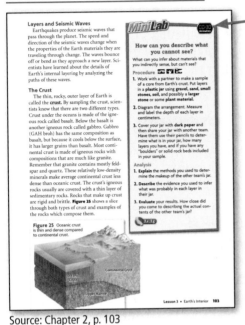

Source: Chapter 2, p. 103

MiniLab These investigations emphasize the lesson's content. MiniLabs are located in either a margin, like the one shown here, or on a full page. The California Science Content Standards that correlate to the material are listed.

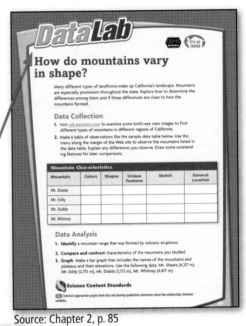

Source: Chapter 2, p. 85

DataLab These investigations emphasize the lesson's content by using mathematical analysis. DataLabs are located in either a margin or on a full page, as shown here. The California Science Content Standards and the California Mathematics Content Standards that correlate to the material are listed.

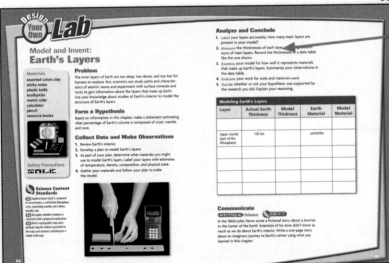

Source: Chapter 2, pp. 110–111

Lab Full-length investigations emphasize the chapter's content. Included are Labs, Design Your Own Labs, or Use the Internet Labs. The California Science Content Standards that correlate to the material are listed.

Previewing Your Textbook

Special Features

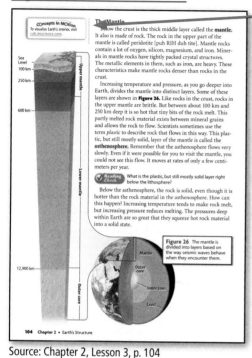

Source: Chapter 2, Lesson 3, p. 104

Concepts in Motion interactive art or diagrams that can be accessed through the Glencoe Web site to help you build understanding of concepts

Real-World Science Four connections with science are made in this feature: Science and Career, Science and Technology, Science and History, and Science and Society. These four connections will help you practice written and oral presentation skills.

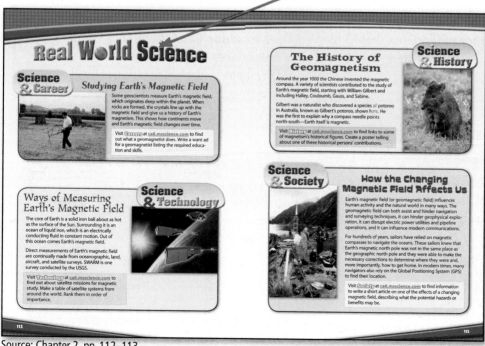

Source: Chapter 2, pp. 112–113

Previewing Your Textbook

Standards Review

Linking Vocabulary and Main Ideas a concept map to assist you in reviewing your vocabulary

Using Vocabulary a variety of questions that will check your understanding of vocabulary definitions

Source: Chapter 2, p. 115

Understanding Main Ideas multiple-choice questions

Applying Science short-answer and extended-response questions to practice higher-level thinking skills

Cumulative Review short-answer questions covering material from earlier in the unit

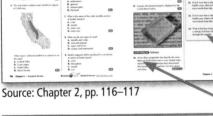

Source: Chapter 2, pp. 116–117

Writing in Science an exercise to practice writing skills; the California English/Language Arts Content Standards that correlate to the material are listed

Applying Math a series of questions that practice math skills related to the chapter; the California Mathematics Content Standards that correlate to the material are listed

Standards Assessment

Standards Assessment multiple-choice questions to review the California Science Content Standards covered in the chapter

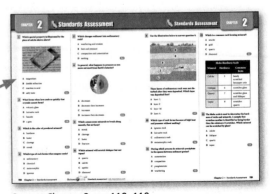

Source: Chapter 2, pp. 118–119

Scavenger Hunt

Focus On Earth Science contains a wealth of information. The secret is to know where to look to learn as much as you can.

As you complete this scavenger hunt, either on your own or with your teachers or family, you will quickly learn how the textbook is organized and how to get the most out of your reading and study time.

1. How many units are in the book? How many chapters?

2. On what page does the glossary begin? What glossary is online?

3. In which Student Resource at the back of your book can you find a listing of Laboratory Safety Symbols?

4. Suppose you want to find a list of all the Launch Labs, MiniLabs, DataLabs, and Labs, where do you look?

5. How can you quickly find the pages that have information about scientist Alfred Wegener?

6. What is the name of the table that summarizes the key concepts and vocabulary of a chapter? On what page in Chapter 4 are these two things located?

7. In which Student Resource at the back of your book can you find information on unit conversion? What are the page numbers?

8. On what page can you find **The BIG Idea** for Chapter 1? On what page can you find the **Main Idea** for Chapter 1, Lesson 2?

9. What feature at the start of each unit provides insight into a scientist's work?

10. What study tool shown at the beginning of a chapter can you make from notebook paper?

11. **Concepts In Motion** are interactive animations. Where do you go to interact with the animation?

12. What activities at the beginning of each chapter will help improve your reading?

Introduction to
Investigation and Experimentation

What is science? Science is the process of studying nature at all levels, from the farthest reaches of space to the smallest particle of matter, and the collection of information that is learned through this process. Every day, scientists ask questions about the natural world and propose explanations based on evidence they gather. This evidence can then be used by other scientists to answer their own questions about the natural world.

What is Earth science?

Earth science describes many fields of study that a to understand the changes that take place on and i Earth and the reasons for those changes. The Earth science topics covered in this book address scientif evidence collected about Earth's rocks, soil, oceans atmosphere, surface features, and how living syste interact with Earth. Familiarize yourself with the to in the Introduction to Investigation and Experimen tion to help you conduct your own investigations topics in Earth science.

Table of Contents

The Branches of Science

There are an infinite number of questions to ask about the natural world. However, these questions are often organized into different fields of study. The chart below lists three areas of science that you will study in middle school.

Earth Science

Volcanologists are Earth scientists that study volcanoes. This team of student volcanologists is studying patterns in cooled volcanic lava. This team of volcanologists is studying a hot volcano lava tube in Kilauea, Hawaii.

Earth scientists ask questions such as:
- What makes the ocean salty?
- What causes an earthquake?
- Why are there more earthquakes in California than in Arizona?
- How are mountains formed?
- What causes a tsunami?

Life Science

Microbiologists are life scientists that ask questions about organisms that are too small to see with the naked eye. This microbiologist is studying the growth of bacteria in order to find out which medicine can treat a disease.

Life scientists ask questions such as:
- What causes plants to grow?
- How do diseases spread in a population?
- Why do some whales beach themselves, but others do not?

Physical Science

Electron microscopists are physical scientists that observe objects at magnifications up to 800,000 times their actual size. This electron microscopist is using a scanning electron microscope at the University of California, Berkeley to observe the structure of an ant's head.

Physical scientists ask questions such as:
- Why does the sunlight melt snow?
- Why are some buildings damaged more than others during earthquakes?
- What makes up stars?
- What causes acid rain to form?

7.e Recognize whether evidence is consistent with a proposed explanation.

7.d Communicate the steps and results from an investigation in written reports and oral presentations.

Scientific Methods

You might think that science is only about facts and discoveries. But, science is also about the skills and thought processes required to make discoveries. There is no one scientific method used by scientists. Instead, scientific methods are based on basic assumptions about the natural world and how humans understand it.

By counting the number of tree seedlings in this field, this restoration ecologist can predict the pattern of growth for a forest.

Assumptions of the Scientific Method

1. There are patterns in nature.

Science assumes that there are patterns in nature. Patterns are characteristics or interactions between things that repeat over and over. Patterns can be observed using the five human senses—sight, hearing, touch, smell, and taste.

2. People can use logic to understand an observation.

Science assumes that an individual can make an observation and then create a series of logical steps in order to find a valid explanation for the observation. This series of steps can then be communicated to others.

3. Scientific discoveries are replicable.

Something that is replicable in science can be repeated over and over again. If a scientist claims to have made a discovery using a certain set of steps in his or her investigation, another scientist should be able to repeat the same steps and get the same result. This ensures that scientists provide reliable evidence to support their claim.

Scientific methods cannot answer all questions.

Questions that deal with your feelings, values, beliefs, and personal opinions cannot be answered using scientific methods. Although people sometimes use scientific evidence to form arguments about these topics, there is no way to find answers for them using scientific methods. Good science is based on carefully crafted questions and objectively collected data.

Questions Science Cannot Answer

The following are examples of questions that cannot be answered by science.

- Which band has the best songs?
- Why do bad things happen?
- What does it mean to be a good person?

What is science?:
Scientific Theories

7.e Recognize whether evidence is consistent with a proposed explanation.

Scientific Theories

Using scientific methods to ask questions about the natural world has led to the formation of scientific theories. A **scientific theory** is explanation of things or events that is based on knowledge gained from many observations and investigations. They are independently tested by many scientists and are objectively verified. However, even the best scientific theory can be rejected if new scientific discoveries reveal new information.

How is a scientific theory different from a common theory?

Scientific Theory	Common Theory
• A scientific theory is an explanation for a observation supported by evidence from many scientific investigations.	• A common theory is a collection of related ideas that one supposes to be true.
• Strength of a scientific theory lies solely in the accuracy of its predictions.	• Strength of a theory is based on the clarity of the explanation, not necessarily objectively obtained evidence.
• A scientific theory is modified or rejected if new evidence makes the theories predictions no longer true.	• A common theory may or may not be modified or rejected when presented with new evidence.
• A scientific theory must be rejectable.	• A common theory does not have to be rejectable.

Scientific Laws

A rule that describes a pattern in nature is a **scientific law.** For an observation to become a scientific law, it must be observed repeatedly. The law then stands until someone makes observations that do not follow the law. A law helps you predict that an apple dropped from arm's length will always fall to Earth. A scientific law, unlike a scientific theory, does not attempt to explain *why* something happens. It simply describes a pattern.

7.b Select and use appropriate tools and technology (including calculators, computers, balances, spring scales, microscopes and binoculars) to perform tests, collect data, and display data.

Lab and Field Study Tools

Lab and field study tools are physical tools that help you make better observations during scientific investigations. These tools enable you to measure the amounts of liquids, measure how much material is in an object, and observe things that are too small or too far away to be seen with the naked eye. Learning how to use them properly will help you when designing your own investigations.

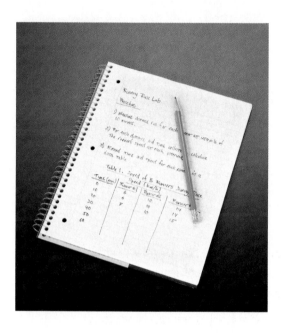

Science Journal Use a Science Journal to record questions, procedures, observations, and conclusions from your investigations.

TIP Your Science Journal can be a spiral-bound binder, a loose-leaf notebook, or anything that will help you record and save information.

TIP It is important that you keep your Science Journal organized. An organized journal will enable you to find information that you have collected in the past.

TIP Write down the date when you are recording information in your Science Journal, and leave extra space to go back to later.

Rulers and Metersticks Use metric rulers and metersticks to measure length or distance.

TIP Metric units of measurement for length include kilometers (km), meters (m), centimeters (cm), and millimeters (mm). The unit of measurement you choose to use will depend on the distance or length you need to measure. For example, a meterstick would be best to measure the length of your classroom. However, a 10-cm ruler would be the best tool to measure the length of a maple leaf.

TIP Estimate one decimal place beyond the markings on the ruler. For a meterstick, measure to the nearest 0.1 cm.

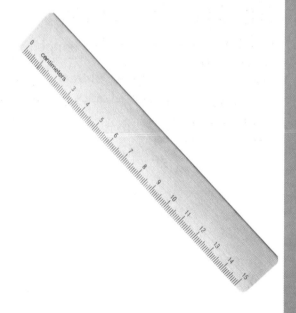

Tools of the Earth Scientist:
Lab and Field Study Tools

7.b Select and use appropriate tools and technology (including calculators, computers, balances, spring scales, microscopes and binoculars) to perform tests, collect data, and display data.

Thermometers Use a thermometer to measure the temperature of a substance.

TIP The metric unit of measurement for temperature is degrees Celsius (°C).

TIP When measuring a liquid that is being heated from the bottom, do not let the thermometer rest on the bottom of the container. This will result in an inaccurate reading.

SAFETY Be careful when transporting a glass thermometer. Glass thermometers are very fragile and are easily broken if dropped or bumped.

Beakers Use a beaker for holding and pouring liquids.

TIP Use a graduated cylinder instead of a beaker to measure the volume of a liquid. The lines on the side of a beaker are not accurate.

SAFETY Use a beaker that holds about twice as much liquid as you are measuring to avoid overflow.

TIP Use a hot plate to keep a substance warmer than room temperature.

SAFETY Use goggles to protect your eyes when working with liquids in the lab.

SAFETY Use gloves to protect your hands when working with liquids in the lab.

 Think Like a Scientist

If you use hot plates to simultaneously heat 200 mL of water and 400 mL of water to 80°C, will one return to room temperature faster than the other? Why?

7.b Select and use appropriate tools and technology (including calculators, computers, balances, spring scales, microscopes and binoculars) to perform tests, collect data, and display data.

Tools of the Earth Scientist:
Lab and Field Study Tools

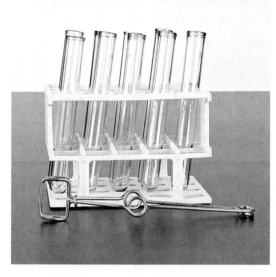

Test Tubes Use a test tube to study small samples of solids, liquids and gases.

TIP Use a test-tube rack to keep your test tubes upright and organized.

SAFETY Since liquids can spill or splash from test tubes, use small amounts of liquids and keep the mouth of the test tube pointed away from you and other people.

SAFETY Use a test-tube holder if you are heating the substance in a test tube or if the substance in the test tube is dangerous to touch.

SAFETY Do not put a stopper in a test tube if you are heating it.

Graduated Cylinder Use a graduated cylinder to measure the volume of a liquid.

Using a Graduated Cylinder

1. Place the graduated cylinder on a level surface so that your measurement will be accurate.

2. To read the scale on a graduated cylinder, make sure to have your eyes at the same level as the surface of the liquid.

3. The surface of the liquid in a graduated cylinder will be curved—this curve is called a meniscus. Read the line at the bottom of the meniscus.

TIP A 10-mL graduated cylinder will measure a small volume of liquid more precisely than a 100 mL graduated cylinder.

TIP Estimate one decimal place beyond the markings on the graduated cylinder. For a 100-mL graduated cylinder, estimate to the nearest 0.1 mL.

TIP You can use a graduated cylinder to find the volume of a solid object by measuring the increase in a liquid's level after the object is submerged in the liquid.

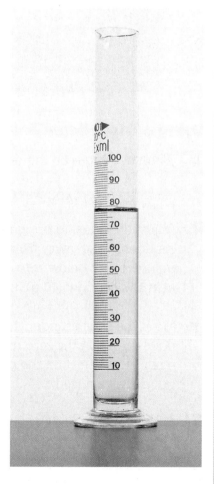

Tools of the Earth Scientist:
Lab and Field Study Tools

7.b Select and use appropriate tools and technology(including calculators, computers, balances, spring scales, microscopes and binoculars) to perform tests, collect data, and display data.

Triple-Beam Balance Use a triple-beam balance to measure the mass, or amount of material contained in an object.

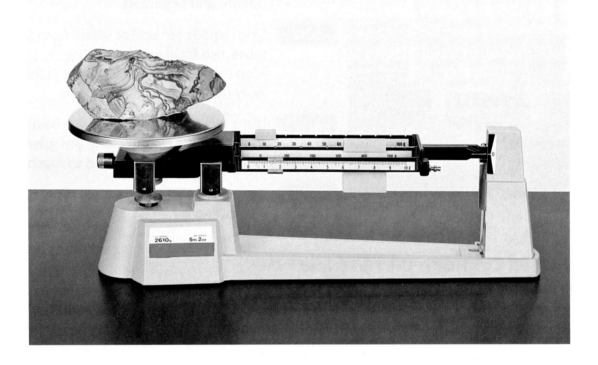

Using a Triple-Beam Balance

1. When nothing is on the pan, make sure the pointer of the balance and the riders are at zero.

2. Place the object you want to measure on the pan. The pointer will rise above the zero mark.

3. Adjust the riders to bring the pointer back down to zero. To do this, start by moving the largest rider (100 g) away from the pan one notch at a time. If moving the largest rider causes the pointer to fall below zero, set the largest rider back at the previous notch. Then, move the next smaller rider (10 g) in the same way.

4. Move the smallest rider (1 g) until the pointer rests at the zero mark. This means the object on the pan and the riders are balanced.

5. Add the measurements from the three beams together to determine the mass of the object.

Think Like a Scientist

Use a triple beam balance to measure the mass of your pencil. If you measured it's mass on the Moon would it be the same? Why or why not? Perhaps you have heard that the terms mass and weight mean different things? Do you know the difference?

7.b Select and use appropriate tools and technology(including calculators, computers, balances, spring scales, microscopes and binoculars) to perform tests, collect data, and display data.

Dissecting Microscope

Use a dissecting microscope to observe small objects you have collected from the field. A dissecting scope will help you study what organisms live in a sample of pond water, look more closely at a rock, and study an insect or leaf in greater detail.

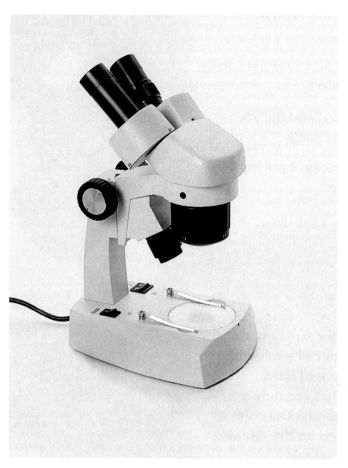

SAFETY When carrying a microscope, always hold one hand under the base and hold the arm with your other hand.

TIP If the light is too bright or too dark, you might not be able to see the sample very well. Adjust the amount of light on the stage plate until you can see the object in detail.

7.b Select and use appropriate tools and technology (including calculators, computers, balances, spring scales, microscopes and binoculars) to perform tests, collect data, and display data.

Binoculars
Use binoculars to make distant objects look bigger. Binoculars are a good tool for observing animals in the field, studying geologic formations, or even star gazing.

SAFETY Never look directly at the Sun with binoculars.

TIP Adjust binoculars until you see only one image.

TIP Fold the eyecups back if you wear glasses.

Streak Plate
Use a streak plate to reveal the powder color of a mineral. A streak plate is a piece of hard, unglazed porcelain that can help you identify types of minerals. The color of the powder left behind on the streak plate is the mineral's streak.

TIP Do not rely on a streak color alone to identify a mineral. Some minerals have the same streak color, and others are too hard to leave a streak.

⊙ Select and use appropriate tools and technology(including calculators, computers, balances, les, microscopes and binoculars) to perform tests, collect data, and display data. **7.f** Read phic map and a geologic map for evidence provided on the maps and construct and interpret a simple scale map.

Tools of the Earth Scientist:
Lab and Field Study Tools

Maps
Use maps when you need information about geographic features and places. Earth scientists use maps to locate places, show the distribution of Earth's physical features, and visually describe weather patterns in a particular area.

Topographic Maps
Typographic maps are used to describe changes in elevation. The lines on the map are called contour lines. The change in elevation between two side-by-side contour lines is called a contour interval. Contour intervals are dependent upon the terrain.

 TIP   For hilly terrain, contour lines are very close together. Lines that are far apart represent relatively flat terrain.

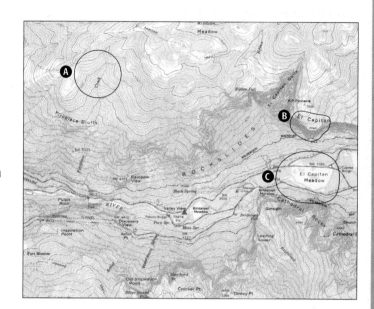

Geologic Maps
Geologic maps are used to describe what types of rock the surface of Earth is made of in a particular area. Each color represents a volume of a certain kind of rock of a given age range. A sandstone of one age might be colored bright orange, while a sandstone of a different age might be colored pale brown.

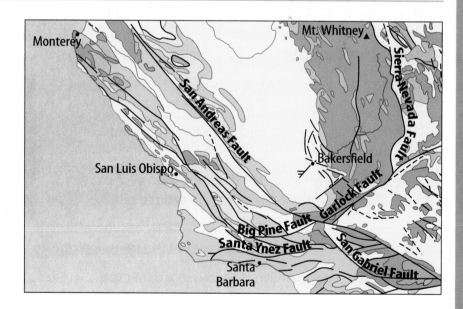

Tools of the Earth Scientist:
Lab and Field Study Tools

7.b Select and use appropriate tools and technology (including calculators, computers, balances, spring scales, microscopes and binoculars) to perform tests, collect data, and display data.

Compass Use a compass to find a known reference direction to help you navigate.

Using a Compass

1. Decide which direction you want to go.

2. Place the compass flat in the palm of your hand with the direction-of-travel arrow pointing in front of you.

3. Turn the compass housing (not the entire compass) until it lines up with the letter corresponding to the direction you chose.

4. Holding the compass, turn your body until the red part of the compass needle lines up with the orienting arrow inside the compass housing.

5. Make sure the red part of the compass needle is pointing north in the compass housing.

6. Walk in the direction the direction-of-travel arrow is pointing.

TIP The red part of the compass needle will always point due north.

TIP Hold the compass flat so the compass needle can turn freely.

Plot Sample Use a plot sample to estimate the population, or number of plants, slow-moving animals, or types of rock you find in an area.

Using a Plot Sample

1. Choose a place to study. For example, you might want to know what plant species are growing near your school.

2. Randomly select a spot and place the plot sample on the ground.

3. Use a field guide to identify different species of plants that appear inside the border of the sample plot.

4. Carefully count the number of each different type of plant that appears inside the border of the sample plot.

5. Record the data you have collected in a data table.

TIP A simple 1 m × 1 m plot sample like the one in the photo can be constructed using wooden stakes and string.

TIP Often you will find different things each time you place your sample plot in a particular area. For example, dandelions might grow near the edge of the school yard but not in the middle.

7.b Select and use appropriate tools and technology (including calculators, computers, balances, spring scales, microscopes and binoculars) to perform tests, collect data, and display data.

Tools of the Earth Scientist:
Lab and Field Study Tools

Rain Gauge Use a rain gauge to measure the amount of rain that falls in a given period of time.

Using a Rain Gauge

1. Insert the funnel into the rain gauge so that it fits into the measuring tube.

2. Place the rain gauge outside and note the time you started rain collection in your Science Journal.

3. Read the marks on the measuring tube to determine how much rain was collected.

4. Record the volume of rain collected and the amount of time spent collecting it in a data table.

5. Empty the measuring tube and the overflow tube after each use.

   TIP To get an accurate measurement of rainfall, locate the rain gauge in a location that is unobstructed by trees or buildings.

Wind Vane and Anemometer Use a wind vane to measure the horizontal direction of moving air. An anemometer, or wind speed gauge, is used to measure the rate at which air is moving.

Using a Wind Vane and Anemometer

1. Place the device in a location that is exposed to wind from all directions.

2. Record the date and time in a data table.

3. Record the direction that the wind vane points in a data table.

4. Mark one of the cups of the anemometer with a pen.

5. Have a lab partner with a stopwatch to keep time. When the time keeper says "Go," count the number of revolutions that the marked cup makes in one minute.

6. Record the number of revolutions per minute in a data table.

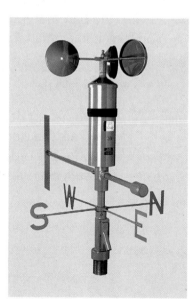

7.b Select and use appropriate tools and technology (including calculators, computers, balances, spring scales, microscopes and binoculars) to perform tests, collect data, and display data.

Hygrometer Use a hygrometer to measure the relative humidity or concentration of water vapor in the air.

Using a Hygrometer

1. Fill the water bottle and attach it to the instrument.

2. Hang the hygrometer on a wall or a tree for 10 min.

3. A hygrometer contains two thermometers: a wet-bulb thermometer, which is submerged in the bottle, and a dry-bulb thermometer. Record the wet-bulb and dry-bulb temperatures in a data table.

4. To determine relative humidity, compare the wet-bulb temperature to the dry-bulb tempterature.

SAFETY Be careful with the instrument. The thermometers are made of glass and can easily break.

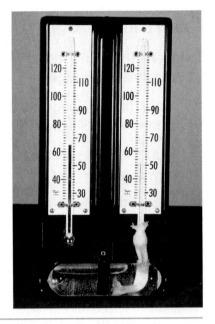

Barometer Use a barometer to measure atmospheric pressure.

Using a Barometer

1. Hang the barometer on a wall.

2. Most barometers have more than one scale. Read the scale that is measured in either millibars or millimeters of mercury.

3. Record the barometer reading and the time you took it in a data table.

TIP The barometer will not measure properly if laid flat.

TIP A barometer needs to be recalibrated if you are measuring at different elevations in order to account for air pressure changes due to altitude.

7.b Select and use appropriate tools and technology (including calculators, computers, balances, spring scales, microscopes and binoculars) to perform tests, collect data, and display data.

Computers and the Internet Use a computer to collect, organize, and store information about a research topic.

Computer Hardware

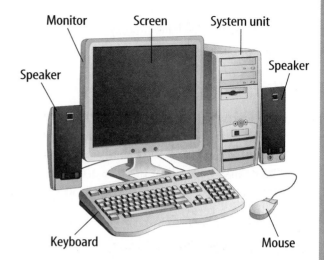

Monitor   Screen   System unit   Speaker   Speaker   Keyboard   Mouse

- Most desktop computers consist of the components shown above. Notebook and tablet computers have the same components in a compact, book-sized unit.

- Electronic probes can be attached to computers in order to automatically record measurements including temperature, sound, and light.

Computer Software

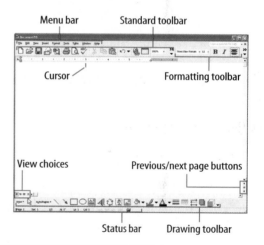

Menu bar   Standard toolbar   Cursor   Formatting toolbar   View choices   Previous/next page buttons   Status bar   Drawing toolbar

- Use a **word processor** to compose, edit, and format lab reports and other written material.

- Use a **spreadsheet program** to create data tables and graphs.

- Use a **presentation program** to organize the results of your research into multimedia presentations.

Using the Internet

Navigation buttons   Address bar   Loading indicator   Link indicator

- You can find information on the Internet using the World Wide Web. Web pages are arranged in collections of related material called "**Web sites.**" A Web site is viewed using a program called a Web browser. Web browsers allow you to browse the Web by clicking on highlighted hyperlinks which move you from Web page to Web page.

- Web content can also be searched by topic by using a search engine. **Search engines** are located on Web sites that catalog keywords on Web pages all over the World Wide Web.

Tools of the Earth Scientist:
Scientific-Thinking Tools

7.h Identify changes in natural phenomena over time without manipulating the phenomena (e.g., a tree limb, a grove of trees, a stream, a hillslope).

Tools of Scientific Thinking

Scientific thinking tools are techniques that help you to refine your questions, make useful observations, and think critically about scientific information. As you work in the lab, refer to this guide to help you understand the nature of science.

Observations

An **observation** is an act of watching something and taking note of what occurs. Although observing often refers to vision, all five human senses can be used to make observations. You make observations in order to ask questions and make informed hypotheses. You also make observations when you collect data in an experiment. There are two types of observations:

A researcher makes observations by tracing the tracks of a Bengal tiger in India.

Qualitative Observations

Qualitative observations are descriptions of the natural world using words.

Quantitative Observations

Quantitative observations are expressed as numbers.

 TIP Record quantitative observations in addition to qualitative observations. Quantitative observations describe what you have observed more specifically.

EXAMPLE

A student studying weather conditions made qualitative and quantitative observations during two different seasons.

Observations of Weather Conditions		Qualitative Observation	Quantitative Observation
Day 1		The air temperature was warm.	The air temperature measured 23°C.
Day 2		The air temperature was cool.	The air temperature measured 0°C.

7.a Develop a hypothesis.   **7.e** Recognize whether evidence is consistent with a proposed explanation.

Tools of the Earth Scientist:
Scientific-Thinking Tools

Hypotheses and Predictions

A **hypothesis** is a tentative explanation or an answer to a question that can be tested with a scientific investigation to describe *what* will happen and *why* it will happen.

A **prediction** is a forecast of what will happen next in a sequence of events, but it does not explain why something happens.

EXAMPLE

Imagine you have two daisies in your classroom. One looks healthy while the other is turning brown. You notice that the healthy-looking daisy receives a lot of sunlight, and the unhealthy daisy receives less sunlight. You know both plants are given the same amount of water every day.

What is one hypothesis that could be used to investigate why one daisy is healthy and the other is not?

1. Start by asking a question.	**1.** Question: Why is one daisy healthy and the other is not?
2. Document what you already know from prior observations.	**2.** Observations: The healthy-looking daisy receives a lot of sunlight. The unhealthy daisy receives little sunlight.
3. Write a hypothesis which tentatively explains your observation.	**3.** Hypothesis: The daisy is not healthy because it is not receiving enough light to grow.
4. Write a prediction that can be used to test your hypothesis.	**4.** Prediction: If I provide the unhealthy daisy with the same amount of sunlight as the healthy daisy, it will become healthier.

TIP   The results of an experiment do not *prove* that a hypothesis is correct. Instead, the results of an experiment either *support or do not support* the hypothesis. This is because scientific inquiry is uncertain. You cannot be sure that you are aware of everything that could have affected the results of your experiment.

TIP   An experiment is not a failure if the results do not support your hypothesis. In the experiment above, if the unhealthy plant does not improve after providing it with more light, you can eliminate that as the cause of the problem and revise your hypothesis.

Tools of the Earth Scientist:
Scientific-Thinking Tools

 7.e Recognize whether evidence is consistent with a proposed explanation.

Inferring

An **inference** is a logical conclusion based on your prior knowledge and observations. To make a good inference, analyze all of your observations. Then, based on everything you know, explain or interpret what you have observed.

EXAMPLE

It is easy to trick yourself into making an inference without adequate evidence to support it. Look at the two photographs shown below. Based on what you see, you might infer that the cat knocked over the plant. However, you need to carefully note your observations first.

Photo #1 Observations	Photo #2 Observations
• The plant is upright. • The cat is near the plant.	• The plant lying sideways on the floor. • The cat is absent from the image.

Even though your past experience with cat behavior might make you think that the cat knocked the plant off the table, you did not observe the cat knocking the plant off the table. Therefore, you have collected no evidence that supports that inference. The only way to know is to investigate the question further and gather more evidence.

TIP It is important to distinguish between observations and inferences. Descriptions of color, temperature, length, and weight are all examples of observations. However, when you offer an explanation for that observation, you are making an inference.

 **Think Like a Scientist**

Carefully examine the image. What do you see? Is the white vapor in the photo a cloud, steam from a nearby factory, smoke, or something else? What evidence do you see in the image? What can you infer from the image?

7.e Recognize whether evidence is consistent with a proposed explanation.

Tools of the Earth Scientist:
Scientific-Thinking Tools

Evaluating Evidence and Explanations

Whether you are reading science articles and lab reports or drawing conclusions from data you have collected in a lab, it is essential to think critically about the data and the scientific explanations presented to you. **Critical thinking** means comparing what you already know with the explanation you are given in order to decide if you agree with it or not.

Evaluating Scientific Evidence

Start by evaluating the quality of the evidence presented to you. Valid scientific investigations contain quantitative or qualitative evidence called **data**. Data can be descriptions, tables, graphs, or labeled drawings. Data are used to support or refute the investigation's hypothesis. When evaluating data from an investigation ask the following questions:

- **Does the journal article or lab report contain data?** A proper scientific investigation always contains data to support an explanation.

- **Are the data precise?** Data used to support an explanation should be exact. Quantitative observations or detailed descriptions and drawings of events are much better than vague descriptions of events. Imprecise phrases such as "a lot" and "a little" do not accurately describe an event because it's impossible to know to what that description is being compared. Vague descriptions lead to incorrect explanations.

- **Have the results of the experiment been repeated?** If a friend told you he could hit a home run, but he was unable to do it while you watched, would you believe him? Probably not. Likewise, scientific data are more reliable when the investigator has repeated an experiment several times and consistently produced the same results. Scientific evidence is considered to be even more reliable when multiple investigators try the same experiment and get the same results.

Why do you think scientific evidence is more reliable when different investigators try the same experiment rather than the same investigator performing the experiment multiple times?

Evaluating Scientific Explanations

Having good data is the first step to providing a good explanation for the data. However, it's easy to make a mistake and accidentally arrive at the wrong conclusion. When evaluating an inference or a conclusion, ask yourself the following questions:

- **Does the explanation make sense?** Be skeptical! There need to be logical connections between the investigator's question, hypothesis, predictions, data, and conclusions. Read the information carefully. Can the investigator reasonably draw his or her conclusion from the results of the experiment?

- **Are there any other possible explanations?** Since it is virtually impossible to control every variable that could affect the outcome of an experiment, it's important to think of other explanations for the results of an experiment. This is particularly true when the data are unusual or unexpected.

Tools of the Earth Scientist:
Scientific-Thinking Tools

7.d Communicate the steps and results from an investigation in written reports and oral presentations.

Writing a Lab Report

How would you know what your friends are thinking if you didn't communicate? Likewise, scientists share the results of their investigations. Part of being a science student involves learning to communicate the outcome of your investigations clearly and effectively.

Written Reports

Written reports use text, data tables, graphs, and drawings in an organized manner to state clearly the topic you investigated and to explain your findings. Written reports can be used to communicate the results of a lab or a more involved investigation such as a Design Your Own Lab or a science fair project.

EXAMPLE

The format below shows the basic organization of a good lab report. Depending on the complexity of your investigation, you might need to add more information.

Problem
State exactly what problem the investigation is attempting to solve or what question the investigation is attempting to answer.

Form a Hypothesis
Describe what you thought would happen and why.

Procedure
Describe the steps you took in a logical manner.

Results
Organize your data into a table so that the information is easy to read and analyze.

Analyze and Conclude
The conclusions you draw from the investigation should be based on your analysis of the data. Also, include notes about measuring errors or problems you encountered.

7.d Communicate the steps and results from an investigation in written reports and oral presentations.

Tools of the Earth Scientist:
Scientific·Thinking Tools

Making an Oral Presentation

Oral presentations are a convenient way to communicate the results of your investigation to several people at once.

Organize your presentation.

Problem State exactly what problem the investigation is attempting to solve or what question the investigation is attempting to answer.

Hypothesis Describe what you thought would happen and why.

Procedure Proceed in a very logical manner, telling what you did step by step.

Results Explain both your controls and variables. Use graphs to summarize your data.

Conclusion State the conclusions you can validly draw from the investigation and support them with data from the investigation. Describe any measuring errors or problems you encountered that could affect your conclusions.

Questions When you are finished, ask your audience if they have any questions about your investigation. If you don't know the answer to a question, admit that you don't know and indicate that you will check into the matter.

TIP Use visuals.

Visuals illustrate points so they are easier to understand, reinforce major ideas in your presentation, and keep the audience focused. They also will help guide you as you speak.

What are visuals?

- graphs and charts
- maps, photos, and drawings
- models
- video/film

What media are used?

- poster display
- transparencies/slides
- video projection/projector
- blackboard/whiteboard
- computer presentation programs

TIP Practice your presentation.

Be sure to rehearse your presentation before you face your audience. Rehearsing will help you remember your presentation when you are presenting to the class, and it will help you to relax. Dress neatly and stand up straight.

- Don't chew gum or candy.
- Stand to the side of your display.
- Speak loudly enough to be heard by the audience.
- Make eye contact with the audience.
- If you have a partner, be sure you share equally in the presentation.

Tools of the Earth Scientist:
Data Analysis Tools

7.c Construct appropriate graphs from data and develop qualitative statements about the relationships between variables.

Data Analysis Tools

Use data analysis tools to help you organize your data and display patterns in your results.

Making Data Tables

Data tables help you organize and record the measurements you make. A data table displays information in rows and columns so that it is easier to read and understand.

EXAMPLE

Suppose you were competing in a 50-km bicycle race. You planned to keep a pace of 10 km/h. In order to know if you stayed on pace or not, you had a friend record your time at every 10 km.

Construct the Data Table

Step 1.	Think about the variables you plan to investigate. Then, organize the data table into columns and rows.
Step 2.	Create headings that describe the variable and the corresponding unit of measurement.
Step 3.	Give the data table a title and a number.

Your data can be organized like this:

Table 1 Bicycle Race Data

Distance (km)	Time (h)
0	0
10	0.75
20	2
30	3.5
40	4
50	5

Or like this:

Table 2 Bicycle Race Data

Distance (km)	0	10	20	30	40	50
Time (h)	0	0.75	2	3.5	4	5

 Think Like a Scientist

Study the types of graphs discussed in the pages ahead. Which type of graph would be appropriate for displaying the bicycle race data—a line graph, bar graph or circle graph? Why?

7.c Construct appropriate graphs from data and develop qualitative statements about the relationships between variables.

Making Line Graphs

A line graph shows a relationship between two variables that change continuously.

- Line graphs are good for showing how an independent variable affects a dependent variable or showing how a variable changes over time.
- Both variables in a line graph must be numbers.

EXAMPLE

Suppose you measured the temperature of the air from sea level to the top of the highest mountain in the world. You measured air temperature in 1,000-m increments from 0 m to 7,000 m. What would a graph of these data look like?

Air Temperature and Altitude	
Air Temperature (°C)	**Altitude (m)**
15	0
7.5	1,000
0	2,000
−7.5	3,000
−15	4,000
−22.5	5,000
−30	6,000
−37.5	7,000

Construct the Graph

Step 1. Use the horizontal x-axis for the independent variable (altitude) and the vertical y-axis for the dependent variable (air temperature).

Step 2. Draw the x-axis and y-axis using a scale that contains the smallest and largest values for each variable. Label each axis.

Step 3. To plot the first data point, find the x-value (0) on the x-axis. Imagine a line rising vertically from that place on the axis. Then, find the corresponding y-value (15) on the y-axis. Imagine a line moving horizontally from that place on the axis. Make a data point where the two imaginary lines intersect. Repeat this process for the remaining data.

Step 4. Connect the data points with lines.

Step 5. Title the graph.

Interpreting Line Graphs

What can you say about the relationship between altitude and air temperature?

- As altitude increased, temperature decreased.
- Air temperature changed about the same amount every 1,000 m.

TIP Be sure to examine the scale of a graph carefully. The scale of a graph can give a distorted picture of the data.

Altitude v. Air Temperature

Tools of the Earth Scientist:
Data Analysis Tools

7.c Construct appropriate graphs from data and develop qualitative statements about the relationships between variables.

Making Bar Graphs

A bar graph uses rectangular blocks, or bars, of varying sizes to represent and compare quantitative data. The length of each bar is determined by the amount of the variable you are measuring.

EXAMPLE

Suppose you measured the total rainfall for each month of the year in Los Angeles. You collected the rainfall data in the following data table.

Monthly Rainfall in Los Angeles

Month	Rain (cm)
January	6.1
February	6.4
March	5.1
April	1.8
May	0.3
June	0
July	0
August	0.5
September	0.8
October	0.8
November	4.6
December	4.3

Constructing the Graph

Step 1. Use the horizontal *x*-axis for the category (month) and the vertical *y*-axis for the measured variable (rain).

Step 2. Draw the *x*-axis and *y*-axis. Evenly space the category names below the *x*-axis. Use a scale that contains the smallest and largest values for the measured variable. Then, label each axis.

Step 3. To draw the first bar, find the first category name (January) below the *x*-axis. Then, draw a bar up to the measured value for that category (6.1) on the *y*-axis. Repeat this process for the remaining data.

Step 4. Title the graph.

Interpreting Bar Graphs

- Los Angeles received the most rainfall in February.

- Los Angeles received no measurable rain in both June and July.

- Question to ask: *Does Los Angeles usually receive no rain in June and July?* To find out, compare rainfall amounts for June and July to the recorded rainfall for those months in the past several years.

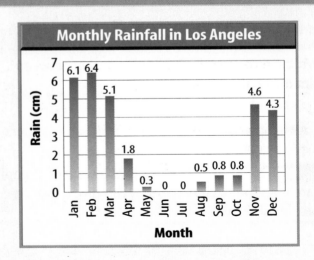

7.c Construct appropriate graphs from data and develop qualitative statements about the relationships between variables.

Tools of the Earth Scientist:
Data Analysis Tools

Making Circle Graphs

A circle graph, or pie graph, is used to show some fixed quantity is broken down into parts. The circular pie represents the total. The slices represent the parts and usually are presented as percentages

EXAMPLE

Suppose you want to find out what portion of the recyclable waste your family recycles is aluminum, glass, plastic and paper in a single month. Every week before the recyclables are collected you measure the mass of aluminum, glass, plastic and paper in the bin. You add up the total mass for each category and create the following data table. Now, what portion does each category represent?

Recyclable Material Used in One Month	
Material Type	**Mass (kg)**
Aluminum	1
Glass	3
Plastic	2
Paper	6

Constructing the Graph

Step 1.	Find the total of the measured variable (mass) by adding together the values for all of the categories. **1 kg + 3 kg + 2 kg + 6 kg = 12 kg**
Step 2.	Calculate the number of degrees of a circle that the first category's value (1) represents. To do this, write a fraction comparing the measured value with the total for all categories (1/12). Then multiply this fraction by 360° **(1/12) × 360° = 30°.** Repeat this process for the remaining categories.
Step 3.	Draw a circle. Use a protractor to draw the angle (number of degrees) for each category.
Step 4.	Color and label each section of the graph.
Step 5.	Title the graph.

Interpreting Circle Graphs

- In the time measured, 50 percent of the total mass of your family's recycled waste was paper.

- Questions to ask: *What factors could have caused this outcome? Did your family recycle more paper than usual? Is this amount typical?* Only further investigation will tell you.

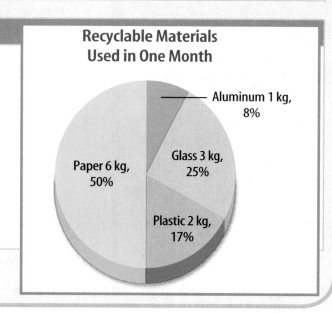

Recyclable Materials Used in One Month

Aluminum 1 kg, 8%
Glass 3 kg, 25%
Plastic 2 kg, 17%
Paper 6 kg, 50%

Designing a Controlled Experiment

In this section you will apply your lab skills, scientific thinking skills, and data analysis skills to the task of designing your own controlled experiment. A controlled experiment is a type of scientific investigation that tests how one thing affects another. Use this section to help you with Design Your Own Labs and science projects.

Asking Scientific Questions

Scientific investigations often begin when someone observes an event in nature and wonders why or how it occurs. To begin designing an experiment, questions need to be refined into specific questions that can be answered with the time and resources available to you.

EXAMPLE

You have observed that fewer young maple trees have been growing in the forest. You have also noticed a sharp increase in the deer population in the same forest. You decide to find out if there might be a relationship between these observations.

Question:

Why are there so many fewer young maple trees in the forest?

Observations:

- Few maple tree seedlings are becoming saplings.
- Deer populations in eastern forests are increasing.
- Maple seedlings often appear to have been bitten off at the stalk.

Refined Question:

Are deer causing maple tree populations to decline by feeding on maple seedlings?

7.d Communicate the steps and results from an investigation in written reports and oral presentations.

Tools of the Earth Scientist:
Designing a Controlled Experiment

Writing a Hypothesis and Prediction

A **hypothesis** is a tentative explanation that can be tested with a scientific investigation. It uses your prior knowledge and observations to predict what will happen and why. A **prediction** is a statement of what will happen next in a sequence of events—in this case, your experiment. If the results of the experiment match the prediction, the hypothesis is considered to be supported.

Hypothesis:
There are fewer young maple trees in eastern forests because deer are eating the seedlings.

Prediction:
If deer are prevented from feeding on maple tree seedlings in part of the forest, then more maple tree seedlings will grow to become saplings there than in the rest of the forest.

Defining Variables and Constants

To test a prediction, you need to identify variables and constants you want to use in your experiment. Variables and constants are factors you think could affect the outcome of your experiment.

Variables

A **variable** is any factor that can have more than one value. In controlled experiments, there are two types of variables—independent variables and dependent variables.

The **independent variable** is the factor you want to test. It is manipulated or changed by the investigator to observe how it affects a dependent variable.

A **dependent variable** is the factor you measure or observe during an experiment.

Constants

To test how the independent variable affects the dependent variable, you need to keep all other factors the same for each test. The factors that remain the same are called **constants**. Without constants, two independent variables could change at the same time and you won't know which variable affected the dependent variable.

Independent Variable: presence or absence of deer in the forest

Dependent Variable: number of maple seedlings that survive to become saplings

Constants: same forest, same weather conditions, same density of new maple seedlings at start of experiment, same amount of time seedlings are studied

Tools of the Earth Scientist:
Designing a Controlled Experiment

7.d Communicate the steps and results from an investigation in written reports and oral presentations.

Experimental Group and Control Group

A controlled experiment has at least two groups—a control group and an experimental group. The **experimental group** is used to study the effect of a change in the independent variable on the dependent variable. The **control group** contains the same factors as the experimental group, but the independent variable is not changed. Without a control, it is impossible to know if your experimental observations result from the variable you are testing or some other factor.

Experimental group:
10 m × 10 m square plots enclosed by 10 m × 10 m fences that prevent deer from feeding on maple seedlings

Control group:
10 m × 10 m square plots marked off with poles, but no fence

Measuring the Dependent Variable

Before you write a procedure, think about what kind of data you need to gather from the dependent variable to know how it relates to the changes you make to the independent variable. Dependent variables can be measured qualitatively or quantitatively.

Qualitative Measurement

Qualitative measurements of the dependent variable use words to describe what you observe in your experiment. Qualitative measurements are easy to make. For some investigations, qualitative data might be the only kind of data you can collect.

Qualitative Measurement of Maple Seedlings

Independent Variable
Presence or absence of deer

Dependent Variable
Presence or absence of maple seedlings in sample plots

Quantitative Measurement

Quantitative measurements of the dependent variable use numbers to describe what you observe in your experiment. In most experiments, quantitative measurements will provide you with greater precision in your data than qualitative measurements.

Quantitative Measurement of Maple Seedlings

Independent Variable
Presence or absence of deer

Dependent Variable
Number of maple seedlings counted in sample plots

○ Select and use appropriate tools and technology (including calculators, computers, balances,
es, microscopes, and binoculars) to perform tests, collect data, and display data. **7.d**
ate the steps and results from an investigation in written reports and oral presentations.

Tools of the Earth Scientist:
Designing a Controlled Experiment

Writing a Procedure

A procedure is a set of instructions that you use to gather the data you
need to answer your question. Each step in the experiment's procedure
should be clear and easy to follow. Record your procedure in your
Science Journal so you can execute it with precision.

Procedure

Step 1: Randomly select three places in the
forest to set up pairs of test and control plots.
Each control plot will be located next to the
test plot.

Step 2: In late winter, set up plots in the forest.
Test plots will consist of four poles with a fence
around them. They will have the same poles, but
no fence. In both cases, the area enclosed will
be 100 m².

Step 3: Allow plots to remain in the forest
for three months, but check plots monthly for
repair.

Step 4: After three months have passed, count
the number of maple seedlings in test plots and
control plots.

Step 5: Subtract the number of seedlings in
the test plot from the number of seedlings in
the neighboring control plot.

Determining Materials

Carefully examine each step in your procedures. Determine what
materials and tools are required to complete each step.

1. metric tape measure, 24 metal stakes (4 per sample area),
large roll of nylon fencing, hammer for driving stakes into
ground, scissors for cutting nylon fencing

2. pencil, eraser and Science Journal for recording observations

3. calculator for calculating seedling counts

Tools of the Earth Scientist:
Designing a Controlled Experiment

7.c Construct appropriate graphs from data and develop qualitative statements about the relationships between variables. **7.d** Communicate the steps and results from an investigation in written reports and oral presentations.

Recording Observations

Once your procedure has been approved, follow the steps in your procedure and record your data. As you make observations, note anything that differs from your intended procedure. If you change a material or have to adjust the amount of time you observe something, write that down in your Science Journal.

Seedlings Counted After 3 Months

Plot Type	Area 1	Area 2	Area 3
Control plot (with deer)	6	6	2
Test plot (without deer)	21	8	15
Difference	−15	−2	−13

Analyzing Results

To summarize your data, look at all of your observations together. Look for meaningful ways to present your observations. Presenting your data in the form of a graph is a powerful tool to communicate patterns in your data.

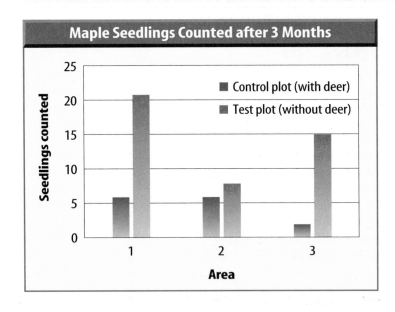

7.e Recognize whether evidence is consistent with a proposed explanation.

Tools of the Earth Scientist:
Designing a Controlled Experiment

Drawing Conclusions

To draw conclusions from your experiment, examine the data tables and graphs you have created. Describe trends you see in the data. Then, compare the results to your prediction and hypothesis. Determine if the results support or do not support your hypothesis. Use evidence in your results to support your determination.

> In all three areas sampled, more seedlings were counted in test plots than in control plots. This matches the prediction that if deer are responsible for maple decreases, then test plots should have more seedlings than control plots. Therefore, I conclude that the data support the hypothesis that deer decrease maple numbers by eating seedlings.

Analyzing Error

Error is a part of any scientific research. It's important to document anything that you changed in your procedure or could have caused uncertainty in your measurements. Be sure to include unanticipated factors or accidents that may have influenced your results and offer alternative explanations for your results.

> 1. There was less of a difference in seedling counts for area 2 than area 1 or area 3. Since the test plot for area 2 was found damaged, it's possible that deer were able to access the plot and eat some of the seedlings.
>
> 2. An alternate explanation for the results could be that deer may have been attracted to the fence. Not being able to get in, however, they instead fed in the area, eating a lot more seedlings in the control area than they would normally have eaten in the forest without fences. Further testing would be required to determine if this was a factor or not.

Case Study: The Next Big Tsunami

Mahyudin Jamil was on his fishing boat a couple of hundred yards off the northern Sumatra coast when the first wave arrived. As he watched from his boat, a 20-m sea wave crashed into the shore and obliterated his village. Ashore, he found that the wave had carried away his wife and five of his seven children. The tsunami left more than 283,000 dead in Indonesia, Sri Lanka, India, and Thailand, making it one of the deadliest disasters in modern history.

The Indian Ocean Tsunami of 2004

The December 26, 2004, earthquake that produced the wave was the largest quake in 40 years. Centered about 100 km off the coast of Sumatra, the magnitude-9 earthquake lasted 9 minutes. It released as much energy as all other earthquakes combined over the past 15 years.

The earthquake resulted from the collision of two sections of Earth's outer, rigid layers, called tectonic plates. In a process known as subduction, one of the plates was forced beneath the other one. This caused the seafloor to rise suddenly, forcing water toward the surface. When the water collapsed, it produced the tsunami. The sea wave caused more deaths than any other tsunami in recorded history.

What does tsunami mean?

Tsunami is the Japanese name for "harbor wave." It is so named because these waves were only observed when they reached their maximum height in bays or harbors.

How does a tsunami form?

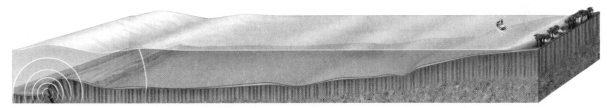

Sudden up or down movement along an underwater fault causes powerful waves that are transferred to and spread across the water's surface. Tsunami waves are less than 1 m high in deep water.

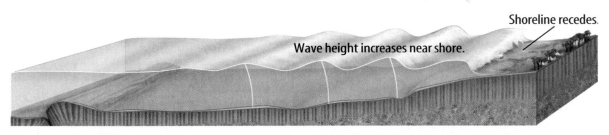

Shoreline recedes.

Wave height increases near shore.

When a tsunami wave reaches shallow water, friction slows it down and causes it to roll up into a wall—sometimes 30 m high. Just before the tsunami crashes to shore, the water near a shoreline may move rapidly toward the sea.

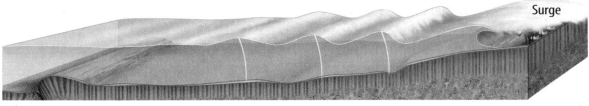

Surge

The tsunami wave crashes into the shoreline, causing a surge of water that is capable of traveling hundreds of meters inland.

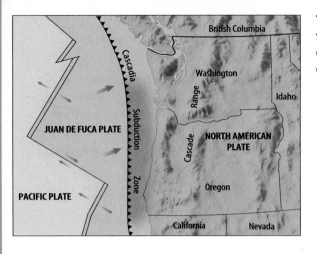

The Cascadia subduction is formed by the intersection of the Juan de Fuca oceanic plate and the North American continental plate.

Could it happen in California?

Could an equally strong tsunami strike the coast of California? Until a few years ago, the idea was practically unthinkable. Despite California's history of earthquake activity, only the 1964 tsunami that struck Crescent City, California was serious enough to cause property damage and life loss—and the origin of that tsunami was Alaska. But scientists now know that a fault along the Pacific Northwest coast of the United States has produced 9.0-magnitude earthquakes. The last one occurred in A.D. 1700. It produced a tsunami that was as powerful as the one that struck Sumatra in 2004.

The Cascadia Fault

As with Sumatra, two tectonic plates intersect just off the Pacific Northwest coast. These are the North American continental plate and the Juan de Fuca seafloor plate. Their intersection forms the Cascadia subduction zone. It extends from British Columbia to Northern California. Although the two plates are now locked, the Juan de Fuca plate has slowly been forcing its way into the edge of the North American Plate. A major slippage along the Cascadia fault would cause the underwater portion of the North American plate to spring upwards. The movement would then set in motion a tsunami.

No Written Record on the West Coast

In the 1980s, few scientists imagined that a major earthquake could strike the Pacific Northwest coast. Researchers were studying the movement of the Juan de Fuca and North American Plates. Were they sliding harmlessly past each other, or did they occasionally lurch past each other? Written records of the Pacific Northwest did not tell of any great earthquakes there. But those records only went back a little more than 200 years.

U.S.G.S. geologist, Brian Atwater points to a thin, dark grey line of beach sand he found digging in inland marshes—evidence that a tsunami occurred here.

Evidence of Falling Land

Brian Atwater is a geologist with the U.S. Geological Survey in Seattle. In the mid-1980s, he knew that subduction earthquakes in other parts of the world had caused land to rise or fall by several meters. He wondered if evidence of suddenly rising or falling land could be found on the Pacific Northwest coast. So he decided to examine the marshes along the Washington coast for similar geological evidence. He gathered data from these marshes because he expected they would serve as natural recorders of earthquakes if any had occurred.

Evidence from Soil Layers

Atwater reported his results in 1987. His research revealed that parts of the Washington coast had indeed fallen as many as six times over the last several thousand years. He knew that coastal land may rise or fall due to causes other than earthquakes. But he determined the evidence could be best explained by a large earthquake.

Mud Atwater observed, for example, mud from estuaries deposited directly on top of soil layers. This suggested to him that the sunken areas had very quickly become covered by mud. If these areas had fallen more slowly, he reasoned, there would have been some evidence of a gradual transition between the mud and the underlying soil.

One way geologists collect data is by studying layers of soil and sediment called soil horizons.

Peat Soils record the environment of the lowlands as sediment layers. At low tide, the soil layers can be seen in riverbanks. After land drops suddenly during a great earthquake, the killed freshwater coastal plants are transformed into dark soils, called peat.

Sand Another key piece of evidence best explained by a tsunami was found in a layer of sand. The sand, in the form of a sheet, rested directly on the buried soil and contained marine diatoms. Atwater suspected that the same earthquake that lowered the land triggered a tsunami that washed over the freshly down-dropped coast.

Evidence of Widespread Tree Deaths

Atwater next formed a team to look for further evidence of ground sinking farther down the Washington coast. There they found Sitka spruce stumps in coastal areas. The trees had died centuries earlier, after being engulfed by tidal mudflats. Radiocarbon dates of the stumps yielded tree-death ages that were indistinguishable in two areas 55 km apart. This radiocarbon dates suggested the trees had died about 300 years earlier. This result supported the hypothesis that a single great earthquake had caused the land to fall.

Victims of sudden, earthquake induced lowering of the land, these tree stumps along the Washington coast contain a lasting record of an earthquake's occurrence. By studying the preserved tree rings, geologists can determine when the earthquake occurred.

Before earthquake | Several months after earthquake | Several decades after earthquake

Soil | Ocean water | Silt

When the land is suddenly lowered, sea water rushes in. Trees cannot tolerate the sea water and die. Eventually, ocean water is replaced with mud, and the dead trees rot. Often only their stumps are preserved in the mud.

When did the trees die?

David Yamaguchi of the University of Washington decided to look at groves of weather-beaten western red cedar tree trunks in coastal marshes. Like the Sitka spruce trees, the cedars had died centuries earlier. However, the cedars were resistant to rotting so were better preserved. Yamaguchi compared the growth rings from the western cedars with those of cedars that were still alive. He attempted to date the year of the trees' deaths, but the bark and outer rings had worn away after centuries of exposure to the elements. Still, Yamaguchi's rough estimates were consistent with the radiocarbon estimates for the spruce trees.

A team led by Alan Nelson of the U.S. Geological Survey in Denver next performed radiocarbon dating on the earthquake-killed trees. In 1995, they reported that the trees all along the Pacific Northwest coast had died between 1695 and 1720. The Cascadia event had apparently affected most of the Pacific Northwest. But was there a single giant earthquake or a series of smaller earthquakes?

Japanese scientists had been following this research closely. They had reason to believe that the Cascadia event was linked to a tsunami that struck the Japanese coast in 1700.

Radiocarbon Dating

When cosmic rays enter the atmosphere, they convert nitrogen into a radioactive form of carbon. This material is taken up by living plants. After the plants die, this exchange no longer takes place. Over time, the radioactive carbon decays. But the ratio of radioactive to nonradioactive carbon in Earth's atmosphere remains constant. Therefore, the age of plant material can be estimated from the ratio of the two types of carbon.

Historical Evidence from Japan

A large tsunami struck the Japanese coast January 27–28, 1700. Earthquake historians discovered accounts of this event when they were reviewing old documents. They found no mention of an accompanying earthquake. There was no explanation for its origin. By the 1990s, there was much interest in this orphan tsunami.

In 1996, Kenji Satake and his colleagues at the Geological Survey of Japan published an astonishing paper. They proposed that a single earthquake off the Pacific Northwest coast had produced the Japanese tsunami. An account of their research appears in *The Orphan Tsunami of 1700.*

Historical accounts suggested that the waves from the 1700 tsunami reached 2–5 m in height in northern Japan. The researchers compared these heights with those produced by other foreign tsunamis. The comparisons suggested the parent earthquake of 1700 had a magnitude of 9.0.

Historical records and geological evidence allowed the researchers to rule out quake origins other than the Pacific Northwest. Given its magnitude, they concluded the quake had ruptured the entire length of the Cascadia fault. Calculating backward from the time the tsunami arrived, they estimated that the earthquake struck at 9:00 P.M. on January 26, 1700.

Japanese Tsunamis

Approximately 22 percent of all documented tsunamis around the world have occurred in Japan. The Japanese coastline has the most extensive tsunami documentation anywhere. Records of tsunami there go back to A.D. 684. Most of these ocean waves have resulted from local earthquakes, but some have had foreign origins.

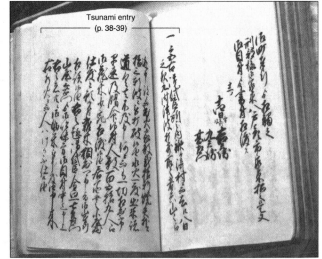

These historic administrative records document a tsunami of unknown origin washing up on Japan's coastline in the year 1700.

Evidence Confirmed with Tree Rings

Scientists on the Pacific Northwest coast next set out to verify the January 1700 date. They had already established that an earthquake had killed trees in North America sometime between 1695 and 1720 based on radiocarbon techniques. But now they would need more refined methods to pinpoint the actual date.

David Yamaguchi and his coworkers decided to take another look at tree rings. They would try to locate the final growth rings in the western red cedars that had been studied earlier. Since the outer layers of the trunks of these trees were badly weathered, they would have to dig up the bark-covered roots.

Yamaguchi's team compared the rings in the trunk with those in the roots. They found that in all but one case, the trees had died sometime after the 1699 growing season but before the 1700 growing season. That finding placed the date of the earthquake somewhere between August 1699 and May 1700.

Cascadia Today

Geologists place the chances of another earthquake occurring on the Pacific Northwest coast at 100 percent. Although it is not possible to prevent a tsunami, California has taken measures to detect tsunamis and rapidly evacuate people from tsunami prone areas in the event that one occurs. Recent seafloor studies have shown that subduction earthquakes on the Cascadia fault occur roughly every 500 to 600 years. But the intervals between those earthquakes have ranged from 200 to 1,000 years. No one can say when the next one will occur. Nor does anyone know whether the entire fault will rupture as it did in 1700 and produce another mega-tsunami.

Meanwhile, a second team led by Gordon Jacoby of the Lamont-Doherty Earth Observatory in New York looked at the dead Sitka spruce in Washington and northern Oregon. They hoped to find evidence of stress produced by post-earthquake tides. The evidence was there. In half the trees examined, the shape of the annual rings had changed between 1700 and 1710.

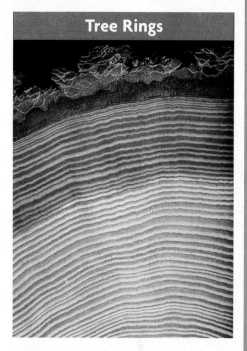

Tree Rings

Each year, living trees form new rings of tissue in their trunks. These rings are called annual growth rings. This photo shows the cross section of a western red cedar that was killed by the 1700 Cascadia tsunami. Scientists were able to determine that the tree died between August 1699 and May 1700 by counting the tree rings.

Tsunami hazard zone and tsunami evacuation route signs help direct you to high ground in the event of a tsunami.

The Unique Planet Earth
The structure of our planet, both inside and out, makes it the only place in our solar system that harbors intelligent life.

West-Coast Events

30 Million Years Ago
North American Plate first touches the Pacific Plate, eventually causes the San Andreas Fault.

10 Million Years Ago
San Andreas Fault first moves.

1562
Diego Gutierrez makes first map of Baja California.

1777
San José was established as the first city in California.

A.D. 1   1500   1700   1750

World Events

c. 150
Claudius Ptolemy writes book on geography that includes color maps based on knowledge of Earth's surface at the time.

1746
Jean Etienne Guettard presents first map of minerals in France to French Academy of Sciences.

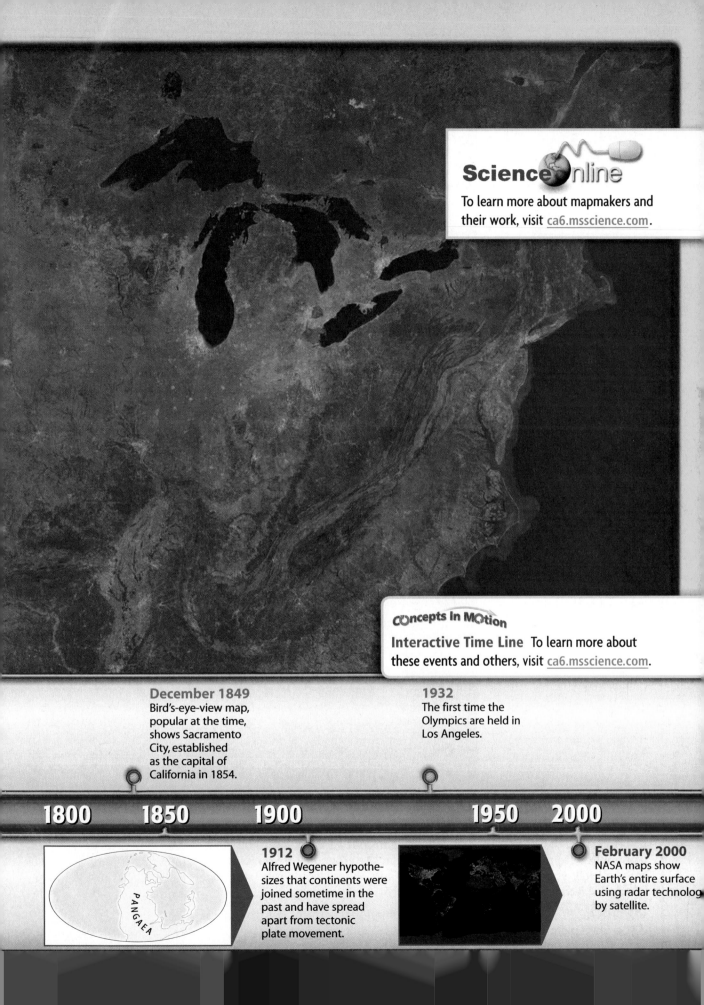

Science Online
To learn more about mapmakers and their work, visit ca6.msscience.com.

Concepts In Motion
Interactive Time Line To learn more about these events and others, visit ca6.msscience.com.

December 1849
Bird's-eye-view map, popular at the time, shows Sacramento City, established as the capital of California in 1854.

1932
The first time the Olympics are held in Los Angeles.

1800 1850 1900 1950 2000

1912
Alfred Wegener hypothesizes that continents were joined sometime in the past and have spread apart from tectonic plate movement.

PANGAEA

February 2000
NASA maps show Earth's entire surface using radar technolog by satellite.

CHAPTER 1

Mapping Earth's Surface

The BIG Idea

Earth's surface can be represented in many ways. Maps provide two-dimensional views of Earth's three-dimensional surface.

LESSON 1 7.f
Reading Maps

Main Idea Maps represent large areas of Earth at a size we can easily see and study.

LESSON 2 7.b, 7.c, 7.f, 7.h
Topographic and Geologic Maps

Main Idea Specialty maps are used to show specific features such as changes in elevation and geologic characteristics.

Where would you build your home?

Do you wish to live close to the beach? A single centimeter on this satellite image of the San Francisco Bay area might represent a long distance on land. Can you tell from this image how steep or flat the ground is? If you look more closely, you can see the San Andreas Fault. Where would it be safe to build your home?

Science *Journal* List some kinds of information you might get from maps if you were planning to build a new home.

Launch Lab

00:20 minutes

How might you map your neighborhood?

What would your neighborhood look like from the air? How would you draw it on paper?

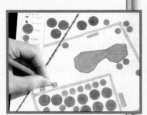

Think About This

- **Consider** What would you see on the ground if you flew in a plane over your neighborhood?

- **Determine** Think about where objects are in relation to other objects.

Procedure

1. Make an empty two-column table. In the left column, list ten objects you would include in a map of your neighborhood.

2. In the right column, draw each object as it might appear from above.

3. Make a list of six symbols to represent objects such as trees, houses, and roads.

4. Draw a map of your neighborhood using a **ruler, grid paper,** and **colored pencils.**

 7.f

Science Online

Visit ca6.msscience.com to:

▶ view **Concepts in Motion**

▶ explore Virtual Labs

▶ access content-related Web links

▶ take the Standards Check

Types of Maps Make the following Foldable to organize information about different types of maps.

STEP 1 **Fold** a sheet of paper in half from top to bottom and then in half from side to side.

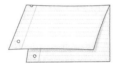

STEP 2 **Unfold** the paper once. **Cut** along the fold of the top flap to make two flaps.

STEP 3 **Label** the flaps as shown.

Topographic Maps | Geologic Maps

Reading Skill **ELA6: R 2.4**

Monitoring Your Comprehension As you read this chapter, list details about each type of map. Include information about how each map looks, what information it contains, and how it is used.

Get Ready to Read

Preview

① Learn It! If you know what to expect before reading, it will be easier to understand ideas and relationships presented in the text. Follow these steps to preview your reading assignments.

1. Look at the title and any illustrations that are included.
2. Read the headings, subheadings, and anything in bold letters.
3. Skim over the passage to see how it is organized. Is it divided into many parts?
4. Look at the graphics—pictures, maps, or diagrams. Read their titles, labels, and captions.
5. Set a purpose for your reading. Are you reading to learn something new? Are you reading to find specific information?

② Practice It! Take some time to preview this chapter. Skim all the main headings and subheadings. With a partner, discuss your answers to these questions.

- Which part of this chapter looks most interesting to you?
- Are there any words in the headings that are unfamiliar to you?
- Choose one of the lesson review questions to discuss with a partner.

③ Apply It! Now that you have skimmed the chapter, write a short paragraph describing one thing you want to learn from this chapter.

Target Your Reading

Use this to focus on the main ideas as you read the chapter.

Reading Tip

As you preview this chapter, be sure to scan the illustrations, tables, and graphs. Skim the captions.

1 **Before you read** the chapter, respond to the statements below on your worksheet or on a numbered sheet of paper.

- Write an **A** if you **agree** with the statement.
- Write a **D** if you **disagree** with the statement.

2 **After you read** the chapter, look back to this page to see if you've changed your mind about any of the statements.

- If any of your answers changed, explain why.
- Change any false statements into true statements.
- Use your revised statements as a study guide.

Before You Read A or D	Statement	After You Read A or D
	1 Latitude lines run north to south.	
	2 Degrees, minutes, and seconds can be used to measure distance on maps.	
	3 A map legend is a historic map.	
	4 Longitude lines run north to south.	
	5 A meridian is a longitude line that forms a semicircle.	
	6 Contour lines run up and down on hillsides.	
	7 Contour intervals indicate horizontal distance on topographic maps.	
	8 A map scale is used to measure the weight of heavy maps.	
	9 Geologic maps use color to indicate the different ages of rocks.	
	10 Geologic cross sections can be used to visualize the slope of geologic formations beneath Earth's surface.	

Sciencenline

Print a worksheet of this page at
ca6.msscience.com.

Science Content Standards

7.f Read a topographic map and a geologic map for evidence provided on the maps and construct and interpret a simple scale map.

Reading Guide

What *You'll Learn*

▶ **Define** *latitude* and *longitude.*

▶ **Explain** how latitude and longitude are used to determine a location on Earth.

▶ **Demonstrate** how map scales are used.

Why *It's Important*

Being able to get from one place to another or to determine your exact location is an important life skill.

New Vocabulary

longitude
latitude
map view
profile view
map legend

Review Vocabulary

pole: either end of an axis of a sphere, especially of Earth's axis (Grade 5)

Reading Maps

Main Idea Maps represent large areas of Earth at a size we can easily see and study.

Real-World Reading Connection Have you ever pretended to find a map to lost treasure? Was the treasure buried somewhere on a deserted island? Look at **Figure 1,** and use your imagination to hypothesize why pirates would bury their treasure and make treasure maps.

Understanding Maps

Maps have been used for many centuries, and there are countless types of maps. All show where things are on Earth or where things are in relation to each other. For example, a street map shows the locations of streets in relation to other streets, towns, and landmarks. Other maps may show the position of a weather system.

A single map is like a picture of the location of things at a given time. However, a series of maps drawn over many years can show how Earth's surface and interior change over time.

Because Earth is large, humans need a way to determine where they are located on the planet. Imagine telling someone your exact position on the snow-covered continent of Antarctica. It would be difficult to describe. Ships' captains and airplane pilots experience the same problems as they plot their courses across Earth.

Figure 1 How might pirates have indicated the locations of their buried treasure?

A Grid System for Plotting Locations

Mapmakers created a system for identifying locations on Earth. This system uses an imaginary grid of lines that encircles the globe. The intersection, or crossing, of these lines can be used to pinpoint a location.

Latitude and Longitude Two sets of lines, called latitude and longitude, make up this imaginary grid. **Longitude** is the distance in degrees east or west of the prime **meridian.** The prime meridian, shown in **Figure 2,** is a longitude line that runs from the north pole to the south pole and passes through Greenwich, England. This line represents zero degrees longitude. If you were to travel 180° east or west from the prime meridian, you would reach another longitude line called the 180° meridian. It lies exactly opposite of the prime meridian, on the other side of Earth. Separately, each of these lines is a semicircle. Together, they form a complete circle that divides Earth into two imaginary halves—the eastern and western hemispheres. The 180° meridian also forms part of the International Date Line.

Latitude is the distance in degrees north or south of the equator. Earth's equator, shown in **Figure 2,** represents zero degrees latitude. It is a latitude line that runs perpendicular to the longitude lines. The equator also divides Earth into two halves—the northern and southern hemispheres. Latitude lines form complete circles. The equator forms the largest circle. All remaining latitude lines form smaller and smaller circles as they approach Earth's poles.

Degrees Lines of latitude and longitude are labeled in units of degrees (°). Because Earth is spherical, each set of the lines of latitude and longitude make up 360°. The north pole is located at 90° north latitude (90°N) and the south pole is located at 90° south latitude (90°S). There are an infinite number of latitude and longitude lines on Earth. However, on a globe, they are generally labeled every 10°.

Minutes and Seconds Because Earth is so large, it is necessary to make this location system more precise. So, the lines of latitude and longitude are divided into smaller units. Each degree of latitude or longitude is divided into 60 minutes ('). Each minute is divided into 60 seconds ("). The degrees, minutes, and seconds of a line of latitude or longitude can be used to pinpoint a precise location on a map.

WORD ORIGIN

meridian

meri– prefix; from pre-Latin; means *mid*

–dian from pre-Latin *dies;* means *day*

Figure 2 Latitude lines run east to west, but longitude lines run north to south.

Observe Why do you think latitude lines are called parallels, but longitude lines are called meridians?

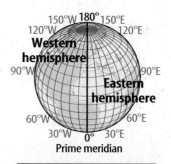

Longitude

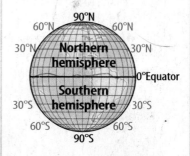

Latitude

Figure 3 The tiny marks on the *x*-axis and *y*-axis are called tic marks. California occupies land from 32°N to 42°N, and from 114°W to 124°W.

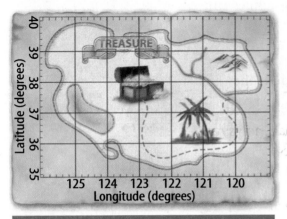

Treasure Map

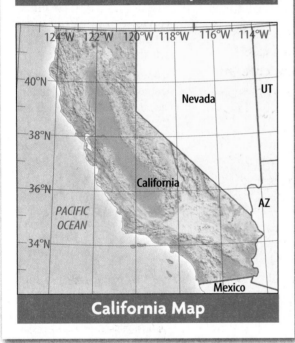

California Map

Figure 4 The terms *map view* and *plan view* mean the same thing.

Infer Why do you think geographers use the term *map view* but architects use the term *plan view*?

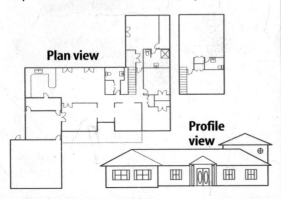

Plotting Locations The lines of latitude and longitude intersect, or cross each other at an infinite number of places on Earth. These intersections describe precise locations on Earth. A location on Earth's surface is described by the intersection of the nearest lines of latitude and longitude. The combination of the latitude number and the longitude number is referred to as a coordinate. Latitude always is listed first when describing a location. For example, Sacramento, California is located near 38°N latitude by 121°W longitude. The coordinates of California's State Capitol Building in Sacramento, however, is 38°34'33" north (N) latitude by 121°29'29" west (W) longitude. Practice finding locations by describing the location of the treasure in **Figure 3.** Then try to determine the latitude and longitude of your hometown.

 Reading Check Which is listed first—latitude or longitude—when describing a location?

Different Views

Most maps are drawn in **map view,** which means they are drawn as if you were looking down on an area from above Earth's surface. Map view may also be referred to as plan view. Lines of latitude and longitude usually are drawn on a map-view map.

Cross sections are drawn in **profile view,** which is a drawing showing a vertical section of the ground. A profile view is like a side view of a house. To help you visualize this concept, a map view and a profile view of a house are shown in **Figure 4.** Map views and profile views will be used to describe topographic maps and geologic maps in Lesson 2. Also, you will use profile views when you study models of the inner structures of volcanoes in Chapter 7.

 **Reading Check** How is a profile view related to Earth's surface?

Map Scales and Legends

Maps have two features to help you read and understand the map. They are a map legend and and a map scale.

Map Legends

Maps use specific symbols to represent certain features on Earth's surface. These symbols allow mapmakers to fit neatly many details on a map. All maps include a key, called a map legend, so you can interpret the symbols. A **map legend** lists all the symbols used on the map and explains what each symbol means. **Figure 5** shows an example of a map legend.

Map Scales

When mapmakers draw a map, they need to decide how big or small to make the map. They need to decide on the map's scale.

The map's scale tells you the relationship between a distance on the map and the actual distance on the surface being mapped. The scale can be written as a phrase such as, "1 centimeter is equal to 1 kilometer." The scale also can be written as a **ratio,** such as 1:100. Because this is a ratio, there are no units. Verbally, you would say, "Every unit on the map is equal to 100 units on the ground." If your unit were 1 cm on the map, it would be equal to 100 cm on the ground. If you drew a map of your school on a scale of 1:1, your map would be as large as your school.

Scale bars also are shown on maps to help determine distance. First, you use a ruler to measure the distance on the map. Then you compare that distance to the scale bar. There are many kinds of scale bars. Some examples of scale bars are shown in **Figure 5.**

SCIENCE USE V. COMMON USE

legend

Science Use an explanatory list of the symbols on a map or chart. *You can use the legend to interpret the different symbols.*
Common Use a story that is popularly regarded as historical but not verifiable; a popular myth. *Everyone loved the many legends of the western frontier.*

ACADEMIC VOCABULARY

ratio

(noun) the relationship in quantity or size between two or more things
The ratio of boys to girls in Ms. Smith's class was 2:1.

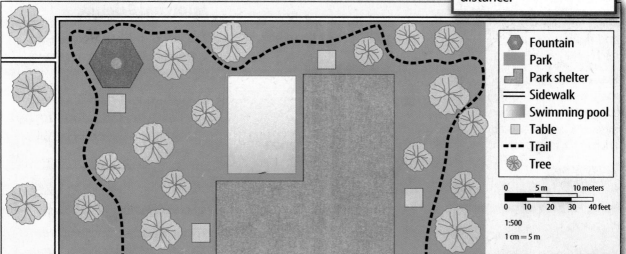

Figure 5 On a map scale, the map distance is listed before the actual distance.

Fountain
Park
Park shelter
Sidewalk
Swimming pool
Table
Trail
Tree

0	5 m	10 meters

0	10	20	30	40 feet

1:500

1 cm = 5 m

The Usefulness of Maps

Maps are used to help you locate places on Earth. Latitude and longitude are important for plotting locations on Earth. Subdividing latitude and longitude into degrees, minutes, and seconds makes it possible to pinpoint exact locations. In Chapter 6, you will use latitude and longitude to plot locations of earthquakes and faults. Map scales are included on maps to represent the actual surface distance that is being represented. Map legends often provide a key that explains different symbols that can be used to interpret maps. Depending on your purpose, map views or profile views can be used to help you visualize your surroundings and to find your way around. As populations grow, the development of maps and what they show will help humankind plan for the future.

LESSON 1 Review

Summarize

Create your own lesson summary as you organize an **outline**.

1. **Scan** the lesson. Find and list the first **red** main heading.

2. **Review** the text after the heading and list 2–3 details about the heading.

3. **Find** and list each **blue** subheading that follows the **red** main heading.

4. **List** 2–3 details, key terms, and definitions under each **blue** subheading.

5. **Review** additional **red** main headings and their supporting **blue** subheadings. List 2–3 details about each.

 ELA6: R 2.4

Standards Check

Using Vocabulary

1. **Differentiate** between latitude and longitude. `7.f`

2. **Define** *map view* in your own words. `7.f`

Understanding Main Ideas

3. **Explain** why coordinates used to determine locations include degrees, minutes, and seconds. `7.f`

4. **Translate** the information in the figure below into words. `7.f`

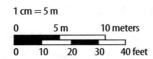
1 cm = 5 m

5. **Diagram** a map view and a profile view of a turkey sandwich. `7.f`

6. Which is the correct way to write location coordinates?

 A. 35° 16' 12" N latitude by 160° 18' 25" W longitude

 B. 12" 16' 36° N latitude by 25" 18' 160° W longitude

 C. 160° 18' 25" W longitude by 35° 16' 12" N latitude

 D. 35° 16' 12" N latitude by 25" 18' 160° W longitude `7.f`

Applying Science

7. **Critique** the usefulness of a road map that has no scale. `7.f`

8. **Outline** the changes a map of a city might show as it grew from the first settlers until today. `7.f`

Science nline

For more practice, visit **Standards Check** at ca6.msscience.com.

MiniLab

Can you map a classroom?

Making a simple scale map of your classroom is easier than mapping your neighborhood. It also is easy to make exact measurements.

Procedure

1. Make a simple sketch of your classroom.

2. Use a **meterstick** to measure the:
   - length and width of the classroom
   - size of the doors and windows
   - locations of the desks, tables, shelves, and other objects

3. Mark your measurements on your sketch. Refer to the figure above as an example.

4. Calculate your map scale by dividing the actual distance by the size of your map area. For more help, see *Applying Math* at the end of Lesson 2.

5. Use your map scale to calculate the scaled distance for each measured distance on your sketch. Draw all the objects in their correct locations according to your scale.

6. Draw a map legend. Refer to the figure at the right to get an idea about what your final drawing might look like.

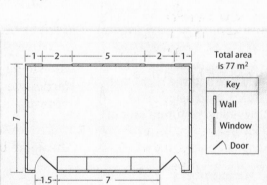

Analysis

1. **Describe** how you made your sketch. Explain any problems you had when making your measurements and sketch.

2. **Analyze** any errors you made in your scale calculations. Did you make the same mistake over again? If you repeat certain math errors, think of a strategy to avoid each type of error.

 Science Content Standards

7.f Read a topographic map and a geologic map for evidence provided on the maps and construct and interpret a simple scale map.

Science Content Standards

7.c Construct appropriate graphs from data and develop qualitative statements about the relationships between variables.

7.f Read a topographic map and a geologic map for evidence provided on the maps and construct and interpret a simple scale map.

7.h Identify changes in natural phenomena over time without manipulating the phenomena (e.g., a tree limb, a grove of trees, a stream, a hillslope).

Also covers: 7.b

Reading Guide

What *You'll Learn*

▶ **Understand** the function of topographic maps.

▶ **Explain** how topographic maps show changes in elevation.

▶ **Describe** geologic maps.

▶ **Compare** topographic maps and geologic maps.

Why *It's Important*

Knowing the shape of Earth's surface and the rocks that make it up helps in the planning for future uses of an area.

Vocabulary
topographic map
contour line
geologic map
geologic formation
contact

Review Vocabulary
geology: the study of Earth

Topographic and Geologic Maps

(Main Idea) Specialty maps are used to show specific features such as changes in elevation and geologic characteristics.

Real-World Reading Connection The geology of California is a complex mix of different rock types and many faults. Geologists and engineers use the information contained in topographic and geologic maps to help them decide where to build homes and businesses. These maps help them determine where it might be unsafe or undesirable to build.

Topographic Maps

Topography refers to the shape of a surface, including its elevation and the position of its features. A special characteristic of **topographic maps** is that they use lines of equal elevation to show the shape of Earth's surface. They show natural features, such as mountains, valleys, rivers, lakes, and coastlines. They also show cultural features—those created by people—such as buildings, roads, and towns. Topographic maps are used for planning, engineering, military, science, and recreation purposes. **Figure 6** shows some of the tools that are used to collect the field data that is used to make these maps.

Figure 6 These surveyors are using equipment to help them calculate land elevation.
Infer What type of map might they be making?

Contour Lines

The lines drawn on maps to join points of equal elevation are called **contour lines.** On topographic maps, contour lines indicate elevation as the distance above sea level. Contour lines make it possible to measure such things as the height of a mountain, the depth of the ocean bottom, and the steepness of a slope. The map view in **Figure 7** illustrates that each contour line represents equal elevation above sea level.

The contour **interval** is the difference in elevation between contour lines that are next to each other. If the contour interval were 20 feet (6.1 m), then the change in elevation after five contour lines would be 100 feet (30.5 m). The contour interval is selected to show the general shape of the terrain without overcrowding the map with too many lines.

Some contour lines are printed with a darker or wider line than others. These are called index contours. Generally, it is only the index contour line that has the elevation written on it.

 **Reading Check** What is the difference between a contour interval and an index contour?

Topographic Profiles

Sometimes, the map view of an area is not enough. You can get a better feeling for the topography of Earth's surface from a profile view, or cross section. A profile view looks as though you were standing on Earth's surface and looking toward the horizon. Examine the steepness of the slope in **Figure 7** by comparing the map view and profile view.

Figure 7 Each contour line shows the same elevation everywhere on the map.

Explain If you walked along a contour line, would you be going uphill, downhill, or neither? Why?

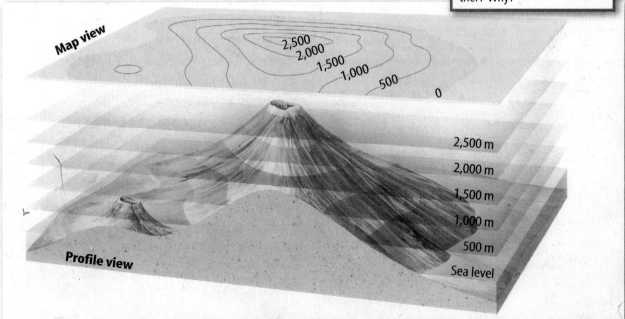

Map view

2,500
2,000
1,500
1,000
500
0

2,500 m
2,000 m
1,500 m
1,000 m
500 m
Sea level

Profile view

C⊙ncepts in M⊙tion

Interactive Table To organize information about topographic map symbols, visit Tables at ca6.msscience.com.

Table 1 USGS Topographic Map Symbols

Description	Type	Symbol
Primary highway	Artificial	
Secondary highway	Artificial	
Unimproved road	Artificial	
Railroad	Artificial	
Buildings	Artificial	
Urban area	Artificial	
Index contour	Landscape	～100～
Intermediate contour	Landscape	
Perennial streams	Landscape	
Intermittent streams	Landscape	
Wooded marsh	Landscape	
Woods or brushwood	Landscape	
Scales		
Ratio	1:24,000	
English	1 ½ 0 1 Mile	
Metric	1 0.5 0 1 Kilometer	

Symbols on Topographic Maps

Interpreting the colors and symbols on topographic maps is an important step in understanding and using topographic maps. For example, contour lines are shown in brown on United States Geological Survey (USGS) topographic maps. Lakes, streams, irrigation ditches, and other water-related features are shown in blue. Vegetation is always green. Land grids and important roads are red. Smaller roads, trails, railroads, boundaries, and other cultural features are shown in black. On some USGS topographic maps, the color purple is used to indicate updated information or changes that were made to an area after the map was originally printed.

 Reading Check What does USGS mean?

USGS topographic maps use a wide variety of symbols to display information. For example, individual houses may be shown as small black squares. But, for large buildings, such as the Rose Bowl in Pasadena, the actual shapes are drawn. Only a tint of color, such as pink, is used to show large cities or areas of high population. **Table 1** shows part of the long legend that accompanies every USGS topographic map.

Figure 8 shows a portion of a topographic map that you might use if you were visiting Yosemite National Park. See if you can find the name of the river that runs through Yosemite Valley. The north direction is at the top of the page on standard USGS topographic maps. Also notice that closely spaced contour lines indicate steep slopes. Widely spaced lines show areas where the land is relatively flat. Contour lines that form a V-shape often indicate the locations of streams or erosion channels.

 Visual Check **Figure 8** Which direction is El Capitan from the location of the large red triangle?

Visualizing Topography

Figure 8

Planning a hike? A topographic map will show you changes in elevation. With such a map, you can see at a glance how steep a mountain trail is, as well as its location relative to rivers, lakes, and roads nearby.

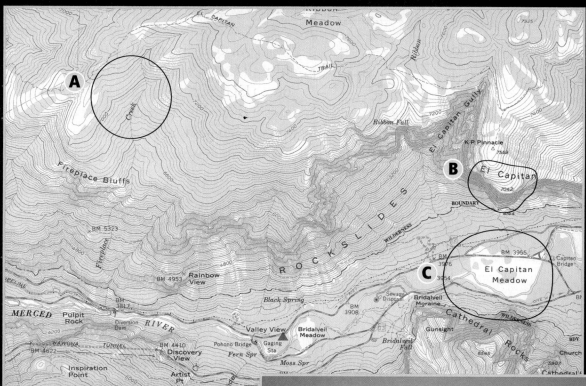

A V-shaped contour lines can indicate the location of a stream or dry gulch. The bottom of the V points upstream.

B Where contour lines on a topographic map are close together, elevation is changing rapidly—and the trail is very steep!

C Widely spaced contour lines indicate a gentle slope—in some places the terrain might be almost flat!

Yosemite Valley as it is viewed from the red triangle in the topographic map.

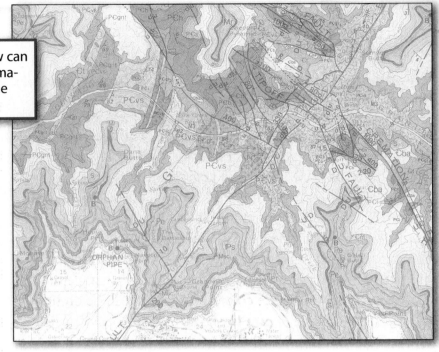

Figure 9 The map key below can be used to locate geologic formations in the geologic map of the Grand Canyon to the right.

QUATERNARY
- S — Landslides and rockfalls
- r — River sediments

PERMIAN
- Pk — Kaibab Limestone
- Pt — Toroweap Formation
- Pc — Coconino Sandstone
- Ph — Hermit Shale
- Pe — Esplanade Sandstone

PENNSYLVANIAN
- IPs — Supai Formation

MISSISSIPPIAN
- Mr — Redwall Limestone

DEVONIAN
- Dtb — Temple Butte Limestone

CAMBRIAN
- Cm — Muav Limestone
- Cba — Bright Angel Shale
- Ct — Tapeats Sandstone

YOUNGER PRECAMBRIAN
- PCi — Diabase sills and dikes
- PCs — Shinumo Quartzite
- PCh — Hakatai Shale
- PCb — Bass Formation

OLDER PRECAMBRIAN
- PCgr₁ — Zoroaster Granite
- PCgnt — Trinity Gneiss
- PCvs — Vishnu Schist

Geologic Maps

Geology has an important effect on life around us, from the likelihood of landslides to the availability of groundwater in wells. It affects the type of soil on the land and the kinds of plants that grow there. **Geologic maps** represent the geology of an area. They can be used for many things, such as locating valuable minerals or understanding earthquakes.

Geologic Formations

The most striking feature of geologic maps is the display of colors. This is shown on the geologic map in **Figure 9.** The colors make it easier to read the maps. Each color represents a different geologic formation or rock unit. A **geologic formation** is a three dimensional body, or volume, of a certain kind of rock of a given age range. For example, sandstone of one age might be represented by light yellow, while sandstone of a different age might be colored bright orange.

The colors on a geologic map are not intended to show actual rock colors. The colors are used only to separate the rocks into different formations. Some formations span millions of years. In these cases, variations of the same color are generally used. For example, in **Figure 9,** rocks of Precambrian age are shades of pinks and orange. A map key, like the one shown in **Figure 9,** is always included on a geologic map.

Visual Check

Figure 9 Use this geologic map with its map key to locate the oldest rocks in the Grand Canyon.

Contacts

All rock formations are formed over, under, or beside other formations. The place where two rock formations occur next to each other is called a **contact.** There are two main types of contacts—depositional contacts and fault contacts. Depositional contacts occur between rocks that formed when sediments were deposited on other sediments. The rock layers of the Grand Canyon are a good example. A depositional contact is shown by a thin line on a geologic map.

There are many places in California where rock formations are broken up and moved. When rock formations that are next to one another have been moved, the contact is called a fault contact. Fault contacts are usually shown as thick black lines on geologic maps. Faults can cut through a single rock unit. This means the same colored rock unit will be on both sides of the fault line. **Figure 10** shows examples of depositional contacts and fault contacts.

 What is the difference between a depositional contact and a fault contact?

Making Geologic Maps

Geologists collect and study many rock samples from Earth's surface. They use their data to create geologic maps. Many rock units are named according to where they are viewed best, or where they were studied first. For example, the Briones Sandstone was first described by geologists studying rocks in Briones Valley, California.

The colored areas on geologic maps represent the rock units that are nearest to the surface. Usually, the soil on top of rock units is not mapped. So, you might have to dig through a lot of soil in some places to reach the rocks.

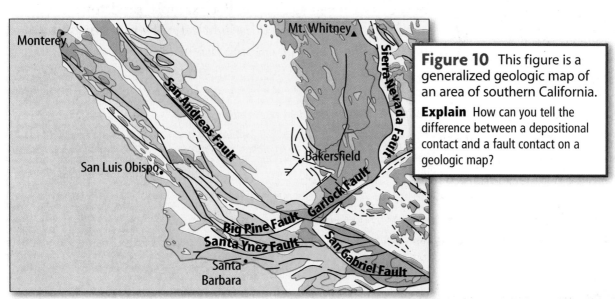

Figure 10 This figure is a generalized geologic map of an area of southern California.

Explain How can you tell the difference between a depositional contact and a fault contact on a geologic map?

SCIENCE USE v. COMMON USE

bed

Science Use a unit of material commonly arranged or deposited in layers. *A bed of sand was laid down before the workers poured the concrete.*

Common Use a piece of furniture on which to sleep. *The children fell asleep shortly after they went to bed.*

Geology Beneath the Surface

Often, it is necessary to know about the geology below Earth's surface. So, how do geologists draw maps of what is underground? Whenever possible, geologists study outcrops or cliffs where the rocks are exposed. However, sometimes they need to drill deep into the ground and retrieve samples of rock. These samples are shaped like long tubes and are called cores. The core samples are studied and the data is recorded for scientific research.

Geologic Cross Sections

What does a core sample or an outcrop show? Imagine yourself at a birthday party where they have just cut the cake. The cake might have several different layers. Many kinds of rocks form in broad, flat layers, called beds. The rock beds stack up just like the layers of a cake. In some parts of California, thick stacks of rock beds remain in their original flat orientation. In other places, however, forces from within Earth have bent and tilted the rocks.

A geologic cross section shows how the rocks are stacked under the surface. It helps geologists interpret what is beneath the surface of the land. **Figure 11** shows an example of a geologic map and a geologic cross section.

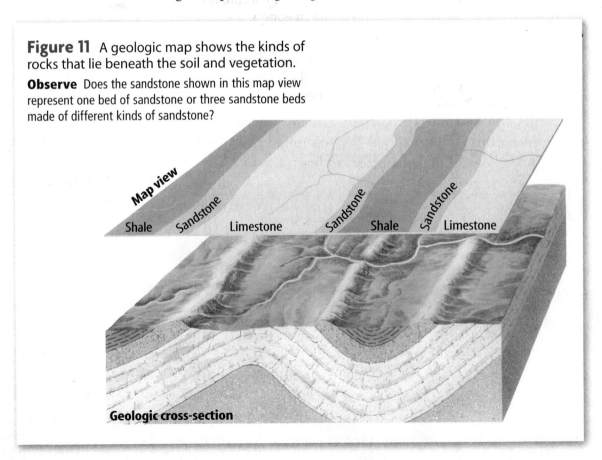

Figure 11 A geologic map shows the kinds of rocks that lie beneath the soil and vegetation.

Observe Does the sandstone shown in this map view represent one bed of sandstone or three sandstone beds made of different kinds of sandstone?

Map view

Shale Sandstone Limestone Sandstone Shale Sandstone Limestone

Geologic cross-section

Geology and Maps

Topographic maps and geologic maps are essential tools for geologists. Topographic maps show changes in elevation and surface features, along with improvements such as roads and buildings. Geologic maps show the distribution of surface rock formations. Together these maps provide a complete picture of the geology of a particular area.

LESSON 2 Review

Standards Check

Using Vocabulary

1. **Compare and contrast** road maps and topographic maps. **7.f**

2. **List** the characteristics of a geologic formation. **7.f**

Understanding Main Ideas

3. **Relate** the spacing of contour lines to the steepness of an area. **7.f**

4. **Explain** how topographic maps show the features of Earth's surface. **7.f**

5. **Describe** the difference between a topographic profile and a geologic cross section. **7.f**

6. **Draw and label** four contour lines that illustrate a hill on a map. **7.f**

7. **Copy and fill** in this graphic organizer to list the features that are commonly included on topographic maps. **7.f**

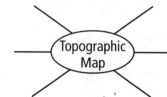

Topographic Map

8. What does the contour interval on a topographic map represent?

   A. distance from a mountain peak

   B. volume of land beneath the surface

   C. distance above or below sea level

   D. lengths of the contour lines **7.f**

Applying Science

9. **Explain** why the contour interval on a topographic map of a mountainous area usually is large. **7.f**

10. **Infer** why soils are not included on a geologic map. **7.f**

Science nline

For more practice, visit **Standards Check** at ca6.msscience.com.

DataLab

How does a landscape change over time?

Landscape can change over time. Humans build cities, dams, and farms. In this lab, you will look at maps of the same area over a period of time and see how the landscape has changed.

Data Collection

Look at two topographic maps provided by your teacher. Record the year of the oldest map as Year 1. Record the year of the most recent map as Year 2.

Data Analysis

1. **List** the features you can identify on the maps. This might include roads, buildings, streams, wetlands, fences, and so on.

2. **Count** the number of each type of feature you can find on each map. **Record** these numbers in your data table.

3. **Draw** line graphs with data points to show the change in the number of features over time. Which features increased in number over the years? Which features decreased?

4. **Write** descriptive observations of any other changes you see in the maps, such as stream paths being changed or features moving as a result of earthquakes.

Year 1

Year 2

 Science Content Standards

7.c Construct appropriate graphs from data and develop qualitative statements about the relationships between variables.

7.f Read a topographic map and a geologic map for evidence provided on the maps and construct and interpret a simple scale map.

7.h Identify changes in natural phenomena over time without manipulating the phenomena (e.g., a tree limb, a grove of trees, a stream, a hillslope).

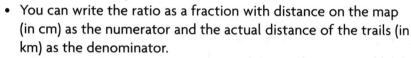

Applying Math

Using Ratios

Ratios show how one quantity is related to another quantity in fraction form. If a map of a park has a key showing that 1 cm on the map equals 5 km of trails, you can write two different ratios of km of trails to cm on the map.

- You can write the ratio as a fraction with distance on the map (in cm) as the numerator and the actual distance of the trails (in km) as the denominator.

$$\frac{\text{distance on the map}}{\text{actual distance of trails}} \quad \text{or} \quad \frac{1 \text{ cm}}{5 \text{ km}}$$

- You can also write a fraction the opposite way, as shown below.

$$\frac{\text{actual distance of trails}}{\text{distance on the map}} \quad \text{or} \quad \frac{5 \text{ km}}{1 \text{ cm}}$$

0   5   10   15 km

Map key
1 cm = 5 km

Map Key
1 cm = 5 km of trails

Example

Which ratio would you need to find how many kilometers of trails are equal to 2.5 cm on the map of the park?

What you know: 1 cm on the map = 5 km of trails in the park

What you need to find: how many kilometers of trails = 2.5 cm on the map

1 Write a fraction for what you need to find: $\dfrac{? \text{ km of trails}}{2.5 \text{ cm on the map}}$

2 Select the ratio to relate kilometers of trails to cenimeters on the map: $\dfrac{5 \text{ km}}{1 \text{ cm}}$

3 Multiply to find your answer:

To change from 1 cm to 2.5 cm, multiply by 2.5. To find the number of kilometers of trails, multiply 5 by 2.5.

$$\frac{5 \text{ km}}{1 \text{ cm}} \xrightarrow[\times 2.5]{\times 2.5} \frac{12.5 \text{ km}}{2.5 \text{ cm}}$$

Answer: 2.5 cm on the map equals 12.5 km of trails.

Practice Problems

1. Write a ratio for the map key that shows how many cm are equal to how many km of bike trails.

   Map Key
   2 cm = 9 km of trails

2. How many kilometers of trails would it be for 8 cm on the map?

Science Online
For more math practice, visit **Math Practice** at ca6.msscience.com.

Model & Invent:
Mapping a Race Route

Materials

topographic map
clear plastic tub
modeling clay
poster board or
    cardboard
rulers
scissors

Safety Precautions

WARNING: *Always wear goggles in the lab. Wear lab coats and gloves to portect your clothing and hands when using stains.*

Science Content Standards

7.b Select and use appropriate tools and technology (including calculators, computers, balances, spring scales, microscopes, and binoculars) to perform tests, collect data, and display data.
7.f Read a topographic map and a geologic map for evidence provided on the maps and construct and interpret a simple scale map.

Problem

Your committee must choose a route for a cross-country bike race. You decide to make three-dimensional models of topographic maps to help choose the race route. The route should go up and down some steep slopes and over a trail that zigzags along a dried-up stream valley.

Form a Hypothesis

Do you think a three-dimensional map will be helpful in choosing a race route? Why or why not? Predict how a three-dimensional map might be helpful to the committee.

Collect Data and Make Observations

1. Read and complete the lab safety form.
2. **Interpret Data** Read a topographic map and figure out how to cut shapes from poster board to stack them into a three-dimensional map.
3. **Create** Can you think of other ways to create a three-dimensional contour map? If you want to use a method of your own, get approval for your idea and materials list before you begin.
4. **Use Models** When you have completed the three-dimensional map, discuss with your committee what the best race route would be, and why.

Analyze and Conclude

1. **Identify** the elevation interval between contour lines on your map.
2. **Locate** the area of highest elevation. What is the altitude? How high is it on your map?
3. **Describe** how you know where water is likely to flow. **Describe** these places on the topographic map and on your three-dimensional map.
4. **Locate** the areas with the steepest slope. **Explain** how you know by looking at the topographic map. **Describe** what the steepest slope looks like on your three-dimensional map.
5. **Communicate** the usefulness of the topographic map in mapping a race route. What factors were easier to assess with the three-dimensional model? What ideas did you get for the race route that you might not have had by looking at the flat topographic map?
6. **Error Analysis** Critique the procedure and tools you used to build the three-dimensional map. What would you do differently next time?

Communicate

WRITING in Science   **ELA6:** W 1.2

Share Your Data How do you think your three-dimensional area map could be useful to other members or groups in your community? Think of organizations that may be helped by your three-dimensional map. Write a newspaper article describing the rationale, tools, materials, and methods you used to make your three-dimensional map. If possible, donate your map to one of those organizations.

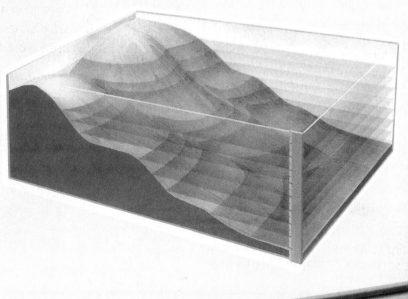

Real World Science

You can map Earth's surface

Geodesy is the science of studying the exact shape and curvature of Earth's surface and the position of landmarks on the surface. Scientists measure small changes in the shape and motions of Earth. They set up observation points on the surface and then measure the exact positions of these points using Global Positioning System (GPS) satellites.

Visit **Careers** at <u>ca6.msscience.com</u> to find out additional information on what these scientists do. Create and present a poster to the class that shows an example of one type of project where a geodesist might work.

Finding Your Exact Location

The science of geodesy is based on fixed reference points on Earth's surface. These points are marked by metal disks (called benchmarks), or very stable structures called monuments. Geodesists can find a location using both the benchmarks and coordinates from the Global Positioning System (GPS). The GPS is a group of 24 satellites that transmit signals to receivers throughout the world.

Visit **Technology** at <u>ca6.msscience.com</u> to find out more about the technologies being used today. Write a short article giving an example of where accurate positioning is vital in construction or navigation.

 **ELA6: W 1.2**

The Importance of Mapping in Early America

The National Geodetic Survey, our nation's first civilian scientific agency, was established by President Thomas Jefferson in 1807 as the Survey of the Coast. This agency was critical in mapping and discovering the expanding territories and boundaries of the early United States.

In 1878 this agency became the Coast and Geodetic Survey (C&GS). In 1970, a reorganization of agencies created the National Oceanic and Atmospheric Admin-istration (NOAA). The part of NOAA responsible for geodetic functions was named the National Geodetic Survey.

Research the Louisiana Territory and the surveying equipment used during Thomas Jefferson's time. Divide into small groups to create a week-long journal of a mapping survey conducted during the early 1800s.

GIS as a Tool to Understanding Earth Changes

GIS is a computer system used for capturing, storing, checking, and analyzing data related to positions on Earth's surface. A GIS assembles information from a wide variety of sources, including ground surveys, maps, and aerial photos. As the scientific community recognizes the environmental consequences of human activity, GIS technology is becoming an essential tool in the effort to understand the process of global change. When the data are inputted over time, the changes can be seen.

Visit Society at ca6.msscience.com to find out more about GIS. Create a table listing some param-eters that can be used to create a GIS map.

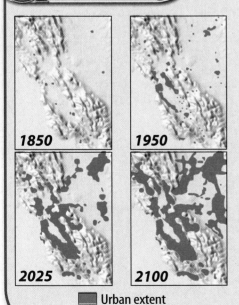

1850

1950

2025

2100

■ Urban extent

The BIG Idea Earth's surface can be represented in many ways. Maps provide two-dimensional views of Earth's three-dimensional surface.

Lesson 1 Reading Maps  7.f

Main Idea Maps represent large areas of Earth at a size we can easily see and study.

- Latitude and longitude can be used to determine the location of everything on Earth's surface.

- Map legends, or keys, commonly are found on maps.

- Map scale indicates the amount of detail shown on maps.

- **latitude** (p. 49)
- **longitude** (p. 49)
- **map legend** (p. 51)
- **map view** (p. 50)
- **profile view** (p. 50)

Lesson 2 Topographic and Geologic Maps 7.b, 7.c, 7.f, 7.h

Main Idea Specialty maps are used to show specific features such as changes in elevation and geologic characteristics.

- Topographic maps show the relief, or topography, of Earth's surface—its canyons, rivers, mountains, plains, plateaus, and other landforms.

- Contour lines show the elevation of Earth's surface above or below sea level.

- Topographic profiles show the land from the perspective of a person standing on Earth's surface.

- Geologic maps show the type, the age, and the characteristics of rocks at Earth's surface.

- Geologic cross sections illustrate the subsurface from a sideways view as if Earth were cut with a giant knife, revealing what is underneath.

- **contact** (p. 59)
- **contour line** (p. 55)
- **geologic formation** (p. 58)
- **geologic map** (p. 58)
- **topographic map** (p. 54)

 **STUDY TO GO** Download quizzes, key terms, and flash cards from ca6.msscience.com.

Linking Vocabulary and Main Ideas

Use the vocabulary terms on page 68 to complete the concept map.

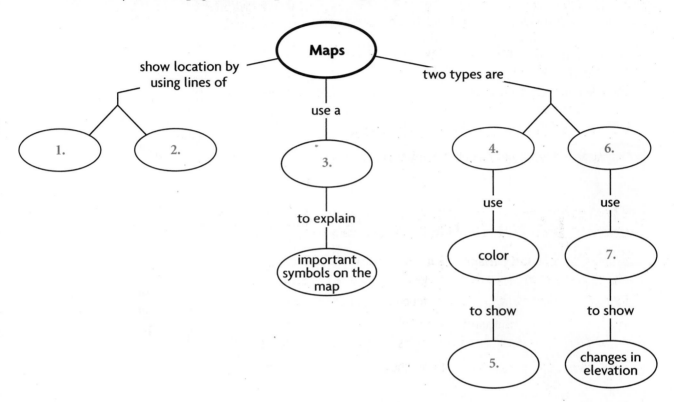

Maps

show location by using lines of

1.

2.

use a

3.

to explain

important symbols on the map

two types are

4.

6.

use

use

color

7.

to show

to show

5.

changes in elevation

Science Online

Visit ca6.msscience.com for:
▶ Vocabulary PuzzleMaker
▶ Vocabulary eFlashcards
▶ Multilingual Glossary

Using Vocabulary

Fill in the blanks with the correct vocabulary words.

8. A(n) _____ appears as if you were looking down on it from above.

9. A(n) _____ explains the symbols on a map.

10. _____ lines are used to determine a location north or south of the equator.

11. On a map, _____ are used to indicate elevation.

12. The shape of Earth's surface is represented on a(n) _____ map.

13. The distribution of rocks on Earth's surface is shown on a(n) _____ map.

Understanding Main Ideas

Choose the word or phrase that best answers the question.

1. The equator divides Earth into which two halves?
   A. northern and southern hemispheres
   B. top and bottom hemispheres
   C. eastern and western hemispheres
   D. right and left hemispheres **7.f**

Use the figure below to answer questions 2 and 3.

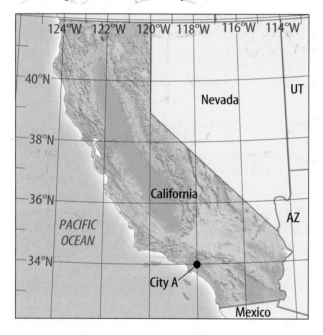

2. Which type of view is shown?
   A. cross section
   B. profile view
   C. map view
   D. topographic view **7.f**

3. What is the correct location of city A?
   A. 34°N, 118°W
   B. 34°W, 118°N
   C. 118°N, 34°W
   D. 118°W, 34°N **7.f**

4. What is the benefit of a profile view?
   A. It shows the vertical shape of the surface. **7.f**
   B. It shows what is beneath the surface.
   C. It explains the symbols on the map view.
   D. It uses colors to indicate geologic formations.

5. Which defines the difference in elevation between contour lines?
   A. bed contour
   B. contact line
   C. contour interval
   D. index contour **7.f**

6. Which indicates a certain body of rock of a specific age?
   A. rock strata
   B. geologic formation
   C. outcrop
   D. rock layer **7.f**

7. What do colors represent on geologic maps?
   A. actual rock colors
   B. position of each rock on the map
   C. age of the rocks
   D. the type of rock **7.f**

8. Rock formations that are broken up and moved meet at which location?
   A. depositional contact
   B. fault contact
   C. bedding plane
   D. geologic formation **7.f**

9. An illustration of a geologic cross section is shown below.

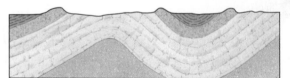

What does a geologic cross-section primarily show?
   A. how steep the surface is
   B. locations of streams, rivers and mountains
   C. rock formations below the surface
   D. distribution of rock formations on the surface **7.f**

Science Online Standards Review ca6.msscience.com

Applying Science

10. Explain why degrees of latitude and longitude are subdivided into smaller units such as minutes and seconds. **7.f**

11. Infer why lines of longitude on a globe get closer together toward the poles as illustrated below. **7.f**

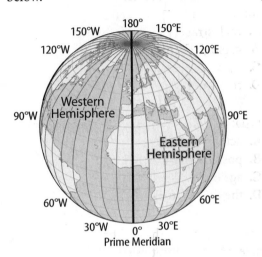

12. Describe in your own words the scale shown below. **7.f**

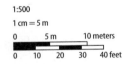

13. Describe how an index contour would be illustrated on a USGS topographic map. **7.f**

14. Evaluate the benefit of adding contour lines to geologic maps. **7.f**

15. Assess the wisdom of the following scenario: A geologist is mapping rock formations and following trails on a topographic map. The trail she is following ends at the edge of her map and appears to continue on. She decides to continue on the trail. **7.f**

WRITING in Science

16. Write a newspaper article describing the discovery of a set of maps dating from 1850–2000 for a fictitious town. Include a description of at least three of the maps and the changes that they record. **ELA6: W 1.2**

Applying Math

Use the illustration below to answer questions 17 through 22.

Map key
1 cm = 10 m

17. Calculate a ratio from the map key that shows how many meters of bike trails are equal to one centimeter on the map. **MA6: NS 1.2**

18. Calculate how many meters of bike trails would be represented by 5 cm on the map. **MA6: NS 1.2**

19. Calculate how many centimeters on the map would represent 480 m of bike trails. **MA6: NS 1.2**

20. Calculate a ratio from the map key that shows how many kilometers of bike trails are equal to one centimeter on the map. **MA6: NS 1.2**

21. Calculate how many kilometers of bike trails would be represented by 5 cm on the map. **MA6: NS 1.2**

22. Calculate how many centimeters on the map would represent 0.6 km of bike trails. **MA6: NS 1.2**

1 Which are useful for measuring position north or south of the equator?

A index contours

B lines of latitude

C lines of longitude

D map legends   7.f

2 Where is 90°S located?

A equator

B north pole

C south pole

D prime meridian   7.f

3 The illustration below includes a map legend and a portion of a map.

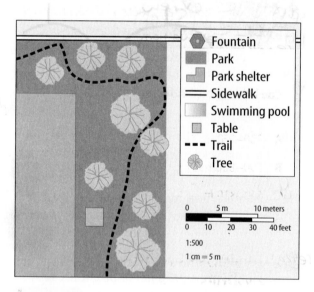

⬡	Fountain
▨	Park
⬓	Park shelter
═	Sidewalk
▦	Swimming pool
▩	Table
- - -	Trail
✿	Tree

0   5 m   10 meters
0 10 20 30 40 feet
1:500
1 cm = 5 m

Which is the best estimate of the actual diameters of the trees?

A 1 cm

B 5 cm

C 5 m

D 10 m   7.f

Use the illustration below to answer questions 4–6. The numbers on the drawing represent meters above sea level.

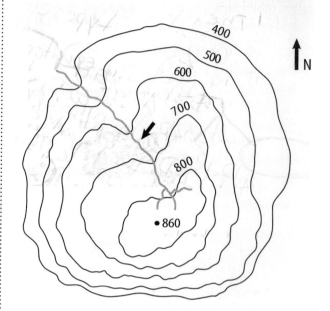

400
500
600
700
800
•860
N

4 Which side of the feature has the steepest slope?

A east side

B north side

C south side

D west side   7.f

5 What is the highest elevation on the feature?

A 400 m

B 800 m

C 860 m

D 960 m   7.f

6 What does the line that is marked by the short arrow represent?

A a contour line

B an index contour

C a high ridge

D a stream   7.f

Use the illustration below to answer questions 7–8.

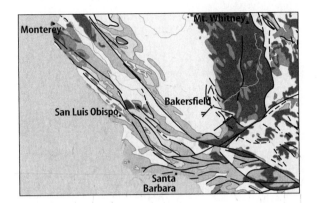

7 **Which is shown above?**

  A geologic map

  B geologic cross section

  C topographic map

  D topographic profile   7.f

8 **What do the thick, black lines represent?**

  A contour intervals

  B index contours

  C depositional contacts

  D fault contacts   7.f

9 **Which type of map would you use to find the location of a layer of coal at Earth's surface?**

  A geologic map

  B geologic cross section

  C topographic map

  D topographic profile   7.f

10 **Which type of map would you use to find the orientation of a layer of coal below Earth's surface?**

  A geologic map

  B geologic cross section

  C topographic map

  D topographic profile   7.f

Use the illustration below to answer questions 11–12. The numbers on the drawing represent meters above sea level.

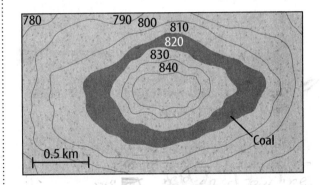

11 **The layer of coal is located between what elevation?**

  A 780–790 m

  B 790–800 m

  C 810–820 m

  D 830–840 m   7.f

12 **Approximately how wide is the actual layer of coal at Earth's surface?**

  A 0.2 cm

  B 0.5 cm

  C 0.2 km

  D 0.5 km   7.f

CHAPTER 2

Earth's Structure

The BIG Idea

Heat escaping from Earth's internal layers constantly changes the planet's surface.

LESSON 1 (1.e, 1.f, 2.a, 7.c)
Landforms

Main Idea Forces inside and outside Earth produce Earth's diverse landforms.

LESSON 2 (2.c, 6.b, 6.c, 7.e)
Minerals and Rocks

Main Idea The solid Earth is made of minerals and rocks.

LESSON 3
(1.b, 4.c, 7.e, 7.f, 7.g)

Earth's Interior

Main Idea Earth's interior has a layered structure.

Now how did *that* happen?

Imagine the results of a fender bender between two cars. The fenders of each are a crumpled mass of metal. When two continents collide, the results are similar—the rocks become crumpled and broken. The photo shows folded rock layers near Lulworth in the United Kingdom. They are the result of a collision between the African and European plates hundreds of kilometers away.

Science *Journal* Describe what an auto collision might look like in slow motion.

How can you model landscapes?

Imagine you are hiking through a natural area such as Yosemite Valley, California. Make a list of the landscape features you think you would see.

Procedure

1. Identify features on your list that are the highest and the lowest in elevation.

2. What makes each feature unique? Were some flat, or peaked on the top?

3. Stack several pieces of **artfoam** in layers, one on top of another. Put your hands on both ends of the stack, and shape the layered artfoam into different terrains.

Think About This

- **Explain** What did you do to the artfoam that might indicate how a landscape would form in nature?

- **Examine** the side of the model you made. What might the layers represent?

Science Online

Visit ca6.msscience.com to:

▶ view **Concepts in Motion**

▶ explore Virtual Labs

▶ access content-related Web links

▶ take the Standards Check

 **Earth's Layers** Make the following Foldable to show Earth's layers.

STEP 1 **Fold** a sheet of paper in half lengthwise. Make the back edge about 2 cm longer than the front edge.

STEP 2 **Fold** into thirds.

STEP 3 **Unfold** and **cut** along the folds of the top flap to make three flaps.

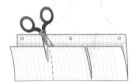

STEP 4 **Label** as shown.

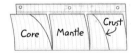

Clarify As you read this chapter, identify Earth's layers on the tabs. Under each tab, explain the features and describe the energy in that layer.

Get Ready to Read

Identify the Main Idea

 ELA6: R 2.3

①Learn It! Main ideas are the most important ideas in a paragraph, a lesson, or a chapter. Supporting details are facts or examples that explain the main idea. Understanding the main idea allows you to grasp the whole picture.

②Practice It! Read the following paragraph. Draw a graphic organizer like the one below to show the main idea and supporting details.

> The wearing away of soil and rock is called erosion. Water does most of this work. Rivers and streams carry rock fragments as the water flows downhill. Over long periods of time, this action changes the landscape. Mountains are worn down to flat plains. As rivers flow toward lakes or oceans, they carve valleys and steep-sided canyons.
>
> —from page 80

Main Idea

③Apply It! Pick a paragraph from another lesson of this chapter and diagram the main idea as you did above.

Target Your Reading

Reading Tip

The main idea is often the first sentence in a paragraph but not always.

Use this to focus on the main ideas as you read the chapter.

1 Before you read the chapter, respond to the statements below on your worksheet or on a numbered sheet of paper.

- Write an **A** if you **agree** with the statement.
- Write a **D** if you **disagree** with the statement.

2 After you read the chapter, look back to this page to see if you've changed your mind about any of the statements.

- If any of your answers changed, explain why.
- Change any false statements into true statements.
- Use your revised statements as a study guide.

Science Online

Print a worksheet of this page at ca6.msscience.com .

Before You Read A or D	Statement	After You Read A or D
	1 Energy from the Sun changes Earth's landscapes.	
	2 Earth's internal energy pushes up the land; surface processes wear it down.	
	3 Most of Earth, including its interior, is composed of rock.	
	4 Hardness and color are the two main characteristics of gems used in jewelry.	
	5 Matter and energy move from Earth's interior toward the surface.	
	6 Heat is always escaping from Earth's interior.	
	7 Humans have drilled holes and collected samples to about 500 km deep in Earth.	
	8 There is one type of crust near Earth's surface, and it is found on the continents.	
	9 The thickest of Earth's layers is the core.	
	10 Seismic waves do not penetrate Earth's layers.	

Reading Guide

What *You'll Learn*

▶ **Classify** landforms.

▶ **Explain** how landforms are produced.

▶ **Relate** your knowledge of landforms to California landscapes.

Why *It's Important*

You'll appreciate landforms around you as you discover how they form and change.

Vocabulary

landform
uplift
erosion

Review Vocabulary

weather: current condition of the atmosphere; temperature, wind speed and direction, humidity, and air pressure (Grade 5)

Landforms

Main Idea Forces inside and outside Earth produce Earth's diverse landforms.

Real-World Reading Connection Imagine you're making a sculpture by piling up sand near the shore. Suddenly, a wave comes and washes away part of your new artwork. Through different and slower processes, landforms are constantly being built up and worn down on Earth's surface.

How do landscapes form?

You live on the surface of Earth. Look out the window at this surface, or look at a photograph or drawing of a landscape. **Figure 1** is an example. There are tall mountains, deep valleys, and flat plains. Why does the landscape have different shapes and forms?

An endless interaction of forces reshapes Earth's topography. The transfer of matter and energy from Earth's interior builds mountains. Forces on the surface continuously wear down the mountains. These forces are caused by uneven heating of the surface by the Sun. In turn, this energy is transferred to the atmosphere. This makes weather that constantly bombards surface material and erodes it away, especially in higher areas. Without these eroding forces, the planet's surface would be more mountainous.

 What is the source of energy for Earth's weather?

Figure 1 Earth's landscape is the result of internal and external forces constantly acting upon the surface.

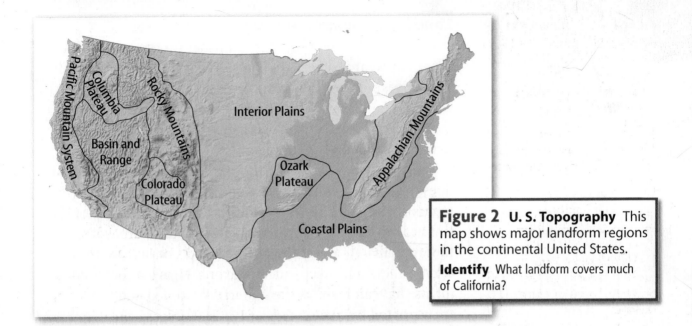

Figure 2 **U. S. Topography** This map shows major landform regions in the continental United States.

Identify What landform covers much of California?

Landforms

Features sculpted by processes both on Earth's surface and resulting from forces within Earth are called **landforms.** They can cover large regions or be smaller, local features. **Figure 2** shows the landform regions of the continental United States. These are large areas with similar topography. Find your location on the landform map in **Figure 2.**

Three main types of landforms are shown on the landform map. These examples are mountains, plateaus, and plains. Mountains and plateaus are areas with high elevations. Plains are low, flat areas.

Landforms Made by Uplift

Uplift is any process that moves the surface of Earth to a higher elevation. Both mountains and plateaus are formed by uplift. If a large flat area is uplifted, a plateau is formed. If the uplifted area is not flat, but has many steep slopes, it is called a mountain.

Earth's internal energy produces uplift. As thermal energy from Earth's interior moves toward the surface, it also causes matter in the interior to move upward. An example of a landform moved by uplift is shown in **Figure 3.** Sometimes Earth's internal heat energy melts rocks. If this melted rock moves to the surface, a mountain called a volcano can form. More often, the heat does not melt the rocks but makes mountains by pushing solid rocks upward. Scientists call the forces that can push solid rocks upward plate tectonics.

Figure 3 **Uplifted Landforms** Mountains and plateaus are made by uplift.

ACADEMIC VOCABULARY ···
transport (trans PORT)
(verb) to carry from one place
to another
*A large truck was needed to
transport the cargo.*

Landforms Shaped by Surface Processes

While Earth's internal energy pushes up the land, surface processes wear it down. As you read earlier, energy from the Sun drives some of these processes on the surface. Water, wind, ice, and gravity break apart the rocks that make up mountains. These broken fragments are carried downhill, making the mountains smaller.

The wearing away of soil and rock is called **erosion.** Water does most of this work. Rivers and streams carry rock fragments as the water flows downhill. Over long periods of time, this action changes the landscape. Mountains are worn down to flat plains. As rivers flow toward lakes or oceans, they carve valleys and steep-sided canyons. **Figure 4** shows landforms that can form as the material is eroded and **transported** by rivers.

When rivers eventually slow, they deposit some of their load of rock fragments. The fragments are distributed by the water to build other landforms, like the beach shown in **Figure 4.** Wave action from the ocean moves fragments of rocks, such as the sand on this beach, along the coastline.

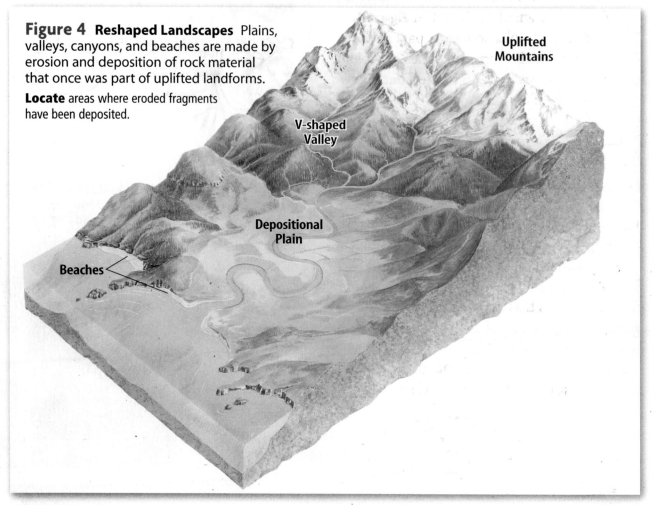

Figure 4 Reshaped Landscapes Plains, valleys, canyons, and beaches are made by erosion and deposition of rock material that once was part of uplifted landforms.

Locate areas where eroded fragments have been deposited.

Uplifted
Mountains

V-shaped
Valley

Depositional
Plain

Beaches

California Landforms

California has many types of landforms. Some are so spectacular that they are preserved in state or national parks. Maybe you have taken a trip to visit one of these parks.

Yosemite Valley

For example, the U-shaped surface of the valley in California's Yosemite National Park is shown in **Figure 5.** Glaciers carved this shape into the valley as they moved across its surface about one million years ago. In contrast, rivers usually carve sharper, V-shaped valleys as they cut through and erode rock.

 **Reading Check** How do valleys carved by glaciers differ in shape from valleys carved by rivers?

Lassen Peak

Another national park with landforms is Lassen Volcanic National Park. It features an active volcano, which is shown in **Figure 5.** Lassen Peak is a volcano that is part of the Cascade Mountain Range. A series of violent volcanic eruptions in 1915 blasted out a new crater at Lassen Peak's summit. The explosion expelled melted rock, gas, and ash that dramatically changed the landscape around the volcano. Volcanic ash mixed with snow and ice. This caused a rapid flow of mud down the sides of Lassen Peak and into river valleys below. Residents living in the vicinity of the eruptions lost their homes.

These California landforms show how different forces can act to change the landscape. External forces that caused precipitation for glacial ice to accumulate shaped the landscape of Yosemite Valley. Internal forces caused volcanic eruptions that altered the landscape surrounding Lassen Peak.

Yosemite Valley

Lassen Peak

Figure 5 Glaciers and Volcanoes
Yosemite Valley and Lassen Peak show how diverse the California landscape can be.

Mountains

California's major landforms are shown in **Figure 6.** This is a shaded relief map of the state. Find the Sierra Nevada and the Coastal Ranges. These are examples of mountains formed by the forces of plate tectonics. Solid rock was pushed up, forming high peaks. Because the ranges are long and narrow, they sometimes are called mountain belts.

 Figure 6 Identify two landform regions to the north of the Transverse ranges.

Now find Mount Shasta in **Figure 6.** It looks different from the other mountains. In fact, Mount Shasta looks like a distinct circle on the map. Mount Shasta is a volcano. It did not form by uplift of solid rock, as did most of the mountains in California. Mount Shasta's cone-shape formed when melted rock poured out from its center onto the land surface.

California's mountains continue to grow upward. Most often they grow so slowly you don't even realize this uplift is happening. Other times a volcanic eruption or an earthquake causes sudden uplift.

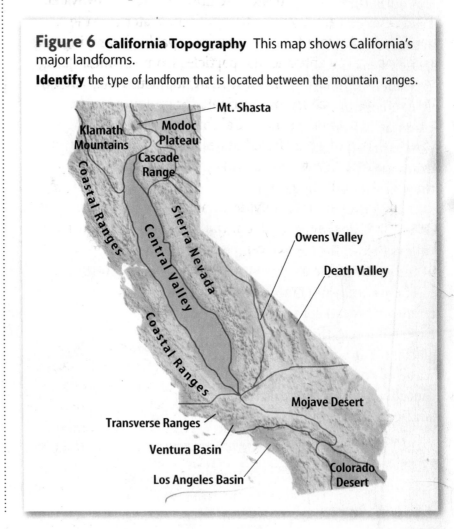

Figure 6 **California Topography** This map shows California's major landforms.

Identify the type of landform that is located between the mountain ranges.

Figure 7 The particles eroded from the mountain ranges surrounding the Central Valley have provided the soil base for producing most of California's agricultural products.

California Agriculture Statistics
• California has been the top agricultural state for more than 50 years.
• Agriculture generates almost $26.7 billion per year.
• Almost one-third of California's land area is used for farming.
• California produces more than 350 crops.
• California grows more than half of the United States' fruits, vegetables, and nuts.

Source: *USDA Agriculture in the Classroom*

Valleys

Next to the California mountain ranges are flat, open valleys. As the mountain peaks rise upward, erosion by water, wind, ice, and gravity wear them down. Water flowing downhill has a lot of kinetic energy, capable of carrying loosened rock fragments and soil particles from the mountains down to the valleys. This loose material helps make the valley's farmland rich in soil nutrients for growing plants.

These fertile valleys make California a top-ranked agricultural producer in the United States. **Figure 7** shows a farm located in the Great Central Valley. What is being produced on the farm shown here?

California also has many deep, narrow valleys. Rivers carve these valleys as they flow from the mountains toward the Pacific Ocean. The water carries loosened rock fragments from the west side of the Sierra Nevada, down toward the Central Valley, and eventually to the Pacific coast.

Beaches

Sand-sized grains of rock loosened from mountains toward the east provide material for beaches along the Pacific coast. Beaches are temporary features that must have sediment added constantly in order to exist. This is because sand is constantly washed away by ocean currents moving parallel to the shore. Without rivers continuously adding more sand, beaches would disappear. Material that has been transported by a creek and deposited along the Pacific shore is shown in **Figure 8.**

Figure 8 As moving water slows down, its sediment is deposited in sandbars and on the beaches.

Changing Landforms

Although they might seem like permanent features, landforms in your surroundings change continuously. Heat energy from the Sun and from Earth's interior provides the energy to change these landscapes. The constant movement of energy from Earth's interior to the surface results in forces that uplift the land into mountains and plateaus. At the same time, thermal energy from the Sun provides the energy for weather that includes precipitation, which wears down the uplifted landforms. At times, these changes are abrupt and dramatic, as when volcanoes erupt. Most often though, the changes are slow and steady, but endlessly sculpt Earth's landforms.

LESSON 1 Review

Summarize

Create your own lesson summary as you design a **visual aid.**

1. **Write** the lesson title, number, and page numbers at the top of your poster.

2. **Scan** the lesson to find the **red** main headings. Organize these headings on your poster, leaving space between each.

3. **Design** an information box beneath each **red** heading. In the box, list 2–3 details, key terms, and definitions from each **blue** subheading.

4. **Illustrate** your poster with diagrams of important structures or processes next to each information box.

 **ELA6:** R 2.4

Standards Check

Using Vocabulary

1. A glacier scraping sediment and rock from the sides of a mountain is an example of _____. **2.a**

2. In your own words, write a definition for *landform.* **1.e**

Understanding Main Ideas

3. How did the landform shown above most likely form? **1.e**

   A. when a block of rock uplifted

   B. when sediment was piled up by a river

   C. when a volcano erupted

   D. when a glacier passed over a valley

4. **Identify** a landform you have seen that was made by erosion. **2.a**

5. **Compare and contrast** the ways that internal and external forces produce surface landforms. **1.f**

6. **Compare and contrast** the formation of Lassen Peak with the formation of the Sierra Nevada. **1.e**

Applying Science

7. **Predict** what would happen to Earth's surface if all of Earth's internal heat escaped. **1.e**

8. **Decide** if a constantly changing landscape is beneficial for people. **1.e**

Landscape	Benefit	Harm

 **Science** **Online**

For more practice, visit **Standards Check** at ca6.msscience.com.

00:45 minutes

Try at Home

How do mountains vary in shape?

Many different types of landforms make up California's landscape. Mountains are especially prominent throughout the state. Explore how to determine the differences among them and if these differences are clues to how the mountains formed.

Data Collection

1. Visit ca6.msscience.com to examine some bird's-eye view images to find different types of mountains in different regions of California.

2. Make a table of observations like the sample data table below. Use the menu along the margin of the Web site to observe the mountains listed in the data table. Explain any differences you observe. Draw some outstanding features for later comparisons.

Mountain Characteristics

Mountain	Colors	Shapes	Unique Features	Sketch	General Location
Mt. Shasta					
Mt. Eddy					
Mt. Diablo					
Mt. Whitney					

Data Analysis

1. **Identify** a mountain range that was formed by volcanic eruptions.

2. **Compare and contrast** characteristics of the mountains you studied.

3. **Graph** Make a bar graph that includes the names of the mountains and plateaus and their elevations. Use the following data: Mt. Shasta (4,317 m), Mt. Eddy (2,751 m), Mt. Diablo (1,173 m), Mt. Whitney (4,417 m).

 Science Content Standards

7.c Construct appropriate graphs from data and develop qualitative statements about the relationships between variables.

Reading Guide

What *You'll Learn*

▶ **Identify** minerals by observing their properties.

▶ **Explain** the value of minerals in your life.

▶ **Classify** rocks according to how they form.

▶ **Illustrate** how the rock cycle continuously recycles Earth materials.

Why *It's Important*
The majority of Earth materials, even those in the deep interior, are solid rock.

Vocabulary

mineral	lava
density	sediment
rock	rock cycle
magma	

Review Vocabulary

igneous rock: rock that forms from magma or lava (Grade 4)

Minerals and Rocks

(Main Idea) The solid Earth is made of minerals and rocks.

Real-World Reading Connection You stand on the bank of a creek and throw rocks in the water. Rocks seem to be everywhere. But in your yard there are hardly any rocks. What are rocks? What are they made from? Where do they come from?

What is Earth made of?

The solid part of Earth is made up of minerals and rocks. People use them to build homes and roads. Minerals and rocks break down to form the soil in which farmers grow food. Some rocks and minerals are even used as jewelry because they are so beautiful. Minerals and rocks are such a common part of the environment that you might not realize they are all around you. **Figure 9** shows some common items made from mineral and rock resources.

Minerals are the substances that make up rocks. Scientists have identified about 3,800 distinct minerals, but most of these are rare. There are only about 30 common minerals. Minerals form when crystals grow in nature. For example, they can grow in melted rock material or from material dissolved in water.

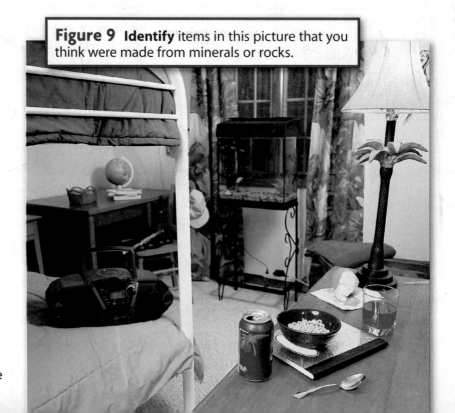

Figure 9 **Identify** items in this picture that you think were made from minerals or rocks.

What is a mineral?

The word *mineral* has several common meanings. You might drink mineral water, or someone might tell you to eat healthful food, so that you get all the vitamins and minerals that you need to be healthy. In Earth science, the word *mineral* has a specific definition. A **mineral** is a naturally occurring, generally inorganic solid that has a crystal structure and a definite chemical composition. How can you tell if something you are looking at is a mineral? Materials classified as minerals have the following properties.

Naturally Occurring To be considered a mineral, a substance must be found in the natural world. Anything manufactured by people, such as one of the gemstones in **Figure 10,** are not minerals. For example, diamonds mined from Earth are minerals, but synthetic diamonds made in laboratories are not.

Generally Inorganic Most minerals are formed by processes that do not involve living things. But, there are some minerals made by living things. The mineral aragonite is found in pearls, which are made by oysters, and the mineral apatite is found in your bones and teeth.

Solid Substances that are liquids or gases are not considered minerals. Therefore, natural emeralds like the ones shown in **Figure 10** are minerals, but the liquid that would form if they were to melt is not a mineral.

WORD ORIGIN · · · · · · · · · · · ·
mineral
minera- Latin; means *mine* or *ore*
mineralis- Latin; means *of* or *from the mine*

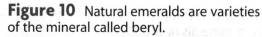
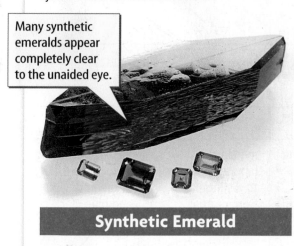

Figure 10 Natural emeralds are varieties of the mineral called beryl.
Compare and contrast the appearances of the synthetic and the natural emeralds.

Many synthetic emeralds appear completely clear to the unaided eye.

Most natural emeralds contain defects such as fractures or mineral impurities.

Synthetic Emerald

Natural Emerald

Figure 11 The cubic nature of the halite crystal is one property used to identify it.

Crystal Structure The atoms in a mineral are arranged in orderly, repeating patterns. This regular atomic pattern is called a crystal structure. The smooth flat surfaces on a crystal represent a well-organized, internal structure of atoms. Observe the crystal structure of the mineral halite shown in **Figure 11.** Notice that the outer, smooth faces of the halite crystal make the same shape as its internal atomic structure.

Definite Composition A mineral is made of specific elements. Not only must a mineral have certain elements, but the elements also must be in definite proportions. A common example is the mineral quartz. It is made of the elements silicon (Si) and oxygen (O). The chemical formula for quartz is SiO_2. The formula tells you there are two oxygen atoms for every silicon atom in quartz. The chemical formula shows both the elements and their proportions.

Concepts In Motion
Interactive Table Organize information about minerals at ca6.msscience.com.

Table 1 Is it a mineral?

	Amber	Rock Candy	Synthetic Ruby	Fluorite
Did it form in nature?	Yes	No	No	Yes
Is it inorganic?	No	No	Yes	Yes
Does it have a crystal structure?	No	Yes	Yes	Yes
Does it have a definite chemical composition?	No	Yes	Yes	Yes
Is it a mineral?	No	No	No	Yes
Comments	common gemstone; made of tree resin; mixture of many organic compounds	organic compound made by humans	made in laboratories, hard to distinguish from natural rubies	gemstone ranging in color from clear or green to violet and blue black

Physical Properties of Minerals

You can tell one mineral from another by its physical properties. Physical properties are characteristices that can be observed or measured without changing the identity of the mineral. If you learn how to test a mineral for these properties, you will be able to use the tests to identify many minerals. Some of the more common physical properties you can use to identify minerals are described next.

Hardness

You can test the hardness of a mineral by observing how easily it is scratched. Any mineral can be scratched by another mineral that is harder. In the early 1800s, Austrian scientist Friedrich Mohs developed a hardness scale with 10 minerals. On this scale, the hardest mineral, diamond, has a hardness of 10. The softest mineral, talc, has a hardness of 1. **Table 2** shows the Mohs' hardness scale. Quartz, feldspar, and calcite are on the scale, and they all are common minerals.

 Table 2 Which minerals can be scratched by feldspar?

Color

A mineral's color can sometimes help you identify it. The mineral malachite, for example, always has a distinctive green color because it contains the metal copper. Most minerals do not have a single distinctive color, as shown by the many colors of quartz in **Figure 12.**

Table 2 Mohs Hardness Scale		
Mineral	**Hardness**	**Common Tests**
Talc	1	rubs off on clothing
Gypsum	2	scratched by fingernail
Calcite	3	barely scratched by copper coin
Fluorite	4	scratches copper coin deeply
Apatite	5	about same hardness as glass
Feldspar	6	scratches glass
Quartz	7	scratches glass and feldspar
Topaz	8	scratches quartz
Corundum	9	scratches most minerals
Diamond	10	scratches all common materials

Uncut Quartz

Cut Quartz

Figure 12 Quartz cannot be identified by color alone.

Figure 13 **Constant Streak** Although the colors of hematite can be different, the streak is always reddish-brown.

Infer which is harder—the porcelain tile or the hematite.

Metallic Luster

Glassy Luster

Figure 14 Galena and quartz have distinctive crystal shapes and lusters.

Streak and Luster

Streak is the color of powder from a mineral. You can look at the powder by scratching the mineral across a tile made of unglazed porcelain. Some minerals that vary in color have distinct streak colors. For example, the color of the mineral hematite can be silver, black, brown, or red. But, notice in **Figure 13** that the two different-colored hematite samples both show a reddish-brown streak.

Luster is the way a mineral's surface reflects light. Geologists use several common words to describe mineral luster. Two of these are shown in **Figure 14.** Galena has a shiny metallic luster. Quartz has a glassy luster. Other terms used to describe luster are *greasy, silky,* and *earthy.* Look again at **Figure 13** and try to use these terms to describe the luster of the hematite samples. Do both hematite samples have the same luster?

Crystal Shape

Every mineral has a unique crystal shape. A crystal that forms on Earth's surface will be small, because the erupting lava flow cools rapidly. Crystals are large and perfect when they form underground where Earth's heat is maintained and the magma source cools slowly. As **Figure 14** illustrates, each crystal has a distinct shape, which sometimes is referred to as crystal habit.

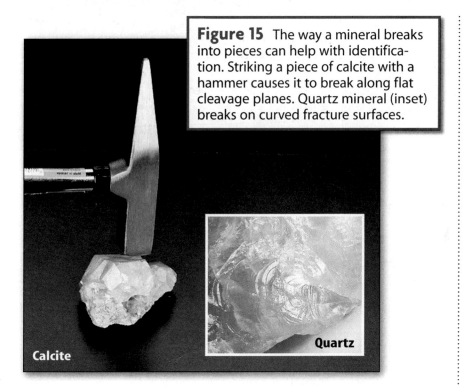

Figure 15 The way a mineral breaks into pieces can help with identification. Striking a piece of calcite with a hammer causes it to break along flat cleavage planes. Quartz mineral (inset) breaks on curved fracture surfaces.

Calcite

Quartz

Cleavage and Fracture

Cleavage and fracture describe the way a mineral breaks. If it breaks along smooth, flat surfaces, it has cleavage. A mineral can have one or more distinct cleavage directions. If a mineral breaks along rough or irregular surfaces, it displays fracture. **Figure 15** shows examples of both cleavage and fracture. The calcite has three distinct cleavage directions. This makes it break into blocks. Quartz does not have cleavage. It breaks along curved surfaces, so it displays fracture.

 How many directions of cleavage does calcite have?

Density

Density is the amount of matter an object has per unit of volume. Some minerals are denser than others. If you pick up a piece of galena and a piece of quartz, and both are about the same size, you can feel that the galena is much heavier. This is because galena is denser than quartz.

Most metals have high densities compared to nonmetals. Minerals with atoms packed closely together also tend to have higher densities. Quartz and feldspar are common minerals with relatively low densities. Olivine, with a closely packed structure of atoms and some iron in its structure, has a relatively high density. When a mineral has an especially high or low density, its density can be used to identify it.

| Magnetism | Double Refraction |

Figure 16 Both magnetite and calcite have noticeable physical properties that help identify them.

Explain how the property of magnetism could help physically separate minerals.

Other Properties

Some minerals have properties that make them easy to identify. For example, magnetite is magnetic. **Figure 16** shows how magnetite attracts a magnet. Calcite reacts chemically to acids. If you place a drop of acid on calcite, it fizzes.

Calcite also shows an interesting property that occurs when light interacts with it. If you look at an object through a clear calcite crystal, you can see two images of the object, as shown in **Figure 16.** This is called double refraction, and it occurs when light splits into two separate rays, each forming its own distinct image of the object.

 What property of calcite produces double images of objects viewed through it?

Many properties of minerals make them ideal to use in industry. For example, quartz can produce an electric current when pressure is applied to it. Graphite can be used to mark on paper. Copper is used in electronic wiring because it is a good conductor of electricity.

Every mineral has properties that can be observed to help identify it. But remember that many minerals have similar properties. You need to test for a combination of properties to find those that are unique to a particular mineral. It can be a challenge to find an unfamiliar mineral and try to figure out what it is.

minutes

Mineral Identification by Property

It can be challenging to identify a mineral correctly, because many of them have similar properties. But, with a few simple tools, you can observe a set of characteristic physical properties for an unknown mineral. This can help you determine what it is.

Procedure

1. Complete a safety worksheet.
2. Obtain three or four unknown numbered mineral samples from your teacher.
3. Use a **field guide for rocks and minerals**, a **magnifying glass**, a **streak plate**, a **copper coin**, a **glass plate**, a **magnet**, a **graduated cylinder**, and a **triple-beam balance** to help you determine the physical properties of each sample.
4. For each sample, observe and record the physical properties, color, streak, luster, hardness, and cleavage or fracture using information in Lesson 2.
5. To determine the density of a sample, place it on the triple-beam balance and measure the mass in grams. Then tie a string around the sample and carefully lower it into the graduated cylinder that has a recorded volume of water in it. Subtract the original volume from the new volume of water. Divide the mass by the volume.

Properties to Identify Minerals							
Mineral Name	Color	Streak	Luster	Hardness	Cleavage/ Fracture	Density (g/mL)	Other Properties

Analysis

1. **Compare** your results to the information in the field guide.
2. **Identify** each mineral using your observations and the guide.
3. **Evaluate** which properties were most helpful for you to identify a mineral. Describe any properties that could help you identify a mineral without testing other properties.

 Science Content Standards

7.e Recognize whether evidence is consistent with a proposed explanation.

Mineral Uses

Some minerals are important because they contain materials that have many uses. Others are important because they have special properties or because they are rare. People **appreciate** some minerals solely for their beauty.

Metallic Ores

Rich deposits of valuable minerals are called ores. The metals you use every day come from these ores. The minerals chalcopyrite and malachite are examples of copper ores. Copper is a common metal used in wires to conduct electricity.

Iron used to make steel comes from hematite and magnetite. Steel is used to manufacture cars, bridges, skyscrapers, and many other things you use every day. Galena is the major ore for producing lead. Most lead is used to manufacture automobile batteries. The minerals gold and silver are considered precious metals. They are used in industry and also in jewelry.

 Reading Check What is the major ore used for producing lead?

Gemstones

People have been collecting minerals for their beauty for thousands of years. These minerals are called gems. Many gems have intense colors, a glassy luster, and are 7 or more on the Mohs hardness scale. Diamonds, rubies, sapphires, and emeralds are among the most valuable gemstones. When these rare minerals are cut and polished, their value can last for hundreds of years. **Figure 17** shows the difference between these minerals before and after they are cut and polished.

ACADEMIC VOCABULARY ⋯
appreciate (uh PRE shee ayt)
(verb) to grasp the nature, quality, worth or significance of
It is difficult for most people to appreciate patience.

Figure 17 The clear diamond, ruby, blue sapphire, and ruby are cut and polished to make jewelry.

Cut ruby on uncut matrix

Uncut sapphire

Cut sapphire

Uncut diamond

Cut diamond

Rocks

A **rock** is a natural, solid mixture of particles. These particles are made mainly of individual mineral crystals, broken bits of minerals, or rock fragments. Sometimes rocks contain the remains of organisms or are made of volcanic glass. Geologists call the particles that make up a rock grains.

Most of Earth is made of rocks. Mountains, valleys, and even the seafloor under the oceans are made of rocks. You might not always notice the rocks under your feet. **Figure 18** shows an example of how rocks and soil are present beneath a landscape's surface.

Rocks are classified, or placed into groups, based on the way they form. There are three major groups of rocks: igneous rocks, metamorphic rocks, and sedimentary rocks.

 Figure 18 What happens to particles eroded from the mountains?

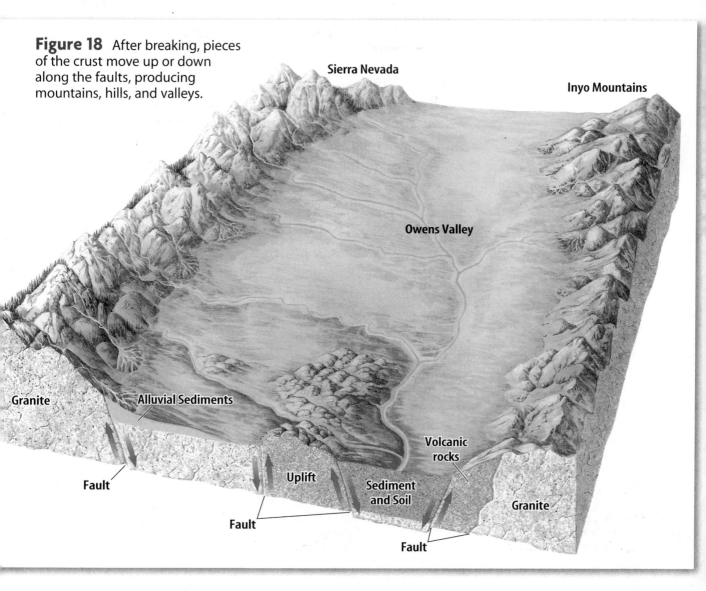

Figure 18 After breaking, pieces of the crust move up or down along the faults, producing mountains, hills, and valleys.

Sierra Nevada

Inyo Mountains

Owens Valley

Granite

Alluvial Sediments

Fault

Uplift

Fault

Sediment and Soil

Volcanic rocks

Granite

Fault

Igneous Rocks

Igneous rocks are formed from molten, or liquid, rock material called **magma.** As the temperature of magma drops, tiny crystals of minerals begin to form. These tiny crystals become the grains in an igneous rock.

Located at Earth's surface, magma, now called **lava,** cools quickly. The crystals in lava do not have much time to grow, so they are small. Volcanic glass forms when lava cools so rapidly that atoms do not form well-organized crystal structures.

Deep within Earth, magma cools slowly because thick layers of rock surround it. There is more time for larger crystals to grow. **Figure 19** shows a cross-section, or slice, through Earth. Notice that the igneous rock called granite in **Figure 19** has larger mineral grains than the igneous rock called basalt. This is because granite cools much more slowly than basalt does.

 Reading Check Why does magma cool slowly?

Like the word *mineral, texture* is a common word. But in Earth science it has a specific definition. The grain size and the way grains fit together in a rock are called texture. Because granite and basalt have different-sized grains, they have different textures. Granite's texture is coarse grained and basalt's texture is fine grained. **Figure 20** shows El Capitan, which is a huge mountain of granite now exposed at the surface by uplift.

The igneous rocks granite and basalt do not differ only in texture. They also differ in mineral composition. Granite contains low-density minerals such as quartz and feldspar. Basalt is made of higher-density minerals than granite, such as olivine and magnetite.

Figure 19 Cooling Rates The grain size of an igneous rock depends in part on how quickly the magma cools.

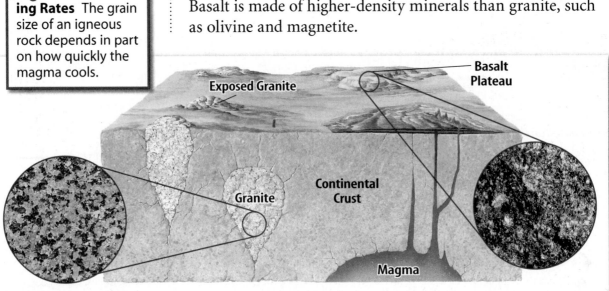

Exposed Granite

Basalt Plateau

Granite

Continental Crust

Magma

Visualizing Igneous Rock Features

Figure 20

Intrusive igneous rocks are formed when a mass of magma is forced upward toward Earth's surface and then cools before emerging. The magma cools in a variety of ways. Eventually the rocks may be uplifted and erosion may expose them at Earth's surface. A selection of these formations is shown here.

▶ This dike in Israel's Negev Desert formed when magma squeezed into cracks that cut across rock layers.

▶ A batholith is a large igneous rock body that forms when rising magma cools below the ground. Towering El Capitan, right, is just one part of a huge batholith. It looms over the entrance to the Yosemite Valley.

▲ Sills such as this one in Death Valley, California, form when magma is forced into spaces that run parallel to rock layers.

▶ Volcanic necks like Shiprock, New Mexico, form when magma hardens inside the vent of a volcano. Because the volcanic rock in the neck is harder than the volcanic rock in the volcano's cone, only the volcanic neck remains after erosion wears the cone away.

Contributed by National Geographic

Metamorphic Rocks

Metamorphic rocks form when solid rocks are squeezed, heated, or exposed to fluids, changing them into new rocks. To be considered metamorphic, rocks must stay solid as they change. If the conditions are correct to melt them, new igneous rocks will form instead of metamorphic rocks.

The original rock that is changed is called the parent rock. Heat, pressure, and hot fluids composed mainly of water and carbon dioxide applied to a parent rock cause the growth of new mineral grains. These new grains may have a different texture and might even have a different mineral composition than the grains in the parent rock.

 Reading Check When exposed to heat, pressure, or fluids, what can happen to mineral grains?

Figure 21 shows changes that can happen when two parent rocks are metamorphosed. Increased pressure and temperature made the grains in the marble bigger and sparkly, compared to the grains in the parent limestone. The grains remain as crystals of calcite, but they are larger than in limestone.

The metamorphic rock, gneiss (NISE), in **Figure 21** shows a more dramatic texture change. Look closely at the parallel layers of dark and light mineral grains. This layering is called foliation. Foliation results from uneven pressure.

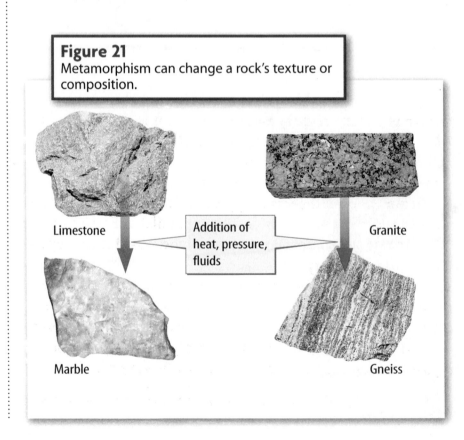

Figure 21
Metamorphism can change a rock's texture or composition.

Limestone

Addition of heat, pressure, fluids

Granite

Marble

Gneiss

Deposition → **Compaction** → **Cementation**

Figure 22 Rocks, gravel, pebbles, and sand are the sediments produced from solid rock. This is the beginning of sedimentary rock formation.

Sedimentary Rocks

Processes at Earth's surface break down rocks, changing them physically and chemically. **Sediment** is rock that is broken down into smaller pieces or that is dissolved in water. Forces that cause erosion, such as water, wind, ice, and gravity move sediment to new locations.

Sediment eventually is dropped, or deposited, in low-lying areas. Sediment usually is deposited parallel to Earth's surface in flat layers. This produces the most obvious characteristic of sedimentary rocks, layering.

Sediment is changed into sedimentary rock as grains are compressed by the weight of the material above them. The sediment grains also are cemented together by dissolved mineral material that crystallizes between grains. **Figure 22** shows possible stages in the formation of sedimentary rock.

 **Reading Check** What two things change sediment grains into rock?

The Rock Cycle

When you observe a mountain of rock, it is hard to imagine it can ever change. But rocks are changing all the time. It happens so slowly that you usually do not notice it. The series of processes that change one rock into another is called the **rock cycle.** Forces on Earth's surface and deep within the planet drive this cycle.

Figure 23 shows how the three major rock groups are related through the rock cycle. The circles show the different Earth materials: magma, sediment, and rocks. The arrows in **Figure 23** represent the processes that change one type of material into another. There are many different pathways through the rock cycle. How fast do rocks move through the rock cycle? It varies, but generally it can take many thousands to many millions of years.

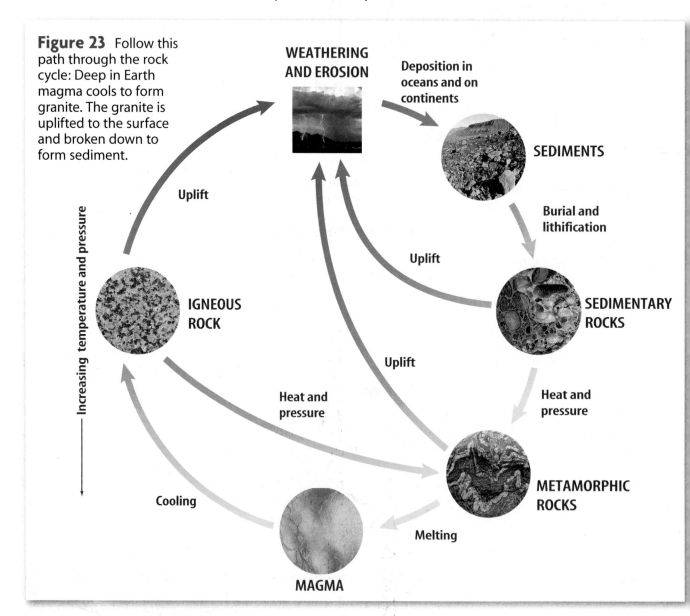

Figure 23 Follow this path through the rock cycle: Deep in Earth magma cools to form granite. The granite is uplifted to the surface and broken down to form sediment.

WEATHERING AND EROSION

Deposition in oceans and on continents

SEDIMENTS

Uplift

Increasing temperature and pressure

IGNEOUS ROCK

Burial and lithification

Uplift

SEDIMENTARY ROCKS

Uplift

Heat and pressure

Heat and pressure

Cooling

METAMORPHIC ROCKS

Melting

MAGMA

Earth Materials

The solid part of Earth is made of minerals and rocks. Scientists have used hardness, luster, streak, color, crystal habit, cleavage, and fracture to identify minerals. Rocks fall into three classes. Igneous rocks are made from melted rock that moves toward the surface where it hardens. Metamorphic rocks are any rocks that are changed while exposed to high pressure and high heat without melting. Sedimentary rocks form when bits and pieces of rocks are pressed and cemented together. As rocks form or change they go through a repeating cycle. Scientists call this recycling pattern the rock cycle. In the next lesson you will learn what scientists think exists where no human has ever been before, inside Earth.

LESSON 2 Review

Summarize

Create your own lesson summary as you design a **study web.**

1. **Write** the lesson title, number, and page numbers at the top of a sheet of paper.

2. **Scan** the lesson to find the **red** main headings.

3. **Organize** these headings clockwise on branches around the lesson title.

4. **Review** the information under each **red** heading to design a branch for each **blue** subheading.

5. **List** 2–3 details, key terms, and definitions from each **blue** subheading on branches extending from the main heading branches.

 ELA6: R 2.4

Standards Check

Using Vocabulary

1. Use the word *sediment* in a sentence. **2.c**

2. Distinguish between the words *magma* and *lava*. **2.c**

Understanding Main Ideas

3. Which best describes the luster of the galena shown here?

A. greasy   **6.b**
B. metallic
C. glassy
D. dull

4. **Compare and contrast** the formation of minerals with the formation of rocks. **6.b**

5. **List** seven minerals that are valuable resources. **6.c**

6. **Summarize** the rock cycle. **2.c**

7. **Compare and contrast** granite and basalt. **6.b**

Applying Science

8. **Illustrate** a path through the rock cycle that changes sedimentary rock to igneous rock. **2.c**

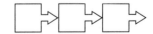

9. **Evaluate** how your life would change if Earth's mineral resources became scarce. **6.c**

Science Online

For more practice, visit **Standards Check** at ca6.msscience.com.

Science Content Standards

1.b Students know Earth is composed of several layers: a cold, brittle lithosphere; a hot, convecting mantle; and a dense, metallic core.

4.c Students know heat from Earth's interior reaches the surface primarily through convection.

7.e Recognize whether evidence is consistent with a proposed explanation.

7.g Interpret events by sequence and time from natural phenomena.

Also covers: 7.f

Reading Guide

What *You'll Learn*

▶ **Explain** how scientists determined that Earth has internal layers.

▶ **Describe** Earth's internal layers.

▶ **Analyze** the role that convection plays inside Earth.

Why *It's Important*

Learning about Earth's interior will help you understand formations and changes on Earth's surface.

Vocabulary

crust
mantle
asthenosphere
lithosphere
core

Review Vocabulary

magnetic field: the region of space surrounding a magnet or magnetized object (Grade 4)

Earth's Interior

Main Idea Earth's interior has a layered structure.

Real-World Reading Connection Maybe you've tried to figure out what was inside a wrapped gift by tapping or shaking it. Without actually opening the gift, you may have figured out what was inside. Scientists can't see deep inside Earth. How might they discover what the planet's interior is like?

Layers

No one can directly sample Earth's deep interior from depths any greater than around 12 km. Because humans cannot see or directly take samples from deep inside Earth, indirect methods are used to determine Earth's layers. Sometimes rock samples from as deep as 200 km are brought to the surface by volcanic eruptions, but these are rare. Most of the evidence for Earth's interior structure comes from the study of seismic waves.

Earth's interior is made up of layers. Each layer has a different composition. Also, the temperatures and pressures within Earth increase as you go deeper. **Figure 24** shows Earth's three basic layers. How did scientists learn so much about the inside of Earth?

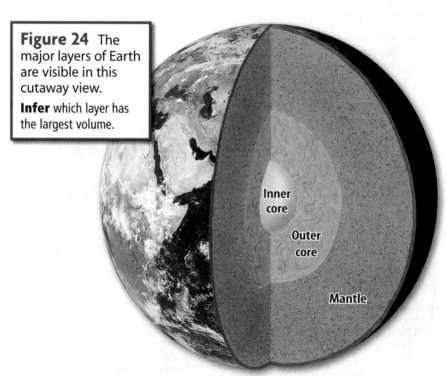

Figure 24 The major layers of Earth are visible in this cutaway view.

Infer which layer has the largest volume.

Layers and Seismic Waves

Earthquakes produce seismic waves that pass through the planet. The speed and direction of the seismic waves change when the properties of the Earth materials they are traveling through change. The waves bounce off or bend as they approach a new layer. Scientists have learned about the details of Earth's internal layering by analyzing the paths of these waves.

The Crust

The thin, rocky, outer layer of Earth is called the **crust.** By sampling the crust, scientists know that there are two different types. Crust under the oceans is made of the igneous rock called basalt. Below the basalt is another igneous rock called gabbro. Gabbro (GAH broh) has the same composition as basalt, but because it cools below the surface, it has larger grains than basalt. Most continental crust is made of igneous rocks with compositions that are much like granite. Remember that granite contains mostly feldspar and quartz. These relatively low-density minerals make average continental crust less dense than oceanic crust. The crust's igneous rocks usually are covered with a thin layer of sedimentary rocks. Rocks that make up crust are rigid and brittle. **Figure 25** shows a slice through both types of crust and examples of the rocks which compose them.

MiniLab

How can you describe what you cannot see?

What can you infer about materials that you indirectly sense, but can't see?

Procedure

1. Work with a partner to make a sample of a core from Earth's crust. Put layers in a **plastic jar** using **gravel, sand, small stones, soil,** and possibly a **larger stone** or some **plant material.**

2. Diagram the arrangement. Measure and label the depth of each layer in centimeters.

3. Cover your jar with **dark paper** and then share your jar with another team. Have them use their pencils to determine what is in your jar, how many layers you have, and if you have any "boulders" or solid rock beds included in your sample.

Analysis

1. **Explain** the methods you used to determine the makeup of the other team's jar.

2. **Describe** the evidence you used to infer what was probably in each layer in their jar.

3. **Evaluate** your results. How close did you come to describing the actual contents of the other team's jar?

 7.e, 7.g

Figure 25 Oceanic crust is thin and dense compared to continental crust.

Continental Crust

Oceanic Crust

Upper Mantle

Concepts In Motion
To visualize Earth's interior, visit
ca6.msscience.com.

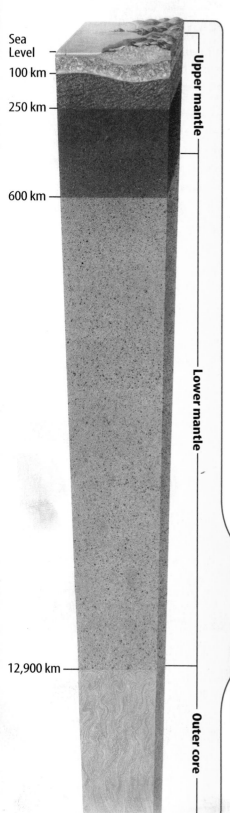

Sea
Level
100 km
250 km

600 km

Upper mantle

Lower mantle

12,900 km

Outer core

The Mantle

Below the crust is the thick middle layer called the **mantle.** It also is made of rock. The rock in the upper part of the mantle is called peridotite [puh RIH duh tite]. Mantle rocks contain a lot of oxygen, silicon, magnesium, and iron. Minerals in mantle rocks have tightly packed crystal structures. The metallic elements in them, such as iron, are heavy. These characteristics make mantle rocks denser than rocks in the crust.

Increasing temperature and pressure, as you go deeper into Earth, divides the mantle into distinct layers. Some of these layers are shown in **Figure 26.** Like rocks in the crust, rocks in the upper mantle are brittle. But between about 100 km and 250 km deep it is so hot that tiny bits of the rock melt. This partly melted rock material exists between mineral grains and allows the rock to flow. Scientists sometimes use the term *plastic* to describe rock that flows in this way. This plastic, but still mostly solid, layer of the mantle is called the **asthenosphere.** Remember that the asthenosphere flows very slowly. Even if it were possible for you to visit the mantle, you could not see this flow. It moves at rates of only a few centimeters per year.

Reading Check What is the plastic, but still mostly solid layer right below the lithosphere?

Below the asthenosphere, the rock is solid, even though it is hotter than the rock material in the asthenosphere. How can this happen? Increasing temperature tends to make rock melt, but increasing pressure reduces melting. The pressures deep within Earth are so great that they squeeze hot rock material into a solid state.

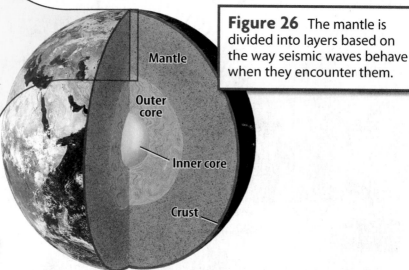

Mantle

Outer core

Inner core

Crust

Figure 26 The mantle is divided into layers based on the way seismic waves behave when they encounter them.

Lithosphere

The crust and the mantle are made of rock. Recall that the crust is cool and brittle and so is the upper 100 km of the mantle. Even though the rocks in the crust and mantle have different compositions, they both are solid and rigid. Together, the crust and the uppermost mantle form the brittle outer layer of Earth called the **lithosphere.**

The Core

The dense metallic center of Earth is called the **core.** It is the densest part of the planet because it is made mainly of metallic elements. The metal is mostly iron with some nickel. The core is divided into two layers. The outer core is a **layer** of molten metal. The metal is liquid because the effects of temperature now outweigh pressure's effects in the outer core. But in the inner core, higher pressures cause the metal to be in the solid state.

Figure 27 shows how Earth's layers are divided into more detailed layers. These divisions are based on the ways that Earth materials within those layers respond to the extreme temperatures and pressures within Earth.

ACADEMIC VOCABULARY
layer (LAY uhr)
(noun) one thickness, course, or fold laid or lying over or under another
The cake had a thin layer of icing covering the top.

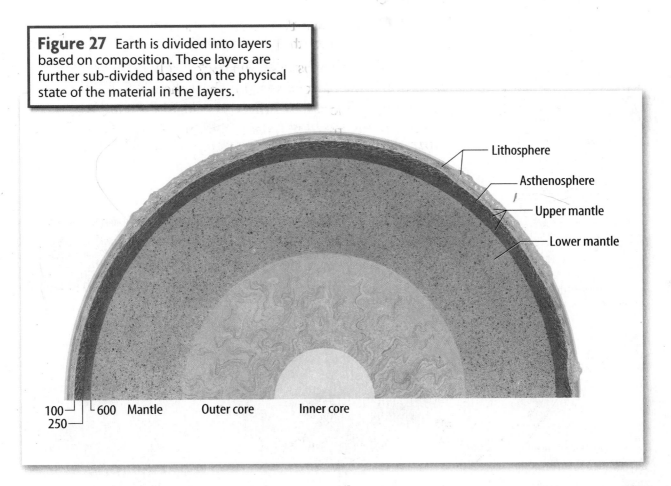

Figure 27 Earth is divided into layers based on composition. These layers are further sub-divided based on the physical state of the material in the layers.

Lithosphere
Asthenosphere
Upper mantle
Lower mantle

100
250
600   Mantle   Outer core   Inner core

Heat Transfer in Earth

In Chapter 3, you will read that heat movement in a fluid is by a process called convection. This type of heat transfer occurs in two of Earth's layers that you just read about. Convection processes transfer heat in the outer core and in the mantle. This transfer process is driven by changes in density.

Density

The density of all Earth materials is not the same. You read in the last lesson that some minerals and rocks are denser than others. This is partly because of their composition. But, there are other factors that can affect density. These factors include temperature and pressure. As the temperature of a material is raised, its density decreases. This happens because material expands when heated, and the volume increases. The amount of material does not change. But, it takes up more space, so it is less dense. As pressure on a material is raised, its density increases. Again, the mass of a material does not change, but that material is squeezed into a smaller space, causing its density to increase.

What do you think density has to do with layering in Earth? The three major layers have distinct compositions, and therefore, they have different densities. The core is metallic, similar to the metorite shown in **Figure 28**. The force of gravity has pulled it to the center of the planet. Most elements that make up mantle and crust rocks are less dense than material in the core, so as the metallic core material sank, mantle and crust matter moved up toward the surface. The rocks in the crust are the least dense of all rocks. This compositional layering is thought to have formed billions of years ago, when Earth was young.

Figure 28 Scientists think that billions of years ago dense metallic elements sank to the center of Earth, forming a core. The lighter elements floated upward, forming the mantle and the crust. Earth would have had to melt for this to occur. This meteorite is made of material similar to Earth's core.

Convection in the Core and Mantle

Thermal energy in Earth's outer core and mantle escapes toward the surface mostly by convection. This is important for two major Earth processes. First, convection in the outer core produces Earth's magnetic field. As Earth spins on its axis, convection currents of molten iron produce a magnetic field around the planet. This causes Earth to act a little like a huge bar magnet. In Chapter 4, you will read about how Earth's magnetic field helps scientists understand plate tectonics.

 Reading Check What produces Earth's magnetic field?

Second, convection in the mantle is important for plate tectonics. It might seem hard to think about convection in the mantle, because it is made mostly of solid rock. But scientists have discovered that even solid rock can flow. In order for this to happen, the rock in some places must be very hot, and it must be cooler in other places. The flow takes place extremely slowly.

Energy and matter from the mantle are transferred to the plates. At one time, most scientists thought the flow of material in Earth's mantle drove the plates, much like items moving along on a conveyor belt below them. But recent studies show that the plates themselves might control the convective flow of the mantle below them. **Figure 29** shows what the convection currents in the outer core and mantle might look like. Remember that there still is much to learn about this movement of material in Earth's interior.

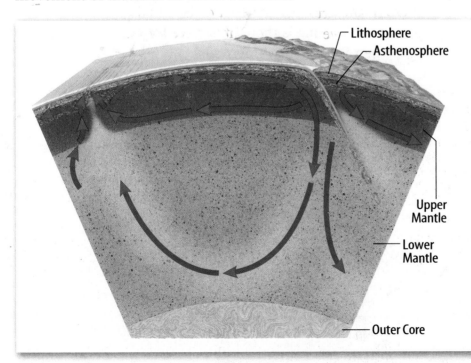

Lithosphere
Asthenosphere
Upper Mantle
Lower Mantle
Outer Core

Figure 29 Matter and energy in Earth's mantle and core move mainly by convection. The blue arrows shown in this sketch suggest general directions of motion but much remains to be learned about this motion.

Dynamic Layers

Now that you've thought about Earth's structure from the surface to the core, you probably realize that Earth is a dynamic planet. Still energized by decay of radioactive elements in the interior, material within Earth continues to move. As long as this movement of matter occurs, heat escapes and changes Earth's surface by uplift in some regions.

Layering of Earth started when Earth first formed billions of years ago. Trying to dig a hole to look at Earth's interior is impossible, so scientists had to rely on other methods to find out what was there. Using earthquakes and other vibrations brought the layers to light. Today we are looking for ways to learn even more about our planet's layers.

LESSON 3 Review

Summarize

Create your own lesson summary as you write a script for a **television news report.**

1. **Review** the text after the red main headings and write one sentence about each. These are the headlines of your broadcast.

2. **Review** the text and write 2–3 sentences about each **blue** subheading. These sentences should tell *who, what, when, where,* and *why* information about each **red** heading.

3. **Include** descriptive details in your report, such as names of reporters and local places and events.

4. **Present** your news report to other classmates alone or with a team.

 **ELA6:** LS 1.4

Standards Check

Using Vocabulary

1. Use the word *asthenosphere* in a sentence. **1.b**

2. In your own words, write a definition for Earth's *core*. **1.b**

Understanding Main Ideas

3. **Give an example** of a common object that has a layered structure. **1.b**

4. **List** the names of Earth's internal layers, starting at the center of the planet. **1.b**

5. **Name** the layers of Earth. Add extra ovals to show how the layers are divided. **1.b**

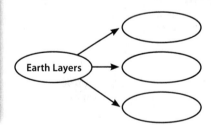

6. **Apply** what you have learned about density to explain why a bar of soap floats in the bathtub. **4.c**

7. **Compare** the materials in the outer core to the materials in the lithosphere. **4.c**

Applying Science

8. **Imagine** Earth's internal heat suddenly increased. Would convection currents flow more quickly or more slowly? **4.c**

 **Science** nline

For more practice, visit **Standards Check** at ca6.msscience.com.

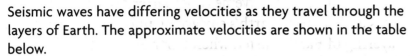

Applying Math

Seismic Wave Velocity

Seismic waves have differing velocities as they travel through the layers of Earth. The approximate velocities are shown in the table below.

Seismic Wave Velocities			
Wave Type	Velocity in Earth's Crust (v_{cr}) (km/s)	Velocity in Earth's Mantle (v_m) (km/s)	Velocity in Earth's Core (v_{co}) (km/s)
P wave	5–7	8	8
S wave	3–4	4.5	N/A

Example

If a P wave has a velocity of 6.2 km/s in Earth's crust, find the increase in velocity as the P wave enters the mantle.

1 **What do you know:** velocity in the crust (v_{cr}): 6.2 km/s
velocity in the mantle (v_m): 8 km/s

2 **What do you need to know:** difference in velocity (v_d)

3 **Use this equation:** $v_d = v_m - v_{cr}$

4 **Subtract the velocities:** $v_d = 8 - 6.2 = 1.8$

Answer: The P wave increases in velocity by 1.8 km/s when it travels from the crust through the mantle.

Practice Problems

1. If a P wave has a velocity of 5.6 km/s in Earth's crust, find the increase in velocity as the P wave enters the mantle?

2. If an S wave has a velocity of 3.7 km/s in Earth's crust, what is the increase in velocity as the wave travels into to the mantle?

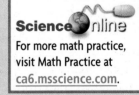

Science Online
For more math practice, visit Math Practice at ca6.msscience.com.

Model and Invent:
Earth's Layers

Materials

assorted colors clay
sticky notes
plastic knife
toothpicks
metric ruler
calculator
pencil
resource books

Safety Precautions

Problem

The inner layers of Earth are too deep, too dense, and too hot for humans to explore. But, scientists can study paths and characteristics of seismic waves and experiment with surface minerals and rocks to gain information about the layers that make up Earth. Use your knowledge about studies of Earth's interior to model the structure of Earth's layers.

Form a Hypothesis

Based on information in this chapter, make a statement estimating what percentage of Earth's volume is composed of crust, mantle, and core.

Collect Data and Make Observations

1. Review Earth's interior.
2. Develop a plan to model Earth's layers.
3. As part of your plan, determine what materials you might use to model Earth's layers. Label your layers with estimates of temperature, density, composition, and physical state.
4. Gather your materials and follow your plan to make the model.

Science Content Standards

1.b Students know Earth is composed of several layers; a cold brittle lithosphere; a hot, convecting mantle; and a dense, metallic core.
7.e Recognize whether evidence is consistent with a proposed explanation
7.f Read a topographic map and a geologic map for evidence provided on the maps and construct and interpret a simple scale map.

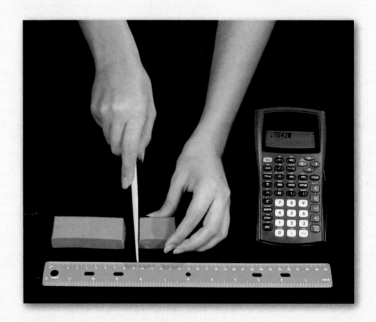

Analyze and Conclude

1. **Label** your layers accurately. How many main layers are present in your model?
2. **Measure** the thicknesses of each layer, including subdivisions of main layers. Record the thicknesses in a data table like the one shown.
3. **Examine** your model for how well it represents materials that make up Earth's layers. Summarize your observations in the data table.
4. **Evaluate** your work for scale and materials used.
5. **Decide** whether or not your hypothesis was supported by the research you did. Explain your reasoning.

Modeling Earth's Layers				
Layer	**Actual Earth Thickness**	**Model Thickness**	**Earth Material**	**Model Material**
Upper mantle (part of the lithosphere)	100 km		peridotite	

Communicate

In the 1860s Jules Verne wrote a fictional story about a *Journey to the Center of the Earth*. Scientists of his time didn't know as much as we do about Earth's interior. Write a one-page story about an imaginary journey to Earth's center using what you learned in this chapter.

Real World Science

Studying Earth's Magnetic Field

Some geoscientists measure Earth's magnetic field, which originates deep within the planet. When rocks are formed, the crystals line up with the magnetic field and give us a history of Earth's magnetism. This shows how continents move and Earth's magnetic field changes over time.

Visit **Careers** at **ca6.msscience.com** to find out what a geomagnetist does. Write a want ad for a geomagnetist listing the required education and skills.

Ways of Measuring Earth's Magnetic Field

The core of Earth is a solid iron ball about as hot as the surface of the Sun. Surrounding it is an ocean of liquid iron, which is an electrically conducting fluid in constant motion. Out of this ocean comes Earth's magnetic field.

Direct measurements of Earth's magnetic field are continually made from oceanographic, land, aircraft, and satellite surveys. SWARM is one survey conducted by the USGS.

Visit **Technology** at **ca6.msscience.com** to find out about satellite missions for magnetic study. Make a table of satellite systems from around the world. Rank them in order of importance.

The History of Geomagnetism

Around the year 1000 the Chinese invented the magnetic compass. A variety of scientists contributed to the study of Earth's magnetic field, starting with William Gilbert and including Halley, Couloumb, Gauss, and Sabine.

Gilbert was a naturalist who discovered a species of potoroo in Australia, known as Gilbert's potoroo, shown here. He was the first to explain why a compass needle points north-south—Earth itself is magnetic.

Visit **History** at ca6.msscience.com to find links to some of magnetism's historical figures. Create a poster telling about one of these historical persons' contributions.

How the Changing Magnetic Field Affects Us

Earth's magnetic field (or geomagnetic field) influences human activity and the natural world in many ways. The geomagnetic field can both assist and hinder navigation and surveying techniques, it can hinder geophysical exploration, it can disrupt electric power utilities and pipeline operations, and it can influence modern communications.

For hundreds of years, sailors have relied on magnetic compasses to navigate the oceans. These sailors knew that Earth's magnetic north pole was not in the same place as the geographic north pole and they were able to make the necessary corrections to determine where they were and, more importantly, how to get home. In modern times, many navigators also rely on the Global Positioning System (GPS) to find their location.

Visit **Society** at ca6.msscience.com to find information to write a short article on one of the effects of a changing magnetic field, describing what the potential hazards or benefits may be.

The BIG Idea Heat escaping from Earth's internal layers constantly changes the planet's surface.

Lesson 1 Landforms
🏴 1.e, 1.f, 2.a, 7.c

Main Idea Forces inside and outside Earth produce Earth's diverse landforms.

- Uplift produces elevated landforms, such as mountains and plateaus.
- Erosion produces landforms by removing sediment, which is deposited at another location.
- Valleys and beaches are landforms resulting from erosion and deposition of Earth's surface materials.
- California has many uplifted mountain ranges and volcanoes.
- In California there are large, open valleys parallel to mountain ranges and river valleys running down to the ocean, where beaches form.

- **erosion** (p. 80)
- **landform** (p. 79)
- **uplift** (p. 79)

Lesson 2 Minerals and Rocks
🏴 2.c, 6.b, 6.c, 7.e

Main Idea The solid Earth is made of minerals and rocks.

- The solid Earth is made of rocks, which are made of minerals.
- Each mineral can be identified by testing for a set of physical properties.
- Minerals are valuable resources that are used by humans in many ways.
- There are three major groups of rocks: igneous, metamorphic, and sedimentary.
- Rocks continuously change as they are subject to processes of the rock cycle.

- **density** (p. 91)
- **lava** (p. 96)
- **magma** (p. 96)
- **mineral** (p. 87)
- **rock** (p. 95)
- **rock cycle** (p. 100)
- **sediment** (p. 99)

Lesson 3 Earth's Interior
🏴 1.b, 4.c, 7.e, 7.g

Main Idea Earth's interior has a layered structure.

- Earth is composed of three major layers, which have distinct compositions.
- The three major layers differ in physical state and composition.
- Scientists study the behavior of seismic waves to indirectly determine the details of Earth's layers.
- Convection in the core produces Earth's magnetic field, and convection in the mantle moves matter and energy to Earth's surface.

- **asthenosphere** (p. 104)
- **core** (p. 105)
- **crust** (p. 103)
- **lithosphere** (p. 105)
- **mantle** (p. 104)

STUDY TO GO Download quizzes, key terms, and flash cards from ca6.msscience.com.

Linking Vocabulary and Main Ideas

Use vocabulary terms from page 114 to complete this concept map.

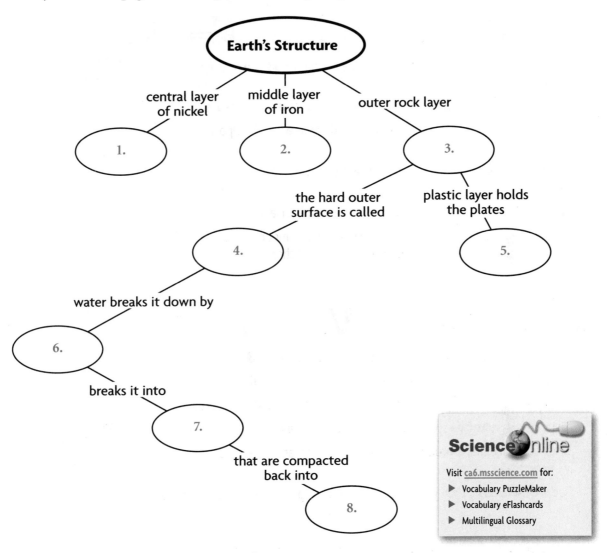

Using Vocabulary

Fill in the blanks with the correct vocabulary words. Then read the paragraph to a partner.

There are more than 3,800 examples of _____ 9. _____, which are the materials that make up rocks. Sometimes, temperature and pressure conditions are just right for rocks to melt beneath Earth's surface to form _____ 10. _____. When this happens, and the molten rock moves to Earth's surface, it can produce a volcanic mountain, which is a _____ 11. _____ that forms by _____ 12. _____, making an area that is elevated compared to its surroundings.

Understanding Main Ideas

Choose the word or phrase that best answers the question.

1. Which California mountain was made by volcanic eruptions?
   A. Lassen Peak
   B. Sierra Nevada
   C. Mt. Fuji
   D. Mt. Baldy   `1.e`

2. What landforms are low and flat?
   A. volcanoes
   B. mountains
   C. plains
   D. plateaus   `1.f`

3. What produces a U-shaped valley?
   A. uplift
   B. glacial erosion
   C. glacial uplift
   D. river deposition   `2.a`

4. The map below outlines major landform regions of California.

   What major California landform is colored in on the map?
   A. Central Valley
   B. Coast ranges
   C. Death Valley
   D. Sierra Nevada   `2.a`

5. The photo below shows a fragment of the mineral rhodochrosite.

   The surfaces of this rhodochrosite sample indicate that it displays which type of breakage?
   A. fracture
   B. luster
   C. cleavage
   D. linear   `2.c`

6. Which type of rock is crystallized from melted rock?
   A. sedimentary
   B. igneous
   C. metamorphic
   D. chemical   `6.c`

7. What is the name of the solid, metallic portion of Earth's interior?
   A. crust
   B. mantle
   C. inner core
   D. outer core   `1.b`

8. What are the two types of crust?
   A. metallic and rocky
   B. rock and mineral
   C. basaltic and organic
   D. oceanic and continental   `1.b`

9. Earth's magnetic field is produced by convection in which of Earth's layers?
   A. crust
   B. lithosphere
   C. mantle
   D. core   `1.b`

Science⌒nline Standards Review ca6.msscience.com

Applying Science

10. Classify these layers of Earth as solid or liquid: inner core, outer core, mantle, lithosphere, and crust.   `1.b`

11. Justify mining for ore minerals. Mining produces large amounts of pollution, which is harmful to people's health. Justify the continued extraction of ores considering the environmental problems associated with it.   `6.b`

12. Predict what the texture of an igneous rock would be like if the following happened:
A. The magma started to cool and crystallize deep within Earth.
B. Next, the molten rock with crystals in it suddenly was forced to the surface and erupted from a volcano.   `1.b`

13. Describe the characteristics of the asthenosphere that allow the plates to ride on it.   `4.c`

14. Sketch a graph that shows, in general, how temperature changes with increasing depth in Earth.   `4.c`

15. Explain the physical property displayed by the crystal shown below.   `2.c`

pment along with computers enables the
rd to monitor the vo... ...each vessel. Cut
onsible for offshor... ...an ar ...scue. Many
r ships are ...ped with ...ding p...
opters to ... rapid emergency se...es. In
years ... ...ives have been save... ...quick res...
men a... ...women of the C... ...uard.
esides sear... ...and rescue ...ations the Coast
in the enforce... ...customs and imm...
and the prevention... ...illegal smuggling. Sin

WRITING in ▶ Science

16. Write three paragraphs that describe the main layers of Earth from crust to core. Include information about how scientists have determined this layered structure and list a few facts about each layer.   `ELA6: W 1.2`

Cumulative Review

17. Identify a type of map that accurately displays landforms.   `2.a`

18. Name the kind of map you would use to show rock structures that are underground.   `2.a`

Applying Math

Use the table on page 109 to answer questions 19–23.

19. What is the loss of speed as a P-wave travels at a velocity of 6.3 km/s through Earth's crust through the mantle?   `MA6: NS 2.0`

20. If an S-wave has a velocity of 2.9 km/s in Earth's core, what is the loss in velocity as the wave travels from the mantle to the core?   `MA6: NS 2.0`

21. If an S-wave has a velocity of 3.7 km/s in Earth's crust, what is the gain in velocity as the wave travels from the crust to the mantle?   `MA6: NS 2.0`

22. If an S-wave has a velocity of 2.5 km/s in Earth's core, what is the loss in velocity as the wave travels from the mantle to the core?   `MA6: NS 2.0`

23. What is the loss of speed as a P-wave travels at a velocity of 8 km/s through Earth's core through the mantle?   `MA6: NS 2.0`

1 Which special property is illustrated by the piece of calcite shown below?

A magnetism

B double refraction

C reaction to acid

D salty taste `2.c`

2 What forms when lava cools so quickly that crystals cannot form?

A volcanic glass

B intrusive rock

C bauxite

D a gem `1.b`

3 Which property describes the color of powdered mineral?

A hardness

B luster

C cleavage

D streak `2.c`

4 Which type of rock forms when magma cools?

A sedimentary

B chemical

C metamorphic

D igneous `2.c`

5 Which changes sediment into sedimentary rock?

A weathering and erosion

B heat and pressure

C compaction and cementation

D melting `2.c`

6 In general, what happens to pressure as you move outward from Earth's interior?

A decreases

B decreases then increases

C increases

D increases then decreases `4.c`

7 Which causes some minerals to break along smooth, flat surfaces?

A streak

B cleavage

C luster

D fracture `2.c`

8 Which mineral will scratch feldspar but not topaz?

A quartz

B calcite

C apatite

D diamond `2.c`

9   Use the illustration below to answer question 9.

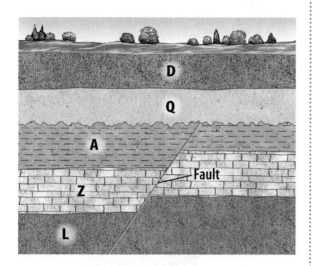

These layers of sedimentary rock were disturbed after they were deposited. Which layer was deposited first after the fault occurred?

A   layer L

B   layer Z

C   layer Q

D   layer A   **1.f**

10   Which type of rock forms because of high heat and pressure without melting?

A   igneous rock

B   intrusive rock

C   sedimentary rock

D   metamorphic rock   **1.e**

11   During which process do minerals precipitate in the spaces between sediment grains?

A   cementation

B   compaction

C   conglomerate

D   weathering   **1.b**

12   Which is a common rock forming mineral?

A   azurite

B   gold

C   quartz

D   diamond   **2.c**

Use the table below to answer question 13.

Portion of Mohs Hardness Scale		
Mineral	**Hardness**	**Common Tests**
Calcite	3	barely scratched by copper coin
Feldspar	6	scratches glass
Quartz	7	scratches glass and feldspar
Topaz	8	scratches quartz

13   The Mohs scale is used to determine the hardness of rocks and minerals. A sample that scratches another is identified as being harder than the substance it scratches. Which mineral can be scratched by glass?

A   calcite

B   feldspar

C   quartz

D   topaz   **2.c**

Thermal Energy and Heat

The BIG Idea

Thermal energy moves from warmer to cooler materials until the materials have the same temperature.

LESSON 1 3.a
Forms of Energy

Main Idea Energy exists in many forms.

LESSON 2 3.a, 3.b
Energy Transfer

Main Idea Energy can be transferred from one place to another.

LESSON 3 3.a, 7.c
Temperature, Thermal Energy, and Heat

Main Idea Thermal energy flows from areas of higher temperature to areas of lower temperature.

LESSON 4 3.c, 3.d, 7.a
Conduction, Convection, and Radiation

Main Idea Thermal energy is transferred by conduction, convection, and radiation.

Feeling The Burn

This raging forest fire glows red as it burns the trees and other vegetation in its path. The changes caused by a forest fire are due to the release of thermal energy. The thermal energy released by the fire causes the high temperatures that help keep the fire going.

Science Journal List three changes that occur when you light a match.

How cold is it?

Hot and cold are words you often use. How accurate is your sense of hot and cold?

Procedure

1. Complete a lab safety form.

2. Fill a **pan** with **lukewarm water.** Fill a second **pan** with **cold water** and **ice.** Fill a third **pan** with **warm tap water.**

3. Put one hand into the cold water and the other hand into the warm water. Keep your hands in the water for 15 s.

4. Quickly remove both hands from the pans and put them both into the pan of luke-warm water.

Think about This

- **Identify** which hand felt warmer when you placed both hands in the lukewarm water.

- **Explain** whether your sense of warm and cold would make a useful thermometer.

3.a

Science Online

Visit ca6.msscience.com to:

▶ view **Concepts in Motion**

▶ explore Virtual Labs

▶ access content-related Web links

▶ take the Standards Check

Energy and Change
Make the following Foldable to record the types of energy and examples of changes caused by the energy.

▷ **STEP 1** **Fold** a sheet of paper into thirds lengthwise.

▷ **STEP 2** **Unfold** and **draw** vertical lines along the folds. Draw three horizontal lines to divide the paper into four rows. Label as shown.

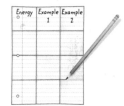

Reading *Skill*   **ELA6:** R 2.4

Inferring

As you read this chapter, list in the first column the types of energy discussed. In the second column, list an example from the text of a change caused by that energy. In the third column, describe a different change that you know is caused by that type of energy.

Get Ready to Read

New Vocabulary

①Learn It! What should you do if you find a word you don't know or understand? Here are some suggested strategies:

1. Use context clues (from the sentence or the paragraph) to help you define it.
2. Look for prefixes, suffixes, or root words that you already know.
3. Write it down and ask for help with the meaning.
4. Guess at its meaning.
5. Look it up in the glossary or a dictionary.

②Practice It! Look at the word *bonds* in the following passage. See how context clues can help you understand its meaning.

Context Clue
Bonds occur between atoms.

Context Clue
Bonds join atoms together.

Context Clue
Bonds can be broken.

When you eat, you take in another type of potential energy. Chemical potential energy is the energy stored in bonds between the atoms that make up matter. Remember that atoms are joined together by chemical bonds to form molecules, as shown in **Figure 6.** Chemical potential energy can be released during chemical reactions when these bonds are broken and new bonds are formed.

—*from page 128*

③Apply It! Make a vocabulary bookmark with a strip of paper. As you read, keep track of words you do not know or want to learn more about.

Target Your Reading

Use this to focus on the main ideas as you read the chapter.

Reading Tip

Read a paragraph containing a vocabulary word from beginning to end. Then, go back to determine the meaning of the word.

1 **Before you read** the chapter, respond to the statements below on your worksheet or on a numbered sheet of paper.

- Write an **A** if you **agree** with the statement.
- Write a **D** if you **disagree** with the statement.

2 **After you read** the chapter, look back to this page to see if you've changed your mind about any of the statements.

- If any of your answers changed, explain why.
- Change any false statements into true statements.
- Use your revised statements as a study guide.

Science Online

Print a worksheet of this page at ca6.msscience.com.

Before You Read A or D	Statement	After You Read A or D
	1 Energy is the ability to cause change.	
	2 There is only one type of energy.	
	3 Thermal energy always flows from larger objects to smaller objects.	
	4 Only waves can transfer energy from place to place.	
	5 Thermal energy from the Sun travels to Earth as waves.	
	6 Energy can be stored in a stretched rubber band.	
	7 Objects must be touching each other in order for energy to flow from one to another.	
	8 A baseball player can transfer energy to a baseball by hitting the ball with a bat.	
	9 Most materials get smaller when they are heated.	
	10 The atoms and molecules in an object move slower as the object cools.	

Science Content Standards

3.a Students know energy can be carried from one place to another by heat flow or by waves, including water, light and sound waves, or by moving objects.

Reading Guide

What *You'll Learn*

▶ **Define** energy.

▶ **Describe** different forms of energy.

▶ **Distinguish** between kinetic energy and potential energy.

Why *It's Important*

Energy is the cause of all the changes you observe in the world around you.

Vocabulary

energy
kinetic energy
potential energy
elastic potential energy
thermal energy

Review Vocabulary

gravity: attractive force between two objects that depends on the masses of the objects and the distance between them (Grade 5)

Forms of Energy

Main Idea Energy exists in many forms.

Real-World Reading Connection A pizza fresh from the oven sure smells good! When you put it in the oven, it was just soft dough covered with cold tomato sauce, cheese, and uncooked vegetables. Now, the dough is crisp and golden, the sauce is hot, and the vegetables are toasted. What caused these changes?

What is energy?

Think about the changes you see and feel every day. You might have seen cars moving and felt the wind. All the changes around you are caused by energy. **Energy** (EN ur jee) is the ability to cause change. There are different forms of energy.

The ball shown in **Figure 1** changes the window from a solid sheet of glass to many smaller glass pieces. These changes occur because the moving ball has energy. **Kinetic** (kuh NEH tihk) **energy** is the energy an object has because it is moving. The amount of kinetic energy a moving object has depends on two things. One is the object's mass. The other is the object's speed. The kinetic energies of some moving objects are given in **Figure 2.**

Figure 1 The kinetic energy carried by the moving baseball caused this window to change.

Visualizing Kinetic Energy

Figure 2

▶ The amount of kinetic energy of a moving object depends on the mass and speed of the object. Energy is measured in units called joules (J). For example, the fastest measured speed a baseball has been thrown is about 45 m/s. The kinetic energy of a baseball traveling at that speed is about 150 J.

▼ A 600-kg race car traveling at about 50 m/s has about 5,000 times the kinetic energy of the baseball.

▲ There is evidence that a meteorite 10 km in diameter collided with Earth about 65 million years ago and might have caused the extinction of dinosaurs. The meteorite might have been moving 400 times faster than the baseball and would have a tremendous amount of kinetic energy due to its enormous mass and high speed—about a trillion trillion joules.

▼ Earth's atmosphere is continually bombarded by particles called cosmic rays, which are mainly high-speed protons. The mass of a proton is about a 100 trillion trillion trillion times smaller than the mass of the baseball. Yet some of these particles travel so fast, they have nearly the same kinetic energy as the baseball.

◀ A sprinter with a mass of about 55 kg running at 9 m/s has kinetic energy about 15 times greater than the baseball.

Contributed by National Geographic

Kinetic Energy and Speed

In a game of pool, the cue ball hits the rack of balls and causes these balls to move. The kinetic energy of the moving cue ball changes the positions and the motions of these other balls when it hits the rack. A greater change **occurs** when the cue ball is moving faster. Then the balls in the rack move faster and become more spread out. This means that the cue ball has more kinetic energy when it is moving faster. The faster an object moves, the more kinetic energy it has.

ACADEMIC VOCABULARY
occur (uh KUHR)
(verb) to happen
Something exciting occurred today.

Kinetic Energy and Mass

The kinetic energy a moving object has increases as the mass of the object increases. Suppose you roll a volleyball down a bowling alley instead of a bowling ball. If the two balls have the same speed, the volleyball will knock down fewer pins than the bowling ball. Even though the balls have the same speed, the volleyball has less kinetic energy because it has less mass.

Figure 3 shows how the kinetic energy of moving objects depends on their mass and speed. The two cars have the same mass, but they don't have the same kinetic energy. The blue car has more kinetic energy because it is moving faster. The truck and the blue car have the same speed. However, the truck has more kinetic energy because its mass is greater.

Units of Energy

Energy is measured in units of joules, abbreviated as J. If you dropped a softball from a height of about 0.5 m, it would have about 1 J of kinetic energy before it hit the floor. All forms of energy are measured in units of joules.

Figure 3 The kinetic energy of each vehicle depends on its mass and speed. The truck has more kinetic energy than the blue car because it has more mass. The blue car has more kinetic energy than the green car because it is moving faster.
Compare the kinetic energy of the two cars if they have the same speed.

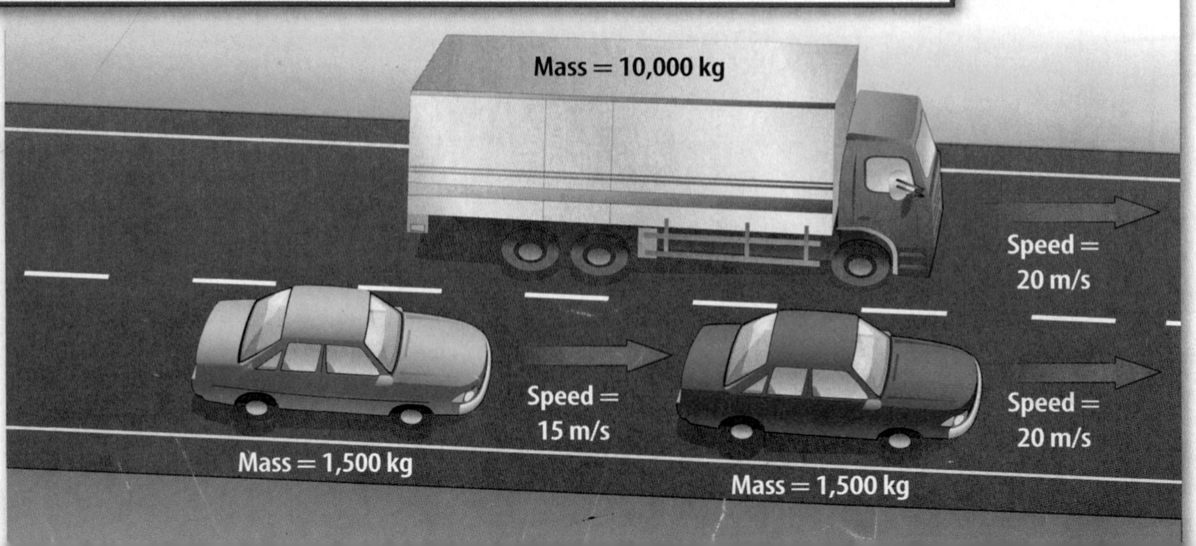
Mass = 10,000 kg
Speed = 20 m/s
Speed = 15 m/s
Speed = 20 m/s
Mass = 1,500 kg
Mass = 1,500 kg

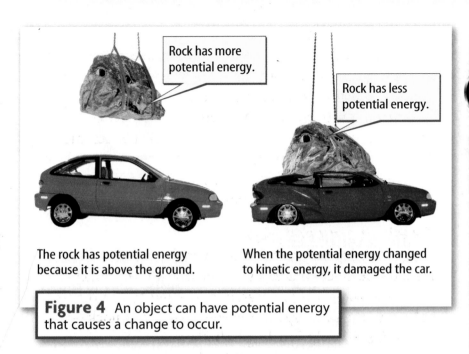

Rock has more potential energy.

Rock has less potential energy.

The rock has potential energy because it is above the ground.

When the potential energy changed to kinetic energy, it damaged the car.

Figure 4 An object can have potential energy that causes a change to occur.

Potential Energy—Stored Energy

An object can have energy even if it is not moving. Look at the hanging rock in **Figure 4.** Would you say it has energy? Even though no changes are occurring, the rock still has energy. Remember that energy is the ability to cause change. When the rock falls, it causes a change. Because the rock had the ability to cause change before it fell, it had energy as it was hanging above the car. The hanging rock has stored energy, called **potential** (puh TEN chul) **energy.** There are different forms of potential energy.

Gravitational Potential Energy

The rock hanging above the ground has a form of stored energy called gravitational potential energy. This form of energy is due to the downward pull of Earth's gravity. Gravitational potential energy depends on an object's mass and its height above the ground. The hanging rock in **Figure 4** has gravitational potential energy due to its height above the ground.

The higher an object is above the ground or the greater its mass, the more gravitational potential energy it has. For example, the hanging rock in **Figure 4** would cause even more damage if it fell from a greater height. More damage also would be caused if a rock with more mass fell from the same height. The gravitational potential energy of an object increases if its mass or height above the ground increases.

Reading Check What are two ways to increase an object's gravitational potential energy?

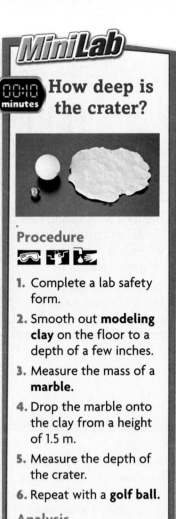

MiniLab

00:10 minutes

How deep is the crater?

Procedure

1. Complete a lab safety form.
2. Smooth out **modeling clay** on the floor to a depth of a few inches.
3. Measure the mass of a **marble.**
4. Drop the marble onto the clay from a height of 1.5 m.
5. Measure the depth of the crater.
6. Repeat with a **golf ball.**

Analysis

1. **Relate** the depth of the crater to the mass of the balls.
2. **Infer** how the potential energy of the balls before they fell depended on their masses.

3.a

WORD ORIGIN
potential
from Latin *potens,* means *power*

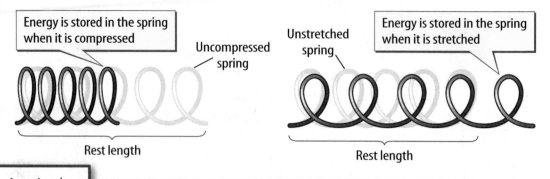

Energy is stored in the spring when it is compressed

Uncompressed spring

Unstretched spring

Energy is stored in the spring when it is stretched

Rest length

Rest length

Figure 5 A spring has elastic potential energy when it is compressed or stretched.

Identify two ways a spring can store elastic potential energy.

SCIENCE USE V. COMMON USE
matter

Science Use something that has mass and occupies space. *The weight of an object depends on the amount of matter it contains.*

Common Use trouble or difficulty. *What's the matter with your CD player?*

Elastic Potential Energy

If you stretch a rubber band and then let it go, you know about another type of stored energy. **Elastic** (ih LAS tik) **potential energy** is the energy stored when an object is squeezed or stretched. When you stretch a rubber band, the elastic potential energy of the rubber band increases. This stored energy then can cause the rubber band to fly across the room when you let it go.

Figure 5 shows the two ways that a spring can store elastic potential energy. If the spring is squeezed together, or compressed, it has a tendency to change back to its rest length. The spring also will return to its rest length if it is stretched. Either by compression or stretching, stored elastic potential energy gives an object the ability to change.

Chemical Potential Energy

When you eat, you take in another type of potential energy. Chemical potential energy is the energy stored in bonds between the atoms that make up matter. Remember that atoms are joined together by chemical bonds to form molecules, as shown in **Figure 6.** Chemical potential energy can be released during chemical reactions when these bonds are broken and new bonds are formed. You get energy by eating because food contains chemical potential energy. Fossil fuels, such as oil and coal, are energy sources because they contain chemical potential energy.

Figure 6 The atoms in molecules are held together by chemical bonds that store chemical potential energy.

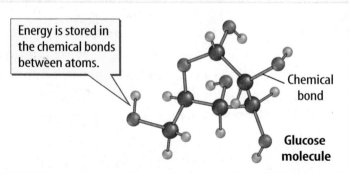

Energy is stored in the chemical bonds between atoms.

Chemical bond

Glucose molecule

Light Energy and Thermal Energy

When you turn on a lamp, a change occurs. Light from the lamp enables you to see objects in the room more clearly. When you turn on a stove to heat a pot of water, a change occurs. Heating the pot causes the temperature of the water to increase. These changes are caused by light energy and thermal energy.

Light Energy

What causes the changes you observe in plants? You may know that sunlight is needed for plants to grow. Sunlight causes plants to grow because sunlight contains a form of energy called light energy. Light energy is the energy carried by light waves. When you turn on a lamp, the light spreads out to make a room seem bright. **Figure 7** shows why light energy also is sometimes called radiant energy. It spreads out, or radiates, in all directions from its source.

 Reading Check Why is light energy sometimes called radiant energy?

Thermal Energy

When you put your hands against a warm cup, as in **Figure 8,** you've felt another form of energy. **Thermal** (THUR mul) **energy** is energy that moves from one place to another because of differences in temperature. Thermal energy is sometimes called heat energy. Like all forms of energy, thermal energy can cause changes. For example, a cup of hot cocoa is warmer than your hands. When you hold a cup of hot cocoa, thermal energy moves from the hot cocoa to your hands. This causes a change to occur. Your hands become warmer, and the cocoa becomes cooler. Thermal energy always moves from warmer objects to colder objects.

Figure 7 Light energy spreads out in all directions from a light source.

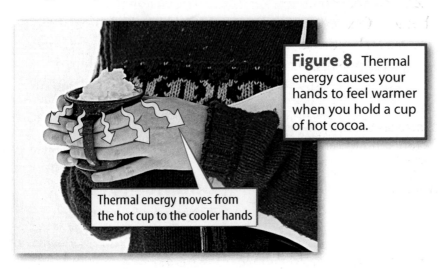

Figure 8 Thermal energy causes your hands to feel warmer when you hold a cup of hot cocoa.

Thermal energy moves from the hot cup to the cooler hands

The Different Forms of Energy

Energy is the ability to cause change. There are different forms of energy, but all forms of energy can cause something to change. A moving object has a form of energy called kinetic energy. Kinetic energy increases if the speed or mass of the object increases.

Potential energy is stored energy. There are different kinds of potential energy. In this lesson you read about gravitational potential energy, elastic potential energy, and chemical potential energy.

Finally, thermal energy is energy that moves from place to place because of differences in temperature. You will read more about thermal energy in Lesson 3.

LESSON 1 Review

Summarize

Create your own lesson summary as you organize an **outline.**

1. **Scan** the lesson. Find and list the first **red** main heading.

2. **Review** the text after the heading and list 2–3 details about the heading.

3. **Find** and list each **blue** subheading that follows the **red** main heading.

4. **List** 2–3 details, key terms, and definitions under each **blue** subheading.

5. **Review** additional **red** main headings and their supporting **blue** subheadings. List 2–3 details about each.

 ELA6: R 2.4

Standards Check

Using Vocabulary

1. A car traveling along the highway has _____ energy due to its motion. **3.a**

2. Define *thermal energy* in your own words. **3.a**

Understanding Main Ideas

3. **Identify** the form of energy the rock has due to its height above Earth. **3.a**

4. **Describe** the two ways a spring can store energy. **3.a**

5. **Explain** why your hands feel warm when you hold a cup of hot cocoa. **3.a**

6. **Organize Information** Copy and fill in the graphic organizer below with information about the different forms of potential energy. **3.a**

Applying Science

7. **Imagine** a bowling ball and a golf ball that have the same kinetic energy. Which one is moving faster? If they move at the same speed, which one has more kinetic energy? **3.a**

8. **Identify** the two forms of potential energy contained in an orange hanging from the branch of an orange tree. **3.a**

 **Science** **Online**

For more practice, visit **Standards Check** at ca6.msscience.com.

Forms of Energy ca6.msscience.com Brain POP

Science Content Standards

3.a Students know energy can be carried from one place to another by heat flow or by waves, including water, light and sound waves, or by moving objects.

3.b Students know that when fuel is consumed, most of the energy released becomes heat energy.

Reading Guide

What *You'll Learn*

▶ **Recognize** how moving objects transfer energy from one place to another.

▶ **Describe** how waves transfer energy from one place to another.

▶ **Explain** ways that energy can change from one form to another.

Why *It's Important*

All sports involve the transfer of energy and the conversion of energy from one form to another.

Vocabulary

work
wave
fuel
friction

Review Vocabulary

force: a push or a pull that one object exerts on another object (Grade 2)

Energy Transfer

(**Main Idea**) Energy can be transferred from one place to another.

Real-World Reading Connection What sports do you like? Do you like shooting baskets? Maybe you like running, swimming, or riding a bicycle. In all of these activities, energy moves from one place to another.

Moving Objects Transfer Energy

A moving object transfers energy from one place to another. During a baseball game, a pitcher's moving arm transfers energy to a baseball and makes it move. The moving ball then has kinetic energy. When the ball hits the catcher's mitt, this kinetic energy is transferred to the catcher's mitt.

You can see another example in **Figure 9.** The bowler is able to knock down the pins by using the bowling ball to transfer energy. Some of the kinetic energy of the ball is transferred to the pins when the ball hits the pins. The energy that is transferred causes the pins to move.

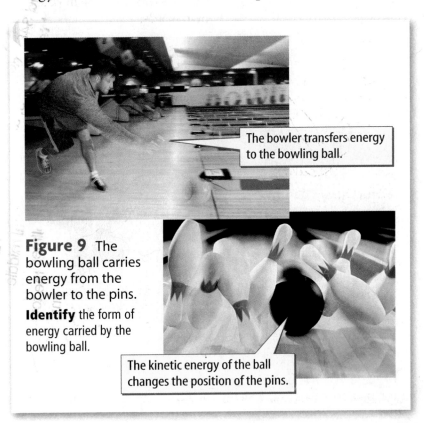

The bowler transfers energy to the bowling ball.

Figure 9 The bowling ball carries energy from the bowler to the pins.
Identify the form of energy carried by the bowling ball.

The kinetic energy of the ball changes the position of the pins.

Doing Work Transfers Energy

Studying for a test may seem like hard work, but in science, it isn't work at all. Scientists define **work** as the transfer of energy that occurs when a push or a pull makes an object move. Recall that a push or a pull also is called a force. **Figure 10** shows that pushing on something doesn't mean that you are doing work. Pushing on a wall might make you feel tired. However, because the wall doesn't move, no work is being done. Work is done only when an object moves in the same direction as the force applied to the object.

If you pick up a box, however, you do work. You pull on the box and cause the box to move upward. Your pull is a force that makes the box move, so you have done work. Because you increase the height of the box above the ground, the gravitational potential energy of the box increases. As you lift the box, you transfer energy to the box. The energy you transfer increases the box's gravitational potential energy. By doing work on the box, you transferred energy to the box.

 **Reading Check** Why is no work done when you push on a wall and it does not move?

Waves Transfer Energy

Have you ever been floating in a wave pool? The waves cause you to move up and down as they move beneath you. The waves cause your position and motion to change because they carry energy. A **wave** is a disturbance that transfers energy without transferring matter. In a wave pool, some of the energy carried by a water wave is transferred to you as the wave moves past you. The energy transferred by the wave causes you to move upward. Like moving objects, waves transfer energy from one place to another.

The student hasn't done work because the wall doesn't move in the direction of the force.

Force

No work is done.

The student does work because the box moves in the direction of the force he applies.

Force

Motion

Work is done.

Figure 10 The boy only does work when the object moves in the direction of his force.

Water Waves

The water wave in **Figure 11** transfers energy as it moves horizontally along the surface of the water. However, the wave does not transfer matter from place to place. The water wave causes the bobber to move up and down as it passes. After the wave passes, the bobber is in the same place. The bobber and the water beneath it have not been carried along with the wave. Instead, only the energy carried by the wave moves along the surface of the water. Like all waves, water waves transfer energy from one place to another, but they don't transfer matter.

Sound Waves

You can see water waves as they move along the surface of a pool. Energy also can be transferred by waves you can't see, such as sound waves. Sound waves are caused by the back-and-forth movement, or vibration, of an object. **Figure 12** shows how a vibrating drum causes sound waves.

When the drummer in **Figure 12** hits the drum, the head of the drum moves back and forth many times each second. Each time the drum head moves, it hits nearby air particles and transfers kinetic energy to them. These particles, in turn, hit other air particles and transfer kinetic energy. When the drum head moves outward, it causes the particles in air nearby to be bunched together. When the drum head moves inward, it causes air particles outside the drum to spread apart. In this way, energy travels from the drum through the air. When this energy strikes your ears, your hear the sound of the drum.

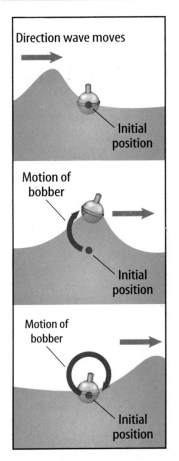

Figure 11 The bobber is not carried along with the wave. It returns to its initial position after the wave passes.

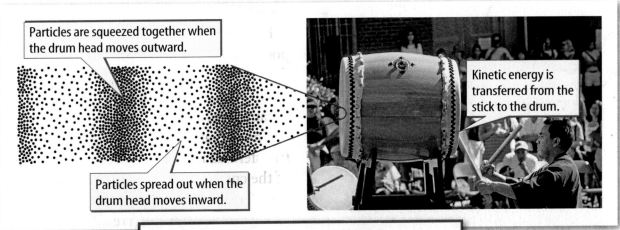

Particles are squeezed together when the drum head moves outward.

Particles spread out when the drum head moves inward.

Kinetic energy is transferred from the stick to the drum.

Figure 12 Energy is passed through the air by sound waves.

Infer what causes the sound waves shown here.

Figure 13 Radiant energy moves from the Sun to Earth as electromagnetic waves.

Electromagnetic Waves

Like sound waves and water waves, light is also a type of wave. However, unlike water waves and sound waves that can travel only in matter, light waves also can travel in empty space. For example, the Sun gives off light waves that travel almost 150 million km to Earth through empty space, as shown in **Figure 13.** Light waves are a type of wave called electromagnetic waves. Electromagnetic waves are waves that can **transfer** energy through matter or empty space. The energy carried by electromagnetic waves also is called radiant energy.

ACADEMIC VOCABULARY
transfer (TRANS fur)
(verb) to move something from one place to another
The doctor will transfer the patient to a different hospital.

Types of Electromagnetic Waves

There are other types of electromagnetic waves besides light waves, as **Figure 14** shows. Radio waves are used to carry the signals you hear when you listen to the radio. Microwaves are electromagnetic waves that heat food in a microwave oven. Microwaves also are used to carry signals to cell phones. The warmth you feel when you sit in sunlight is caused by infrared waves that are emitted by the Sun. X rays are used by doctors to diagnose broken bones.

Figure 14 The energy carried by electromagnetic waves is used in different ways. **Identify** the type of electromagnetic wave being used in each photo.

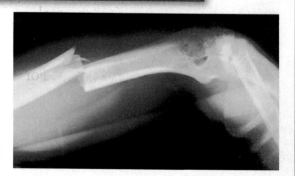

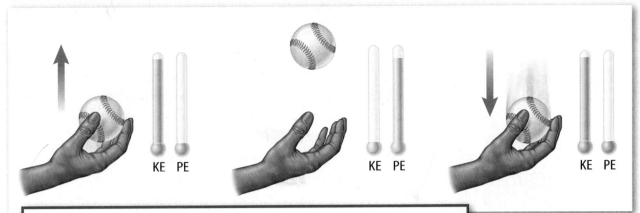

Figure 15 Kinetic energy is converted into potential energy as the ball rises. Potential energy is converted into kinetic energy as the ball falls.

Energy Conversions

Energy comes in different forms and can be converted from one form to another.

Converting Potential Energy and Kinetic Energy

When you throw a ball upward, as in **Figure 15,** energy changes form. As the ball moves upward, the ball's kinetic energy changes into potential energy. When the ball reaches its highest point, all its initial kinetic energy has been converted to potential energy. Then, as the ball falls down, potential energy is converted back into kinetic energy.

Converting Chemical Potential Energy

As the wood burns in **Figure 16,** it gives off light energy and thermal energy. Chemical potential energy is energy stored in the bonds between the atoms that make up wood. As a result, the molecules in the wood contain chemical potential energy. When the wood burns, chemical reactions occur that change chemical potential energy into thermal energy and light energy.

Wood is an example of a **fuel,** a material that can be burned to release energy. For example, when wood is burned, most of the chemical energy changes form. Only a small fraction of the wood's original chemical potential energy remains in the ashes, like those shown in **Figure 16.**

Figure 16 Burning wood converts chemical potential energy to heat and light energy.

Light energy   Thermal energy

The wood contains chemical potential energy.

The ashes contain much less chemical potential energy.

MiniLab

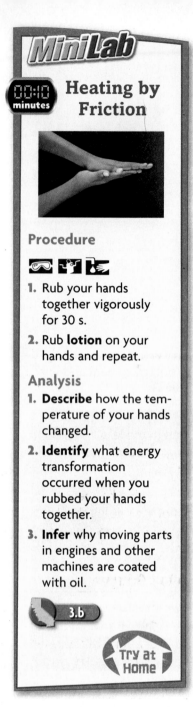

00:10 minutes

Heating by Friction

Procedure

1. Rub your hands together vigorously for 30 s.
2. Rub **lotion** on your hands and repeat.

Analysis

1. **Describe** how the temperature of your hands changed.
2. **Identify** what energy transformation occurred when you rubbed your hands together.
3. **Infer** why moving parts in engines and other machines are coated with oil.

3.b

Try at Home

WORD ORIGIN

friction
from Latin *fricare;* means *to rub*

Converting Thermal Energy to Kinetic Energy

Thermal energy from burning fuels can be used to perform work. As the gasoline in a car's engine burns, most of its stored chemical potential energy changes to thermal energy. The thermal energy produced by the burning gasoline causes forces to be exerted on parts of the engine. These forces make various parts in the car's engine move. These moving parts do work by making the car's wheels turn.

However, in a car's engine, only about one-fourth of the thermal energy produced by burning gasoline is converted into kinetic energy. The other three-fourths of the energy is transferred to the engine and the surrounding air. This causes the engine and other parts of the car to become hot.

Reading Check What happens to most of the chemical potential energy in gasoline when it is burned in an engine?

Converting Kinetic Energy to Thermal Energy

A car's engine changes some of the chemical potential energy in burning fuel to the kinetic energy of the moving car. But as the car moves, some of this kinetic energy is converted back into thermal energy. The conversion of kinetic energy to thermal energy in a car is due to friction between the moving parts in the car. **Friction** (FRIK shun) is the force that resists the sliding of two surfaces in contact.

It is friction that causes a bicycle to stop when you apply the brakes, as shown in **Figure 16.** When you apply a bicycle's brakes, the brake pads rub against the wheels. Friction between the brake pads and the wheels changes the kinetic energy of the wheel to thermal energy. As a result, the bicycle stops. In a similar way, a car's brakes use friction to change the kinetic energy of the car to thermal energy.

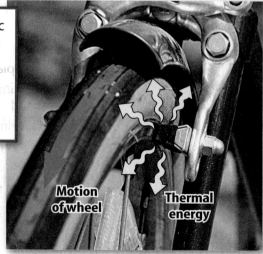

Figure 17 The kinetic energy of the bicycle is transformed into heat energy due to friction.

Explain why applying the brakes brings the bicycle to a stop.

Motion of wheel

Thermal energy

What have you learned?

Energy can be transferred in different ways. Moving objects can transfer energy when they collide with other objects. Water waves, sound waves, and electromagnetic waves can also transfer energy as they move from place to place.

You also read about converting energy from one form to another. One example of energy conversion is the burning of fuels. Fuel contains chemical potential energy stored in the bonds between atoms and molecules. When this energy is released, it can be converted into thermal energy and radiant energy.

LESSON 2 Review

Summarize

Create your own lesson summary as you write a **newsletter.**

1. **Write** this lesson title, number, and page numbers at the top of a sheet of paper.

2. **Review** the text after the **red** main headings and write one sentence about each. These will be the headlines of your newsletter.

3. **Review** the text and write 2–3 sentences about each **blue** subheading. These sentences should tell *who, what, when, where,* and *why* information about each headline.

4. **Illustrate** your newsletter with diagrams of important structures and processes next to each headline.

 ELA6: W 1.2

Standards Check

Using Vocabulary

1. Write a sentence using the word *friction*. **3.b**

2. Gasoline in a car is the car's _____, because the car's engine converts the chemical potential energy of the gasoline to thermal energy and kinetic energy. **3.b**

Understanding Main Ideas

3. **List** three types of energy conversions. **3.b**

4. In which of the following is work NOT done on the box? **3.a**

   A. lifting a box from the floor and placing it on a table

   B. holding a box above your head

   C. carrying a box up the stairs

   D. lowering a box from a table to the floor

5. **Describe** three ways energy can be transferred by waves. **3.a**

6. **Give an example** not mentioned in the lesson of energy changing from one form to another. **3.b**

7. **Analyze** what happens to most of the energy that is released when firewood is burned. **3.b**

Applying Science

8. **Determine Cause and Effect** Copy the graphic organizer below and list the results of applying the brakes on a bicycle. **3.b**

9. **Suggest** reasons why scientists might try to design cars that can convert light energy from the Sun to kinetic energy of the car. **3.a**

For more practice, visit Standards Check at ca6.msscience.com.

Applying Math

Thermal Expansion

Thermal expansion of materials is an increase in the volume of the substance resulting from an increase in temperature. The linear thermal expansion coefficient is change in length of a bar per degree of temperature change. It is measured in parts per million (ppm) per °C. The table below shows linear thermal expansion coefficients for some common substances. To find the volumetric thermal expansion coefficient, multiply the linear thermal expansion coefficient times three.

Linear Thermal Expansion Coefficients	
Substance	Linear Thermal Expansion Coefficients (ppm/°C)
Aluminum	23
Copper	17
Diamond	1
Gold	14
Iron	12
Lead	29
Platinum	9

Example

Find the volumetric thermal expansion coefficient for aluminum.

What you know:	linear thermal expansion coefficient: 23 ppm/°C
What you want to know:	volumetric thermal expansion coefficient
Multiply by 3:	23 ppm/°C × 3 = 69 ppm/°C
Answer: The volumetric thermal expansion coefficient is 69 ppm/°C.	

Practice Problems

1. Find the volumetric thermal expansion coefficient for gold.
2. Find the difference in the volumetric thermal expansion coefficient for aluminum and the volumetric thermal expansion coefficient of gold.

Science Online

For more math practice, visit **Math Practice** at ca6.msscience.com.

Reading Guide

What *You'll Learn*

▶ **Recognize** that thermal energy flows from a warmer object to a cooler object.

▶ **Explain** how temperature depends on particle motion.

▶ **Compare** different temperature scales.

Why *It's Important*

You use thermal energy every time you cook food.

Vocabulary

temperature
thermal expansion
heat

Review Vocabulary

speed: a measure of how quickly an object changes its position (Grade 2)

Temperature, Thermal Energy, and Heat

Main Idea Thermal energy flows from areas of higher temperature to areas of lower temperature.

Real-World Reading Connection Before you cook pancakes, you first have to heat up the pan. Turning on the stove's burner under the pan makes the pan hotter. What happens when the temperature of the pan increases?

What is temperature?

You probably use the word *temperature* often. On a cold or hot day, you may ask a friend what the temperature is. You know that cooking changes the temperature of food. But what does the word *temperature* really mean?

Particles in Motion

Look around you at objects that are sitting still. These objects, and all matter, contain particles such as atoms and molecules that are always moving. **Figure 18** illustrates that even though the object is not moving, the particles that make it up are constantly in motion.

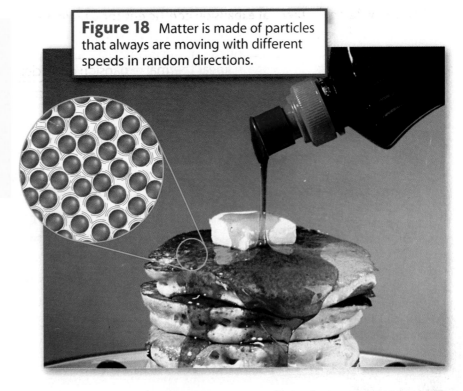

Figure 18 Matter is made of particles that always are moving with different speeds in random directions.

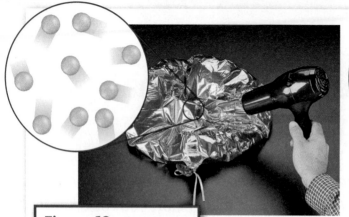
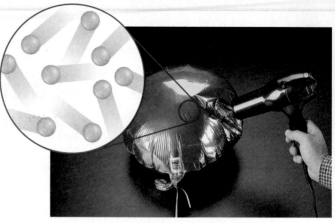

Figure 19 Heating the balloon causes the particles to move faster and take up more space.

Temperature and Particles in Motion

The temperature of an object depends on how fast the particles in the object are moving. The balloon in **Figure 19** contains particles of air that always are moving. Because these particles are moving they have kinetic energy. However, these particles don't move with the same speed. Some are moving faster and some are moving more slowly. Recall that kinetic energy increases as the object moves faster. This means the particles that are moving faster have more kinetic energy.

Because particles in the balloon are moving at different speeds, they have different amounts of kinetic energy. **Temperature** is a measure of the average kinetic energy of the particles in a material. When the particles in a material move faster, they have more kinetic energy. As a result, the average kinetic energy of the particles increases and the temperature of the material increases.

 What is temperature?

Thermal Expansion

Something happens to the balloon in **Figure 19** as it's heated by the hair dryer—it gets larger. Heating the balloon changes the motion of the particles in the balloon.

As the temperature of the air in the balloon increases, the particles in the balloon move faster. As the particles move faster, they tend to move farther apart. This causes the air in the balloon to expand. **Thermal expansion** (THUR mul • ihk SPAN shun) is an increase in the **volume** of a substance when the temperature increases. Most materials expand when their temperature increases. Usually, the greater the increase in temperature, the more the material expands.

WORD ORIGIN · · · · · · · · · · · ·
temperature
from Latin *temperare;* means
moderation

ACADEMIC VOCABULARY · · ·
volume (VAWL yewm)
(noun) the amount of space
occupied by an object or a
region of space
*A basketball has a larger
volume than a tennis ball.*

Measuring Temperature

Temperature is a measure of the average kinetic energy of the particles in a material. However, these particles are so small that it is impossible to measure their kinetic energies. Instead, a practical way to measure temperature is with a thermometer.

Thermometers

You probably have used a thermometer, such as the ones in **Figure 20,** to measure temperature. This type of thermometer contains a red liquid inside a thin glass tube. As the temperature of the liquid increases, it expands, and its volume increases. Then the height of the liquid in the tube increases. When the liquid temperature decreases, the liquid shrinks. Then the height of the liquid in the tube decreases. A scale on the side of tube indicates the temperature. The numbers on this scale depend on the temperature scale being used.

Temperature Scales

There are three common temperature scales—Fahrenheit, Celsius, and Kelvin. These three temperature scales are shown in **Figure 20.** Temperature values on the Kelvin scale are 273 degrees more than temperatures on the Celsius scale. The Fahrenheit scale most often is used in the United States. Other countries use the Celsius scale. The Celsius and Kelvin scales both are used in science.

Figure 20 Fahrenheit, Celsius, and Kelvin are three common temperature scales.
Identify the temperature at which water boils on the Fahrenheit, Celsius, and Kelvin scales.

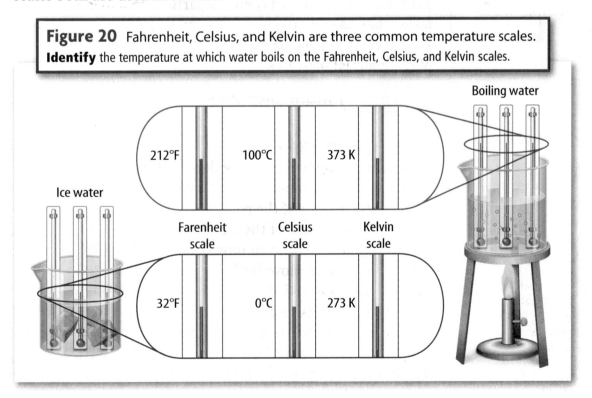

Boiling water

Ice water

212°F   100°C   373 K

Farenheit scale   Celsius scale   Kelvin scale

32°F   0°C   273 K

Air temperature

Thermal energy

Soup temperature

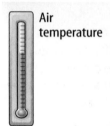

Air temperature

Soup temperature

Figure 21 Thermal energy transfers from areas of higher temperature to areas of lower temperature.

Compare the flow of thermal energy in the top photo to the flow in the bottom photo.

Heat

When you put an ice cube in a glass of water, the water becomes colder. The water becomes colder because thermal energy moves from the warmer water to the colder ice cube. The movement of thermal energy from an object at a higher temperature to an object at lower temperature is called **heat.**

 What is heat?

Heat and Temperature Differences

Heat always transfers energy from something at a higher temperature to something at a lower temperature. The bowl of hot soup sitting on a table in **Figure 21** is warmer than the air around it. As a result, thermal energy flows from the hot soup to the surrounding air. This causes the particles in the air to move faster and the particles in the soup to move more slowly. The kinetic energy of the air particles increases, so the temperature of the air increases. The kinetic energy of the particles in the soup decreases, so the temperature of the soup decreases. As thermal energy continues to move from the soup to the air, the air becomes warmer, and the soup becomes cooler.

Reaching the Same Temperature

Suppose you measured the temperature of the soup and the air as the soup cooled. You would find that after a while, the temperature of the soup and the air were the same. If you continued to measure the air and soup temperatures, you would find that these temperatures did not change.

Thermal energy keeps moving from a warmer object to a cooler object until both objects reach the same temperature. Then there is no difference in temperature between the objects. As a result, there is no longer any transfer of thermal energy. The temperature of the two objects stays constant.

What have you learned?

Temperature is a measure of the average kinetic energy of the particles in a material. In Lesson 1 you read that kinetic energy increases as the speed of the object increases. If a material's temperature increases, the particles in the material move faster. As the particles move faster, the material usually expands and its volume increases.

Heat is the movement of thermal energy due to a difference in temperature. Thermal energy moves from an object at a higher temperature to an object at a lower temperature. The movement of thermal energy stops when the objects reach the same temperature.

LESSON 3 Review

Summarize

Create your own lesson summary as you design a **visual aid.**

1. **Write** the lesson title, number, and page numbers at the top of your poster.

2. **Scan** the lesson to find the **red** main headings. Organize these headings on your poster, leaving space between each.

3. **Design** an information box beneath each **red** heading. In the box, list 2–3 details, key terms, and definitions from each **blue** subheading.

4. **Illustrate** your poster with diagrams of important structures or processes next to each information box.

 ELA6: R 2.4

Standards Check

Using Vocabulary

1. Distinguish between temperature and thermal energy. **3.a**

2. Define *thermal expansion* in your own words. **3.a**

Understanding Main Ideas

3. **Identify** three temperature scales. State the boiling and freezing points of water on the three scales. **3.a**

4. **Explain** why the balloon in the photo below expands when its temperature increases. **3.a**

5. **Distinguish** between temperature and kinetic energy. **3.a**

6. **Determine Cause and Effect** Copy and fill in the graphic organizer below to describe two things that happen when an object is heated. **3.a**

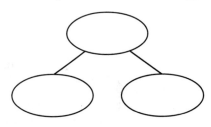

Applying Science

7. **Predict** how the motion of water molecules changes as the water is heated from 10°C to 95°C. **3.a**

8. **Calculate** the temperature on the Kelvin temperature scale of an object with a temperature of 50°C. Use **Figure 20**. **3.a**

Science nline
For more practice, visit **Standards Check** at ca6.msscience.com.

DataLab

How are temperature scales related?

Almost all countries in the world and science laboratories use the Celsius scale. However, in the United States the Fahrenheit temperature scale is used. Using a graph is one way to convert Celsius temperatures into Fahrenheit temperatures.

Fahrenheit and Celsius Temperatures		
Data Point	Fahrenheit Temperature (°F)	Celsius Temperature (°C)
1	120.0	48.9
2	60.0	15.6
3	0.0	−17.8
4	−60.0	−51.1

Data

1. Using **graph paper,** create a graph comparing Fahrenheit temperatures to Celsius temperatures. Use the data table provided. Plot the temperature value in °F along the *x*-axis and the temperature value in °C along the *y*-axis for each data point.

2. Your graph should have:
   - a title
   - a scale for each axis
   - units on each axis
   - a title for each axis

Data Analysis

1. **Describe** the shape of your plotted line.

2. **Make use of** the graph to determine what Celsius temperature is the same as 70°F.

3. **Identify** the temperature at which the Fahrenheit and Celsius scales are the same.

 Science Content Standards

7.c Construct appropriate graphs from data and develop qualitative statements about the relationships between variables.

Science Content Standards

3.c Students know heat flows in solids by conduction (which involves no flow of matter) and in fluids by conduction and by convection (which involves flow of matter).
3.d Students know heat energy is also transferred between objects by radiation (radiation can travel through space).
7.a Develop a hypothesis.

Reading Guide

What *You'll Learn*

▶ **Describe** how thermal energy is transferred by collisions between particles.

▶ **Explain** how thermal energy is transferred by the movement of matter from one place to another.

▶ **Describe** thermal energy transfer by electromagnetic waves.

Why *It's Important*

The transfer of thermal energy is involved in cooking food, staying warm in a winter coat, and the warming of Earth by the Sun.

Vocabulary

conduction
conductor
convection
fluid
convection current
radiation

Review Vocabulary

density: the amount of matter in a unit volume (p. 27)

Conduction, Convection, and Radiation

Main Idea Thermal energy is transferred by conduction, convection, and radiation.

Real-Life Reading Connection When it is cold outside, you might put on a coat or sweatshirt. Why does putting on another layer of clothes help you stay warm?

Conduction

You can burn your fingers if you try to pick up a hot pan. Your fingers burn because thermal energy moves from the hot pan into your cooler skin. One way that thermal energy moves is by conduction. **Conduction** (kuhn DUK shun) is the transfer of thermal energy by collisions between particles in matter.

Collisions Between Particles

In **Figure 22,** thermal energy is transferred throughout the pot by conduction. The particles in the pot close to the fire gain thermal energy from the hot fire. This causes the particles to move faster and collide with nearby particles. In this way, thermal energy moves upward through the pot. Eventually, the entire pot becomes hot.

Figure 22 Thermal energy moves by conduction throughout the pot.

Thermal energy

Is metal a good conductor?

Safety Precautions

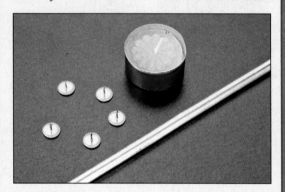

Procedure

1. Read and complete a lab safety form.

2. Clamp a **30 cm metal rod** or **thick wire** horizontally about 20 cm above the lab table on a **ring stand.**

3. Light a **candle** and drip the melted candle wax into the metal rod about 10 cm from the end. Stick a **metal thumb tack** to the wax before it hardens.

4. Repeat step 3 to place 4 more thumb tacks at 2-cm intervals along the rod.

5. Heat the rod at one end with the candle. Record the time when heat is first applied.

6. Record the time when each tack falls off the rod.

Analysis

1. **Create** a graph to show the time that was required for each thumb tack to melt away from the rod.

2. **Infer** how thermal energy moved through the rod to melt the wax.

3. **Infer** how your data would change if the metal rod were replaced by a material that was an insulator.

Conduction in Solids

Conduction transfers thermal energy by collisions of particles. In solids, collisions occur between neighboring particles. **Figure 22** on the preceding page shows how particles in a solid are arranged. In a solid, particles are packed closely together. The particles can move back and forth slightly, but they stay in one place. As a result, collisions can occur only between particles that are next to each other. Thermal energy is conducted through a solid object as energy is passed from one particle to the next.

Reading Check How is thermal energy transferred through a solid?

Conductors

Conduction occurs in all materials. However, thermal energy moves by conduction more quickly in some materials than in others. For example, thermal energy moves more than 15,000 times faster in copper than in air. Materials in which thermal energy moves quickly are **conductors.** Solids usually are better conductors than liquids and gases. Metals are the best conductors, which is why cooking pans usually are made of metal. **Table 1** lists some common materials that are good conductors.

Table 1 Examples of Conductors and Insulators	
Conductors	**Insulators**
Silver	Air
Copper	Plastic foam
Aluminum	Fiberglass
Steel	Cork
Brass	Wood

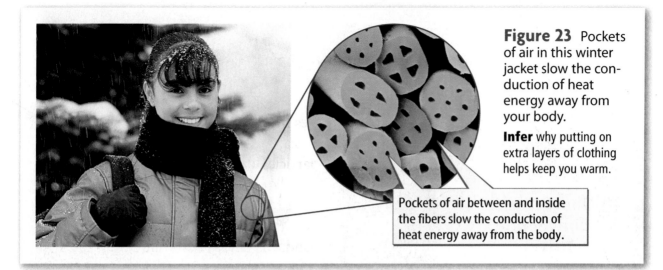

Figure 23 Pockets of air in this winter jacket slow the conduction of heat energy away from your body.

Infer why putting on extra layers of clothing helps keep you warm.

Pockets of air between and inside the fibers slow the conduction of heat energy away from the body.

Insulators

Thermal energy moves more slowly in gases than in solids and liquids. Thermal energy moves slowly in gases because the particles in a gas are so spread apart that collisions occur less often. A material in which thermal energy moves slowly is called an insulator. Air is a gas and is a good insulator. Winter jackets like the one in **Figure 23** are made of materials that trap pockets of air. The jacket keeps you warm because the air slows the rate of thermal energy flowing from your body.

Convection

Another way for thermal energy to move is by convection. **Convection** (kuhn VEK shun) is the transfer of thermal energy by the movement of matter from one place to another. For convection to take place, the particles of the material must be able to move easily from place to place. In solids, particles cannot move from one place to another. However, a **fluid** is a material made of particles that easily can change their locations. Liquids and gases are fluids, so convection occurs only in liquids and gases.

WORD ORIGIN

convection
from Latin *convehere;* means *to bring together*

Density

During convection, parts of a fluid that have a higher temperature move to a region where the temperature of the fluid is lower. Convection occurs because of differences in density in a fluid. Recall that density is the amount of matter, or mass, contained in a unit volume of a material. Think about picking up a full 2-L bottle of soda versus picking up a 2-L bottle that is full of air. Both bottles have the same volume. But because the density of soda is greater than the density of air, the bottle of soda has more mass and is heavier.

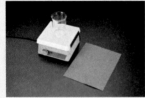

Figure 24 Burners heat the air in the balloons, causing the density of the balloons to decrease.

Density and Temperature

Different materials have different densities, but even samples of the same material can have different densities. Recall that most materials expand, or take up more space, if their temperature increases. Since the mass of the material doesn't change, and the volume increases, its density decreases when its temperature increases.

Density and Floating

Why does a stone sink when you throw it into a pond? An object sinks if its density is greater than the density of the fluid that surrounds it. An object will float if its density is less than the density of the fluid that surrounds it. So the stone sinks because its density is greater than the density of water. A stick floats because it is less dense than water.

Density differences cause the hot-air balloon, shown in **Figure 24,** to rise. When the pilot heats the air in the balloon, the air in the balloon expands. This makes the air inside the balloon less dense than the air around the balloon. As a result, the balloon will rise off the ground and float into the air.

 Reading Check Why does a hot-air balloon float in the air?

While the balloon is in the air, thermal energy flows from the balloon to the surrounding air. As the air in the balloon cools, its volume decreases, so its density increases. When the balloon becomes denser than the surrounding air, it sinks downward.

Convection Currents

A hot-air balloon rises as the air inside is heated. The balloon sinks when the air inside cools. In a similar way, when a fluid is heated, the warmer parts of the fluid are less dense than the surrounding fluid. As **Figure 25** shows, the warmer fluid then can float upward. When the warm fluid cools enough, it can become more dense than the surrounding fluid and sink.

As the warm water rises, it cools and its density increases. The cooler, denser water sinks along the sides of the beaker. At the bottom of the beaker, the water may again be heated. The circular motion that results in a fluid heated from below is called a **convection current.**

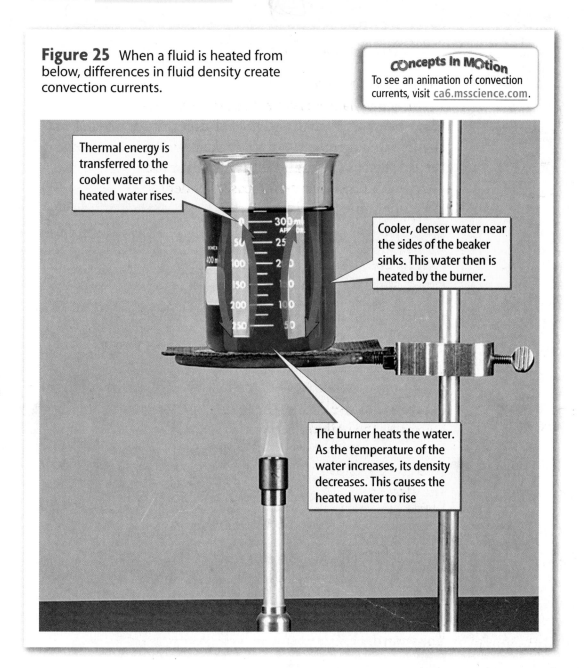

Figure 25 When a fluid is heated from below, differences in fluid density create convection currents.

COncepts In MOtion
To see an animation of convection currents, visit ca6.msscience.com.

Thermal energy is transferred to the cooler water as the heated water rises.

Cooler, denser water near the sides of the beaker sinks. This water then is heated by the burner.

The burner heats the water. As the temperature of the water increases, its density decreases. This causes the heated water to rise

Thermal energy

Figure 26 The lamp gives off electromagnetic waves that transfer thermal energy to the food, keeping it warm.

ACADEMIC VOCABULARY

summary (SUH muh ree)
(noun) a presentation of the content of a text in a condensed form or by reducing it to its main points
The newspaper included a summary of the day's top stories.

Radiation

A third type of thermal energy transfer is **radiation** (ray dee AY shun), which is the transfer of thermal energy by electromagnetic waves. Conduction and convection cause thermal energy to move in matter. However, electromagnetic waves can travel in matter and in empty space. As a result, thermal energy can be transferred by radiation within matter and through space. Radiation can cause thermal energy to be transferred between objects that aren't touching.

For example, the lamp in **Figure 26** gives off electromagnetic waves that travel to the food. The atoms and molecules in the food absorb the energy carried by these waves. This absorbed energy causes the particles in the food to move faster. As a result, the temperature of the food increases.

The Sun gives off enormous amounts of energy. Radiation transfers some of this energy from the Sun to Earth. You can feel this energy when the Sun shines on your skin. Life on Earth depends on the energy emitted from the Sun. The atmosphere and Earth's surface absorb this energy and become warm enough for life to survive. **Table 2** summarizes the three ways thermal energy is transferred.

CONcepts In MOtion
Interactive Table Organize information about the transfer of thermal energy at ca6.msscience.com.

Table 2 Thermal Energy Transfer		
Type of Transfer	**How Transfer Occurs**	**Example**
Conduction	Particles with more thermal energy collide with nearby particles that have less thermal energy. Conduction transfers thermal energy only through matter.	
Convection	Part of a fluid that is at a higher temperature moves to where the fluid is cooler. Convection transfers thermal energy only through matter.	
Radiation	Electromagnetic waves given off by objects at a higher temperature are absorbed by objects at a lower temperature. Radiation transfers thermal energy through matter and space.	

What have you learned?

In this lesson you read about three methods of transferring thermal energy from place to place. Conduction is the transfer of thermal energy by collisions between particles in matter. Particles in the material bounce back and forth against one another, transferring energy throughout the material.

Convection is the transfer of thermal energy by the movement of matter from one place to another. Convection currents arise due to differing densities caused by differing temperatures.

Radiation is the transfer of thermal energy by electromagnetic waves. Radiation is the process that transfers energy from the Sun to Earth.

LESSON 4   Review

Summarize

Create your own lesson summary as you design a **study web.**

1. **Write** the lesson title, number, and page numbers at the top of a sheet of paper.

2. **Scan** the lesson to find the **red** main headings.

3. **Organize** these headings clockwise on branches around the lesson title.

4. **Review** the information under each **red** heading to design a branch for each **blue** subheading.

5. **List** 2–3 details, key terms, and definitions from each **blue** subheading on branches extending from the main heading branches.

 ELA6: R 2.4

Standards Check

Using Vocabulary

1. Distinguish between conduction and convection. **3.c**

2. Define *radiation* in your own words. **3.d**

Understanding Main Ideas

3. **State** which types of heat transfer take place only in matter. **3.c**

4. Which of the following is an example of conduction? **3.c**
   A. water moving in a pot of boiling water
   B. warm air rising from hot pavement
   C. warmth you feel standing near a fire
   D. warmth you feel holding a cup of hot cocoa

5. **Give examples** of heat transfer by conduction, convection, and radiation that you have experienced. **3.c**

6. **Illustrate** how a convection current depends on density differences. **3.c**

7. **Sequence** Draw a graphic organizer like the one below showing the sequence of processes that produce a convection current. **3.c**

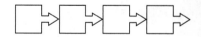

Applying Science

8. **Suggest** a way to reduce the rate of heat loss if you take a warm slice of pizza from home for your school lunch. **3.c**

9. **Hypothesize** why windows in some homes have two panes of glass separated by a layer of air. **3.c**

For more practice, visit **Standards Check** at ca6.msscience.com.

Create a Thermos

Materials

heated water
graduated cylinder
alcohol thermometers
   (3)
foam, metal, ceramic,
   and/or plastic
   containers
10 cm × 10 cm card-
   board (3 pieces)
stopwatch or clock
tongs
thermal mitts

Safety Precautions

Science Content Standards

3.c Students know heat flows in solids by conduction (which involves no flow of matter) and in fluids by conduction and by convection (which involves flow of matter).

7.a Develop a hypothesis.

Problem

Have you ever used a thermos bottle to keep hot soup hot or cold water cold? The food or liquid inside a thermos bottle changes its temperature slowly. Thermal energy moves slowly across the walls of the bottle. What materials are best for making a thermos bottle?

Form a Hypothesis

➢ **Review** the results from this chapter's laboratory investigations.

➢ **Make a prediction** about what types of materials are the best insulators.

Collect Data and Make Observations

1. Read and complete the lab safety form.
2. Make a data table like the one shown below.
3. Pour 80 mL of heated water into three different kinds of containers.
4. Make a small hole in each piece of cardboard and insert a thermometer through each hole.
5. Place each piece of cardboard on a container. Make sure the thermometer does not touch the sides or bottom of the container.

6. For each container, measure the temperature of the water every minute and record your results in your data table. Continue measuring for 10 min.

Data Table			
Time (min)	Container 1 Temperature (°C)	Container 2 Temperature (°C)	Container 3 Temperature (°C)
0			
1			
2			
3			
4			
5			

7. Graph your data. Make a graph with time as the horizontal axis and temperature as the vertical axis. Using a different color pencil for each container, plot the data you collected. Connect the plotted points for each container.

8. Make sure that any water that was spilled is wiped up.

Analyze and Conclude

1. Identify at least three factors that you were careful to keep the same in each trial for the lab.

2. Describe the shapes of your graphs.

3. Identify which material is the best insulator. Use the data to support your answer.

4. Identify which material is the best conductor. Use the data to support your answer.

5. Diagram the flow of thermal energy between the water, the container, and the surrounding air.

6. Identify any experimental errors that could affect your data. How could you improve your procedure?

Communicate

 **Science** ELA6: W 2.2

Write a marketing proposal for an actual thermos to sell in stores. Research thermal containers that are available for sale. What are the advertising points? What design features are used? How can you improve on the design of the thermal containers for sale? Explain.

Real World Science

Science & Career

Heating and Cooling

Almost everywhere in the United States, homes and other buildings require heating and cooling systems to keep them at a comfortable temperature throughout the year. The installation and maintenance of these systems is done by technicians called HVAC technicians. *HVAC* stands for heating, ventilation, and air conditioning. To become an HVAC technician usually requires a high-school degree and on-the-job training.

Imagine you are interviewing an applicant for a HVAC technician position. List five questions you would ask the person.

Science & Technology

Hybrid Cars

Automobile engines have been developed that use an electric motor along with a gasoline-burning engine to power a car. The electric motor usually provides a power boost when the car is speeding up. Cars that use these engines are called hybrid cars. Hybrid cars usually get better gas mileage than conventional cars and emit less carbon dioxide and other gases. As a result, hybrid cars can help conserve gasoline and improve air quality.

Visit **Technology** at <u>ca6.msscience.com</u> to find information about hybrid cars. **Describe** three ways a hybrid engine is different from a conventional automobile engine.

From the Caloric Theory to the Kinetic Theory

In the eighteenth and nineteenth centuries scientists thought objects contained an invisible fluid called caloric. This fluid couldn't be created or destroyed. The temperature of an object changed when the object lost or gained caloric. However, observations showed that friction could make objects hotter, implying that caloric could be created. In the late nineteenth century, the caloric theory was replaced by the kinetic theory, which relates temperature to the motion of atoms and molecules.

Visit **History** at **ca6.msscience.com** to find out more about the caloric theory. Make a time line showing the events in the development and replacement of the caloric theory.

Steam Engines in the Industrial Revolution

In the late 1700s, James Watt developed an improved version of the steam engine. The steam engine provided the source of energy for factories to manufacture various goods, such as cloth, that used to be made by hand. The steam engine was also used to power trains, which could travel more quickly than horse-drawn vehicles. Machines began to replace manual labor, a change that is called the industrial revolution.

Visit **Society** at **ca6.msscience.com** to research the history of the steam engine. **Write** 500–700 words about how the steam engine helped change society in the eighteenth and nineteenth centuries.

ELA6: W 1.2

The BIG Idea Thermal energy moves from warmer to cooler materials until the materials have the same temperature.

Lesson 1 Forms of Energy 3.a

Main Idea **Energy exists in many forms.**

- Energy is the ability to cause change.
- Kinetic energy is energy of motion.
- Potential energy is stored energy.
- Thermal energy moves because of a temperature difference.

- **elastic potential energy** (p. 128)
- **energy** (p. 124)
- **kinetic energy** (p. 124)
- **potential energy** (p. 127)
- **thermal energy** (p. 129)

Lesson 2 Energy Transfer 3.a, 3.b

Main Idea **Energy can be transferred from one place to another.**

- Moving objects transfer kinetic energy from place to place.
- Water waves and sound waves transfer energy through matter.
- Electromagnetic waves transfer energy through matter or empty space.
- Fuels store energy that can be changed into other forms of energy.

- **friction** (p. 136)
- **fuel** (p. 135)
- **wave** (p. 132)
- **work** (p. 132)

Lesson 3 Temperature, Thermal Energy, and Heat 3.a, 7.c

Main Idea **Thermal energy flows from areas of high temperature to areas of low temperature.**

- Temperature is a measure of the average kinetic energy of the particles in matter.
- Materials usually expand if their temperature increases.
- Heat is the flow of thermal energy from a higher to a lower temperature.

- **heat** (p. 142)
- **temperature** (p. 140)
- **thermal expansion** (p. 140)

 STUDY TO GO Download quizzes, key terms, and flash cards from ca6.msscience.com.

Lesson 4 Conduction, Convection, and Radiation 3.c, 3.d, 7.a

Main Idea **Thermal energy is transferred by conduction, convection, and radiation.**

- Conduction occurs when thermal energy is transferred by collisions between the particles in matter.
- Convection is the flow of thermal energy due to the movement of matter from one place to another.
- Radiation is the transfer of thermal energy by electromagnetic waves.

- **conduction** (p. 145)
- **conductor** (p. 146)
- **convection** (p. 147)
- **convection current** (p. 149)
- **fluid** (p. 147)
- **radiation** (p. 150)

Linking Vocabulary and Main Ideas

Use vocabulary terms from page 156 to complete this concept map.

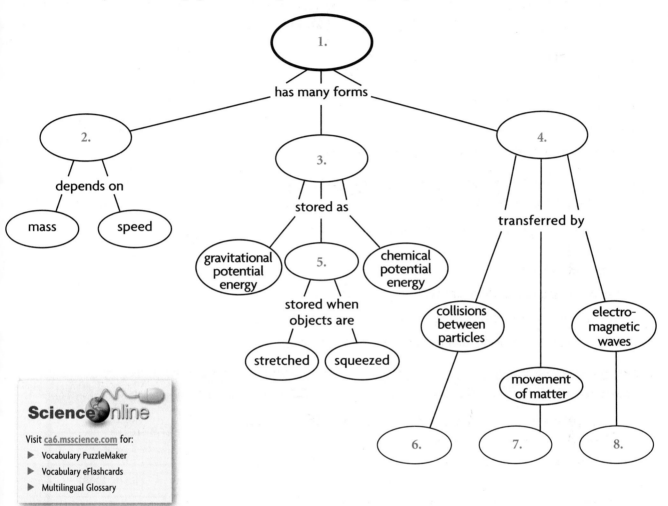

Science Online

Visit ca6.msscience.com for:
▶ Vocabulary PuzzleMaker
▶ Vocabulary eFlashcards
▶ Multilingual Glossary

Using Vocabulary

Match a vocabulary term to each definition below.

9. an increase in the size of a substance when the temperature is increased

10. energy that can move from one place to another because of differences in temperature

11. energy of motion

12. transfer of energy from warmer to cooler matter

13. measure of the average kinetic energy of the particles in a material

14. stored energy

15. transfer of thermal energy by collisions between particles in matter

16. a material in which thermal energy moves easily

17. thermal energy transfer by the movement of matter from one place to another

Understanding Main Ideas

Choose the word or phrase that best answers the question.

1. Which is an example of chemical potential energy changing to kinetic energy?
   A. a flag flapping in the wind   `3.b`
   B. logs burning on a fire
   C. using a battery in a flashlight
   D. burning fuel to move a car

2. The photo below shows a boy pushing on a wall.

   Why is no work being done by the boy?
   A. The boy is not pushing hard enough.   `3.b`
   B. The wall is pushing back on the student.
   C. There is no motion in the direction of the force.
   D. The gravitational potential energy of the wall is not changing.

3. You place a pan on a hot stove. Which is true?
   A. The random motion of the water particles decreases.
   B. The average kinetic energy of the water particles increases.
   C. The water particles are closer together and are better able to transfer energy.
   D. The volume of the water increases as the water particle speed decreases.   `3.c`

4. Milk at 19°C is mixed with hot chocolate at 55°C. What could be the temperature of the mixture?
   A. 17°C   `3.a`
   B. 42°C
   C. 59°C
   D. 74°C

5. Which of the following can cause a convection current?
   A. heating a liquid from above   `3.c`
   B. heating a liquid from below
   C. cooling a solid from above
   D. cooling a solid from below

6. The kinetic energy of an object depends on which?
   A. the object's chemical bonds   `3.a`
   B. the object's mass and speed
   C. the object's temperature and mass
   D. height of the object above the ground

7. The table below shows some conductors and insulators.

Conductors and Insulators	
Conductors	**Insulators**
Silver	Air
Copper	Plastic foam
Aluminum	Fiberglass

   Through which material does thermal energy flow the slowest?
   A. Aluminum
   B. Copper
   C. Fiberglass
   D. Silver   `3.c`

8. What happens to aluminum if you increase its temperature?
   A. volume increases, density increases
   B. volume increases, density decreases
   C. volume decreases, density increases
   D. volume decreases, density decreases   `3.c`

Applying Science

9. **Explain** the energy transformations and transfers that occur when a candle is used to heat a beaker of water. **3.b**

10. **Infer** why you can keep cooler on a sunny day if you sit under a tree or umbrella. **3.d**

11. The figure below shows sound waves produced by a drum.

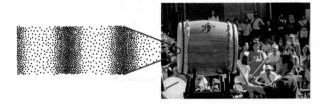

Describe how energy is transferred by sound waves. **3.a**

12. **Design** an experiment to test which colors of fabric are the most efficient absorbers of radiation. Design the experiment to test the fabrics by placing them in direct sunlight for a period of time. **3.d**

13. **Describe** three methods of thermal energy transfer. **3.c**

14. **Explain** how a hot cup of tea and a cold glass of lemonade reach room temperature when both are placed on a kitchen table. In which direction does thermal energy flow in each case? **3.a**

15. **Describe** how the potential energy and the kinetic energy of a ball change as the ball falls. **3.a**

16. **Explain** why materials that contain many small pockets of air are good insulators. **3.c**

17. **Explain** whether your body gains or loses thermal energy if your body temperature is 37°C and the temperature of the air surrounding your body is 25°C. **3.a**

WRITING in ▶ Science

18. **Write** a paragraph that describes the relationship between thermal energy, temperature, and heat. **ELA6: W 1.2**

Cumulative Review

19. **Explain** how Earth's temperature changes with depth, if thermal energy flows from Earth's core to Earth's surface. **1.b, 3.a**

20. **Diagram** how a convection current transfers thermal energy within Earth's mantle. **1.b, 3.c**

21. **Determine** the method of heat transfer that occurs in Earth's solid inner core. **1.b, 3.c**

Applying Math

Use the table below to answer questions 22 through 26.

Linear Thermal Expansion Coefficients	
Material	**Linear Thermal Expansion Coefficient (ppm/°C)**
Aluminum	23
Brass	20
Diamond	1
Steel	12

22. **Find** the volumetric thermal expansion coefficient for brass. **MA6: NS 2.0**

23. **Find** the difference in the volumetric thermal expansion coefficient for brass and the volumetric thermal expansion coefficient of aluminum. **MA6: NS 2.0**

24. **Find** the volumetric thermal expansion coefficient for steel. **MA6: NS 2.0**

25. **Find** the difference in the volumetric thermal expansion coefficient for brass and the volumetric thermal expansion coefficient of steel. **MA6: NS 2.0**

26. **Find** the volumetric thermal expansion coefficient for diamond. **MA6: NS 2.0**

1 Which is a true statement?

A Waves can transfer energy through matter and empty space. **3.a**

B Waves can transfer energy through matter, but not through empty space.

C Waves can transfer energy through liquids and solids, but not through gases or empty space.

D Waves can transfer matter, but not energy.

2 The illustration below shows an ice cube in a glass of room-temperature water.

Which occurred when the ice was added to the water?

A The temperature of the water increased.

B The temperature of the ice decreased.

C Thermal energy moved from the ice to the water.

D Thermal energy moved from the water to the ice. **3.a**

3 The transfer of thermal energy from the Sun to Earth is an example of which process?

A conduction

B convection

C radiation

D thermal expansion **3.d**

4 The illustration below shows the movement of air in a room heated by a furnace.

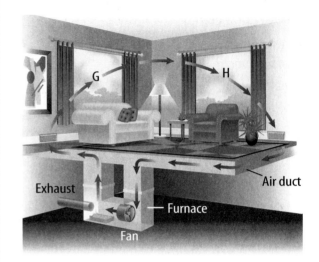

How does the air at point G compare to the air at point H?

A The air at point G is warmer and denser. **3.c**

B The air at point G is warmer and less dense.

C The air at point G is cooler and denser.

D The air at point G is cooler and less dense.

5 How does thermal energy move as you hold an ice cube in your hand?

A by convection from the ice to your hand

B by convection from your hand to the ice

C by conduction from your hand to the ice

D by conduction from the ice to your hand **3.c**

6 What must happen for work to be done?

A An object must speed up.

B A force must be exerted on an object.

C The direction in which an object is moving must change.

D A force must move an object in the same direction as the force. **3.b**

7 Which is the best conductor of thermal energy?

A copper

B rubber

C wood

D air **3.c**

8 The illustration below shows the movement of air near a shoreline.

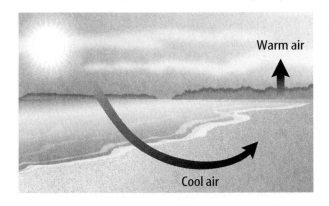

Warm air

Cool air

Which best describes this process?

A contraction

B expansion

C conduction

D convection **3.c**

9 The illustration below shows a wave on a rope.

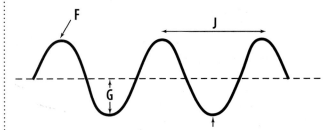

F

J

G

Which property of the wave is shown at J?

A amplitude

B frequency

C trough

D wavelength **3.a**

10 Which process causes thermal energy to move through a metal spoon placed in a pot of boiling water?

A conduction

B convection

C radiation

D insulation **3.c**

11 As a pan of water is heated, how does the motion of the water molecules change?

A They move more slowly.

B They move more quickly.

C Their speed doesn't change and they move farther apart.

D Their speed doesn't change and they move closer together. **3.c**

CHAPTER 4

Plate Tectonics

The BIG Idea

Plate tectonics explains the formation of many of Earth's features and geologic events.

LESSON 1 · 1.a, 7.e
Continental Drift

Main Idea Despite the evidence that supported continental drift, it was rejected by most scientists.

LESSON 2 · 1.a, 7.g
Seafloor Spreading

Main Idea New discoveries led to seafloor spreading as an explanation for continental drift.

LESSON 3
1.b, 1.c, 4.c, 7.a, 7.e

Theory of Plate Tectonics

Main Idea Earth's lithosphere is broken into large brittle pieces, which move as a result of forces acting on them.

A Growing Country

Pingvellir, Iceland, is located on the Mid-Atlantic Ridge, where the North American Plate and the Eurasian Plate are slowly being pulled apart. This process causes Iceland's intense earthquakes and volcanic activity. Iceland is one of the few places where the Mid-Atlantic Ridge can be seen above sea level.

Science Journal Write three questions you would ask a geologist about plate tectonics.

00:20 minutes

Can you put it back together?

Earth's plates are not in the same places as they used to be. Can you match the plates from an orange if someone scrambles them up?

Procedure

1. Read and complete a lab safety form.

2. Make ocean basins in an **orange** by gently carving away some of the top layer of the skin with a **citrus peeler.**

3. Draw continents on the orange with a **ballpoint pen.**

4. Use the pen tip to cut the skin into six or seven irregularly-shaped plates.

5. Peel the plates away from the orange.

6. Trade oranges with a classmate, and try to put each other's oranges back together.

Think About This

List the clues you used to put the plates back together.

 1.a, 7.e

Science Online

Visit ca6.msscience.com to:

▶ view **Concepts in Motion**

▶ explore Virtual Labs

▶ access content-related Web links

▶ take the Standards Check

Plate Tectonics Make the following Foldable to help you monitor your understanding of plate tectonics.

STEP 1 Fold a sheet of paper in half lengthwise. Make the back edge about 2 cm longer than the front edge.

STEP 2 Fold into thirds.

STEP 3 Unfold and **cut** along the folds of the top flap to make three flaps.

STEP 4 Label the flaps as shown.

Plate Tectonics

| Continental Drift | Seafloor Spreading | Theory of Plate Tectonics |

Reading Skill  ELA6: R 1.4

Monitoring

As you read this chapter, use the Reading Checks to help you monitor your understanding of what you are reading. Write the Reading Check questions and answers for each lesson under its tab.

163

Get Ready to Read

Monitor

① Learn It! An important strategy to help you improve your reading is monitoring, or finding your reading strengths and weaknesses. As you read, monitor yourself to make sure the text makes sense. Discover different monitoring techniques you can use at different times, depending on the type of test and situation.

② Practice It! The paragraph below appears in Lesson 1. Read the passage and answer the questions that follow. Discuss your answers with other students to see how they monitor their reading.

> Fossils are the remains, imprints, or traces of once-living organisms. If an organism dies and is buried in sediment, then it can become preserved in various ways. Eventually, the fossil becomes part of a sedimentary rock. Fossils help scientists learn about species from past times. Wegener collected fossil evidence to support his continental drift hypothesis.
>
> —*from page 169*

- What questions do you still have after reading?
- Do you understand all of the words in the passage?
- Did you have to stop reading often? Is the reading level appropriate for you?

③ Apply It! Identify one paragraph that is difficult to understand. Discuss it with a partner to improve your understanding.

Target Your Reading

Use this to focus on the main ideas as you read the chapter.

1 **Before you read** the chapter, respond to the statements below on your worksheet or on a numbered sheet of paper.

- Write an **A** if you **agree** with the statement.
- Write a **D** if you **disagree** with the statement.

2 **After you read** the chapter, look back to this page to see if you've changed your mind about any of the statements.

- If any of your answers changed, explain why.
- Change any false statements into true statements.
- Use your revised statements as a study guide.

Reading Tip

Monitor your reading by slowing down or speeding up depending on your understanding of the text.

Before You Read A or D	Statement	After You Read A or D
	1 Most oceanic crust is made of granite.	
	2 The density of rock increases as its temperature increases.	
	3 Earth's lithosphere is broken into 100 large pieces called plates.	
	4 A slab is less dense than continental crust.	
	5 Fossils of sharks provide evidence for Pangaea.	
	6 Harry Hess proposed the continental drift hypothesis in the mid-1950s.	
	7 Earthquakes and volcanic eruptions occur at boundaries of lithospheric plates.	
	8 Heat is currently escaping from the interior of Earth.	
	9 Seafloor spreading provided part of an explanation of how continents could move on Earth's surface.	
	10 The theory of plate tectonics is well established, so scientists no longer study it.	

Science Online

Print a worksheet of this page at ca6.msscience.com.

Reading Guide

What *You'll Learn*

▶ **Explain** Alfred Wegener's controversial hypothesis.

▶ **Summarize** the evidence used to support continental drift.

▶ **Justify** why most scientists rejected the continental drift hypothesis.

Why *It's Important*

The continental drift hypothesis led to the development of plate tectonics—a theory that explains many of Earth's features and events.

Vocabulary

continental drift
Pangaea

Review Vocabulary

rock: a natural, solid mixture of mineral crystal particles (p. 95)

Continental Drift

Main Idea Despite the evidence that supported continental drift, it was rejected by most scientists.

Real-World Reading Connection Maybe you've had an idea that was really outrageous and exciting. Because your idea seemed so impossible, your friends might have rejected it. You still might have tried hard to convince them that it was a great idea. This is what happened to Alfred Wegener (VAY guh nur) when he tried to convince other scientists that continents slowly drift parallel to Earth's surface.

Drifting Continents

About five hundred years ago, during the age of exploration, European explorers sailed across the Atlantic Ocean. They discovered continents they had never seen before. These continents were North and South America. New maps that included the Americas were drawn.

People who studied these maps, such as the one shown in **Figure 1,** observed something strange. The edges of the American continents look as if they might fit into the edges of Europe and Africa. This observation inspired Alfred Wegener's controversial idea.

Figure 1 Antique Maps This map was published in 1680. Maps like this made people question why the edges of continents appeared as if they could fit together.
Identify the east coast of South America and the west coast of Africa.

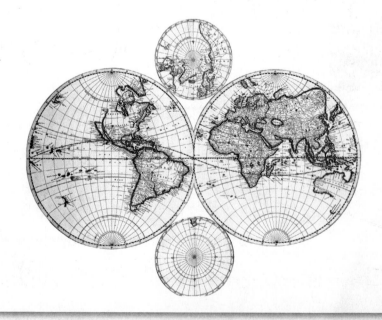

A Controversial Idea

Alfred Wegener thought the edges of continents looked like they might fit together because they once had been attached as one huge landmass.

In the early 1900s, he proposed a hypothesis to explain this. Wegener's hypothesis, **continental drift,** is the idea that the continents move very slowly, over millions of years, parallel to Earth's surface.

Pangaea Breaks Apart

Wegener's continental drift hypothesis proposed that the continents have slowly drifted to their present-day locations. **Figure 2** shows how scientists think the continents broke into pieces as they slowly drifted apart.

1 Wegener proposed that millions of years ago, the continents formed one huge landmass. He named this ancient supercontinent **Pangaea** (pan JEE uh). The top panel of **Figure 2** shows how Pangaea might have appeared about 255 million years ago. According to Wegener, Pangaea started to break apart about 200 million years ago.

Reading Check What is Pangaea?

2 About 152 million years ago, the Atlantic Ocean began to open up between North America and Africa. The southern continents of Pangaea were still mostly intact.

3 India moved toward the ancient Asian continent about 66 million years ago. Oceans widened, and much of the southern continents of Pangaea broke apart. The landmass positions appear much as they do today.

4 The world as you know it is presented here.

Concepts in Motion
To see an animation of Pangaea, visit ca6.msscience.com.

Figure 2 Fragmenting Landmass These maps show the way scientists think Pangaea broke into pieces and drifted apart millions of years ago.

255 Million Years Ago

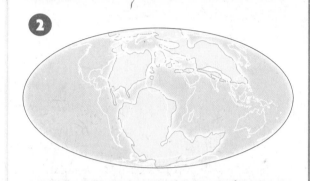

152 Million Years Ago

66 Million Years Ago

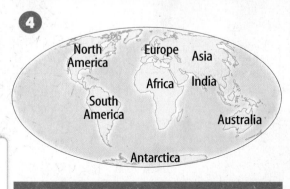

Present-Day

Evidence for Continental Drift

In order to support his continental drift hypothesis Wegener collected **data** from different scientific fields. In 1915, he published this information in a book called *The Origin of Continents and Oceans.* In his book, Wegener presented four major types of evidence for his hypothesis. This evidence included the geographic fit of the continents, fossils, rocks and mountain ranges, and ancient climate records.

Fit of the Continents

The most obvious evidence for continental drift is the geographic fit of the continents. If you were to remove the present-day Atlantic Ocean, the continents would fit back together. The east coast of South America fits into the notch on the west coast of Africa. And, the bulge on northwest Africa fits into the space between North and South America. This is shown in **Figure 3.**

Visual Check **Figure 3** List the continents on which *Glossopteris* lived during the time of Pangaea.

This geographic fit of the continents suggests ways to look for even more evidence for Pangaea. Imagine the continents pieced back together, like pieces of a puzzle. Some rock types and fossils are the same because the continents were connected at the time of Pangaea.

ACADEMIC VOCABULARY

data (DAY ta)

(noun) factual information used as a basis for reading, discussion, or calculation

Data were collected by the accountants to help complete Mr. Smith's tax return.

Figure 3 To support his continental drift hypothesis, Wegener collected fossils from the time of Pangaea.

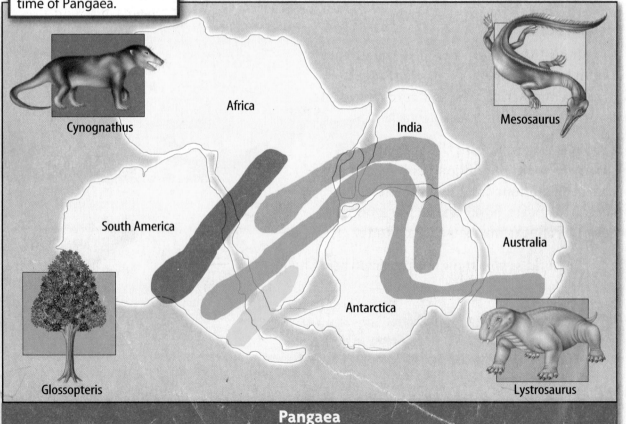

Cynognathus

Africa

India

Mesosaurus

South America

Australia

Antarctica

Glossopteris

Lystrosaurus

Pangaea

Fossil Evidence

Fossils are the remains, imprints, or traces of once-living organisms. If an organism dies and is buried in sediment, then it can become preserved in various ways. Eventually, the fossil becomes part of a sedimentary rock. Fossils help scientists learn about species from past times. Wegener collected fossil evidence to support his continental drift hypothesis. He wanted to learn where the plants and animals from the time of Pangaea lived.

Glossopteris One plant Wegener studied was *Glossopteris* (glahs AHP tur us), a seed fern. Fossils of this fern have been discovered in South America, Africa, India, Australia, and Antarctica. The heavy seeds could not have been blown by the wind, nor could they have floated, across the wide oceans separating these continents.

 What is *Glossopteris*?

So, Wegener concluded that all those continents must have been attached when *Glossopteris* was alive. As shown in **Figure 3,** *Glossopteris* was not the only species that lived on several continents. Wegener used the present-day locations of these various fossils to support the idea that there was a supercontinent when the animals and plants were alive.

Figure 3 cont. Fossils of various species that lived during the time of Pangaea have been found on more than one continent.

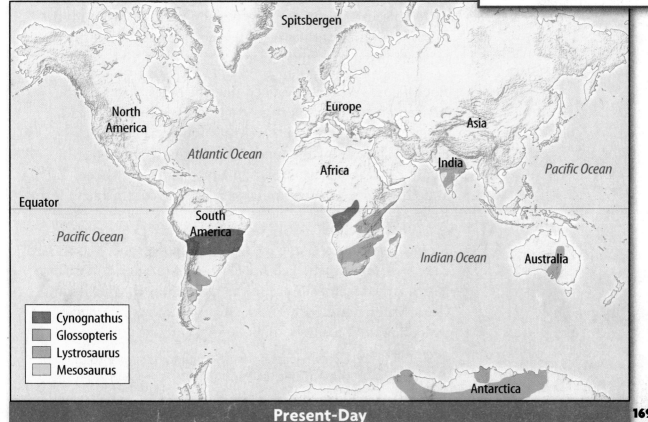

Present-Day

169

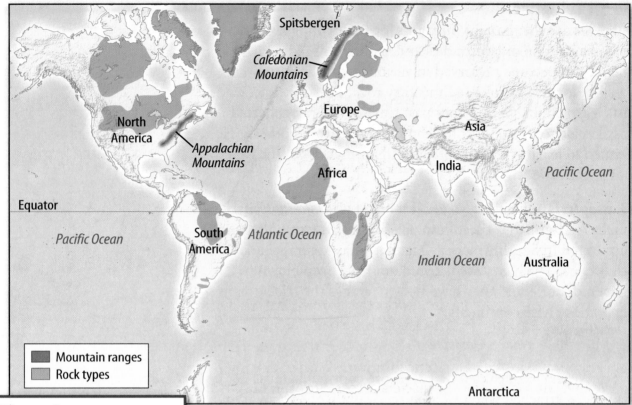

Figure 4 Connecting Landforms Rock types and mountain ranges match up across the continents when they are arranged to form Pangaea.

Rock Types and Mountain Ranges

The locations of rock types and mountain ranges from the time of Pangaea also provide evidence for continental drift. Geologists can identify groups of rocks, much like you can match pieces of a puzzle. Wegener showed that certain types of rocks on the continents would match up if the continents were arranged to form Pangaea.

Rock Types Wegener realized that the oldest rocks on the African and South American continents were next to each other when the continents were assembled as Pangaea. **Figure 4** shows how the types of rocks match up across the Atlantic Ocean. Ancient rocks in North America, Greenland, and Europe also match up if you move the continents to form Pangaea.

Mountain Ranges Some mountain ranges also look as if they were once connected. The Appalachian Mountains in eastern North America are similar to the mountains in Greenland, Great Britain, and Scandinavia. **Figure 4** shows how they would look like a single, long mountain range.

 **Reading Check** List two locations with mountains similar to those of the Appalachian Mountains.

Ancient Climate Evidence

Wegener was a meteorologist. Meteorologists are scientists who study weather and climate. Wegener traveled the planet looking for rocks that contained evidence of past climates.

Recording Climate When sedimentary rocks form, clues about the climate are preserved within the rock. Hot, wet climates produce lots of plants. As plants die, they form coal deposits in the rocks. Tropical seas leave behind fossil reefs. Hot, dry climates produce rocks with preserved sand dunes. Glaciers form in cold climates, leaving ancient glacial formations. Rocks often indicate an ancient climate that is very different from the present-day climate.

Changing Climate Spitsbergen is currently located above the arctic circle, east of Greenland. Rocks that formed during the time of Pangaea show that this island once had a tropical climate. Wegener suggested that the island drifted from the warm tropics to its current arctic location. Wegener also found ancient rocks made by glaciers across Africa, India, and Australia. These places are now too warm to have glaciers. **Figure 5** shows evidence of ancient glaciers in South America, Africa, India, and Australia. The ancient climate evidence supports the existence of Pangaea.

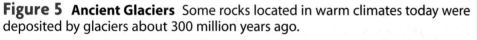

Figure 5 Ancient Glaciers Some rocks located in warm climates today were deposited by glaciers about 300 million years ago.

Explain why rocks formed in tropical climates in Spitsbergen suggest that this island has moved to its present-day location.

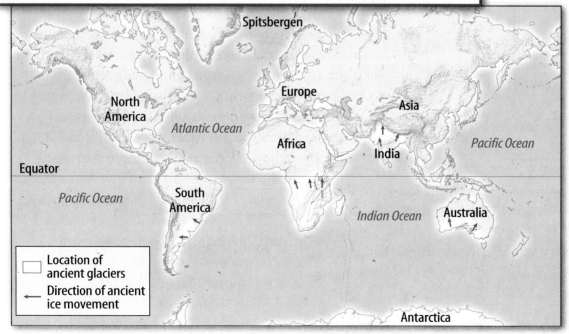

Drifting Continents!

Imagine one huge landmass. This ancient supercontinent began to break apart about 200 million years ago. These pieces very slowly drifted to their present-day locations. Can you model the past, present, and future locations of Earth's continents?

Procedure

1. Complete a lab safety form.

2. Obtain a **map of Pangaea**, a **map of the present-day continents, glue**, and **scissors**.

3. Cut out the present-day continents.

4. Place the pieces in the appropriate locations on the map of Pangaea.

5. Take the pieces and move them to their present-day locations. Refer to a map of the world for help. Think about how far and in what direction each continent has moved.

6. Place the continents where you think they might be millions of years from now.

7. Glue the continents in their future locations.

Analysis

1. **Determine** which continents moved the farthest from the time of Pangaea to the present.

2. **Explain** whether you think there could be another supercontinent in the future.

A Hypothesis Rejected

Wegener presented this evidence for continental drift to other scientists. Wegener had difficulty explaining how, when, or why the continents slowly drifted across Earth's surface.

He proposed that the continents drifted by plowing through the seafloor. He thought the same forces of gravity that produced tides in the ocean had moved the continents.

 What did Wegener think caused the continents to drift?

Wegener knew these forces were not very strong. But, he thought that over millions of years, they could cause the continents to drift. Most other scientists did not accept this explanation. Because these scientists could not think of any forces strong enough to make continents drift, Wegener's hypothesis was rejected.

 Why wasn't continental drift accepted by the scientific community?

Alfred Wegener did not give up when his hypothesis was rejected. He continued to search for evidence to support his continental drift hypothesis.

Wegener died in 1930 with little recognition for his accomplishments. He disappeared in a storm while on an expedition studying the weather in Greenland. The controversy over his hypothesis remained for several decades after his death. He did not live long enough to see the new evidence that made scientists reconsider his controversial idea.

Scientists reconsidered Wegener's controversial idea because of advances in technology, such as sonar and deep-sea drilling. These technological advances helped scientists develop new ideas and evidence that related to continental drift.

Continental Drift Hypothesis

Alfred Wegener thought that the edges of the continents looked like they fit together because they had once been attached as an entire landmass. Wegener's continental drift hypothesis is the idea that the continents move very slowly across Earth's surface. Wegener's evidence included the geographic fit of the continents, fossils, rocks and mountain ranges, and ancient climate records.

Wegener presented this evidence for continental drift to other scientists. Scientists could not think of forces strong enough to make continents drift, so Wegener's hypothesis was rejected.

LESSON 1 Review

Summarize

Create your own lesson summary as you write a script for a **television news report.**

1. **Review** the text after the red main headings and write one sentence about each. These are the headlines of your broadcast.

2. **Review** the text and write 2–3 sentences about each **blue** subheading. These sentences should tell *who, what, when, where,* and *why* information about each **red** heading.

3. **Include** descriptive details in your report, such as names of reporters and local places and events.

4. **Present** your news report to other classmates alone or with a team.

 **ELA6:** LS 1.4

Standards Check

Using Vocabulary

Complete the sentences using the correct term.

Pangaea   continental drift

1. Mesosaurus is a fossil that supports the _____ hypothesis. **1.a**

2. A supercontinent that existed about 200 million years ago is _____. **1.a**

Understanding Main Ideas

3. Why is *Glossopteris* evidence for continental drift?
   A. Its leaves produced coal.
   B. It was exceptionally large.
   C. Its seeds were heavy.
   D. It was found only in Antarctica. **1.a**

4. **Explain** how rocks can preserve a record of ancient climates. **1.a**

5. **Decide** whether or not continental drift would have been accepted if Wegener had collected more evidence. **1.a**

6. **Organize** Draw a diagram like the one below. List evidence for continental drift into two categories. **1.a**

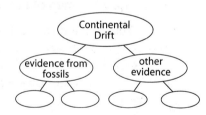

Applying Science

7. **Imagine** a fossil organism that might indicate an ancient tropical reef deposit. **1.a**

8. **Decide** whether scientists were justified in rejecting continental drift. **1.a**

Science online
For more practice, visit **Standards Check** at ca6.msscience.com.

Science Content Standards

1.a Students know evidence of plate tectonics is derived from the fit of the continents; the location of earthquakes, volcanoes, and mid-ocean ridges; and the distribution of fossils, rock types, and ancient climatic zones.
7.g Interpret events by sequence and time from natural phenomena (e.g., the relative ages of rocks and intrusions).

Reading Guide

What *You'll Learn*

▶ **Describe** new discoveries that led to the seafloor spreading hypothesis.

▶ **Explain** how seafloor spreading works.

▶ **Compare and contrast** evidence for seafloor spreading with evidence for continental drift.

Why *It's Important*

The seafloor spreading hypothesis explained continental drift.

Vocabulary
mid-ocean ridge
seafloor spreading

Review Vocabulary
magma: molten, liquid rock material found underground (p. 96)

Seafloor Spreading

Main Idea New discoveries led to seafloor spreading as an explanation for continental drift.

Real-World Reading Connection Do you know how to do a magic trick? When you first see a good trick, it seems impossible. Then, when you learn how the trick works, it doesn't seem impossible any more. In the decades after continental drift was rejected, scientists discovered new technology that helped explain how continents could move.

Investigating the Seafloor

Wegener collected most of his evidence for continental drift at Earth's surface. But, there is also evidence on the seafloor. Scientists began investigating the seafloor by collecting samples of rocks. They knew that most rocks on the seafloor are made of basalt. Recall from Chapter 2 that basalt is an igneous rock that is made of highly dense minerals such as olivine and magnetite.

Scientists wondered why rocks on the seafloor were so different from rocks on land. By the 1950s, new technologies were being developed to explore the seafloor. An example of this technology is shown in **Figure 6.**

Figure 6 The bottom of the ocean is complicated. In this colorized image of the seafloor off the central California coast, the coastline is outlined in white.
Determine whether features colored yellow are above or under water.

Mapping the Seafloor

During World War II, a new method was developed for mapping the seafloor. This new method used technology called sonar. **Figure 7** shows how sonar works. Scientists emit sound waves from a boat. The sound waves bounce off the seafloor. Then, a receiver records the time it takes for the waves to return. Because scientists know the speed of sound waves in water, they can use the data to calculate the depth of water. With this new technology, the topography of the seafloor was mapped.

Mid-Ocean Ridges

Figure 8 shows what scientists discovered when they mapped the topography of the seafloor. Hidden under ocean waters are the longest mountain ranges on Earth. These mountain ranges, in the middle of the seafloor, are called **mid-ocean ridges.** The mountains wrap around Earth much like seams wrap around a baseball.

Maps of the seafloor made scientists want to learn even more about it. They studied temperatures on the seafloor. They discovered that there is more heat escaping from Earth at the mid-ocean ridges than at other locations in the oceans. The closer you move toward a mid-ocean ridge, the more heat flows from the mantle, as shown in **Figure 9.**

Figure 7 Seafloor Mapping
Sonar uses sound waves bounced off the seafloor to measure ocean depths.

Name an animal that uses sound waves to navigate.

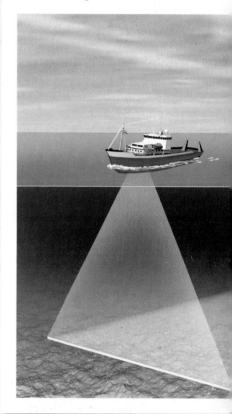

Figure 8 **Depth Changes** The light-blue color on the map shows locations with shallow water.

East Pacific Rise

Mid-Atlantic Ridge

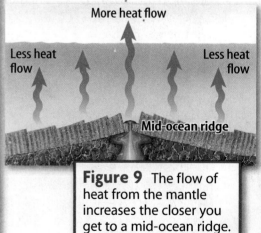

More heat flow

Less heat flow

Less heat flow

Mid-ocean ridge

Figure 9 The flow of heat from the mantle increases the closer you get to a mid-ocean ridge.

The Seafloor Moves

Harry Hess was an American geologist. He studied the seafloor, trying to understand how mid-ocean ridges were formed. He proposed it was hot beneath the mid-ocean ridges because lava erupted there and made new seafloor. Hess suggested a new **hypothesis** describing this process.

Seafloor spreading is the process by which new seafloor is continuously made at the mid-ocean ridges. Convection brings hot material in the mantle toward the surface, causing magma to form. The magma flows out as lava through cracks along the ridge. When the lava cools, it forms new seafloor. Then, the seafloor moves sideways, away from the center of the mid-ocean ridge.

 Where does new seafloor form?

Seafloor spreading seemed to explain continental drift. **Figure 10** shows seafloor moving away from the mid-ocean ridge as new oceanic crust is formed. Notice how the seafloor becomes older as the distance from the mid-ocean ridge increases. Adding new seafloor makes the ocean wider. As a result, continents drift apart as the ocean grows. Scientists looked for evidence that could test the new seafloor spreading hypothesis. Studies of mid-ocean ridges continue today, as shown in **Figure 11.**

ACADEMIC VOCABULARY

hypothesis
(hi PAH thuh sus)
(noun) a tentative explanation that can be tested with a scientific investigation
Michael made a hypothesis that he would have no cavities because he did a good job of brushing and flossing his teeth.

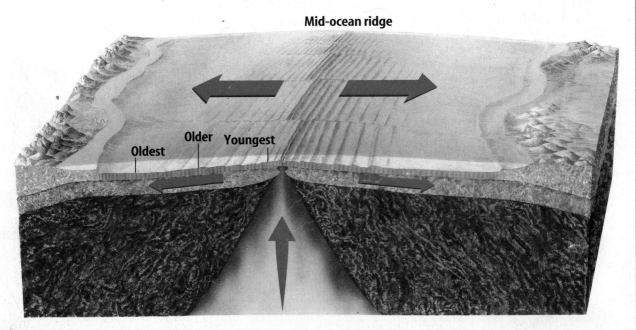

Figure 10 Seafloor spreading forms new oceanic crust. The older oceanic crust moves away from the ridge as new oceanic crust forms.

Mid-ocean ridge

Oldest Older Youngest

Visualizing Mid-Ocean Ridges

Figure 11

Mid-ocean ridges are vast, underwater mountains that form the longest continuous mountain ranges on Earth. Earthquakes and volcanoes commonly occur along the ridges. An example of a mid-ocean ridge is the Mid-Atlantic Ridge. The Mid-Atlantic Ridge was formed when the North and South American Plates pulled apart from the Eurasian and African Plates.

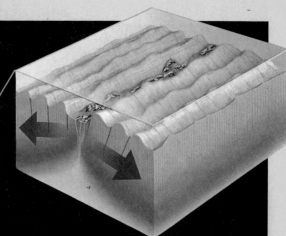

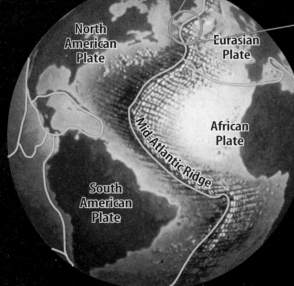

North American Plate

Eurasian Plate

African Plate

Mid-Atlantic Ridge

South American Plate

▲ New oceanic crust is formed as seafloor moves away from the mid-ocean ridge. The seafloor becomes older as the distance from the mid-ocean ridge increases.

▼ Scientists have made many new discoveries on the seafloor. Hydrothermal vents, also known as black smokers, form along mid-ocean ridges. The "smoke" that rises from the hydrothermal vent is actually a hot fluid that is rich in metals.

Some species, such as these giant tube ▲ worms, live next to the hydrothermal vents. The heat and minerals allow them to survive without sunlight.

Contributed by National Geographic

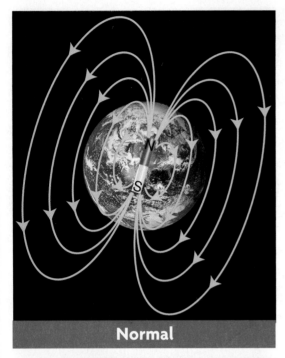

Figure 12A Earth's magnetic poles have reversed many times over many millions of years.

Normal

Reversed

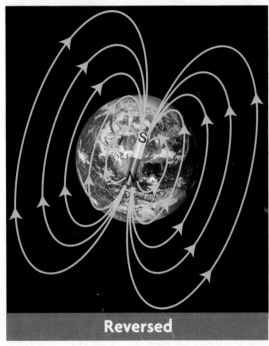

Evidence for Spreading

New evidence connected the ages of seafloor rocks to how Earth's magnetic field was oriented at those times.

Magnetic Polarity Reversals

Whenever you use a compass, the north-seeking end of the needle points to Earth's magnetic north pole. But, Earth's magnetic field has not always had the same orientation. Sometimes the magnetic poles reverse. If you happened to be living at a time after the magnetic poles switched, your compass needle would point south instead of north.

Orientation The top diagram of **Figure 12A** shows the orientation of the magnetic field the way it is today. This is called *normal*. When it points in the opposite direction, it is called *reversed*. Scientists learned the ages of each of these reversals. They used this information to produce a magnetic time scale, which is like a calendar for part of Earth's history.

Recording Reversals Igneous rocks can record these reversals, as illustrated below in **Figure 12B.** This happens along a mid-ocean ridge as oceanic crust forms from lava and cools. Tiny crystals record the magnetic field orientation that existed when the crust cooled.

Figure 12B Igneous rocks that form on both sides of mid-ocean ridges can preserve changes in Earth's magnetic field.

Explain why magnetic polarity reversals are evidence of seafloor spreading.

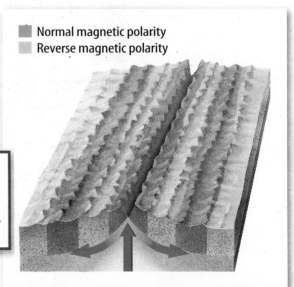

■ Normal magnetic polarity
■ Reverse magnetic polarity

Magnetic Stripes on the Seafloor

As shown in **Figure 13,** scientists can measure Earth's magnetic field with instruments called magnetometers. These instruments can travel over large areas of Earth's surface by ship, plane, and satellite. As they move over the ocean, they measure the strength of the magnetic field. The oceanic crust makes a striped pattern when graphed because it contains alternating strips of rock with normal and reversed polarity.

These magnetic stripes are shown in **Figure 12B.** Just as Hess hypothesized, the seafloor is youngest at the mid-ocean ridge. By measuring the distance of a stripe of rock from the mid-ocean ridge and determining its age, scientists can calculate the speed of seafloor movement.

 **Reading Check** How is the speed of seafloor movement calculated using magnetic polarity reversals?

The seafloor and continents move slowly, only centimeters per year. Learning about seafloor spreading was like learning how a magic trick is done. Scientists finally understood how the continents could move and accepted Wegener's continental drift hypothesis.

Figure 13 Scientists use magnetometers to collect data about Earth's magnetic field.

Figure 14 Drill pipes up to 6 km long are used by scientists in order to reach the seafloor in the deep ocean.

Seafloor Drilling

Not long after scientists learned how to determine the age of the seafloor, they developed deep-sea drilling. They designed a boat that could drill and collect samples from the seafloor. This boat, named the *Glomar Challenger,* made its first voyage in 1968.

Scientists used drill pipes several kilometers long to cut through rock at the bottom of the sea and bring up samples. **Figure 14** shows how the drill pipe extended all the way from the ship to the seafloor. The photo in **Figure 14** shows how the drill bit, with diamonds glued in it, was attached to the bottom of the drill pipe. Recall from Chapter 2 that diamond is the hardest mineral. A diamond-tipped drill can cut through the hardest rock.

 Why are diamonds used in drill bits?

The ages of the samples showed that the oldest rocks were farthest from the mid-ocean ridge. And, the youngest rocks are found in the center of the mid-ocean ridge. This seafloor drilling supported the seafloor-spreading hypothesis.

The small bumps on this modern drill bit are bunches of tiny diamonds glued together. As the bit spins around, they cut through the rocks of the seafloor.

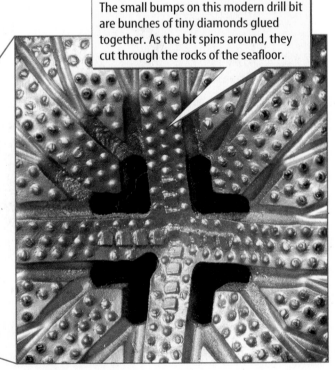

Seafloor Spreading Hypothesis

By the 1950s, new methods and technologies, such as sonar, were being developed to map and explore the seafloor. When scientists mapped the topography of the seafloor they discovered underwater mountain ranges known as mid-ocean ridges. Harry Hess studied the seafloor trying to understand how mid-ocean ridges were formed. He proposed the seafloor spreading hypothesis, which is the process by which new seafloor is continuously made at the mid-ocean ridges. New evidence from around the world showed that the seafloor was spreading, as Hess had thought. Seafloor spreading seemed to explain continental drift. Studies of mid-ocean ridges continue today.

LESSON 2 Review

Summarize

Create your own lesson summary as you organize an **outline.**

1. **Scan** the lesson. Find and list the first **red** main heading.

2. **Review** the text after the heading and list 2–3 details about the heading.

3. **Find** and list each **blue** subheading that follows the **red** main heading.

4. **List** 2–3 details, key terms, and definitions under each **blue** subheading.

5. **Review** additional **red** main headings and their supporting **blue** subheadings. List 2–3 details about each.

 ELA6: R 2.4

Standards Check

Using Vocabulary

1. Use the terms *mid-ocean ridge* and *seafloor spreading* in the same sentence. **1.a**

2. Write a definition for the term *mid-ocean ridge* in your own words. **1.a**

Understanding Main Ideas

3. **Sequence** Draw a diagram like the one below. List the process of seafloor spreading beginning with convection bringing hot material in the mantle toward the surface. **1.a**

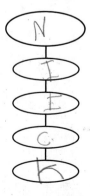

4. **Illustrate** the symmetry of magnetic polarity stripes on the seafloor. **1.a**

5. **Assess** how new data supported the seafloor spreading hypothesis. **1.a**

Applying Science

6. **Suggest** what scientists' reactions to the continental drift hypothesis might have been if data from the seafloor were available in the 1910s. **1.a**

7. **Relate** the high temperatures measured at mid-ocean ridges to the formation of basalt at the ridges. **1.a**

 Science Online

For more practice, visit **Standards Check** at ca6.msscience.com.

How fast does seafloor spread?

Scientists use their knowledge of seafloor spreading and magnetic polarity reversals to estimate the rate of seafloor spreading.

Data

1. Study the magnetic polarity graph.

2. Place a **ruler** vertically on the graph so that it lines up with the center of peak 1 west of the Mid-Atlantic Ridge.

3. Determine and record the distance and age. Repeat this process for peak 1 east of the ridge.

4. Calculate the average distance and age for this pair of peaks.

Graph of Normal and Reverse Polarity

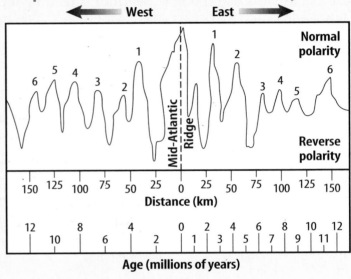

5. Repeat steps 2 through 4 for the remaining pairs of normal polarity peaks.

6. Calculate the rate of movement for the six pairs of peaks. Use the formula rate = distance/time. Convert kilometers to centimeters. For example, to calculate a rate using normal polarity peak 5, west of the ridge:

rate = 125 km/10 million years = 12.5 km/1 million years = 1,250,000 cm/1,000,000 years = 1.25 cm/year

Data Analysis

1. **Compare** the age of igneous rock found near the ridge with that of igneous rock found farther away from the ridge.

2. **Calculate** how long ago a point on the coast of Africa, now 2,400 km away from the ridge, was at or near the Mid-Atlantic Ridge.

 ### Science Content Standards

1.a Students know evidence of plate tectonics is derived from the fit of the continents; the location of earthquakes, volcanoes, and mid-ocean ridges; and the distribution of fossils, rock types, and ancient climatic zones.

7.g Interpret events by sequence and time from natural phenomena (e.g., the relative ages of rocks and intrusions).

 MA6: AF 2.2, AF 2.3

Science Content Standards

1.b Students know Earth is composed of several layers: a cold, brittle lithosphere; a hot, convecting mantle; and a dense, metallic core.

1.c Students know lithospheric plates the size of continents and oceans move at rates of centimeters per year in response to movement in the mantle.

4.c Students know heat from Earth's interior reaches the surface primarily through convection.

Also covers: 7.a, 7.e

Reading Guide

What *You'll Learn*

▶ **Summarize** the theory of plate tectonics.

▶ **Determine** common locations of earthquakes, volcanoes, ocean trenches, and mid-ocean ridges.

▶ **Compare and contrast** oceanic and continental lithosphere.

Why *It's Important*

Plate tectonics cause major geologic features of Earth's crust and contribute to the recycling of material.

Vocabulary

lithospheric plate
plate tectonics
ocean trench
slab
Global Positioning System (GPS)

Review Vocabulary

convection: heat transfer by the movement of matter from one place to another (p. 147)

Theory of Plate Tectonics

Main Idea Earth's lithosphere is broken into large, brittle pieces, which move as a result of forces acting on them.

Real-World Reading Connection The next time you eat a hard-boiled egg, hit it on the table. Even though the shell breaks up, it stays on the egg. Gently slide one of the broken pieces of shell along the surface of the egg. As the pieces move on top of the softer layer of egg, they bump into each other.

Earth's Plates

Canadian geologist J. Tuzo Wilson first used the term *plates* to describe the large pieces of Earth's crust that move horizontally. Much like the pieces of the broken eggshell, Wilson thought the plates were brittle and outlined by cracks. A model of Earth's brittle crust is shown in **Figure 15.** The large brittle pieces of Earth's outer shell are called **lithospheric plates. Figure 16** shows scientists' current mapping of Earth's lithospheric plates.

Figure 15 Earth's brittle crust is cracked and broken into pieces. The red lines show about where major cracks are located on Earth.

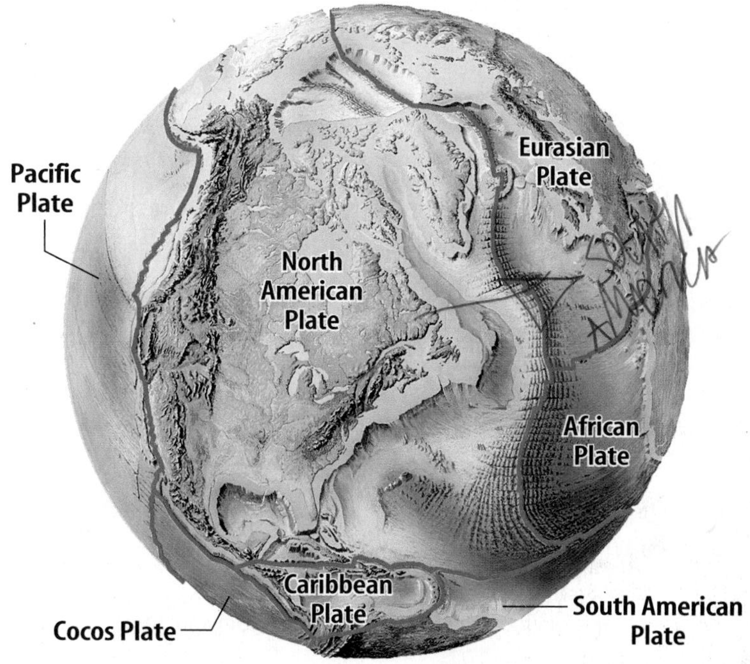

Pacific Plate
Eurasian Plate
North American Plate
African Plate
Caribbean Plate
Cocos Plate
South American Plate

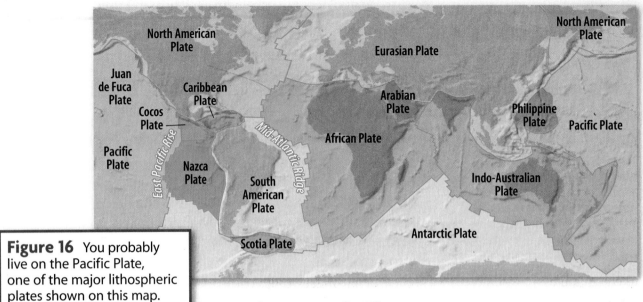

Development of a Theory

The discoveries you have read about and many more were combined in a new theory. The theory of **plate tectonics** explains how lithospheric plates move and cause major geologic features and events on Earth's surface. This theory includes ideas from continental drift and seafloor spreading.

 What does the theory of plate tectonics explain?

Scientists from many countries developed the theory of plate tectonics. **Figure 17** summarizes some important studies that contributed to the development of the theory of plate tectonics. Both successes and failures were valuable in developing the theory of plate tectonics. By the end of the 1960s, there was so much evidence supporting it that the theory gained acceptance by most scientists. Aspects of the theory are still being tested and modified today.

Figure 17 Important studies from scientists around the world contributed to the development of the theory of plate tectonics.

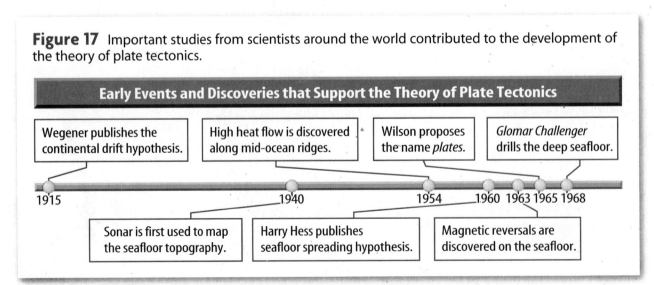

Early Events and Discoveries that Support the Theory of Plate Tectonics

Wegener publishes the continental drift hypothesis.

High heat flow is discovered along mid-ocean ridges.

Wilson proposes the name *plates*.

Glomar Challenger drills the deep seafloor.

1915   1940   1954   1960 1963 1965 1968

Sonar is first used to map the seafloor topography.

Harry Hess publishes seafloor spreading hypothesis.

Magnetic reversals are discovered on the seafloor.

Boundaries of Lithospheric Plates

How can you tell where one lithospheric plate ends and another begins? Mid-ocean ridges show the boundaries of some lithospheric plates. Where would you go to find other boundaries of lithospheric plates?

You would need to find more evidence of forces in Earth such as earthquakes and volcanic eruptions. These geologic features and events occur where the edges of plates are pushed together, pulled apart, or scraped sideways past each other as they move.

Examine **Figure 18** to find the locations of earthquakes and volcanic eruptions. Notice that they are not evenly distributed around the world. There are some places that have many earthquakes and volcanoes. There are other places that have almost none.

 Figure 18 Identify five locations of earthquakes or active volcanoes in North America.

Thin lines of earthquakes and volcanic eruptions **define** the mid-ocean ridges. But, there are thick bands of both earthquakes and volcanoes in other places. Around the edges of the Pacific Ocean, there are some thick bands of earthquakes and volcanoes. These earthquakes and volcanoes are located near long, deep parts of the seafloor called **ocean trenches.** Seafloor that is formed at mid-ocean ridges is destroyed at ocean trenches.

ACADEMIC VOCABULARY
define (de FINE)
(verb) to fix or mark the limits of
The surveyor defined the limits of the property before the house was sold.

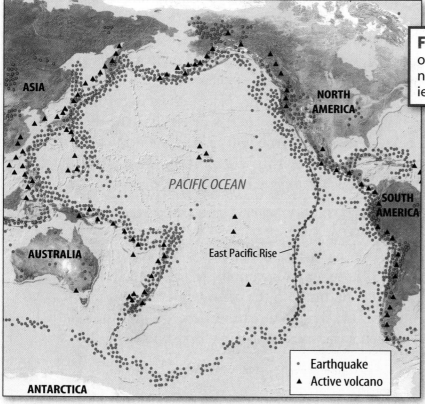

Figure 18 The locations of earthquakes and volcanoes indicate the boundaries of lithospheric plates.

ASIA

NORTH AMERICA

PACIFIC OCEAN

SOUTH AMERICA

East Pacific Rise

AUSTRALIA

ANTARCTICA

• Earthquake
▲ Active volcano

Types of Lithosphere

The word *lithosphere* is based on the Greek word *lithos,* which means "rock"—which is what makes up the lithosphere. In Chapter 2, you read that the lithosphere is made of both crust and the upper mantle. Recall that both the crust and the upper mantle are made of rigid and brittle rocks. They do not flow as much as the warmer, weaker rocks below them in the asthenosphere.

Table 1 shows a slice through a lithospheric plate. Just like there are two different types of crust, there are two different types of lithosphere. These types of lithosphere, like the types of crust, are oceanic and continental.

 What are the two types of lithosphere?

Thickness **Table 1** shows that the thickness of a lithospheric plate varies. Oceanic lithosphere is much thinner than continental lithosphere. When oceanic lithosphere first forms near a mid-ocean ridge, it is about 10 km thick. As the oceanic lithosphere ages and cools after it forms at Earth's surface, it thickens and becomes much denser.

Composition Thickness is not the only difference between oceanic and continental lithosphere. Their compositions also differ, as shown in **Table 1.** Oceanic lithosphere contains oceanic crust, which is made mainly of the dense igneous rocks basalt and gabbro. A relatively thin layer of sediment also occurs on oceanic lithosphere in many places.

In contrast, continental lithosphere contains the continental crust, which is made of igneous rocks such as granite and metamorphic rocks such as gneiss. It also has a covering of sedimentary rocks in many places. Most of the rocks in the continental lithosphere are less dense than the rocks in the oceanic lithosphere. You will read in the next chapter that this difference in density is important when lithospheric plates are pushed together.

 **Table 1** Infer the density of peridotite using the data in Table 1.

Most of the major plates contain both oceanic and continental lithosphere. An example of this is the North American Plate. It contains both the continent of North America and much of the Atlantic Ocean. An exception is the Pacific Plate, which contains mostly oceanic lithosphere.

Table 1 Types of Lithosphere

Type of Lithosphere	Components	Common Rocks	Approximate Densities
Oceanic	• oceanic crust • upper mantle	• crust: basalt, gabbro • upper mantle: peridotite	• crust: 3.0 g/cm^3 • mantle: 3.3 g/cm^3
Continental	• continental crust • upper mantle	• crust: granite, gneiss, sedimentary rocks • upper mantle: peridotite	• crust: 2.65 g/cm^3 • mantle: 3.3 g/cm^3

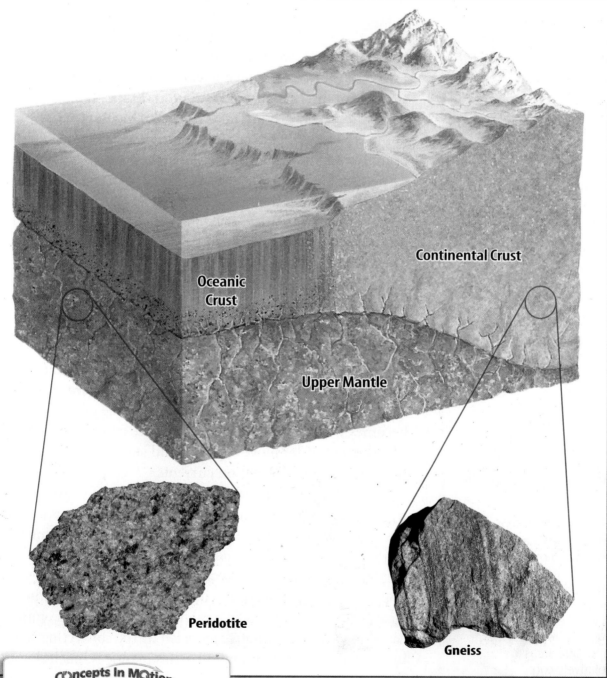

Continental Crust

Oceanic Crust

Upper Mantle

Peridotite

Gneiss

Concepts In Motion
Interactive Table Organize information about the lithosphere at ca6.msscience.com.

What controls plate movement?

In the past, scientists generally agreed that convection currents in Earth's mantle control the movement of lithospheric plates. Convection currents, which you read about in Chapter 3, passively move the lithospheric plates above them. Today, scientists think that other forces might exist that control the movement of lithospheric plates.

Escaping Heat

Heat has been escaping from Earth since it first formed. One way heat is transferred from Earth's deep interior to the surface is by convection. As a result, convection currents in the mantle provide matter and energy for the motion of the plates.

Internal Heat Source

In order for convection currents in the mantle to continue, a supply of internal heat is required in Earth. Temperature in Earth increases with increasing depth. One important source of internal heat is radioactive decay. Some elements, such as uranium and potassium, are radioactive. As radioactive elements decay, they produce heat, along with radiation, in rocks that contain them. The heat increases the temperature of the surrounding rock, causing its density to decrease. This causes movement of both rock and heat from inside Earth toward the surface.

Reading Check Why does the density of the surrounding rock decrease?

Convection

You can't see convection currents directly, but you can see convection in a pot of water, like the one in **Figure 19.** As a flame heats the water from the bottom of the pot, this water becomes warmer and less dense than the cooler water above. The warmer water moves upward and starts to circulate in the pot, forming convection currents.

Figure 19 helps you understand the process of convection in a pot of water. But, keep in mind that this model does not describe mantle convection accurately. Scientists have much work to do before a complete understanding of convection is possible. Compare the convection currents in the pot of water to the warm and cool regions in Earth's interior shown in **Figure 20.** Temperature changes in Earth's interior are complicated and are not accurately modeled by the convecting water in the pot. Also, convection in Earth's mantle is much slower than convection in water.

Figure 19 The flame heats water at the bottom of the pot. The warm water rises, then as it cools, it sinks back down.

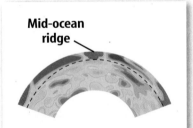

Mid-ocean ridge

Figure 20 Scientists model the locations of warm and cool regions in Earth's interior by studying the behavior of complex waves that travel through them. Red areas shown here are relatively warm, and blue areas are relatively cool.

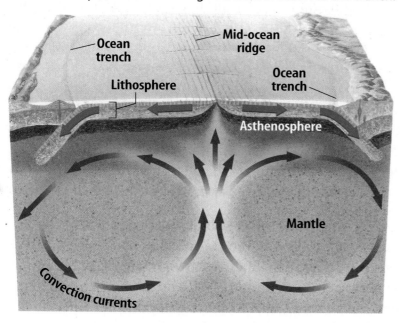

Figure 21 One hypothesis states that convection currents in the mantle move plates from the ridge across the ocean to the trench.

Ocean trench

Mid-ocean ridge

Lithosphere

Ocean trench

Asthenosphere

Mantle

Convection currents

Plate Movement and Convection

Some scientists hypothesize that convection currents in Earth's mantle drive plate movement. The arrows in **Figure 21** show the direction of flow for convection currents. In this model, the brittle lithospheric plates float on top of the weak asthenosphere. This happens in much the same way that a sheet of ice floats along on slowly moving water. **Figure 21** also shows the cooler, denser lithospheric plate sinking down into the mantle. A plate that sinks back into the mantle is called a **slab.** The slab bends and breaks as it sinks down, causing earthquakes.

Density Changes Along a mid-ocean ridge, less dense rock is brought near the surface and melts. At the surface, new oceanic lithosphere forms at the ridge. As the rock cools, it becomes denser. The new plate starts to move away from the ridge with the convection current.

 What happens to the density of new oceanic lithosphere when it cools?

Historical Model This model of convection currents in Earth's mantle convinced scientists to accept the theory of plate tectonics. Convection is still thought to play a role in the movement of plates. Recent studies are making scientists consider new ideas about what makes plates move.

Ridge Push and Slab Pull

Some scientists think that ridge push and slab pull are two forces that might be important in controlling the movement of plates. These forces are related to the lithospheric plates themselves. **Figure 22** shows how these forces are thought to work.

A mid-ocean ridge is higher in elevation than most parts of the seafloor. The force of gravity tends to move things downhill. For example, if you stand on a skateboard, even on a very gentle slope, you will start to roll downhill. That is what scientists think happens to a lithospheric plate along a ridge. The force of gravity moves the plate downward and away from the ridge. This force is known as ridge push.

Sinking Into the Mantle Slab pull occurs when a lithospheric plate sinks into the mantle. The slab is dense and cool as it sinks into the mantle, as shown in **Figure 22.** Again, the force of gravity acts on the plate. The denser slab acts like a sinker on a fishing line. A sinker will pull your line and hook down into the water where a fish might see it. In the same way, the dense slab pulls the plate deep into the mantle.

Future Work Scientists are currently studying individual plates, mid-ocean ridges, slabs, and speeds of complex waves that travel through Earth. They hope to learn which forces are most effective in controlling how plates move. There is still much to discover about the dynamic Earth and plate tectonics.

Figure 22 Ridge push and slab pull are forces that might move plates. Of the two, slab pull is thought to be the more important force.

Ridge Push

Slab Pull

Measuring Plate Movement

Since the 1960s, scientific studies have continued to strengthen the theory of plate tectonics. An example is the ability to directly measure plate movement. Originally, scientists had to estimate the speeds of the plates by using the ages of rocks. Now they use satellites.

Global Positioning System

Global Positioning System (GPS) is a network of satellites used to determine locations on Earth. A receiver on Earth collects radio signals from several satellites that circle the planet as shown in **Figure 23.** Then, a computer inside the receiver calculates the latitude, longitude, and elevation of the receiver's location. Today, airplane pilots, sailors, people driving cars, and even hikers use GPS.

 What is GPS?

Because GPS can be used to accurately measure distances, it can also be used to measure the movement of plates. Scientists have set up a group of receivers to monitor plate movement around the world.

 Figure 23 Infer the advantages of having many GPS satellites measure plate movement.

Satellite Laser Ranging

GPS is not the only satellite system that is used to measure plate movement. Satellite laser ranging (SLR) uses pulses of light instead of radio waves to measure distances. These pulses of light are laser beams, as shown in **Figure 24.**

Rates of Plate Movement

Using satellite data from these methods, scientists have estimated the rates at which lithospheric plates move. The measurements made using GPS and SLR show similar results. The plates are moving only centimeters per year. This is about as fast as your fingernails grow.

Figure 23 This diagram shows the orbits of the 24 GPS satellites that orbit 20,000 km above Earth.

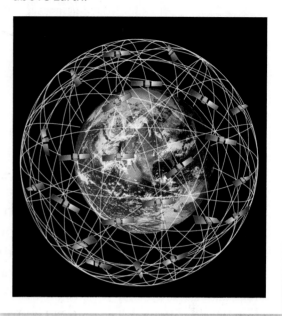

Figure 24 Scientists use pulses of light to measure the rates of plate movements.

Plate Tectonics and the Rock Cycle

Reading about plate tectonics makes it easier to understand how the rock cycle works. **Figure 25** shows the rock cycle from Chapter 2. Imagine where you might find magma, sediment, and the three rock types in a plate tectonic setting.

 Visual Check

Figure 25 Explain why magma is shown at the bottom of the rock cycle.

Rocks are always moving through the rock cycle. Magma will rise up to Earth's surface to become igneous rock at a mid-ocean ridge. A plate will slowly move away from the ridge and cool. It will carry the igneous rock with it. Eventually the plate will sink into the mantle.

While some rocks erode on mountaintops, others form on the seafloor. Earth's lithospheric plates move slowly, but they have millions of years to travel long distances and get where they are going. As plates move, they recycle material by keeping it moving through the rock cycle.

Figure 25 Plate tectonics moves Earth materials through the rock cycle.

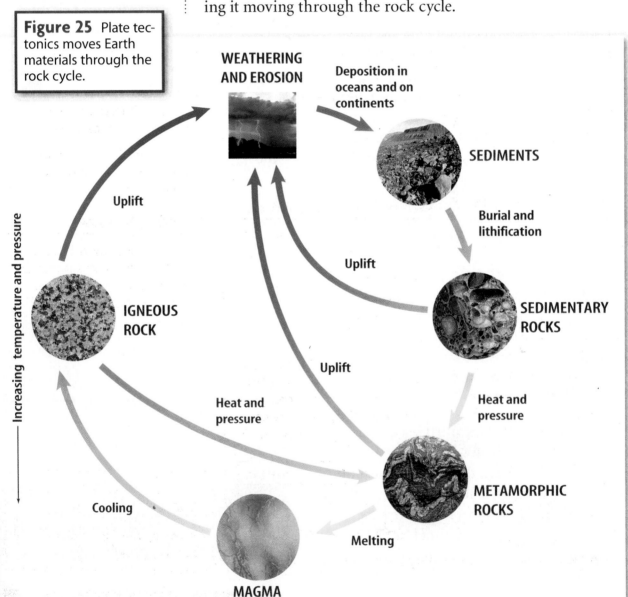

Plate Movements

Earth's outer shell is broken into large, brittle pieces called lithospheric plates. The thickness and composition of a lithospheric plate varies. The thickness and composition depends upon whether the plate is made of oceanic or continental material. Processes in Earth, such as earthquakes and volcanoes, occur along the boundaries of the lithospheric plates.

The theory of plate tectonics explains how lithospheric plates move and cause major geologic features and events on Earth's surface. Some scientists hypothesize that convection currents in Earth's mantle drive the movement of plates. And, some scientists think that ridge push and slab pull are two forces that control the movement of plates.

LESSON 3 Review

Summarize

Create your own lesson summary as you write a **newsletter.**

1. **Write** this lesson title, number, and page numbers at the top of a sheet of paper.

2. **Review** the text after the red main headings and write one sentence about each. These will be the headlines of your newsletter.

3. **Review** the text and write 2–3 sentences about each blue subheading. These sentences should tell *who, what, when, where,* and *why* information about each headline.

4. **Illustrate** your newsletter with diagrams of important structures and processes next to each headline.

 **ELA6:** W 1.2

Standards Check

Using Vocabulary

1. Use the terms *slab* and *ocean trench* in the same sentence. **1.b**

2. A method used to determine the rate of plate movement is _____. **1.c**

Understanding Main Ideas

3. Which rock is found in oceanic lithosphere? **1.c**
   A. granite
   B. gneiss
   C. limestone
   D. basalt

4. **Compare and Contrast** Draw a chart like the one below. Fill in details comparing forces that scientist hypothesize control the movement of plates. **1.c**

Force	Similarities	Differences

5. **Describe** an important source of internal heat in Earth. **1.b**

6. **Compare and contrast** oceanic lithosphere and continental lithosphere. **1.c**

7. **State** three forces that scientists hypothesize move lithospheric plates. **1.c**

Applying Science

8. **Imagine** that there is more internal heat in the planet. Do you think the plates would move faster or more slowly? Explain. **4.c**

 **Science Online**

For more practice, visit **Standards Check** at ca6.msscience.com.

MiniLab

How can you observe convection in water?

Convection in Earth's mantle is difficult to model and cannot be seen directly. However, you can observe movement of convection currents in water.

Procedure 👓 🧤 ✂️ 🧷 🚫 🔥

1. Read and complete a lab safety form.
2. Pour **water** into a **clear, colorless casserole dish** until the water level is about 5 cm from the top of the dish.
3. Center the dish on a **hot plate** and heat it.
4. Add a few drops of **food coloring** to the water above the center of the hot plate.
5. Looking in from one side of the dish, observe what happens in the water.
6. Illustrate your observations by making a labeled sketch.

Analysis

1. **Determine** whether any currents form in the water. Support your answer with the illustration from step 6.
2. **Infer** what caused the water to behave the way it did.
3. **Infer** why convection in Earth's interior is important.

 Science Content Standards

4.c Students know heat from Earth's interior reaches the surface primarily through convection.
7.e Recognize whether evidence is consistent with a proposed explanation.

Applying Math

Percentage of Minerals in Rocks in the Lithosphere

The five major rocks in Earth's oceanic and continental lithosphere are basalt, diorite gneiss, gabbro, granite, and peridotite. The table shows the composition of each type of rock.

Example

How does the percentage of amphibole in basalt differ from the percentage of amphibole in gabbro?

What you know:
- The percentage of amphibole in basalt: 45%
- The percentage of amphibole in gabbro: 15%

What you need to find:
- The difference in the percentages

Subtract:

$45 - 15 = 30\%$

Answer: There is 30% more amphibole in basalt than in gabbro.

Percentage of Minerals in Rocks		
Rock	**Mineral**	**Composition %**
Basalt	amphibole	45
	calcium feldspar	40
	olivine	15
Diorite gneiss	sodium feldspar	50
	amphibole	20
	biotite	10
	orthoclase feldspar	10
	quartz	10
Gabbro	pyroxene	55
	olivine	25
	ampibole	15
	calcium feldspar	5
Granite	orthoclase feldspar	50
	biotite	20
	sodim feldspar	15
	quartz	15
Peridotite	olivine	70
	pyroxene	25
	calcium feldspar	5

Practice Problems

1. How much more calcium feldspar is found in basalt than gabbro?
2. How much more quartz is found in granite than diorite gneiss?

Science nline
For more math practice, visit ca6.msscience.com.

Use the Internet:
Inferring Plate Tectonic Activity

Materials

metric ruler
pencil
world map
computer with internet access

Problem

The movement of lithospheric plates causes forces that build up energy in rocks. Some of this energy is released as earthquakes. Earthquakes occur every day. Many are too small to be felt by humans, but each event tells scientists something more about Earth. Can you infer plate tectonic activity by plotting locations of recent earthquakes on a world map?

Form a Hypothesis

Think about how earthquakes define the boundaries of lithospheric plates. There are some places that have many earthquakes, and other places that have almost none. Make a hypothesis about whether the locations of earthquakes can be used to infer plate tectonic activity.

Collect Data and Make Observations

1. Make a data table like the one shown below.
2. Visit ca6.msscience.com to collect and record data for earthquake locations from the last two weeks.
3. Plot the locations on a copy of a map of the world. This map should include lines of latitude and longitude to guide your plotting.

Locations of Earthquakes			
Location Description	Latitude	Longitude	Date

Analyze and Conclude

1. **Infer** where plate tectonic activity occurs on Earth today.
2. **Compare and contrast** the active plate tectonic areas to the plate boundary map shown below.
3. **Identify** lithospheric plates that are represented by the earthquake locations you plotted.
4. **Explain** how data from a longer period of time might help you better identify plate tectonic activity.

Communicate

 Science  **ELA6:** W 1.2

Write a Paragraph Select one of the earthquakes you plotted. Research the details of the event, including the geography of the area near the earthquake, and whether the lives of humans or other organisms were impacted by the event.

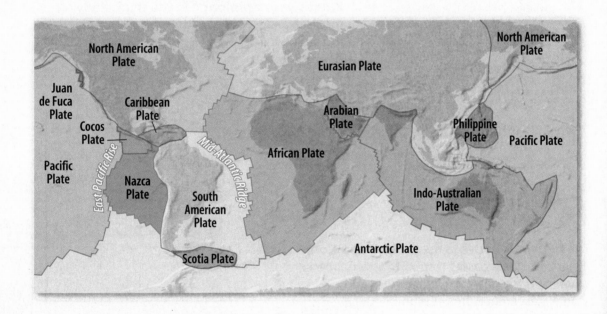

Real World Science

Paleontologists Validate Wegener in Dinosaur Cove

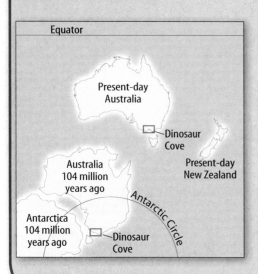

Equator

Present-day Australia

Dinosaur Cove

Australia 104 million years ago

Present-day New Zealand

Antarctic Circle

Antarctica 104 million years ago

Dinosaur Cove

Fossil remains unearthed by paleontologists in Dinosaur Cove are all that remain of dinosaurs that lived in southeastern Australia 100 million years ago. Paleontologists believe these "polar dinosaurs" had keen night vision and were warm-blooded. This allowed them to find food in the dark and below-freezing temperatures. The discovery provides convincing evidence of Australia's northward movement toward the equator during the past 100 million years.

What's your opinion? Do you think studying fossils in Dinosaur Cove is important to understanding life on Earth today? **Prepare** an oral argument describing your opinion.

ELA6: W 2.4

Defining Plate Boundaries

The boundary between the Eurasian and North American Plates passes through Eastern Siberia, where seismic detection equipment and people are scarce. Researchers from Columbia University, Massachusetts Institute of Technology, the University of California, and the Russian Academy of Sciences collaborated to pinpoint the boundary. Using GPS, the team monitored 50 points in the region over six years and defined the boundary.

Analyze the Map **Compare** a plate boundary map to a regional map. **Identify** features in Russia that are near the boundary.

ELA6: RC 2.3

North American Plate

Eurasian Plate

Pacific Plate

Philippine Plate

Behind a Revolutionary Theory

Historians agree that Alfred Wegener was not alone in considering Earth as a shifting, dynamic world. In 1858, 57 years before Wegener published *The Origin of Continents and Oceans,* French scientist Antonio Snider-Pellegrini made maps illustrating how the American and African continents may once have fit together, and then moved apart. Pellegrini supported his hypothesis with evidence from identical plant fossils in Europe and the United States.

Creating a Continental Drift Time Line Visit **History** at **ca6.msscience.com** to research the contributions of Wegener and other scientists. **Create** a time line illustrating key individuals and events.

 ELA6: W 1.3

HOLE 504B

Off the west coast of South America, scientists study Earth through the deepest hole ever drilled into the oceanic crust. Hole 504B bores into the Costa Rica Rift, the zone along which the Cocos and Nazca Plates are pulling apart. First drilled in 1979, the hole is now 2,111 m deep, exposing 6 million years of oceanic crust. Scientists at the drilling site investigate oceanic crust properties at various depths, including thermal conductivity, density, and velocity. Hole 504B is a unique underwater laboratory that allows scientists to study how oceanic crust forms and evolves over time.

Visit **Society** at **ca6.msscience.com** to learn about the history of the Integrated Ocean Drilling Program. Use the information you gather to **write** a brief paper discussing key events in the history of deep ocean drilling.

 ELA6: W 1.2

The BIG Idea Plate tectonics explains the formation of many of Earth's features and geologic events.

Lesson 1 Continental Drift
1.a, 7.e

Main Idea Despite the evidence that supported continental drift, it was rejected by most scientists.

- The continental drift hypothesis states that the continents move slowly across Earth's surface.

- The fit of the continents, fossils, rock types, mountain ranges, and ancient climate evidence support Pangaea's existence.

- Continental drift was rejected by some scientists because Wegener could not explain what forces move continents.

- **continental drift** (p. 167)
- **Pangaea** (p. 167)

Lesson 2 Seafloor Spreading
1.a, 7.g

Main Idea New discoveries led to seafloor spreading as an explanation for continental drift.

- Sonar allowed scientists to map the topography of the seafloor.

- Mid-ocean ridges, the longest mountain ranges on Earth, were found to be regions of high heat flow.

- Harry Hess proposed the seafloor spreading hypothesis, a process by which new seafloor is continuously formed at mid-ocean ridges.

- Magnetic polarity reversals supply data for approximating the ages of rocks on the seafloor.

- Seafloor drilling confirmed the increasing age of the seafloor with distance from mid-ocean ridges.

- **mid-ocean ridge** (p. 175)
- **seafloor spreading** (p. 176)

Lesson 3 Theory of Plate Tectonics
1.b, 1.c, 4.c, 7.a, 7.e

Main Idea Earth's lithosphere is broken into large, brittle pieces, which move as a result of forces acting on them.

- Earth's surface is covered by the brittle lithosphere, which is broken into numerous large plates.

- The locations of earthquakes, volcanoes, and mid-ocean ridges help define the boundaries of plates.

- Plates can be composed of either thin oceanic crust, thick continental crust, or both.

- Scientists use satellites to directly measure the speed and directions of plate movement.

- **Global Positioning System (GPS)** (p. 191)
- **lithospheric plate** (p. 183)
- **ocean trench** (p. 185)
- **plate tectonics** (p. 184)
- **slab** (p. 189)

STUDY TO GO Download quizzes, key terms, and flash cards from ca6.msscience.com.

Linking Vocabulary and Main Ideas

Use vocabulary terms from page 200 to complete this concept map.

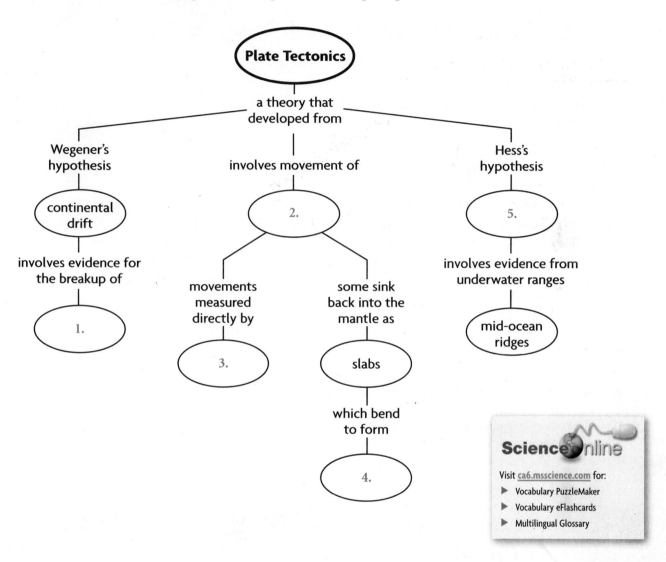

Using Vocabulary

Fill in each blank with the correct vocabulary term.

When scientists discovered _____6._____, which are mountain ranges in the middle of the seafloor, they finally had found a way for _____7._____ to have broken apart. This process, by which continents could separate and drift apart as new seafloor forms, is _____8._____. It was an important discovery that led to the rapid development of the theory of _____9._____.

Understanding Main Ideas

Choose the word or phrase that best answers the question.

1. Which evidence did Alfred Wegener propose to support his continental drift hypothesis?
   1.a
   A. ancient climate belts
   B. seafloor spreading
   C. ancient drift scars on the seafloor
   D. fish fossils

2. Why was Wegener's hypothesis rejected?
   1.a
   A. He collected too little evidence.
   B. He could not find fossils from the time of Pangaea.
   C. He did not publish his ideas.
   D. He could not explain what forces could move the continents.

3. Which modern technology is used to directly measure plate movement?
   A. ATV
   1.c
   B. GPS
   C. MTV
   D. sonar

4. The illustration below shows the distribution of some fossils among southern continents of Pangaea.

   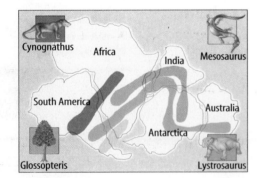

   What are the life-forms shown in the picture?
   A. modern land-based animals
   1.a
   B. fossil sea creatures
   C. fossil land-based species
   D. ancient life-forms migrating to Asia

5. The illustration below shows an area where new seafloor is forming.

   What do the stripes of rock represent?
   A. changes in rock composition
   1.a
   B. changes of rock type
   C. changes in magnetic polarity orientation
   D. changes in ocean-water depth

6. Who developed the theory of plate tectonics?
   A. Alfred Wegener
   1.c
   B. American scientists
   C. Harry Hess
   D. scientists from many countries

7. Which of these is used to map seafloor topography?
   A. magnetic seafloor stripes
   1.a
   B. sonar
   C. GPS
   D. scuba divers

8. Which is **NOT** a major difference between oceanic and continental lithosphere?
   A. density
   1.b
   B. composition
   C. thickness
   D. rigid, or brittle, behavior

9. At what rate do lithospheric plates move?
   A. centimeters per year
   1.c
   B. centimeters per week
   C. kilometers per year
   D. kilometers per week

Science Online Standards Review ca6.msscience.com

Applying Science

10. **Suggest** a reason that explains why there is no continental drift on the Moon. `1.a`

11. **Decide** which was more important in advancing the acceptance of the theory of plate tectonics: GPS or seafloor spreading. `1.c`

12. **Give** an example of something that, like lithospheric plates, moves so slowly you cannot see it move. `1.c`

13. **Compare and contrast** the locations of volcanoes with the locations of ocean trenches. `1.c`

14. **Describe** how heat from Earth's interior reaches the surface. `4.c`

15. **Identify** the continental lithosphere and oceanic crust in the illustration below. `1.c`

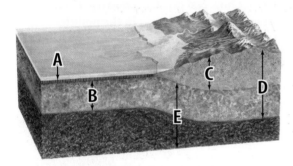

WRITING in ▶ Science

16. **Write** a paragraph summarizing the theory of plate tectonics, including features on the seafloor that provide evidence for the locations of plate boundaries. `ELA6: W 1.2`

Cumulative Review

17. **Evaluate** the usefulness of a road map that has no scale. `7.f`

18. **List** seven minerals that are valuable resources. `6.c`

19. **Infer** why you can keep cooler on a sunny day if you sit under a tree or an umbrella. `3.d`

Applying Math

Percentage of Minerals in Rocks

Rock	Mineral	Composition (%)
Basalt	Amphibole	45
	Calcium feldspar	40
	Olivine	15
Diorite gneiss	Sodium feldspar	50
	Amphibole	20
	Biotite	10
	Orthoclase feldspar	10
	Quartz	10
Gabbro	Pyroxene	55
	Olivine	25
	Amphibole	15
	Calcium feldspar	5
Granite	Orthoclase feldspar	50
	Biotite	20
	Sodium feldspar	15
	Quartz	15
Peridotite	Olivine	70
	Pyroxene	25
	Calcium feldspar	5

20. How much more amphibole is found in basalt than in gabbro? `MA6: NS 1.4, MR 2.4`

21. How much more calcium feldspar is found in basalt than in peridotite? `MA6: NS 1.4, MR 2.4`

22. How much more calcium feldspar is found in gabbro than in peridotite? `MA6: NS 1.4, MR 2.4`

23. How much more olivine is found in gabbro than in basalt? `MA6: NS 1.4, MR 2.4`

1 Which is believed to cause plate movement?

A compression

B convection

C isostasy

D tension `1.c`

2 Which hypothesis states that continents slowly moved to their present-day positions on Earth?

A subduction

B erosion

C continental drift

D seafloor spreading `1.a`

3 The illustration below shows an ancient supercontinent.

What is the name of the ancient supercontinent?

A Pangaea

B Gondwanaland

C Laurasia

D North America `1.a`

4 Who developed the continental drift hypothesis?

A Harry Hess

B J. Tuzo Wilson

C Alfred Wegener

D W. Jason Morgan `1.a`

5 Which term includes Earth's crust and part of the upper mantle?

A asthenosphere

B lithospheric plate

C lower mantle

D core `1.b`

6 About how fast do plates move per year in response to movements in the mantle?

A a few millimeters per year

B a few centimeters per year

C a few meters per year

D a few kilometers per year `1.c`

7 This diagram shows magnetic polarity.

What happens to Earth's magnetic field over time?

A It continually strengthens.

B It changes its polarity.

C It stays the same.

D It weakens and goes away. `1.a`

Science Online Standards Assessment ca6.msscience.com

8 Which feature is evidence that many continents were at one time near Earth's south pole?

A glacial deposits

B earthquakes

C volcanoes

D mid-ocean ridges `1.a`

9 What evidence in rocks supports the seafloor spreading hypothesis?

A plate movement

B magnetic reversals

C subduction

D convergence `1.a`

10 Alfred Wegener described continental drift in 1912. Why weren't Wegener's ideas accepted by scientists until the 1950s?

A Wegener was not respected because he was not a geologist.

B The fossil evidence of the time did not support Wegener's ideas.

C The continents Wegener proposed do not actually fit together well.

D Wegener could not explain how or why the continents moved. `1.a`

11 What is the main cause of convection currents in Earth's mantle?

A Energy from the Sun heats the upper part of the mantle more than the lower part.

B Earthquakes produce cracks in Earth's crust, allowing hot, plasticlike rock to rise up from the mantle.

C Heat released by volcanoes melts rock at the surface, and the molten rock sinks through cracks in the mantle.

D Warmer, less dense rock rises toward the surface and cooler, denser rock sinks back toward the core. `4.c`

12 The graph shows how seafloor depth changes with age.

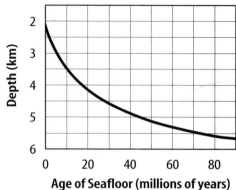

Relationship Between Depth and Age of Seafloor

How does seafloor depth change with age?

A depth decreases

B depth increases

C depth remains the same

D depth varies `1.a`

13 Which observation best supports the continental drift hypothesis?

A Blue whales can be found in every ocean on Earth.

B Limestone often contains fossils of animals that lived in the sea.

C Some mammal species in Australia are found nowhere else on Earth.

D Fossils of the same species of land lizard are found on continents separated by an ocean. `1.a`

14 What theory states that plates move around on the asthenosphere?

A continental drift

B seafloor spreading

C subduction

D plate tectonics `1.a`

Plate Boundaries and California

The BIG Idea

California is located on a plate boundary, where major geologic events occur.

LESSON 1 (1.c, 1.d, 1.e, 7.g)

Interactions at Plate Boundaries

(Main Idea) There are three main types of plate boundaries, where stresses cause rocks to deform.

LESSON 2

(1.e, 1.f, 7.a, 7.b, 7.e)

California Geology

(Main Idea) Many of California's landforms were produced by plate tectonic activity, which continues today.

Whose fault is it?

California and Nevada share more than just a border—they share faults. The fault shown in this photo lies within the Sierra Nevada. About 25 million years ago, the Sierra Nevada started to rise and tilt to the west. Rivers cut deep canyons. Uplift of the Sierra Nevada continues today, especially along its eastern side. This uplift causes faults and large earthquakes.

(Science *Journal*) Imagine you are an explorer and it is 1776. On your expedition, you see the Sierra Nevada for the first time. Write your description of the mountains and your thoughts as you view these mountains.

Launch Lab

How do objects deform?

Depending on what they are made of and how stress is applied to them, solids can change shape, or deform. What will happen to objects when you place a force on them?

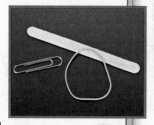

Procedure

1. Complete a lab safety form.

2. Obtain a **small stick,** a **paper clip,** and a **small rubber band.**

3. Observe what happens when you bend or stretch each object with your hands.

4. Experiment with the bending and stretching at different rates.

Think About This

- **Describe** the ways each object deformed.

- **Determine** which objects remained deformed after forces were removed.

 1.d, 7.e

Try at Home

Plate Boundaries Make the following Foldable to help you visualize information about plate boundaries.

STEP 1 Fold a sheet of paper in half lengthwise. Make the back edge about 2 cm longer than the front edge.

STEP 2 Fold into thirds.

STEP 3 Unfold and **cut** along the folds of the top flap to make three flaps.

STEP 4 Label the flaps as shown.

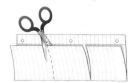

Plate Boundaries

| Divergent | Convergent | Transform |

Reading Skill  ELA6: R 2.4

Visualizing

As you read this chapter, visualize each type of plate boundary and draw it under its tab. Then give examples of places in California where each type of plate boundary can be found.

Science Online

Visit ca6.msscience.com to:

▶ view Concepts in Motion

▶ explore Virtual Labs

▶ access content-related Web links

▶ take the Standards Check

Get Ready to Read

Visualize

1 Learn It! Visualize by forming mental images of the text as you read. Imagine how the text descriptions look, sound, feel, smell, or taste. Look for any pictures or diagrams on the page that may help you add to your understanding.

2 Practice It! Read the following paragraph. As you read, use the underlined details to form a picture in your mind.

> When two plates are moving apart, tension <u>pulls lithosphere apart</u> so that it <u>stretches and becomes thinner</u>. This stretching and thinning is the deformation that results from tension stress.
>
> —from page 211

Based on the description above, try to visualize tension stress. Now look at the diagram on page 211.

- How closely does it match your mental picture?
- Reread the passage and look at the picture again. Did your ideas change?
- Compare your image with what others in your class visualized.

3 Apply It! Read the chapter and list three subjects you were able to visualize. Make a rough sketch showing what you visualized.

Target Your Reading

Reading Tip

Forming your own mental images will help you remember what you read.

Use this to focus on the main ideas as you read the chapter.

1 **Before you read** the chapter, respond to the statements below on your worksheet or on a numbered sheet of paper.

- Write an **A** if you **agree** with the statement.
- Write a **D** if you **disagree** with the statement.

2 **After you read** the chapter, look back to this page to see if you've changed your mind about any of the statements.

- If any of your answers changed, explain why.
- Change any false statements into true statements.
- Use your revised statements as a study guide.

Science Online

Print a worksheet of this page at ca6.mscience.com.

Before You Read A or D	Statement	After You Read A or D
	1 Part of California will eventually break off and fall into the Pacific Ocean.	
	2 The San Andreas Fault is part of a plate boundary.	
	3 Plate boundaries extend deep into Earth's lithosphere.	
	4 Subduction occurs when oceanic and continental lithospheric plates move toward each other.	
	5 Mountains in western South America result from a continent-to-continent convergent plate boundary.	
	6 Faults are surfaces where rocks break and move.	
	7 Los Angeles and San Francisco are moving closer to one another because of a transform plate boundary.	
	8 When rocks are subjected to compression stress, they become thinner.	
	9 The Cascade Range forms on a divergent plate boundary.	

Reading Guide

What *You'll Learn*

▶ **Describe** types of stress that deform rock.

▶ **Relate** geologic features of Earth's surface to types of plate boundaries.

Why *It's Important*

Understanding geologic events that occur at plate boundaries can save lives and prevent damage to property.

Vocabulary

fracture
fault
divergent plate boundary
continental rifting
rift valley
convergent plate boundary
subduction
transform plate boundary

Review Vocabulary

lithospheric plate: large, brittle pieces of Earth's outer shell composed of crust and uppermost mantle (p. 183)

Interactions at Plate Boundaries

Main Idea There are three main types of plate boundaries, where stresses cause rocks to deform.

Real-World Reading Connection Have you ever been stuck in traffic on the freeway? Often, there are bumper-to-bumper cars moving slowly. If all the drivers carefully go the same speed and in the same direction, there are no crashes. But, if a car slows down, speeds up, or turns, there can be a collision, causing crumpled vehicles. Like cars in traffic, Earth's plates can also collide and deform.

Stress and Deformation

In Chapter 4, you read how Earth's lithosphere—made partly of crust and partly of upper mantle—is broken into plates. These plates are packed together more closely than cars in traffic. They also travel at different speeds and in different directions, so there are collisions. Earth's plates move very slowly, not like cars on the freeway. They are so massive that their collisions are very powerful. Such interactions cause stress at plate boundaries. And, like crashing cars, the stresses cause deformation, as shown in **Figure 1.**

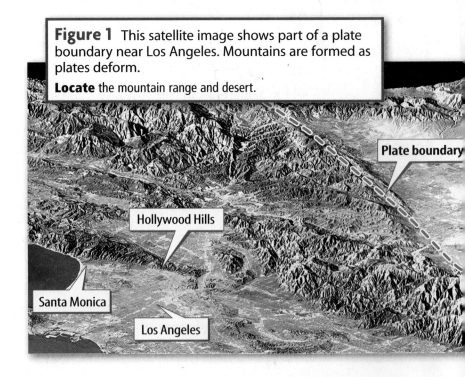

Figure 1 This satellite image shows part of a plate boundary near Los Angeles. Mountains are formed as plates deform.
Locate the mountain range and desert.

Plate boundary

Hollywood Hills

Santa Monica

Los Angeles

Deforming Rocks

It's hard to imagine, but rocks sometimes can bend under stress without breaking. When rocks are stressed at high temperatures and pressures, they can change shape permanently by folding. Scientists call this plastic deformation. Rocks are more likely to deform in a plastic way when stresses are applied to them slowly, or at high temperatures. Sometimes, rocks can snap back to their original shapes after stress is removed, which is called elastic deformation.

Maybe you've found a rock and you wanted to see what it looked like on the inside. If you placed stress on the rock by breaking it with a hammer, you caused the rock to deform in a brittle way. A break, or crack, in rock is called a **fracture.** In nature, if the rocks on one side of a fracture have moved relative to the rocks on the other side, the fracture is a **fault.**

Types of Stress

Three main types of stress can cause faulting. Rocks experience forces that can produce tension, compression, or shear stress. You can explore this in **Figure 2.** It is important to understand that combinations of these stresses are common in nature. Also, a particular type of stress can cause more than one type of fault.

Tension The top diagram in **Figure 2** shows layered rocks that are not deformed. When two plates are moving apart, tension pulls lithosphere apart so that it stretches and becomes thinner. This stretching and thinning is the deformation that results from tension stress.

Compression If rocks are squeezed, as shown in **Figure 2,** the stress is called compression. Where two lithospheric plates are forced together, compression makes the rocks thicker.

Shear When rocks slide horizontally in opposite directions, the stress is called shear. The lithosphere neither thins nor thickens as a result of shear stress.

Figure 2 Three main types of stress are tension, compression, and shear.

Demonstrate tension stress using a rubber band or putty.

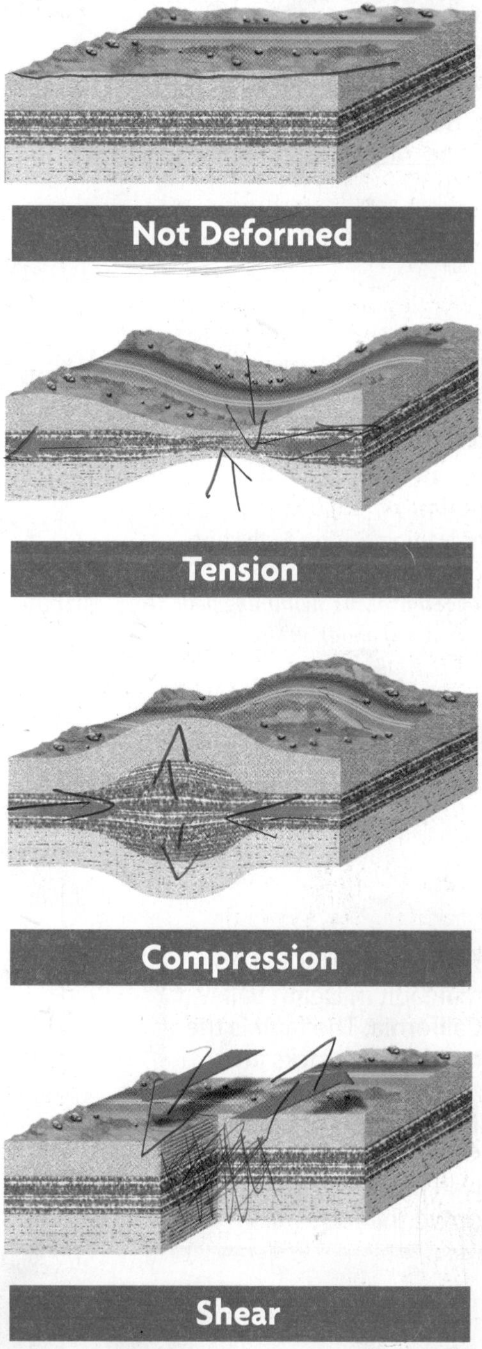

Not Deformed

Tension

Compression

Shear

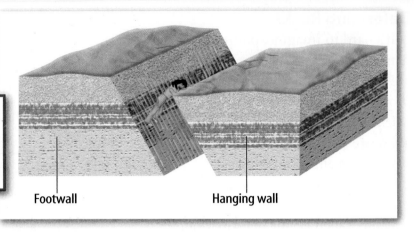

Figure 3 If a fault's surface is inclined, the block of rock above the fault is the hanging wall, and the block of rock below the fault is the footwall.

Footwall   Hanging wall

Types of Faults

Examining a fault helps scientists determine the stresses that caused it. Geologists measure the angle of the fault's surface and try to determine which way the broken sections of the rock have moved. They look for objects that were broken by the fault to determine which direction the rocks moved.

Figure 3 shows an **inclined** fault surface cutting across rocks. Imagine that the rocks were pulled apart at the fault's surface so you could fit between them. If you were to reach up, you could touch the hanging wall. The hanging wall is the block of rock that lies above the fault from which you would be able to hang. The block of rock that lies below the fault is called the footwall. You can imagine stepping on the footwall if the blocks of rock were separated.

 Figure 3 Compare and contrast a hanging wall and a footwall to the ceiling and the floor of your home.

Normal Faults The rocks along faults can move up, down, or sideways. Tension stresses inside Earth pull rock apart, producing normal faults. Normal faults slope at an angle, such as the fault in Death Valley, shown in **Figure 4.** When rock breaks and slips along a normal fault, the hanging wall moves down the footwall.

Figure 4 This is a normal fault in Death Valley, California. The fault is the break in the rocks just above the person's head. The labeled rock layers show how the hanging wall block has moved down the footwall.

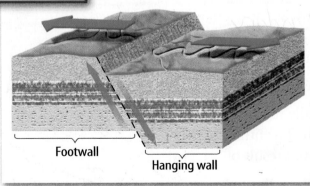

Footwall

Hanging wall

A
A

Footwall

Hanging Wall

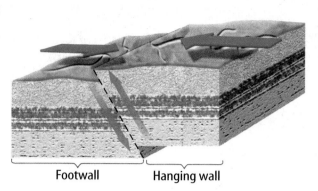

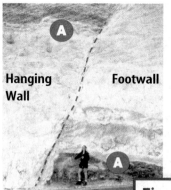

Hanging Wall   Footwall

A

A

Figure 5 Reverse faults look like normal faults, but their motions are different. The labeled rock layers show how the hanging wall block has moved up the footwall.

Explain how the footwall has moved relative to the hanging wall.

Reverse Faults In places where rocks are pushed together, compression stresses produce reverse faults. A reverse fault looks similar to a normal fault, but the blocks of rock move differently. As rocks are pushed together, the hanging wall moves up the footwall. **Figure 5** shows the direction of movement along a reverse fault in the Appalachian Mountains.

 How does the movement of rock along a reverse fault differ from the movement along a normal fault?

Strike-Slip Faults When plates slide horizontally past each other, shearing stresses produce strike-slip faults. Unlike normal and reverse faults, strike-slip faults often are vertical, not inclined. Instead of moving mostly up or down, the rocks slide past each other sideways, or horizontally.

In **Figure 6,** a small strike-slip fault has broken the asphalt on the road. Imagine standing on the white road lines on one side of this fault. The road lines that are across the fault from you have moved to your right compared to the lines you are standing on. Analyzing movement along a fault like this one helps scientists determine the stresses that caused the faulting.

Figure 6 A small strike-slip fault broke the asphalt on this road. The rocks did not move up or down, but slid past each other.

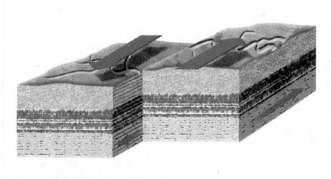

MiniLab

Try at Home

00:25 minutes

How can you model movement of a fault?

A fault is formed when rocks are deformed to the point of breaking, and movement occurs along the break. Scientists observe the movement of faults in nature. This allows them to determine the types of stresses that caused the faulting.

Procedure

1. Read and complete a lab safety form.
2. Cut a **shoe box lid** in half along its width.
3. Turn the **shoe box** over. The bottom of the box will represent the surface of Earth.
4. Use **scissors** to cut the shoe box in half along its width. Cut at an angle to model an inclined fault surface. Examine **Figures 4, 5,** and **6** for examples of how faults look in three dimensions.
5. **Tape** the two halves of the shoe box lid over the shoe box halves to make the fault slope.
6. Model fault movement for a normal, a reverse, and a strike-slip fault.
7. Challenge option: Use **poster paints** to paint rock layers on a side of the shoe box before it is cut to see how the layers move relative to each other.

Analysis

1. **Illustrate** fault movement for each scenario. Use arrows to show how the shoe box halves move relative to each other.
2. **Relate** a type of stress to each of the fault types that you modeled. Use arrows to indicate the directions of stress on your illustrations.

Science Content Standards

1.d Students know that earthquakes are sudden motions along breaks in the crust called faults and that volcanoes and fissures are locations where magma reaches the surface.
7.g Interpret events by sequence and time from natural phenomena (e.g., the relative ages of rocks and intrusions).

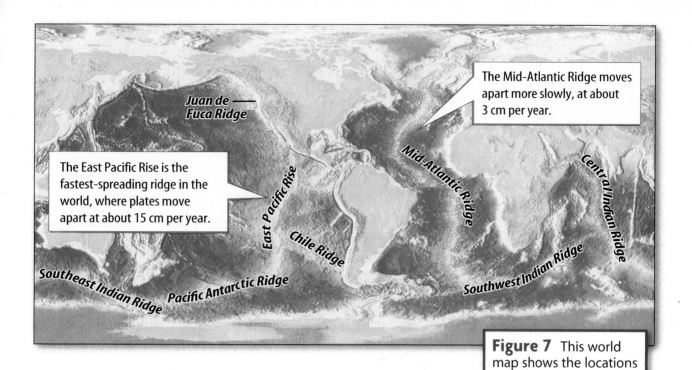

Juan de Fuca Ridge

The East Pacific Rise is the fastest-spreading ridge in the world, where plates move apart at about 15 cm per year.

The Mid-Atlantic Ridge moves apart more slowly, at about 3 cm per year.

East Pacific Rise

Mid-Atlantic Ridge

Central Indian Ridge

Chile Ridge

Southeast Indian Ridge

Pacific Antarctic Ridge

Southwest Indian Ridge

Figure 7 This world map shows the locations of mid-ocean ridges.
Locate a region where a mid-ocean ridge emerges above sea level.

Types of Plate Boundaries

The edges of Earth's plates meet at plate boundaries that extend deep into the lithosphere. Faults form along these boundaries. Like faults, there are three major types of plate boundaries. They are classified according to how rocks on either side of the plate boundary move. The kinds of geologic features that form depend on the type of plate boundary and the stresses generated along the boundary.

Divergent Plate Boundaries

When two lithospheric plates are moving, or pulling apart, it is called a **divergent plate boundary.** Mid-ocean ridges, shown in **Figure 7,** occur along divergent plate boundaries. As the two plates move apart, new seafloor forms. Tension stresses stretch and thin the lithosphere and cause earth-quakes to occur when rocks break and move.

Mid-Ocean Ridges Figure 8 shows an example of a mid-ocean ridge—the East Pacific Rise. Notice that the seafloor located farther from a divergent plate boundary is deeper underwater than the seafloor near the ridge. This is because as rock cools and contracts it becomes denser and moves away from the center of the ridge. It is difficult for scientists to study mid-ocean ridges because they are about 2 km below sea level.

Figure 8 Mid-ocean ridges, such as the East Pacific Rise, are topographic features on the seafloor.

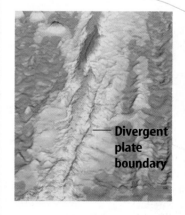

Divergent plate boundary

Continental Rifting Most divergent plate boundaries are located on the seafloor. But, divergent plate boundaries can also form on land when a continent is pulled apart. The process that pulls a continent apart is called **continental rifting.** This is shown in **Figure 9.**

Remember that divergent plate boundaries form where tension stresses cause the lithosphere to stretch and become thinner. Tension stresses in the lithosphere form normal faults. As the hanging wall blocks slip down, a long, flat, narrow **rift valley** forms.

 How does a rift valley form?

Sediment collects on the floor of the rift valley. Oceanic crust made of gabbro and basalt is formed, which is dense and causes the valley to sink. Eventually, ocean water flows into the valley. There are places on Earth today where scientists can directly observe continents rifting.

 Figure 9 Identify the numerous cracks and rift valley of the East African Rift.

Examples of Continental Rifting The East African Rift cuts across the eastern side of Africa for 5,600 km. It is a rift valley where crust is being pulled apart and large slabs of rock are sinking. This generates a rift zone, which is shown in **Figure 10.** If continental rifting continues, then East Africa will eventually part from West Africa.

The Gulf of California is also an example of continental rifting. The bottom of the rift valley has dropped low enough so that ocean water now fills it.

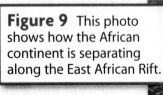

Figure 9 This photo shows how the African continent is separating along the East African Rift.

Visualizing Rift Valleys

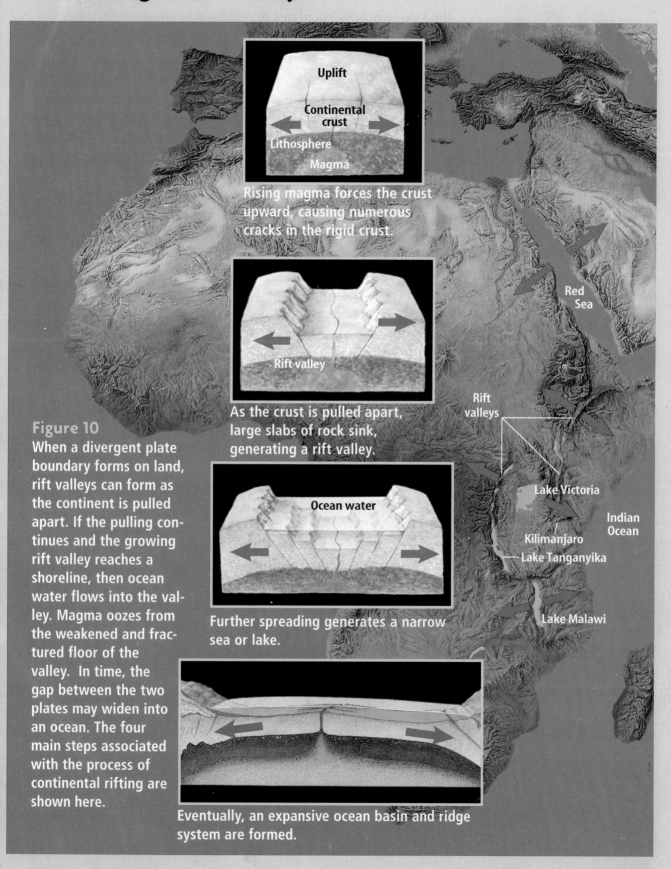

Uplift

Continental crust

Lithosphere

Magma

Rising magma forces the crust upward, causing numerous cracks in the rigid crust.

Rift valley

As the crust is pulled apart, large slabs of rock sink, generating a rift valley.

Ocean water

Further spreading generates a narrow sea or lake.

Eventually, an expansive ocean basin and ridge system are formed.

Red Sea

Rift valleys

Lake Victoria

Kilimanjaro

Lake Tanganyika

Indian Ocean

Lake Malawi

Figure 10
When a divergent plate boundary forms on land, rift valleys can form as the continent is pulled apart. If the pulling continues and the growing rift valley reaches a shoreline, then ocean water flows into the valley. Magma oozes from the weakened and fractured floor of the valley. In time, the gap between the two plates may widen into an ocean. The four main steps associated with the process of continental rifting are shown here.

Convergent Plate Boundaries

A **convergent plate boundary** is formed when two lithospheric plates move toward each other. Interactions between the two lithospheric plates depend upon whether the plates are composed of continental or oceanic lithosphere. There are three possible types of interactions. Both plates can be made of oceanic lithosphere, one plate can be oceanic and the other continental, or two plates with continental lithosphere can converge. In all cases, large earthquakes occur and new geologic features form.

Ocean-to-Ocean Where two oceanic plates move toward each other, one of the plates sinks beneath the other, as shown in the top diagram of **Table 1.** This process, in which one plate is forced down into the mantle beneath another plate, is called **subduction.** The density of the plate determines which plate subducts. Generally, the colder, older, denser slab is forced down into the mantle, forming a deep ocean trench on the seafloor where it bends.

 What forms on the seafloor where a slab bends?

As it sinks deeper into the mantle, high temperatures and pressures release water from minerals in the slab. Where this water rises up into the mantle, it causes mantle rocks to melt. This forms a supply of magma for volcanic eruptions, producing a curved line of volcanoes in the overlying plate.

Ocean-to-Continent If one of the converging plates is oceanic and the other is continental, the oceanic plate always subducts. Most continental rocks are less dense than oceanic rocks. In **Table 1,** the second diagram shows the oceanic plate subducting underneath the continental plate. The melting that results forms a curved string of volcanoes along the leading edge of the overlying continental plate.

Continent-to-Continent Which plate subducts when two plates made from continental rocks collide? Rocks, like granite and shale, aren't dense enough to sink into the mantle, so neither continental plate subducts. Instead, compression stresses force crust to rise up, thicken, and shorten. You can imagine this shortening if you try pushing a stack of paper together from two opposite ends. In nature, the leading edges of colliding continents are uplifted, forming tall mountains.

WORD ORIGIN

boundary
from Latin *boudina;* means
boundary, boundary marker

Table 1 Types of Convergent Plate Boundaries

Concepts In Motion

Interactive Table Organize information about convergent plate boundaries at ca6.msscience.com.

Boundary Type	Description
Ocean-to-Ocean • Older, colder oceanic plate subducts. • Magma forms in mantle above subducted slab. • Curved line of volcanoes forms as magma makes its way to Earth's surface. • example: Marianas Islands	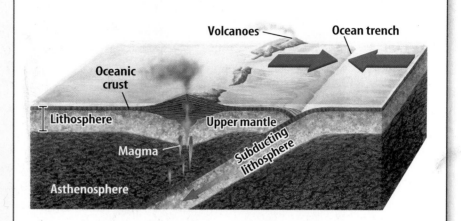
Ocean-to-Continent • Oceanic plate subducts beneath continental plate. • Magma forms in mantle above subducted slab. • A string of volcanoes forms along the leading edge of the continent. • example: Cascade Range	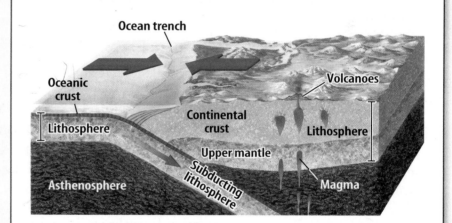
Continent-to-Continent • Neither continental plate subducts. • Continental crust rises up, thickens, and shortens due to compression stresses. • Uplift of lithosphere forms tall mountains. • example: Himalayas	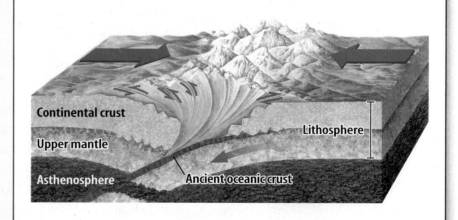

Lesson 1 • Interactions at Plate Boundaries **219**

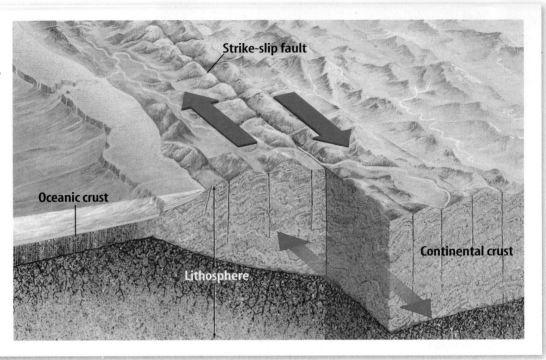

Figure 11 At a transform plate boundary, motion is mostly parallel to Earth's surface.

Labels on figure: Strike-slip fault, Oceanic crust, Lithosphere, Continental crust

SCIENCE USE V. COMMON USE

fault

Science Use a fracture where rocks on one side have moved relative to the rocks on the other side. *The movement along the fault produced a large earthquake.*

Common Use error or mistake. *The accident was the driver's fault.*

Transform Plate Boundaries

Where two plates slide horizontally sideways past one another, a **transform plate boundary** exists. Lithosphere is neither formed nor recycled at these boundaries. **Figure 11** shows how a transform plate boundary is similar to a huge strike-slip fault. When the plates slide sideways past each other and eventually slip, rocks break, causing earthquakes.

Oceanic Oceanic transform plate boundaries connect pieces or segments of the mid-ocean ridges. The ridges are not completely straight but made of many shorter pieces. Most oceanic transform boundaries are relatively short. But, there are a few long transform boundaries on Earth. These are located on the continents.

 Reading Check What connects segments of the mid-ocean ridges?

Continental Some transform plate boundaries slice through continental lithosphere as huge strike-slip faults. Earthquakes resulting from movement along these faults can be very destructive if they occur in populated areas.

The San Andreas Fault in California is the best-studied continental transform plate boundary in the world. Most of California lies on the North American Plate. But, a small portion of California, west of the San Andreas Fault, lies on the Pacific Plate. The continental transform plate boundary separates these two plates.

Deformation and Plate Boundaries

You've read how stresses can cause rocks to deform. Deformation leaves clues for scientists to use when unraveling Earth's complicated history such as fractures and faults. Analyzing how rocks bend and break helps scientists determine the types of stresses exerted on them. From these data, the direction and distance that the lithospheric plates have moved and the way they have interacted at plate boundaries can be determined.

LESSON 1 Review

Summarize

Create your own lesson summary as you design a **visual aid.**

1. **Write** the lesson title, number, and page numbers at the top of your poster.

2. **Scan** the lesson to find the **red** main headings. Organize these headings on your poster, leaving space between each.

3. **Design** an information box beneath each **red** heading. In the box, list 2–3 details, key terms, and definitions from each **blue** subheading.

4. **Illustrate** your poster with diagrams of important structures or processes next to each information box.

 ELA6: R 2.4

Using Vocabulary

1. Distinguish between *fracture* and *fault*. **1.d**

2. In your own words, write a definition for *rift valley*. **1.e**

Understanding Main Ideas

3. Which of the following best describes the direction of movement of rock along a strike-slip fault? **1.d**

   A. downhill
   B. uphill
   C. sideways
   D. vertical

4. **List** three different types of faults. **1.d**

5. **Classify** three types of convergent plate boundaries based on the densities of the plates involved. **1.c**

6. **Relate** the three major types of plate boundaries to types of stress expected at these boundaries. **1.e**

Standards Check

7. **Sequence** Draw a diagram to show the stages of continental rifting. **1.e**

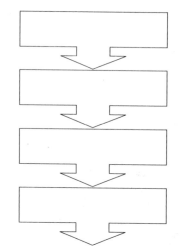

Applying Science

8. **Predict** whether Iceland will become larger or smaller in the future. Defend your prediction. **1.e**

9. **Decide** if you would rather live close to a plate boundary or far away from one. Explain your decision. **1.e**

Science **online**

For more practice, visit **Standards Check** at ca6.msscience.com.

 Mountain Types ca6.msscience.com

Applying Math

Speed of Lithospheric Plates

Earth's lithospheric plates move at rates of centimeters per year. Plate movement can be measured in an absolute speed that is approximated in the table.

Approximate Speed of Lithospheric Plates

Name of Plate	Approximate Speed (cm/yr)
Antarctic	2.05
African	2.15
Arabian	4.65
Caribbean	2.45
Cocos	8.55
Eurasian	0.95
Indian-Australian	6.00
Nazca	7.55
North American	1.15
Pacific	8.10
Philippine	6.35
South American	1.45

How much faster is the Caribbean Plate moving than the Antarctic Plate?

What you know:
- Caribbean Plate velocity: 2.45 cm/yr
- Antarctic Plate velocity: 2.05 cm/yr

What you need to find:
- How much faster is the Caribbean Plate moving than the Antarctic Plate?

Divide the faster speed by the slower speed to find how many times faster:

1 $\frac{2.45}{2.05}$

2 approximately 1.20 times faster

The Caribbean Plate is moving at a rate of about 1.20 times faster than the Antarctic Plate.

Practice Problems

1. Approximately how much faster is the Pacific Plate moving than the Eurasian Plate?

2. Approximately how many more centimeters per year does the Indian-Australian Plate move than the North American Plate?

Science Online
For help visit
ca6.msscience.com.

Science Content Standards

1.e Students know major geologic events, such as earthquakes, volcanic eruptions, and mountain building, result from plate motions.

1.f Students know how to explain major features of California geology (including mountains, faults, volcanoes) in terms of plate tectonics.

7.a Develop a hypothesis.

7.b Select and use appropriate tools and technology (including calculators, computers, balances, spring scales, microscopes, and binoculars) to perform tests, collect data, and display data.

Also covers: 7.e

Reading Guide

What *You'll Learn*

▶ **Describe** California's current plate tectonic setting and how plate movements have produced landforms.

▶ **Predict** future changes in California's tectonic setting and topography.

Why *It's Important*

Interactions between the North American Plate and the Pacific Plate produce California's mountains and basins and cause earthquakes and volcanoes.

Vocabulary

San Andreas Fault

Review Vocabulary

uplift: any process that moves Earth's surface to a higher elevation (p. 79)

California Geology

Main Idea Many of California's landforms were produced by plate tectonic activity, which continues today.

Real-World Reading Connection Have you ever gone back to a former home or an old favorite spot—one that you haven't visited for a long time? Had things changed from the way you remembered them? Similarly, California geology has changed much over millions of years.

Plate Tectonics in California

As shown in **Figure 12,** California's landscape is different now compared to how it was in the past. Movements of plates have changed California dramatically. Seas have disappeared. While old mountains have eroded away, new mountains have been uplifted. Plate boundaries have formed and then disappeared. Today, a continental transform plate boundary cuts across the state. At the northern end of California, a convergent plate boundary sits offshore. This active plate tectonic setting produces earthquakes, volcanoes, and mountains.

Figure 12 This is what scientists think California might have been like 80 million years ago, a time when dinosaurs roamed Earth. Many parts of the state were covered by shallow seas. Tall mountains stood where valleys now exist.

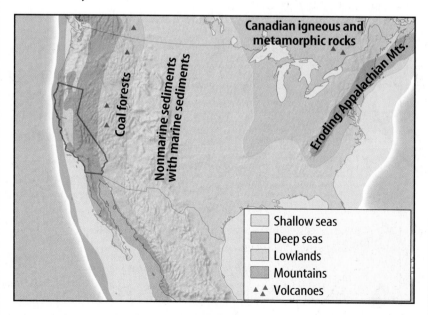

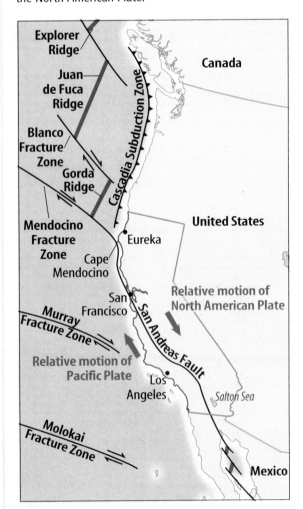

ACADEMIC VOCABULARY
adjacent (ad JAY sunt)
(*adjective*) not distant, nearby
The city and adjacent suburbs were placed under a tornado warning.

Transform Plate Boundary

Most of California is situated on the North American Plate. A small part of California, west of the San Andreas Fault, lies on the **adjacent** Pacific Plate. The Pacific Plate moves northwest, relative to the North American Plate, at a velocity of about 3.4 cm per year. The plate does not slide smoothly, but sticks much of the time and moves in jerks. Each time a jerky movement occurs, an earthquake happens.

The **San Andreas Fault** is a transform plate boundary that is located between the North American Plate and the Pacific Plate. Because the San Andreas Fault is a transform plate boundary, it is also a strike-slip fault. **Figure 13** shows how the San Andreas Fault extends all the way from Cape Mendocino in the north to the Salton Sea in the south.

Figure 13 also shows that the San Andreas Fault is not a straight line. Where there are bends in the fault, blocks of rock get pushed up or drop down, making mountains or basins. The Transverse Ranges and the Coastal Ranges of California have been pushed up as the Pacific Plate moves past the North American Plate. The Los Angeles Basin, the Ventura Basin, and the San Francisco Bay are all blocks of rock that have dropped down.

Convergent Plate Boundary

Just offshore of Northern California, there are two small oceanic lithospheric plates. **Figure 13** shows where these plates, known as the Gorda and Juan de Fuca, are subducted beneath the coast at the Cascadia Subduction Zone. This subduction forms a convergent plate boundary. Melting above this subduction zone produces the volcanic mountains of the Cascade Range.

Figure 13 The San Andreas Fault extends out to sea north of Cape Mendocino. There it becomes an oceanic transform plate boundary.
Describe the relative motion of the Pacific Plate to the North American Plate.

Explorer Ridge

Juan de Fuca Ridge

Blanco Fracture Zone

Gorda Ridge

Cascadia Subduction Zone

Canada

United States

Mendocino Fracture Zone

Cape Mendocino

Eureka

San Francisco

Relative motion of North American Plate

Murray Fracture Zone

Relative motion of Pacific Plate

San Andreas Fault

Los Angeles

Salton Sea

Molokai Fracture Zone

Mexico

California's Mountains

California's mountains often formed from interactions at several plate boundaries. For example, as rocks on one side of a transform plate boundary grind and push against the rocks on the other plate, mountains, such as the Transverse Ranges, can form.

Subduction

California's convergent plate boundaries, both in the past and present, have been important in forming California's mountains. Granitic rocks form under volcanic mountains where plates converge. And, during mountain building, compressive stresses and heat produce metamorphic rocks. The Klamath Mountains, Coastal Ranges, Peninsular Ranges, and Sierra Nevada all contain igneous and metamorphic rocks that formed far below the surface.

 **Reading Check** List two rock types that form far below Earth's surface.

In continental lithosphere above the Cascadia Subduction Zone in northern California, granitic rocks are forming deep in the crust. At the same time, volcanic activity produces the Cascade Range on the surface. Both Lassen Peak and Mount Shasta are active volcanoes in this mountain range.

Rifting

There are even some mountains in California that have formed because of tension stresses. The Panamint Range just west of Death Valley is rising up as the crust in eastern California stretches. Locate all these mountain ranges on **Figure 14.** Mountains help scientists understand processes that are part of California's rich tectonic history.

Figure 14 Interactions at several plate boundaries have formed mountains ranges in California.

Locate a mountain that formed as a result of subduction.

Future Plate Movement

How will California's landscape change in the future? Scientists can estimate the directions and speeds that the plates are moving. They can predict future plate boundary interactions. For example, the small part of California that is on the Pacific Plate, including Los Angeles, will continue moving northwest along the coast relative to the North American Plate. This means that Los Angeles and San Francisco are approaching each other about as fast as your fingernail grows.

LESSON 2 Review

Summarize

Create your own lesson summary as you design a **study web**.

1. **Write** the lesson title, number, and page numbers at the top of a sheet of paper.

2. **Scan** the lesson to find the **red** main headings.

3. **Organize** these headings clockwise on branches around the lesson title.

4. **Review** the information under each **red** heading to design a branch for each **blue** subheading.

5. **List** 2–3 details, key terms, and definitions from each **blue** subheading on branches extending from the main heading branches.

 ELA6: R 2.4

Standards Check

Using Vocabulary

1. The _____ is located between the North American Plate and the Pacific Plate. **1.f**

Understanding Main Ideas

2. In California, which direction is the Pacific Plate moving relative to the North American Plate? **1.f**
   A. east
   B. northeast
   C. northwest
   D. west

3. **Explain** how uplift can occur at all types of plate boundaries. **1.e**

4. **Identify** a California mountain range that formed as a result of tension stress. **1.f**

5. **Relate** the San Andreas Fault to a main type of stress. **1.f**

6. **Determine Cause and Effect** Draw a diagram like the one below. List some effects of subduction. **1.e**

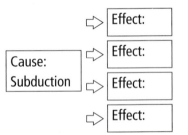

Applying Science

7. **Predict** what type of plate boundary will exist in northern California after the Juan de Fuca and Gorda Plates are completely subducted. Hint: Examine **Figure 13**. **1.f**

Science **n**line

For more practice, visit **Standards Check** at ca6.msscience.com.

DataLab

How do landforms define plate boundaries?

Earth's surface is made up of interacting plates. What landforms occur near plate boundaries? How do these landforms indicate where a plate boundary is?

Data Collection

1. Examine a topographic or raised relief map of the world.

2. Obtain a blank world map from your teacher.

3. Locate some of Earth's mountain ranges on your map. Color them purple.

4. Locate some major fault zones that occur on Earth's surface. Color them red.

5. Locate some areas of recent volcanic activity. Color them orange.

Data Analysis

1. **Compare** your locations of mountain ranges and volcanic activity to a plate boundary map.

2. **Determine** the approximate widths of mountain ranges and fault zones you investigated.

3. **Relate** your answers to questions 1 and 2 to the role of plate boundaries in shaping Earth's topography.

4. **Explain** the existence of some mountain ranges today that aren't located near a plate boundary.

 Science Content Standards

1.e Students know major geologic events, such as earthquakes, volcanic eruptions, and mountain building, result from plate motions.
7.e Recognize whether evidence is consistent with a proposed explanation.

MA6: MR 2.1, MR 2.3

Use the Internet:
Earthquake Depths and Plate Boundaries

Materials

world map with latitude and longitude lines

plate boundary map

graph paper

computer with internet access

Science Content Standards

1.e Students know major geologic events, such as earthquakes, volcanic eruptions, and mountain building, result from plate motions.

7.a Develop a hypothesis.

7.b Select and use appropriate tools and technology (including calculators, computers, balances, spring scales, microscopes, and binoculars) to perform tests, collect data, and display data.

7.e Recognize whether evidence is consistent with a proposed explanation.

Problem

Not all earthquakes occur at the same depth in the lithosphere. Use your knowledge of types of plate boundaries to determine a general relationship between plate boundaries and the depths of earthquakes.

Form a Hypothesis

➤ **Describe** the types of stress that cause earthquakes.

➤ **Review** the basic structure of divergent boundaries, transform boundaries, and convergent boundaries. **Hint:** Refer back to **Figure 10, Figure 11,** and **Table 1.**

➤ Earthquakes range in depth from zero, at Earth's surface, to about 700 kilometers deep. The shallow range is from 0 km to about 70 km deep, intermediate earthquakes are about 70 km to 300 km deep, and deep earthquakes occur from about 300 to 700 km below the surface.

➤ **Make a statement** about which boundaries you think will have shallow, intermediate, or deep earthquakes associated with them.

Collect Data and Make Observations

1. Visit ca6.msscience.com to research data on recent earthquakes.
2. Make a data table like the one shown on page 229.
3. Record latitude, longitude, depth (km), and a location for each earthquake. Leave the plate boundary column blank for now.

Earthquake Locations and Depths				
Latitude	Longitude	Depth (km)	Location	Plate Boundary Type

Analyze and Conclude

1. **Compare** the earthquake locations you plotted with the plate boundary map below.
2. **Specify**, in the table, the type of plate boundary associated with each earthquake or whether an earthquake was not associated with any plate boundary.
3. **Determine** which plate boundaries experience the deepest earthquakes.

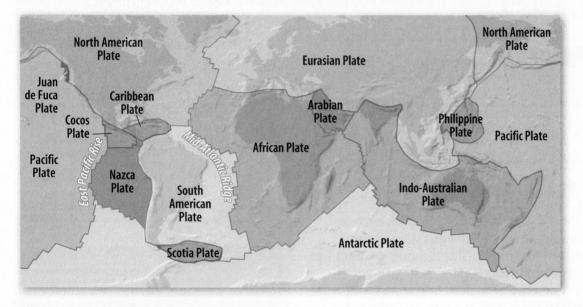

Communicate

WRITING in Science   ELA6: W 1.3

Compare and Contrast Write a paragraph that compares and contrasts the depths of earthquakes that you would expect to occur at transform and at ocean-to-continent convergent plate boundaries. Explain any differences in depth ranges for these types of plate boundaries.

Real World Science

Science
& Career

Studying Earth's Plate Boundaries

As a geophysicist and professor at Stanford University, Dr. Paul Segall is interested in Earth processes. Segall's students benefit from his work with the Plate Boundary Observatory (PBO) project. In this project, Segall and other scientists use data from tools called strainmeters as well as Global Positioning System (GPS) to understand the relationship between plate movement and deformation along plate boundaries. An example of a strainmeter is shown in the photo.

Understanding the Big Picture Visit Careers at **ca6.msscience.com** to record information about the rationale, anticipated results, employment opportunities, and benefits of the EarthScope program. **Prepare** an oral report justifying support for this program which Dr. Segall might present to a funding committee.

ELA6: W 2.5

Finding Faults

Science & Technology

The Quaternary Fault and Fold Database is now available to structural engineers, state disaster planners, and the public. This technology identifies areas in the continental United States where the crust has fractured or folded. The database provides historical details as well as recent fault data.

Faults in Your Region of California Visit Technology at <u>ca6.msscience.com</u> to access the Quaternary Fault and Fold Database. Use the map-based function to find faults nearest your area. **Create** a chart showing available data for these faults. **Analyze** the data to identify the most active fault or faults in your area.

The View from Space

The San Andreas Fault is among California's best-known landforms. Measuring roughly 1,200 km, this strike-slip fault is the longest fault in California. Data from the Shuttle Radar Topography Mission on the Space Shuttle *Endeavor* were used to create this image, which shows the San Andreas Fault along the center, and the North American and Pacific Plates on the right and left.

Online Geologic Fieldtrip Visit **History** at **ca6.msscience.com** to take a geotour with the California Geological Survey. Choose a geological landmark within one of the major geomorphic provinces to explore. **Create** a brochure that a local tourism department might use to advertise this landmark. ELA6: W 1.5

A Shaky History

Parkfield, California sits directly on the San Andreas Fault. Strong earthquakes have occurred regularly in this tiny community. How do Parkfield residents, who number less than 40, deal with living on one of the world's most famous faults? Students have school earthquake drills, and residents are taught to stockpile food and keep fresh batteries in flashlights. Perhaps most importantly, they learn not to panic during emergencies.

Analyzing Seismic Attitudes In a small group, brainstorm two questions that measure the attitude of California residents about living in a seismically active state. **Develop** a brief class survey. Have 5–10 friends or family members complete the survey, then analyze class results.

The BIG Idea — California is located on a plate boundary, where major geologic events occur.

Lesson 1 Interactions at Plate Boundaries

1.c, 1.d, 1.e, 7.e

Main Idea There are three main types of plate boundaries, where stresses cause rocks to deform.

- Stress can break and bend rocks.

- There are three main types of stress that can result in faulting.

- Lithospheric plates move apart at divergent boundaries.

- Divergent boundaries produce mid-ocean ridges and continental rifts.

- Plates move toward each other at convergent boundaries.

- The densities of converging plates determine what type of convergent boundary forms.

- Plates grind past each other at transform boundaries, with the motion mainly sideways, parallel to Earth's surface.

- Most transform boundaries are on the seafloor, but some are on continents.

- **continental rifting** (p. 216)
- **convergent plate boundary** (p. 218)
- **divergent plate boundary** (p. 215)
- **fault** (p. 211)
- **fracture** (p. 211)
- **rift valley** (p. 216)
- **subduction** (p. 218)
- **transform plate boundary** (p. 220)

Lesson 2 California Geology

1.e, 1.f, 7.a, 7.b, 7.e

Main Idea Many of California's landforms were produced by plate tectonic activity, which continues today.

- The San Andreas Fault is a continental transform boundary that runs through California, from Cape Mendocino to the Salton Sea.

- Northern California is located above a subduction zone.

- California's mountains were produced by interactions at plate boundaries.

- Mountains form at all three types of plate boundaries.

- The western sliver of California that is on the Pacific Plate will continue to move northwest, relative to the rest of the North American Plate.

- **San Andreas Fault** (p. 224)

STUDY TO GO Download quizzes, key terms, and flash cards from ca6.msscience.com.

Linking Vocabulary and Main Ideas

Use vocabulary terms from page 232 to complete this concept map.

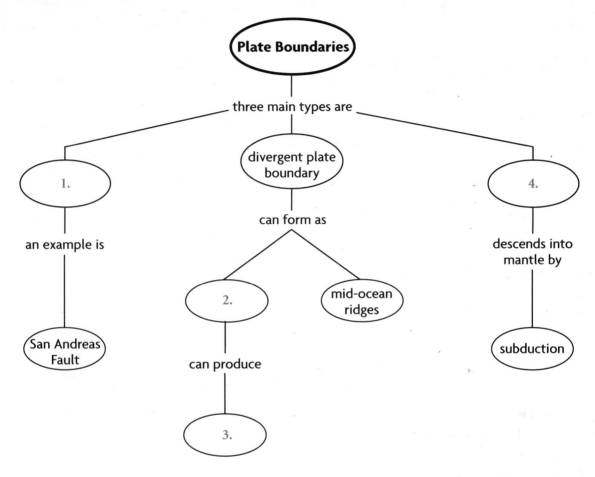

Using Vocabulary

**Fill in the blank with the correct vocabulary words.
Then read the paragraph to a partner.**

The Mariana ocean trench in the Pacific Ocean is an area of
deep ocean water that results from _____5._____. A curved string
of volcanic islands also forms along this ocean-to-ocean
_____6._____. Volcanic eruptions also are common at ocean-to-
continent _____7._____ and _____8._____ boundaries.

Science Online

Visit ca6.msscience.com for:
▶ Vocabulary PuzzleMaker
▶ Vocabulary eFlashcards
▶ Multilingual Glossary

Understanding Main Ideas

Choose the word or phrase that best answers the question.

1. Which form the highest mountains?
   A. mid-ocean ridges
   B. continent-to-continent convergent boundaries
   C. ocean-to-continent convergent boundaries
   D. transform boundaries **1.e**

2. What type of stress is acting on rocks in Death Valley?
   A. tension
   B. compression
   C. convection
   D. shear **1.f**

3. Along which type of plate boundary did the transverse ranges form?
   A. divergent
   B. ocean-to-ocean convergent
   C. continent-to-continent convergent
   D. transform **1.f**

4. The diagram below illustrates a subduction zone.

   Why do volcanoes form above subduction zones?
   A. Tension stresses bring magma to the surface.
   B. Convection currents heat the slab.
   C. Water from the slab causes mantle rocks to melt.
   D. The slab melts when it goes into the mantle. **1.e**

Use the figure below to answer questions 5 and 6.

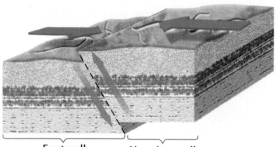

Footwall   Hanging wall

5. What type of fault is shown in this illustration?
   A. normal
   B. reverse
   C. convergent
   D. strike-slip **1.d**

6. In which direction did the hanging wall move relative to the footwall?
   A. down
   B. sideways
   C. up
   D. vertically **1.d**

7. At the Mid-Atlantic Ridge, plates are moving apart at about what rate?
   A. 0.03 cm/year
   B. 0.3 cm/year
   C. 3 cm/year
   D. 30 cm/year **1.c**

8. Where is continental rifting occurring today?
   A. east Africa
   B. west coast of South America
   C. the Himalayas
   D. the Aleutian Islands **1.e**

9. What type of plate boundary occurs along the Mid-Atlantic Ridge?
   A. divergent
   B. continent-to-continent convergent
   C. ocean-to-ocean convergent
   D. transform **1.e**

Science online   Standards Review ca6.msscience.com

Applying Science

10. **Compare and contrast** the types of rocks formed at divergent boundaries to those formed at convergent boundaries. **1.e**

11. **Construct** a diagram showing where different rock types form at a convergent boundary. **1.e**

12. **Imagine** a new continent made from two existing continents. Describe how the new continent forms. **1.e**

13. **Evaluate** the suggestion that people should dispose of hazardous waste by dropping it down oceanic trenches and letting it sink into the mantle with the subducting slab. **1.e**

14. **Illustrate** how chains of volcanoes and deep ocean trenches relate to subduction zones. **1.e**

15. **Identify** the type of stress that is illustrated in the diagram below. **1.e**

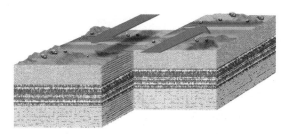

WRITING in Science

16. **Select** a park in California that includes landforms shaped by the interaction of lithospheric plates. Research the park. Prepare a travel brochure that describes how plate boundaries influenced the landforms that a visitor to the park might see. **ELA6: W 1.5**

Cumulative Review

17. **Evaluate** the benefit of a profile view. **7.f**

18. **Compare and contrast** the two types of Earth's crust. **1.b**

19. **Describe** three methods of thermal energy transfer. **3.c**

Applying Math

Use the table below to answer questions 20–24.

Approximate Speed of Lithospheric Plates	
Name of Plate	**Approximate Speed (cm/yr)**
Antarctic	2.05
African	2.15
Arabian	4.65
Caribbean	2.45
Cocos	8.55
Eurasian	0.95
Indian-Australian	6.00
Nazca	7.55
North American	1.15
Pacific	8.10
Philippine	6.35
South American	1.45

20. How much faster is the North American Plate moving than the Eurasian Plate? **MA6: NS 1.2, AF 2.3**

21. Approximately how many more centimeters per year does the North American Plate move than the Eurasian Plate? **MA6: NS 1.2, AF 2.3**

22. How much faster is the Cocos Plate moving than the Nazca Plate? **MA6: NS 1.2, AF 2.3**

23. Approximately how many more centimeters per year does the Cocos Plate move than the Nazca Plate? **MA6: NS 1.2, AF 2.3**

24. How much faster is the Philippine Plate moving than the Arabian Plate? **MA6: NS 1.2, AF 2.3**

1 The illustration below shows two plates moving past each other.

Which type of stress occurs when Earth's lithospheric plates scrape past each other?

A compression `1.d`

B isostasy

C shear

D tension

2 About how fast do plates move?

A a few millimeters per year `1.c`

B a few centimeters per year

C a few meters per year

D a few kilometers per year

3 Earth's lithospheric plates scrape past each other at

A convergent plate boundaries. `1.e`

B divergent plate boundaries.

C transform plate boundaries.

D subduction zones.

4 Which best describes a fault?

A the point where two plates of lithosphere move apart and new seafloor forms `1.d`

B a point inside Earth where movement first occurs during an earthquake

C the surface of a break in a rock along which there is movement

D the snapping back of a rock that has been strained by force

5 In which direction is the Pacific Plate moving?

A northwest `1.f`

B northeast

C southwest

D southeast

6 The illustration below shows a fault in Earth's crust.

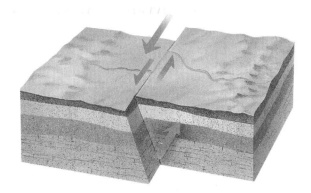

What type of fault is shown?

A normal fault `1.d`

B reverse fault

C strike-slip fault

D thrust fault

7 The table below shows the number of convergent and divergent boundaries for different plates in Earth's crust.

Plate Boundaries		
Plate	Number of Convergent Boundaries	Number of Divergent Boundaries
African	1	4
Antarctic	1	2
Indo-Australian	4	2
Eurasian	4	1
North American	2	1
Pacific	6	2
South American	2	1

Which plate has the most spreading boundaries?

A African   `1.c`

B Antarctic

C Indo-Australian

D Pacific

8 Between the North American Plate and the Pacific Plate lies the San Andreas Fault, which forms a

A continental rift.   `1.f`

B divergent plate boundary.

C mid-ocean ridge.

D transform plate boundary.

9 Which has played an important role in forming California's mountains?

A continental rifting   `1.f`

B a convergent plate boundary

C a divergent plate boundary

D a mid-ocean ridge

10 The map below shows the major plates near South America, their direction of movement, and the type of boundary between them.

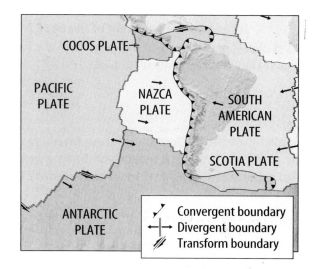

What boundary feature most likely occurs along the Nazca and South American Plates?

A a strike-slip fault   `1.e`

B new oceanic crust

C rift valleys

D volcanoes

11 Which type of fault is produced by tension stresses?

A inclined fault   `1.d`

B normal fault

C reverse fault

D strike-slip fault

Reading on Your Own...

Are you interested in learning more about Earth's structure, its geological features, and the forces that created them? If so, check out these great books.

Nonfiction

Dive to the Deep Ocean, by Deborah Kovacs, a marine scientist, explores both the organisms and the geologic features of the deep ocean. This book provides accurate information about tectonic movement, volcanic action, and undersea technology. *The content of this book is related to* Science Standard 6.1.

Nonfiction

Shaping the Earth, by Dorothy Hinshaw Patent, features full-color photographs highlighting the geological features on Earth's surface. The book explains the forces that created these features. *The content of this book is related to* Science Standard 6.1.

Narrative Nonfiction

The Pebble in My Pocket: A History of Our Earth, by Meredith Hooper, follows a pebble beginning with the cooling of lava from an ancient volcano. The book follows the changes in the formation and development of life on Earth and includes a time line of Earth's history. *The content of this book is related to* Science Standard 6.2.

Narrative Nonfiction

Earth's Fiery Fury, by Sandra Downs, describes the volcanic and geothermal activity of Earth and the features associated with thermal energy. This book helps the reader understand how thermal energy and Earth's inner fire shape Earth. *The content of this book is related to* Science Standard 6.3.

Choose the word or phrase that best answers the question.

1. Which features are evidence that many continents were once near Earth's south pole?
   A. glacial deposits
   B. earthquakes
   C. polar ice caps
   D. mid-ocean ridges   **1.a**

2. What hypothesis states that continents slowly moved to their present positions on Earth?
   A. subduction
   B. erosion
   C. continental drift
   D. seafloor spreading   **1.c**

3. The numbers on the contour map represent meters above sea level.

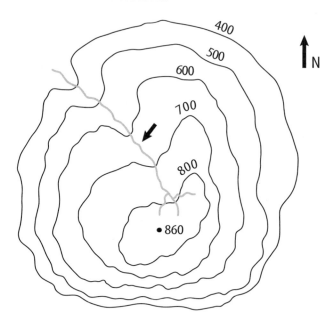

Which side of the feature has the steepest slope?
   A. north side
   B. east side
   C. west side
   D. south side   **7.f**

Write your responses on a sheet of paper.

4. **Compare and contrast** rocks and minerals.   **1.b**

Use the map below to answer questions 5 and 6.

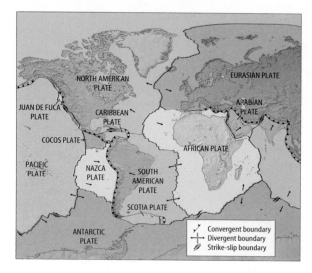

5. **Analyze** why many earthquakes, but only a few volcanic eruptions, occur in the Himalayas.   **1.e**

6. **Explain** the action of the plates along the San Andreas Fault and why volcanoes do not form there.   **1.e**

7. **Analyze** why the fossil of an ocean fish found on two different continents would not be good evidence of continental drift.   **1.a**

8. **Infer** A winter jacket is lined with insulating material that contains air spaces. How do the insulating properties of the jacket change when the insulating material becomes wet?   **3.c**

9. **Apply** When might you use a topographic map instead of a geologic map?   **7.f**

10. **Design an Experiment** Some colors of clothing absorb heat better than other colors. Design an experiment that will test various colors by placing them in the Sun for a period of time. Explain your results.   **3.d**

Shaping Earth's Surface

Captured in Time This plaster cast was made by archaeologists from remains left in Herculaneum by the A.D. 79 Vesuvius eruption.

West-Coast EVENTS

450,000–9,200 Years Ago Mount Shasta volcanic cones are active.

27,000 Years Ago Lassen Peak forms from eruptions.

April 18, 1906 San Francisco earthquake, the largest in America's history, measures 8.3 on the Richter scale.

May 1915 Lassen Peak erupts over several days.

A.D. 1   1800   1900   1920

WORLD EVENTS

August 79 Vesuvius erupts in Italy, burying the towns of Pompeii and Herculaneum.

October 1737 Largest tsunami in recorded history measures 64 m (210 ft.) above sea level.

August 1883 Krakatoa erupts in Indonesia, triggering a tsunami and sending ash 27 km in the air.

August 1914 World War I begins.

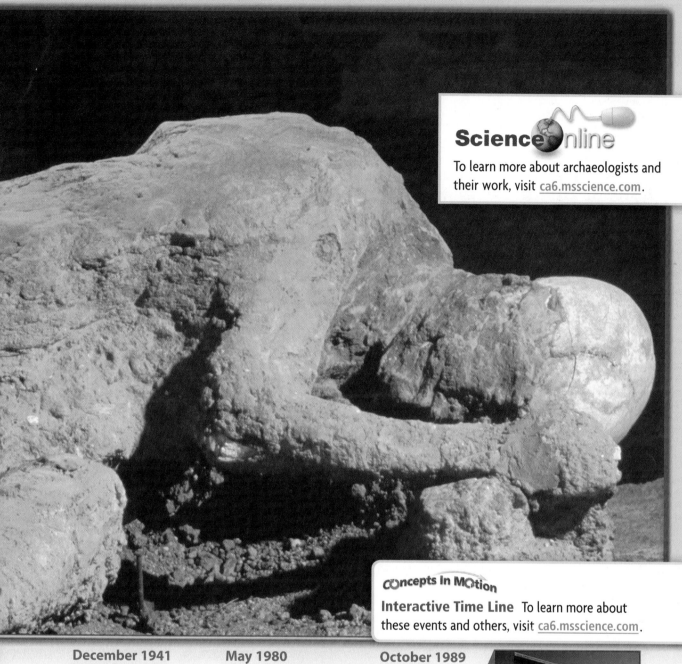

Science Online

To learn more about archaeologists and their work, visit ca6.msscience.com.

Concepts In Motion

Interactive Time Line To learn more about these events and others, visit ca6.msscience.com.

December 1941
Pearl Harbor, on the Hawaiian island of Honolulu, is attacked, bringing the U.S. into World War II.

May 1980
Mt. St. Helens erupts in the state of Washington, triggered by an underground earthquake measuring 5.1 on the Richter scale.

October 1989
Loma Prieta earthquake hits San Francisco, measuring 7.1 on the Richter scale.

1940   **1960**   **1980**   **2000**   **2020**

May 1960
World's biggest earthquake in recorded history—measuring 9.5 on the Richter scale—hits Chile and triggers tsunamis that reach Hawaii.

July 1976
The most devastating earthquake in modern times hits China. Measuring 7.5 on the Richter scale, it claims 240,000 lives.

December 2004
An earthquake measuring 8.9 on the Richter scale causes a tsunami that kills hundreds of thousands in Asia.

Earthquakes

Now how did *that* happen?

On January 17, 1995, at 5:46 A.M., t people of Kobe, Japan, awoke to a major earthquake that toppled building highways, and homes. The Kobe earthquake, also known as the Great Hans earthquake, killed 6,433 people and injured 43,792.

Science *Journal* Have you ever experienced an earthquake? If so, write paragraph about the event. If not, write how you imagine it would feel to experience an earthquake.

Launch Lab

00:10 minutes

Rocks Stretch

When stressed, rock can stretch until it fractures and breaks apart. Can you model the strength of rocks?

Procedure

1. Complete a lab safety form.

2. Lay a **rubber band** in front of you.

3. Mold **modeling clay** into a worm shape.

4. Mold the clay into a spiral shape around half of the rubber band.

5. Put your fingers in the loops of the rubber band and pull gently.

Think About This

- **Describe** what happened to the clay as you pulled and stopped pulling the rubber band.

- **Relate** your observations to how materials near plate boundaries deform in response to stress.

 1.d, 7.e

Science Online

Visit ca6.msscience.com to:

▶ view **Concepts in Motion**

▶ explore **Virtual Labs**

▶ access content-related Web links

▶ take the Standards Check

FOLDABLES™ Study Organizer

Earthquakes Make the following Foldable to organize the causes and effects of earthquakes.

▷ **STEP 1 Fold** a sheet of paper in half lengthwise.

▷ **STEP 2 Fold** the top edge of the paper down from the top as shown.

▷ **STEP 3 Unfold** to form two columns. Label as shown.

Causes Effects

Reading Skill   **ELA6:** R 2.4

Recognizing Cause and Effect

As you read the chapter, explain the causes of earthquakes in the left column. Describe several effects of earthquakes in the right column.

Get Ready to Read

Questioning

① Learn It! Asking questions helps you to understand what you read. As you read, think about the questions you'd like answered. Often you can find the answer in the next paragraph or lesson. Learn to ask good questions by asking *who, what, when, where, why,* and *how.*

② Practice It! Read the following passage from Lesson 1.

> An earthquake is the rupture and sudden movement of rocks along a fault. Remember, a fault is a fracture surface along which rocks can slip. A fault ruptures, or breaks, when rocks are strained so much that they no longer can stretch or bend. This movement causes the release of complex waves that can move objects, as shown in Figure 1.
>
> —from page 246

Here are some questions you might ask about this paragraph:

- What is an earthquake?
- When does a fault rupture?
- What causes the release of complex waves?

③ Apply It! As you read the chapter, look for answers to lesson headings that are in the form of questions.

Target Your Reading

Use this to focus on the main ideas as you read the chapter.

Reading Tip

Test yourself. Create questions and then read to find answers to your own questions.

1 **Before you read** the chapter, respond to the statements below on your worksheet or on a numbered sheet of paper.

- Write an **A** if you **agree** with the statement.
- Write a **D** if you **disagree** with the statement.

2 **After you read** the chapter, look back to this page to see if you've changed your mind about any of the statements.

- If any of your answers changed, explain why.
- Change any false statements into true statements.
- Use your revised statements as a study guide.

Science Online

Print a worksheet of this page at ca6.msscience.com.

Before You Read A or D	Statement	After You Read A or D
	1 Earthquakes are waves of energy that travel across Earth's surface.	
	2 Tsunamis are huge tidal waves.	
	3 Most earthquakes occur in the middle of lithospheric plates.	
	4 Seismic waves are produced at the focus of an earthquake.	
	5 A 4.0 magnitude earthquake releases about twice as much energy as a 3.0 magnitude earthquake.	
	6 Some parts of the United States are at higher risk for earthquakes than others.	
	7 Secondary waves are the fastest seismic waves.	
	8 The San Andreas Fault is a fault zone.	
	9 Fire and landslides are major earthquake hazards.	

Science Content Standards

1.d Students know that earthquakes are sudden motions along breaks in the crust called faults and that volcanoes and fissures are locations where magma reaches the surface.

1.e Students know major geologic events, such as earthquakes, volcanic eruptions, and mountain building, result from plate motions.

7.e Recognize whether evidence is consistent with a proposed explanation.

Reading Guide

What *You'll Learn*

▶ **Explain** what an earthquake is.

▶ **Describe** how faults and earthquakes are related.

▶ **Understand** that most earthquakes occur at plate boundaries.

Why *It's Important*

Understanding what causes earthquakes helps scientists identify where they are likely to occur in the future.

Vocabulary

earthquake
elastic strain
focus

Review Vocabulary

fault: a fracture in rock along which rocks on one side have moved relative to rocks on the other side (p. 211)

Earthquakes and Plate Boundaries

Main Idea Most earthquakes occur at plate boundaries when rocks break and move along faults.

Real-World Reading Connection You're expecting a call. Finally, the cell phone vibrates in your pocket. The shaking stops as you answer the phone. When the ground beneath your feet vibrates during an earthquake, there is no way to stop the shaking.

What is an earthquake?

An **earthquake** is the rupture and sudden movement of rocks along a fault. Remember, a fault is a fracture surface along which rocks can slip. A fault ruptures, or breaks, when rocks are strained so much that they no longer can stretch or bend. This movement causes the release of complex waves that can shake objects, as shown in **Figure 1.**

Most earthquakes occur in Earth's crust, although some happen at great depths where lithospheric plates subduct. Large earthquakes have also occurred in regions far from plate boundaries. Part of the energy released is spread as complex waves that travel through and around Earth.

Figure 1 The shaking during an earthquake is disorienting and frightening. Loose objects that are thrown or that fall down can be dangerous.

Elastic Strain Energy

How can heat from within Earth lead to the shaking people feel during an earthquake? Recall from Chapter 3 that heat in Earth's mantle is a source of energy for plate movement. Some of the heat energy from Earth's interior is transformed into kinetic energy, or energy of motion, for Earth's lithospheric plates. Especially at boundaries between plates, stresses cause strain that occasionally breaks and moves rocks.

 What is kinetic energy?

The plates' kinetic energy is transferred to rocks near the faults. This energy is eventually released as earthquakes, which occur mainly at or near the plate boundaries. This is like the energy stored in a stretched rubber band. The rocks change shape just as the rubber band did. Energy stored as a change in shape is called **elastic strain.** When the rocks cannot stretch to change shape anymore, the faults break and slip as earthquakes.

Faults and Earthquakes

Figure 2 shows how faults and earthquakes are related. The arrows shown in steps 2 and 3 show how rocks slide horizontally past each other. The fault is marked before, during, and after the fault ruptures. As rocks slowly move past each other, elastic strain energy builds up along the strike-slip fault.

Eventually, rocks rupture and slip along the fault, as shown in **Figure 2.** The sudden slip sends complex waves radiating out in all directions into the surrounding rocks. It is the energy in the waves that causes the shaking during an earthquake. Elastic strain energy that was stored in the rocks is partly released by the breaking and moving, and partly released as seismic waves.

 Figure 2 What clues are present in the drawings that show how elastic strain energy is released?

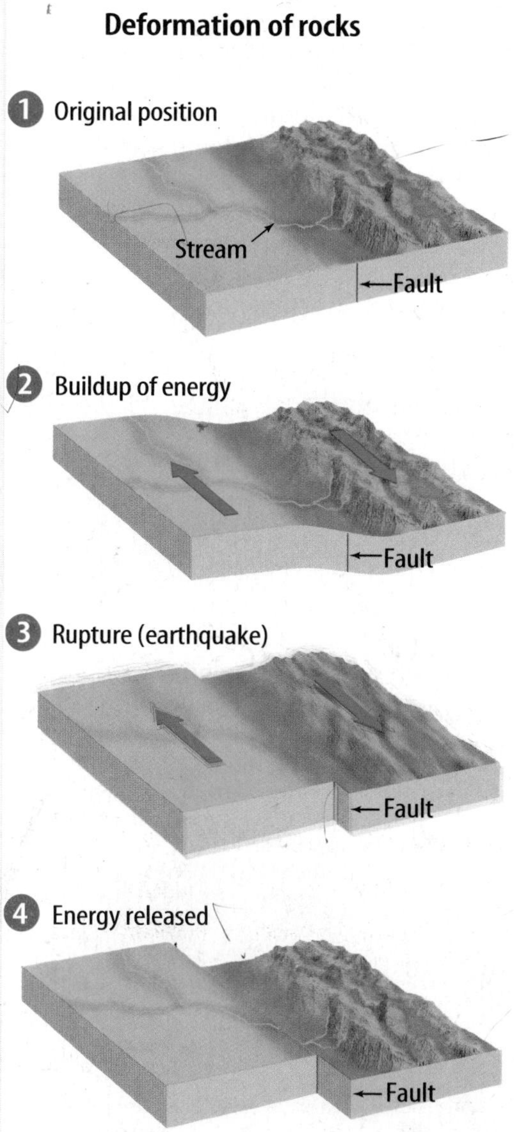

Deformation of rocks

1 Original position

Stream

←Fault

2 Buildup of energy

←Fault

3 Rupture (earthquake)

←Fault

4 Energy released

←Fault

Figure 2 **Strained Rocks** Elastic strain energy builds up in the rocks near the fault. The strength of the rocks is reached, and the fault ruptures, causing an earthquake and releasing the energy.
Identify the type of fault.

Figure 3 Energy and Rupture When elastic strain energy overcomes the strength of the rocks, a rupture begins at the focus. The rupture spreads away from the focus, along the fault, sometimes reaching the ground surface. After the earthquake, most of the elastic strain energy is released.

1 A rupture begins at the focus.

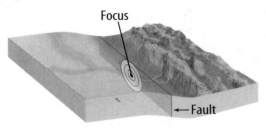

Focus

Fault

2 A rupture spreads away from the fault and more energy is released.

Waves travel outward from the focus

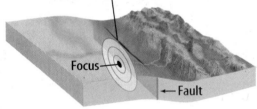

Focus

Fault

3 Most of the elastic strain energy is released, after the earthquake.

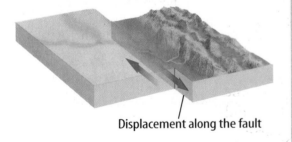

Displacement along the fault

Focus Earthquakes start at the **focus** (plural, foci), which is the location on a fault where rupture and movement begin. **Figure 3** shows the focus, from which a rupture spreads out with time along the fault. As the rupture gets bigger, more and more energy is released into the surrounding rocks.

In general, the closer the focus is to Earth's surface, the stronger the shaking will be. Larger faults can have larger ruptures, which tend to produce larger earthquakes. It takes many small earthquakes to release as much energy as a single, large earthquake.

Fault Zones A plate boundary is often shown as a single line on a map. In reality, plate boundaries are much more complicated. Instead of a single fault, boundaries are usually zones. These fault zones are about 40–200 km wide. The San Andreas Fault is an example of a fault zone. In **Figure 4,** notice that the San Andreas is a group of faults. As a group, these faults result from the plate motion between the Pacific Plate and the North American Plate.

 Visual Check **Figure 4** Identify three faults that are part of the San Andreas Fault zone.

Figure 4 The San Andreas Fault is a zone that contains many faults.

California

Owens Valley Fault

San Francisco

San Andreas Fault

PACIFIC OCEAN

Garlock Fault

Los Angeles

Banning Fault

San Jacinto Fault

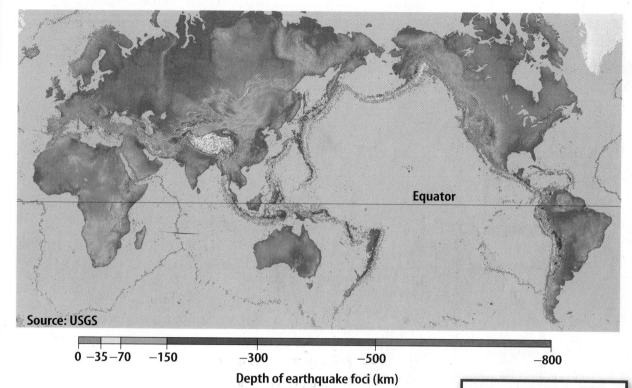

Source: USGS

0 −35 −70   −150   −300   −500   −800

Depth of earthquake foci (km)

Plate Boundaries and Earthquakes

Lithospheric plates **interact** at different boundaries and produce earthquakes. Earthquake size and depth, and the types of faults on which earthquakes occur depend on the type of plate boundary. However, exceptions often occur, and not all earthquakes happen at plate boundaries.

Divergent Plate Boundaries

At divergent plate boundaries, rocks break under tension stress, forming normal faults. **Figure 5** shows that most earthquakes at divergent plate boundaries occur in the crust at relatively shallow depths and are relatively small in size.

Convergent Plate Boundaries

At convergent plate boundaries, rocks break under compression stress, forming reverse faults. **Figure 5** shows that the deepest earthquakes have occurred at convergent plate boundaries. The most devastating earthquakes recorded in Earth's history are associated with convergent plate boundaries.

Transform Plate Boundaries

At transform plate boundaries, rocks slide horizontally past one another, forming strike-slip faults. These earthquakes also occur at relatively shallow depths. However, where transform plate boundaries run through continents, they can cause major earthquakes.

ACADEMIC VOCABULARY
interact (in ter AKT)
(verb) to act on each other
My cat does not interact well with your dog.

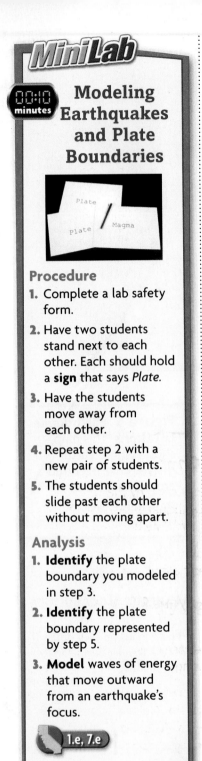

Modeling Earthquakes and Plate Boundaries

Procedure

1. Complete a lab safety form.

2. Have two students stand next to each other. Each should hold a **sign** that says *Plate*.

3. Have the students move away from each other.

4. Repeat step 2 with a new pair of students.

5. The students should slide past each other without moving apart.

Analysis

1. **Identify** the plate boundary you modeled in step 3.

2. **Identify** the plate boundary represented by step 5.

3. **Model** waves of energy that move outward from an earthquake's focus.

1.e, 7.e

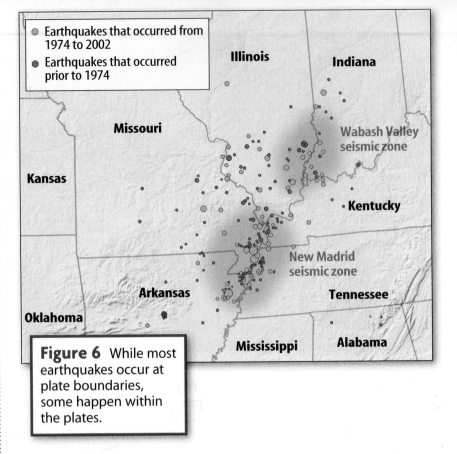

Legend:
- Earthquakes that occurred from 1974 to 2002
- Earthquakes that occurred prior to 1974

Wabash Valley seismic zone

New Madrid seismic zone

Illinois · Indiana · Missouri · Kansas · Kentucky · Arkansas · Oklahoma · Tennessee · Mississippi · Alabama

Figure 6 While most earthquakes occur at plate boundaries, some happen within the plates.

Earthquakes Away from Plate Boundaries

Most earthquakes occur along plate boundaries, where plates are moving relative to one another. However, some earthquakes occur away from plate boundaries, as shown in **Figure 6.** Most of these earthquakes occur in the middle of continents. Even though these earthquakes do not occur often, they can be dangerous. This is partly because people do not expect earthquakes to happen in the middle of continents, and they generally are not prepared for them.

Reading Check Why are earthquakes that occur in the middle of continents dangerous?

In the winter of 1811, three large earthquakes shook New Madrid, Missouri, far from any plate boundary. The largest of the earthquakes changed the course of the Mississippi River, making parts of it flow backward for a while.

Scientists are trying to understand why earthquakes happen so far from plate boundaries. One idea is that there are old, buried faults within the continents. Scientists hypothesize that in the case of the New Madrid earthquakes, millions of years ago, the crust began to pull apart. However, it did not break completely. Instead, a long zone of intense faulting was formed. Today, the crust is being compressed, or squeezed together.

Causes of Earthquakes

Earthquakes occur when elastic strain energy builds up to the point that rocks break and move. This energy is released as earthquakes and complex waves. Boundaries between lithospheric plates are locations where stresses cause rocks to deform, or become strained, as plates slowly move relative to one another. At convergent, divergent, and transform plate boundaries, faults are common. They sometimes rupture and move as earthquakes. Some earthquakes also occur along faults located in the middle of plates, far from present-day plate boundaries. In the next lesson, you'll focus on complex waves—called seismic waves—that release some of the elastic strain energy stored in rocks.

LESSON 1 Review

Summarize

Create your own lesson summary as you write a script for a **television news report.**

1. **Review** the text after the **red** main headings and write one sentence about each. These are the headlines of your broadcast.

2. **Review** the text and write 2–3 sentences about each **blue** subheading. These sentences should tell *who, what, when, where,* and *why* information about each **red** heading.

3. **Include** descriptive details in your report, such as names of reporters and local places and events.

4. **Present** your news report to other classmates alone or with a team.

 **ELA6: LS 1.4**

Standards Check

Using Vocabulary

1. Use the words *focus* and *earthquake* in the same sentence. **1.d**

2. In your own words, write a definition for *elastic strain.* **1.d**

Understanding Main Ideas

3. What is an earthquake? **1.d**

   A. elastic strain stored in rocks

   B. a wave traveling through the crust

   C. rupture and movement along a fault

   D. a fault at a convergent plate boundary

4. **Give an example** of a common object that can store elastic strain energy. **1.d**

5. **Explain** why the deepest earthquakes occur at convergent plate boundaries. **1.e**

6. **Compare and contrast** a fault and a fault zone. **1.e**

Applying Science

7. **Simulate** the buildup and release of elastic strain energy using a wooden stick. **1.d**

8. **Describe** Draw a diagram like the one below. Describe two ways elastic strain energy is released during an earthquake. **1.d**

Elastic Strain Energy		

 **Science** **Online**

For more practice, visit **Standards Check** at ca6.msscience.com.

Science Content Standards

1.g Students know how to determine the epicenter of an earthquake and know that the effects of an earthquake on any region vary, depending on the size of the earthquake, the distance of the region from the epicenter, the local geology, and the type of construction in the region.

7.e Recognize whether evidence is consistent with a proposed explanation.

Reading Guide

What *You'll Learn*

▶ **Explain** how energy released during earthquakes travels in seismic waves.

▶ **Distinguish** among primary, secondary, and surface waves.

▶ **Describe** how seismic waves are used to investigate Earth's interior.

Why *It's Important*

Scientists can locate the epicenter of an earthquake by analyzing seismic waves.

Vocabulary

seismic wave
epicenter
primary wave
secondary wave

Review Vocabulary

wave: a disturbance in a material that transfers energy without transferring matter (p. 132)

Earthquakes and Seismic Waves

(**Main Idea**) Earthquakes cause seismic waves that provide valuable data.

Real-World Reading Connection If you throw or drop a rock into a pond, you might notice that ripples form in the water. Circles of waves move outward from the place where the rock entered the water. In a similar way, an earthquake generates complex waves that move outward through rock.

What are seismic waves?

During an earthquake, the ground moves forward and backward, heaves up and down, and shifts from side to side. Usually this motion is felt as vibrations, or shaking.

Large earthquakes can cause the ground surface to ripple like the waves shown in **Figure 7.** Imagine trying to stand on Earth's surface if it had waves traveling through it. This is what people and structures experience during a strong earthquake. These waves of energy, produced at the focus of an earthquake, are called **seismic** (SIZE mihk) **waves.**

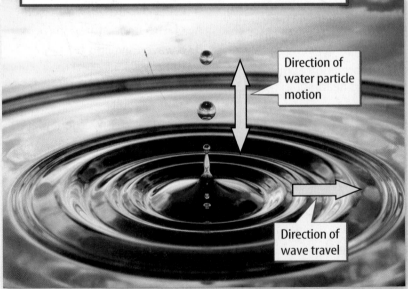

Figure 7 A pebble, dropped in a pond, sends seismic waves outward in all directions. As energy is absorbed by the water, the wave heights decrease.

Direction of water particle motion

Direction of wave travel

How do seismic waves travel?

You read in Lesson 1 that elastic strain energy builds up until it reaches the strength of the rock. Then, the fault ruptures and some of the energy is released in the form of seismic waves. Traveling up to Earth's surface and down deep into the planet, seismic waves move outward from the focus in all directions.

An earthquake's **epicenter** (EH pih sen tur) is the point on Earth's surface directly above the earthquake's focus. Locate the focus and epicenter in **Figure 8.** The shaded spheres show how seismic waves travel outward in all directions from the focus. Rocks absorb some of the energy as the waves move through them. So, the amount of energy in the waves decreases as the waves move farther from the focus.

 Reading Check What happens to the energy of a seismic wave as it travels outward from the focus?

WORD ORIGIN

epicenter

epi– from Greek; means *over*
–center from Latin *centrum;*
means *sharp point, center of
a circle*

SCIENCE USE v. COMMON USE

focus

Science Use the place of origin of an earthquake. *The focus of the earthquake was located five miles off the coast of California.*

Common Use concentrated attention or effort. *Cindy's focus was to help the sick cat recover from its surgery.*

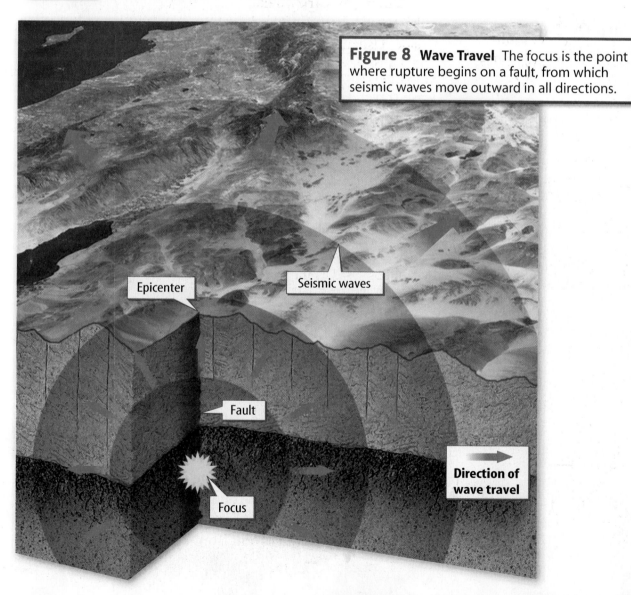

Figure 8 Wave Travel The focus is the point where rupture begins on a fault, from which seismic waves move outward in all directions.

Epicenter

Seismic waves

Fault

Direction of
wave travel

Focus

Types of Seismic Waves

When earthquakes occur, three main types of seismic waves result: primary waves, secondary waves, and surface waves. Each travels differently within Earth.

WORD ORIGIN············

primary
from Latin *primus;* means *first*

Primary Waves (P-waves)

Shown in the first row of **Table 1, primary waves** are compressional waves. When a P-wave moves through rock, particles in the rock move back and forth parallel to the same direction that the wave travels. The energy moves by compressing and expanding the material through which it travels.

Primary waves are the fastest seismic waves. They move between about 5 km/s and 7 km/s, depending on the type of rock they travel through. After an earthquake, primary waves are the first to be detected and recorded by scientific instruments.

WORD ORIGIN············

secondary
from Latin *secundarius;* means
second class, inferior

Secondary Waves (S-waves)

Secondary waves, which are also known as shear waves, cause particles to vibrate perpendicular to the direction of wave travel. For example, when an S-wave moves from left to right through a coiled spring, the spring's vibrations make a 90° angle with the S-wave's direction of travel. This is illustrated in the second row of **Table 1.** This shearing movement changes the shape of rocks. S-waves travel at about 60 percent of the speed of P-waves.

 Table 1 Compare and contrast the motions of P-waves and S-waves.

Surface Waves

When some P-waves and S-waves reach Earth's surface, the energy gets trapped in the upper few kilometers of the crust. This energy forms new types of waves that travel along the surface. Surface waves travel even more slowly than secondary waves.

Surface waves move rock particles in two main ways. Particles move with a side-to-side swaying motion. Particles also move with a rolling motion, as shown in the third row of **Table 1.** Surface waves often vibrate the crust more strongly than P- or S-waves. Their strong shaking damages structures, such as buildings and bridges. These waves usually cause most of the destruction from an earthquake.

Table 1 Types of Seismic Waves

Concepts in Motion
Interactive Table Organize information about seismic waves at ca6.msscience.com.

Seismic Wave	Description
P-Waves  • cause rock particles to vibrate in same direction that waves travel • fastest seismic wave • first to be detected and recorded by scientific instruments • travel through both solids and fluids	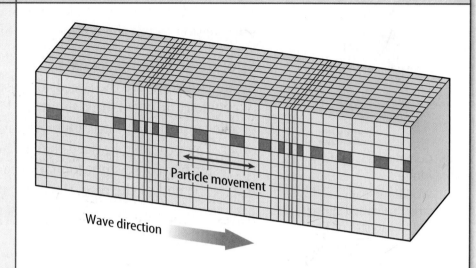
S-Waves • cause rock particles to vibrate perpendicular to direction that waves travel • slower than P-waves • detected and recorded after P-waves • only travel through solids	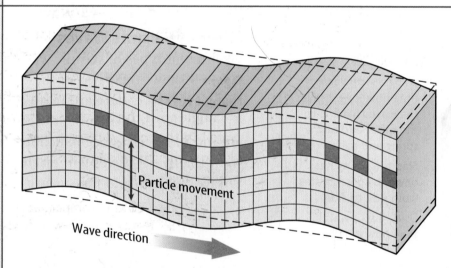
Surface Waves • cause rock particles to move with a side-to-side swaying motion or rolling motion • slowest seismic wave • generally cause the most damage at Earth's surface	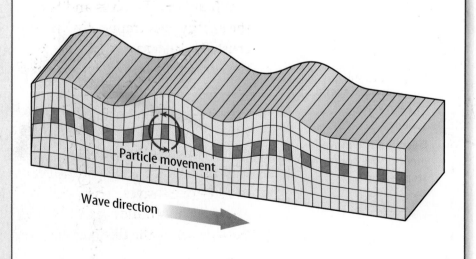

MiniLab

00:15 minutes

Modeling P- and S-Waves

Procedure 🥽 🧤

1. Complete a lab safety form.

2. Working in pairs, stretch a **coiled spring toy** to a length of 3 m.

3. Mark the center of the spring with **tape**.

4. Pinch three or four rings of the spring together at one end and release.

5. Using a **stopwatch**, time how long it takes the wave to move to the other end of the spring and back.

6. Slap the spring near one end so it vibrates from side to side.

7. Using a **stopwatch**, time how long it takes the wave to move to the other end of the spring and back.

Analysis

1. **Identify** which wave represented the P-wave.

2. **Calculate** the speed of each wave.
   speed = $\frac{d}{t}$.

3. **Describe** the movement of the spring in steps 4 and 6.

1.g, 7.e

Using Seismic Wave Data

Different types of seismic waves travel at different speeds. In addition, some seismic waves travel through Earth's interior and some travel along Earth's surface. Seismologists, scientists who study earthquakes and seismic waves, use this information to tell the composition of Earth's interior.

Speeds of Seismic Waves

You can compare the speeds of seismic waves to the speeds of people running. Think about the last time you saw two people running in a race. One person ran faster than the other. You probably noticed that at the beginning of the race, the faster person was not too far ahead of the slower person. But by the end of the race, the faster person was far ahead. Like runners in a race, seismic waves start at the same time. If you are close to the focus of an earthquake, the S-wave is not very far behind the P-wave. If you are far from the focus, the S-wave travels far behind the P-wave.

Paths of Seismic Waves

Different types of seismic waves travel at different speeds. **Figure 9** shows the paths of these seismic waves. Remember that waves travel outward in all directions from the focus. Imagine that an instrument used to measure and record ground motion has been anchored in bedrock. The arrows show that the P-waves will reach the instrument first, the S-waves second, and the surface waves last.

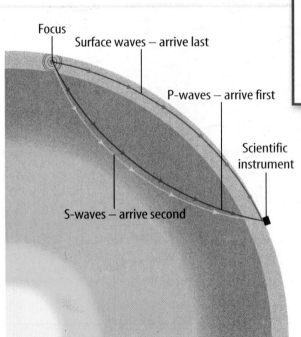

Figure 9 Different types of seismic waves start at the same location, but move at different speeds.

Identify which seismic wave first reached the instrument.

Focus
Surface waves — arrive last
P-waves — arrive first
Scientific instrument
S-waves — arrive second

Mapping Earth's Internal Structure

You read in Chapter 2 that scientists learn the details of Earth's **internal** structure by analyzing the paths of seismic waves. The speed and direction of seismic waves change when properties of materials they travel through change. Scientists use P- and S-waves to investigate this layering because they can travel through Earth's interior. The densities of rocks increase with depth as pressures increase. This makes the paths of the waves change as they pass through Earth.

ACADEMIC VOCABULARY

internal (ihn TUR nul)
(adjective) existing or situated within the limits or surface of something.
The doctor found internal bleeding that required surgery.

 **What makes the paths of seismic waves change as they travel through Earth?**

Early in the twentieth century, scientists discovered that large areas of Earth don't receive any seismic waves from an earthquake. These areas, called shadow zones, are shown in **Figure 10.** Secondary waves can travel only through solids. They stop when they hit the outer core of Earth. The outer core does not stop primary waves, but their paths bend. Because of these observations, scientists think the outer core is liquid. The bending of primary waves and the stopping of secondary waves cause the shadow zone.

Figure 10 Earth's Layers The paths and speeds of P-waves and S-waves help scientists determine the internal structure of Earth.

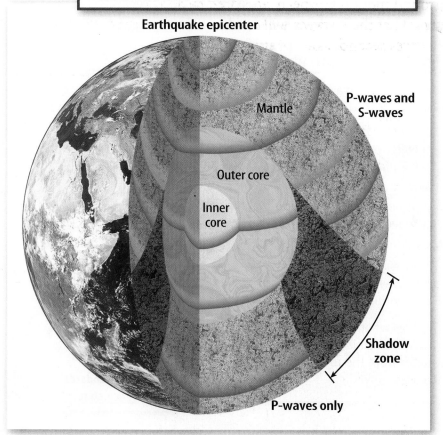

Seismic Waves

As rocks break and move, some of the elastic strain energy that had built up is released as seismic waves. These waves of energy travel outward from the focus. Most of the damage and loss of life from earthquakes results from the seismic waves released during an earthquake.

Primary and secondary waves travel through Earth's interior, although secondary waves do not travel through fluids. Surface waves travel at shallow levels in the crust and cause the most damage to structures. Next, you'll read how seismic waves are measured to determine an earthquakes size and location.

LESSON 2 Review

Summarize

Create your own lesson summary as you organize an **outline.**

1. **Scan** the lesson. Find and list the first **red** main heading.

2. **Review** the text after the heading and list 2–3 details about the heading.

3. **Find** and list each **blue** subheading that follows the **red** main heading.

4. **List** 2–3 details, key terms, and definitions under each **blue** subheading.

5. **Review** additional **red** main headings and their supporting **blue** subheadings. List 2–3 details about each.

 ELA6: R 2.4

Standards Check

Using Vocabulary

1. Distinguish between a *primary wave* and a *secondary wave.* **1.g**

2. In your own words, write a definition for the word *epicenter.* **1.g**

Understanding Main Ideas

3. How do surface waves move rock particles? **1.g**

    A. parallel to direction of wave travel

    B. rolling motion or side-to-side

    C. perpendicular to direction of wave travel

    D. diagonally

4. **Give an example** of how seismic waves provide valuable scientific data. **1.g**

5. **Describe** what happens to the energy of seismic waves as the distance from the focus increases. **1.g**

6. **Sequence** Draw a diagram like the one below. Arrange the types of seismic waves in order of increasing wave speed. **1.g**

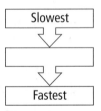

Applying Science

7. **Illustrate** the vibration direction and the direction of travel for an S-wave. **1.g**

8. **Hypothesize** what happens to P-waves and S-waves when they encounter magma. **1.g**

Science nline

For more practice, visit **Standards Check** at ca6.msscience.com.

Applying Math

Speeds of Seismic Waves

Different types of seismic waves travel at different speeds. The difference in the arrival times of P- and S-waves can be used to locate an epicenter of an earthquake. You can use this information to determine the distance from the origin of the seismic wave to your current location. This distance is determined by calculating the difference in arrival times of P- and S-waves and multiplying by 8 km/s.

Example

The data table shows the difference in arrival times for ten different seismic waves. Recall that this number is calculated by finding the difference between the arrival times of P- and S-waves. Use the table to determine the distance from the origin of the first seismic wave to your current location.

Seismic Wave	Difference in Arrival Time
1	4.9
2	8.7
3	12.3
4	17.8
5	18.0
6	19.0
7	24.4
8	42.7
9	51.9
10	52.9

What you know:

- The difference in arrival time for the first seismic wave is 4.9 s.

What you need to find:

- The distance the first seismic wave has traveled from its origin to your current location

Multiply the difference by 8 km/s.

4.9 s × 8 km/s = 39.2 km

Answer: The first seismic wave has traveled a distance of 39.2 km from its origin to your current location.

Practice Problems

1. Determine the distance from the origin of the second seismic wave to your current location.

2. If both the first and second seismic waves occur at the same depth and direction from your location, how far apart are their origins?

Science Online
For more math practice, visit **Math Practice** at ca6.msscience.com.

Science Content Standards

1.g Students know how to determine the epicenter of an earthquake and know that the effects of an earthquake on any region vary, depending on the size of the earthquake, the distance of the region from the epicenter, the local geology, and the type of construction in the region.

7.b Select and use appropriate tools and technology (including calculators, computers, balances, spring scales, microscopes, and binoculars) to perform tests, collect data, and display data.

7.g Interpret events by sequence and time from natural phenomena (e.g., the relative ages of rocks and intrusions).

Reading Guide

What *You'll Learn*

▶ **Explain** how a seismograph records an earthquake.

▶ **Understand** how to locate an earthquake's epicenter.

▶ **Distinguish** among ways earthquakes are measured.

Why *It's Important*

Measuring earthquakes helps scientists understand how and where they occur.

Vocabulary

seismograph
seismogram

Review Vocabulary

sediment: rock material that is broken down into smaller pieces or that is dissolved in water (p. 99)

Measuring Earthquakes

Main Idea Data from seismic waves are recorded and interpreted to determine the location and size of an earthquake.

Real-World Reading Connection You might remember the December 26, 2004, earthquake and tsunami in the Indian Ocean. Scientists described the earthquake as having a magnitude of about 9.0. But, what does this number mean? In what other ways can the size of an earthquake and its effects be described?

How are earthquakes measured?

Compared to most other earthquakes, the December 26, 2004, earthquake in the Indian Ocean was extremely large. Scientists determined its size by measuring how much the rock slipped along the fault. They also analyzed the heights of the seismic waves, which indicate how much energy was released by the earthquake. Because the earthquake occurred under water, the movement of rock caused an ocean wave in the Indian Ocean. A computer model of this ocean wave is shown in **Figure 11.**

Figure 11 The December 26, 2004, earthquake in the Indian Ocean ruptured a fault at a convergent plate boundary. This rupture created an ocean wave. This computer model shows how the wave traveled across the Indian Ocean two hours after the fault ruptured.

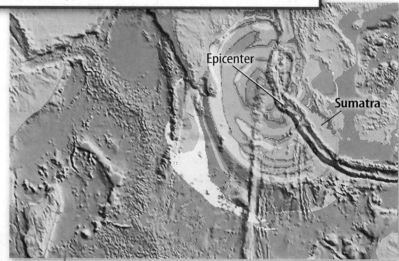

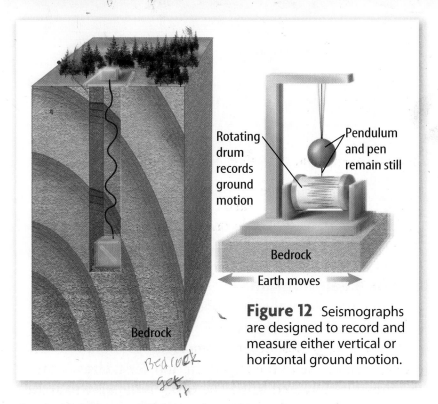

Rotating drum records ground motion

Pendulum and pen remain still

Bedrock

Earth moves

Figure 12 Seismographs are designed to record and measure either vertical or horizontal ground motion.

Bedrock

Recording Seismic Waves

A **seismograph** (SIZE muh graf), shown in **Figure 12,** is an instrument used to record and measure movements of the ground caused by seismic waves. It records the size, direction, and time of the movement. It also records the arrival times of the P- and S-waves. Modern seismographs record the ground motion with electronic signals. They work in much the same way as the older, mechanical seismographs.

Mechanical Seismographs

In order to understand how a seismograph works, consider the parts of a mechanical seismograph. A pen is attached to a weight called a pendulum. When seismic waves shake the ground, the heavy pendulum and the pen remain still. But, the drum moves. This happens because the drum is securely attached to the ground, unlike the freely swinging pendulum. As the ground shakes, the pen records the motion on the paper wrapped around the drum.

Reading Check Which parts of a mechanical seismograph remain still when the ground shakes?

Seismographs record ground motion in two orientations. One orientation is horizontal, or back-and-forth, ground motion. The other is vertical, or up-and-down motion. The record of the seismic waves is called a **seismogram** (SIZE muh gram). Seismograms are used to calculate the size of earthquakes and to determine their locations.

WORD ORIGIN

seismograph
seismogram
seis– from Greek *seismos;* means *earthquake*
–graph from Greek; means *to write*
–gram from Greek; means *written word, a letter*

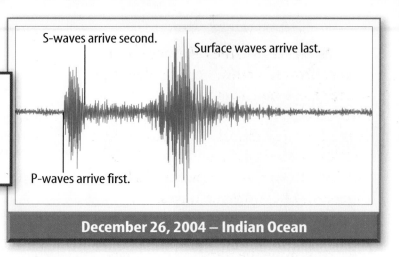

Figure 13 This seismogram is a record of P-waves, S-waves, and surface waves from the December 26, 2004, earthquake in the Indian Ocean that arrived at a seismograph station.

S-waves arrive second.

Surface waves arrive last.

P-waves arrive first.

December 26, 2004 — Indian Ocean

Reading a Seismogram

A seismogram from the December 26, 2004, Indian Ocean earthquake is shown in **Figure 13.** To you, it might look like a bunch of wavy lines. However, seismologists know how to read the lines. You can learn how to do this too. First, observe the *x*-axis. This axis represents time. The first seismic wave to arrive at the seismograph is the fastest wave, the P-wave. As it shakes the ground, it makes wavy lines on the record.

After the P-wave, the S-wave arrives and makes more wavy lines. Finally the surface waves arrive. The heights of the waves on the seismogram **indicate** the sizes of ground motion for each type of wave.

Using average S-wave and P-wave speeds, scientists can plot the difference in arrival times on the *y*-axis of a graph and the distance the waves travel from the epicenter on the *x*-axis. When the arrival times are read from a seismogram, the distance the P- and S-waves have traveled can be determined from a graph. An example of a graph used by scientists is shown in **Figure 14.**

ACADEMIC VOCABULARY

indicate (IHN duh kate)
(verb) to demonstrate or point out with precision
The thermometer indicates that you have a fever.

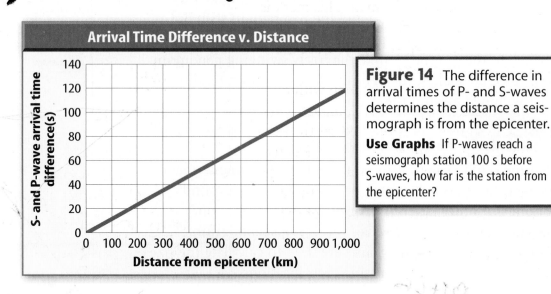

Arrival Time Difference v. Distance

S- and P-wave arrival time difference(s)

Distance from epicenter (km)

Figure 14 The difference in arrival times of P- and S-waves determines the distance a seismograph is from the epicenter.

Use Graphs If P-waves reach a seismograph station 100 s before S-waves, how far is the station from the epicenter?

Locating an Epicenter

Seismologists use the difference in the P- and S-waves' arrival times to determine where an earthquake occurred. If at least three seismographs record distances, the epicenter of the earthquake may be determined by a method called triangulation. This method of locating the epicenter is based on the speeds of the seismic waves.

① Find the arrival time difference.
First, determine the number of seconds between the appearance of the first P-wave and the first S-wave on the seismogram. To do this, use the time scale on the x-axis of the seismogram. Subtract the arrival time of the P-wave from the arrival time of the S-wave.

② Find the distance from the epicenter.
Next, use a graph showing the P- and S-wave arrival time differences plotted against distance. Seismologists can make these graphs because they know the speed that the seismic waves travel inside Earth. Look at the y-axis and find the place on the blue line with the time difference you calculated from the seismogram. Then, read the corresponding distance on the x-axis.

③ Plot the distance on a map.
Next, use a ruler and the scale provided on a map to draw a mark on the map. This mark is the distance away from the seismograph that you just determined. Make sure you use the correct seismograph location.

Finally, draw a circle on the map with a compass. To do this, place the compass point on the seismograph location, and set the pencil at the distance from step 2. The epicenter is located somewhere on the circle. When circles are plotted for data from at least three different seismographs, the location of the epicenter can be found. This location is the point where the circles intersect.

① Find the arrival time difference.

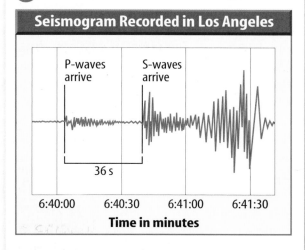
Seismogram Recorded in Los Angeles

② Find the distance from the epicenter.

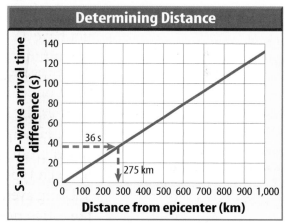
Determining Distance

③ Plot the distance on a map.

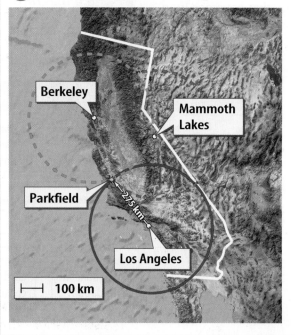

Measuring Earthquake Size

You've probably noticed that some earthquakes are bigger than others. Seismologists use different types of scales to describe the size of an earthquake.

Magnitude Scale

One way to describe an earthquake's size is to measure the heights of the seismic waves recorded on a seismogram. The magnitude scale is based on a seismogram's record of the amplitude, or height, of ground motion. Magnitude is related to the amount of energy released by an earthquake.

 Reading Check What does magnitude measure?

The magnitude of an earthquake is determined by the buildup of elastic strain energy in the crust, at the place where ruptures eventually occur. The magnitude scale does not have an upper or lower limit. However, most measured magnitude values range between about 0 and 9. **Figure 15** shows the magnitude of significant earthquakes from the past 100 years. Each increase of one number on the magnitude scale represents a 10 times increase in ground shaking. But, that same one number increase represents about 30 times more energy released.

Unfortunately, many small earthquakes combined can release only a small fraction of the stored energy. For example, it might take as many as one million 4.0 magnitude earthquakes to release the same amount of energy as a single 8.0 magnitude earthquake.

Figure 15 Several large-magnitude earthquakes have occurred around the world and caused significant damage.

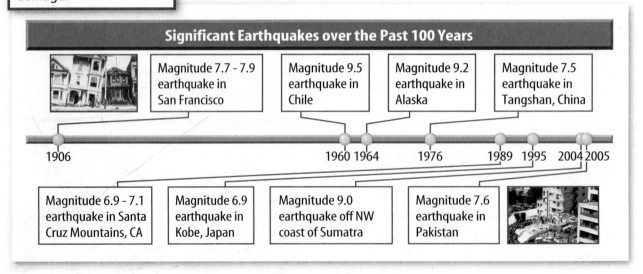

Significant Earthquakes over the Past 100 Years

| Magnitude 7.7 - 7.9 earthquake in San Francisco | Magnitude 9.5 earthquake in Chile | Magnitude 9.2 earthquake in Alaska | Magnitude 7.5 earthquake in Tangshan, China |

1906   1960   1964   1976   1989   1995   2004 2005

| Magnitude 6.9 - 7.1 earthquake in Santa Cruz Mountains, CA | Magnitude 6.9 earthquake in Kobe, Japan | Magnitude 9.0 earthquake off NW coast of Sumatra | Magnitude 7.6 earthquake in Pakistan |

Richter Magnitude Scale

The Richter magnitude scale was the first magnitude scale. It was published in 1935 by Charles Richter. He designed it for use in southern California with the particular type of seismographs that were used there. Richter magnitudes are not accurate for earthquakes that are smaller than about 3.0 in magnitude and larger than about 7.0 in magnitude. Since the 1930s, seismologists have developed magnitude scales to use with more modern seismographs.

Moment Magnitude Scale

Most seismologists today use a more accurate scale for measuring the size of an earthquake. This scale is called the moment magnitude scale. It is based on the amount of energy released during an earthquake. The seismic moment can be calculated by multiplying the area of the fault rupture times the amount that it slips times the strength of the broken rocks. The seismic moment is related to moment magnitude, which is the most widely used measurement scale because it gives a consistent measure of earthquake size—from minor jiggles to devastating jolts. Damage from a devastating jolt during the 1999 earthquake in Turkey is shown below.

Figure 16 The 7.8-magnitude 1999 earthquake in Turkey released about 30 times more energy than the 6.8-magnitude 1994 Northridge earthquake.
Determine what factors caused the damage to the structures in the photo.

Earthquake Intensity

Earthquakes can also be described by the amount of damage they cause. Scientists investigate the effects of an earthquake on structures, the land surface, and people's reactions to the shaking. Based on these data, they assign intensity values for the earthquake.

Effects of Shaking

Earthquake intensity values vary and are based on the effects of ground shaking. They depend on the distance from the epicenter and the local geology. Usually, the maximum intensity is found near the earthquake's epicenter. Therefore, the farther from the epicenter, the less the ground shakes.

Differences in the local rocks or sediment also affect the amount of shaking. Loose sediment or fill tend to shake more violently than rocks.

Plotting Intensity Values

After intensity values are assigned, scientists plot the values on a map. Then, the data are contoured as shown in **Figure 17.** This is done in much the same way that topography is contoured on a topographic map.

Figure 17 The most widely used intensity scale is the modified Mercalli scale. It uses Roman numerals ranging from I (not felt) to XII (damage nearly total) to describe the intensity of the earthquake.

Explain what the shading on this map represents.

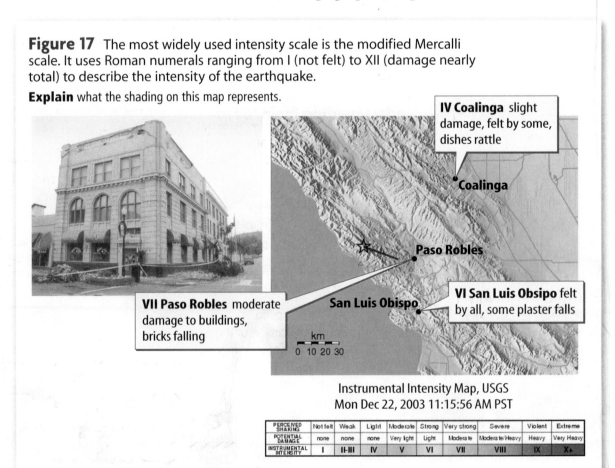

IV Coalinga slight damage, felt by some, dishes rattle

Coalinga

Paso Robles

VI San Luis Obsipo felt by all, some plaster falls

San Luis Obispo

VII Paso Robles moderate damage to buildings, bricks falling

km
0 10 20 30

Instrumental Intensity Map, USGS
Mon Dec 22, 2003 11:15:56 AM PST

PERCEIVED SHAKING	Not felt	Weak	Light	Moderate	Strong	Very strong	Severe	Violent	Extreme
POTENTIAL DAMAGE	none	none	none	Very light	Light	Moderate	Moderate/Heavy	Heavy	Very Heavy
INSTRUMENTAL INTENSITY	I	II-III	IV	V	VI	VII	VIII	IX	X+

Features of Earthquakes

Scientists measure seismic waves to determine where an earthquake occurred and how large it was. Because P- and S-waves travel at different speeds through materials, the time difference at which they arrive at a seismograph indicates how far that seismograph is from an earthquake's epicenter. Magnitude scales are used to measure the energy released by an earthquake. In contrast, the intensity of an earthquake is estimated according to the amount of shaking, and therefore damage, endured in a particular area. Next, you'll read about some hazards this shaking can cause to humans and wildlife habitats.

LESSON 3 Review

Summarize

Create your own lesson summary as you write a **newsletter.**

1. **Write** this lesson title, number, and page numbers at the top of a sheet of paper.

2. **Review** the text after the red main headings and write one sentence about each. These will be the headlines of your newsletter.

3. **Review** the text and write 2–3 sentences about each **blue** subheading. These sentences should tell *who, what, when, where,* and *why* information about each headline.

4. **Illustrate** your newsletter with diagrams of important structures and processes next to each headline.

 ELA6: W 1.2

Standards Check

Using Vocabulary

1. An earthquake's epicenter can be determined from arrival time data recorded on a(n) _____. **1.g**

2. An instrument used to measure ground movement is a(n) _____. **1.g**

Understanding Main Ideas

3. Which scale describes an earthquake's intensity? **1.g**

   A. magnitude scale
   B. modified Mercalli scale
   C. moment magnitude scale
   D. Richter magnitude scale

4. **List** two factors that affect the intensity of an earthquake. **1.g**

5. **Summarize** the steps for determining the location of an earthquake's epicenter. **1.g**

6. **Compare and Contrast** Draw a diagram like the one below. Fill in details comparing an intensity scale with a magnitude scale. **1.g**

Earthquake Scales	Similarities	Differences

Applying Science

7. **Construct** two seismograms that show a difference in distance from an epicenter. **1.g**

8. **Justify** why scientists use data from more than three seismograph locations to determine an epicenter. **1.g**

 **Science Online**

For more practice, visit **Standards Check** at ca6.msscience.com.

Data Lab

Can you locate an earthquake's epicenter?

From the shaking it caused, it might seem like an earthquake happened nearby. But, to determine its epicenter, you need to analyze P- and S-wave data recorded for an earthquake from at least three locations.

Data Collection

1. Read and complete a lab safety form.

2. Obtain a **map of the western United States** from your teacher.

3. Study the three seismograms.

4. Determine the arrival times, to the nearest second, for the P-wave and the S-wave on the first seismogram. Record these data in the table. Also record the seismograph's location.

5. Subtract the P-wave arrival time from the S-wave arrival time. Record this difference, in seconds, in the data table.

6. Use the arrival time difference and the **graph (Figure 14)** to determine the distance between the epicenter and the seismograph.

7. Repeat steps 4–6 for the two other seismograms.

8. Use the map scale to set the **compass** point to a spacing that represents the distance from the first seismograph location to the epicenter. Place the compass point on the seismograph location on your map copy, and draw a circle on the map with radius equal to this distance.

9. Repeat step 8 for the other two seismograph locations.

10. Find the point where all three circles intersect on the map. This is the epicenter of the earthquake.

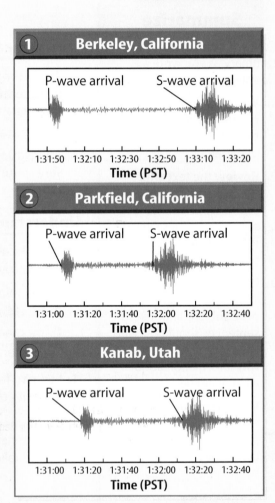

1 **Berkeley, California**

P-wave arrival   S-wave arrival

1:31:50   1:32:10   1:32:30   1:32:50   1:33:10   1:33:20
Time (PST)

2 **Parkfield, California**

P-wave arrival   S-wave arrival

1:31:00   1:31:20   1:31:40   1:32:00   1:32:20   1:32:40
Time (PST)

3 **Kanab, Utah**

P-wave arrival   S-wave arrival

1:31:00   1:31:20   1:31:40   1:32:00   1:32:20   1:32:40
Time (PST)

Earthquake Data Table				
Seismograph Location	P-wave Arrival Time	S-wave Arrival Time	Arrival Time Difference (s)	Distance from Epicenter (km)

Data Analysis

1. **Explain** why you needed to find the difference in time of arrival between P-waves and S-waves for each seismograph location.

2. **Illustrate** how the spacing between P- and S-waves on a seismogram will change as the distance between a seismograph and an earthquake's epicenter decreases.

3. **Summarize** the steps you used to determine the earthquake's epicenter.

4. **Identify** sources of error in the determination of an earthquake's epicenter.

5. **Explain** whether you think data from many seismographs would be useful in determining an earthquake's epicenter.

 Science Content Standards

1.g Students know how to determine the epicenter of an earthquake and know that the effects of an earthquake on any region vary, depending on the size of the earthquake, the distance of the region from the epicenter, the local geology, and the type of construction in the region.
7.b Select and use appropriate tools and technology (including calculators, computers, balances, spring scales, microscopes, and binoculars) to perform tests, collect data, and display data.
7.g Interpret events by sequence and time from natural phenomena (e.g., the relative ages of rocks and intrusions).

 **MA6: NS 2.0**

Science Content Standards

1.g Students know how to determine the epicenter of an earthquake and know that the effects of an earthquake on any region vary, depending on the size of the earthquake, the distance of the region from the epicenter, the local geology, and the type of construction in the region.
2.d Students know earthquakes, volcanic eruptions, landslides, and floods change human and wildlife habitats.
7.a Develop a hypothesis.
7.b Select and use appropriate tools and technology (including calculators, computers, balances, spring scales, microscopes, and binoculars) to perform tests, collect data, and display data.
Also covers: 7.d

Reading Guide

What *You'll Learn*

▸ **Describe** the various hazards from earthquakes.

▸ **Give** examples of ways to reduce earthquake damage.

▸ **List** ways to make your classroom and home more earthquake safe.

Why *It's Important*

Preparing for an earthquake can save lives and reduce damage to property.

Vocabulary

liquefaction
tsunami

Review Vocabulary

San Andreas Fault: fault zone that forms a transform plate boundary between the Pacific Plate and the North American Plate (p. 224)

Earthquake Hazards and Safety

Main Idea Effects of an earthquake depend on its size and the types of structures and geology in a region.

Real-World Reading Connection Sports can be dangerous. People know the hazards involved in playing a particular sport. So, they can design equipment to protect the players. The same is true for designing structures in earthquake-prone regions. The better people understand the hazards of earthquakes, the better they can protect human and wildlife habitats.

Earthquake Hazards

The movement of the ground during an earthquake doesn't directly cause many deaths or injuries. Instead, most injuries result from the collapse of buildings and other structures. It is the shaking of the things people build that does the most damage, as shown in **Figure 18.** Other hazards that might result from an earthquake include fires, landslides, loose sediment, and tsunamis.

Figure 18 It is the damage to buildings and other human-made structures from earthquakes that cause most injuries.

Fire

Fire is the most common hazard that occurs following an earthquake. Fires usually start when earthquakes rupture gas pipes and sever electrical lines. They are especially dangerous if broken water pipes keep firefighters from finding water to fight the fires. Fires that occurred following the San Francisco earthquake of 1906 burned for more than two days and created more damage than the earthquake itself.

Landslides

In areas where there are steep slopes in the landscape, earthquakes may cause landslides. A landslide is the sudden movement of soil and rock down a slope. **Figure 19** shows the damage a landslide can cause. Landslides can block roads, damage and destroy homes, and disrupt gas pipes and electrical lines.

 Reading Check List three ways that landslides can disrupt or destroy human-made structures.

Liquefaction

Have you ever tried to drink a thick milkshake from a cup? Sometimes the milkshake is so thick that it won't flow. How do you make the milkshake flow? You shake it. Something similar can happen to loose, wet sediment during an earthquake.

Wet sediment can be strong most of the time, but the shaking from an earthquake can cause it to act like a liquid. The process by which shaking makes loose sediment move like a liquid is called **liquefaction.**

 **Reading Check** How does shaking from an earthquake influence wet sediment?

When liquefaction occurs in soil under buildings, the buildings can sink into the soil and collapse. This is shown in **Figure 19.** People living in earthquake-prone regions should avoid building on soil or fill made of loose sediment.

Figure 19 In addition to damage caused by shaking, earthquakes also cause fires, landslides, and liquefaction.

Fire

Landslides

Liquefaction

Figure 20
Banda Aceh, Indonesia, sustained great damage from the 2004 tsunami.

Before

After

Tsunamis

Most earthquake damage occurs when seismic waves cause buildings, bridges, and roads to collapse. However, people living on or near a shoreline have another problem. An earthquake that occurs under the ocean causes sudden movement of the seafloor. This sudden movement causes powerful ocean waves that can devaste a region, as shown in **Figure 20.**

Ocean waves caused by earthquakes are called seismic sea waves, or **tsunamis** (soo NAM meez). Far from shore, a tsunami wave is so long and flat that a large ship might ride over it without noticing it. But as one of these waves reaches shallow water, it forms a towering crest that can reach up to about 30 m in height, as shown in **Figure 21.**

Indian Ocean: December 26, 2004 The deadliest tsunami in recorded history occurred on December 26, 2004. The tsunami was triggered by a 9.0-magnitude earthquake off the northwest coast of Sumatra, Indonesia. The series of waves left an estimated 1,126,900 people homeless and killed more than 283,000 people in India, Sri Lanka, Indonesia, and Thailand. **Figure 20** shows an area of Banda Aceh, Indonesia, before and after the 2004 tsunami.

 Figure 20 Identify structures, trees, and shoreline that were destroyed by the 2004 tsunami.

Tsunami Warnings Just before a tsunami crashes onto shore, the water along a shoreline might move back rapidly toward the sea. This exposes a large portion of land that normally is under water. This should be taken as a warning sign that a tsunami could strike soon. You should head for higher ground immediately.

 What is a warning sign that a tsunami could strike soon?

Because of the number of earthquakes that occur around the Pacific Ocean, the threat of tsunamis is constant. To protect lives and property, a warning system has been set up in coastal areas and for the Pacific Islands to alert people if a tsunami is likely to occur. The Pacific Tsunami Warning Center is located near Hilo, Hawaii. It provides warning information including predicted tsunami arrival times.

Currently, no such warning system exists for the Indian Ocean. However, many people in the path of the 2004 tsunami noticed the water receding and large areas of sand exposed. Some people recognized this as a sign of the tsunami and helped alert others to the danger.

Visualizing Tsunamis

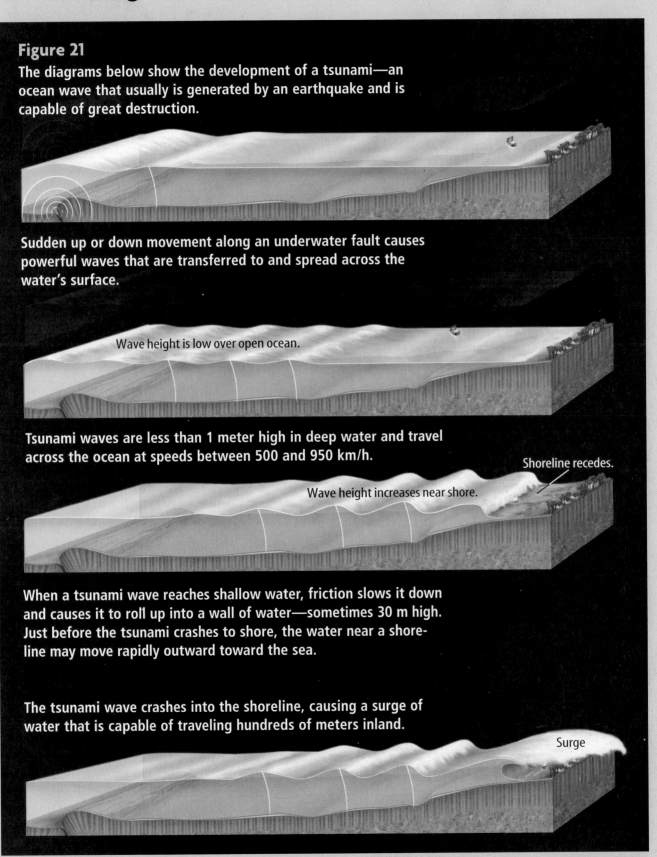

Figure 21

The diagrams below show the development of a tsunami—an ocean wave that usually is generated by an earthquake and is capable of great destruction.

Sudden up or down movement along an underwater fault causes powerful waves that are transferred to and spread across the water's surface.

Wave height is low over open ocean.

Tsunami waves are less than 1 meter high in deep water and travel across the ocean at speeds between 500 and 950 km/h.

Shoreline recedes.

Wave height increases near shore.

When a tsunami wave reaches shallow water, friction slows it down and causes it to roll up into a wall of water—sometimes 30 m high. Just before the tsunami crashes to shore, the water near a shoreline may move rapidly outward toward the sea.

The tsunami wave crashes into the shoreline, causing a surge of water that is capable of traveling hundreds of meters inland.

Surge

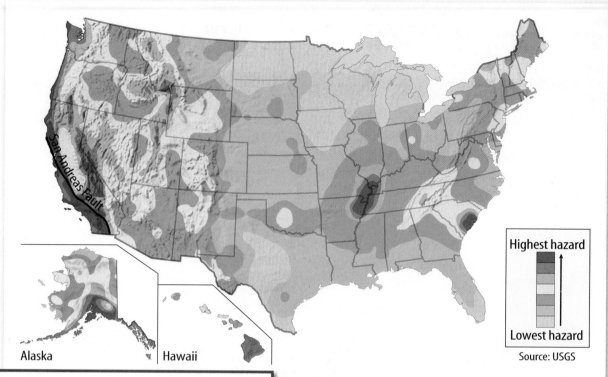

Figure 22 Risk for an earthquake is greater in some locations than in others.

Infer why parts of Alaska and the western continental United States are at high risk for earthquakes.

Avoiding Earthquake Hazards

Recall from Chapter 5 that most earthquakes occur on faults along plate boundaries. The chance of earthquake damage is greater the closer you live to one of these faults. The hazard is also increased in areas where loose sediment is at the surface instead of solid rocks. If you determine where these hazards are the greatest, you can plan ahead to avoid damage and injury.

 List two factors that increase the chance of earthquake damage.

Determining Earthquake Hazards

Scientists can locate active faults. These are places where past earthquakes have occurred. Scientists also have geologic maps that show the locations of rocks and loose sediments at the surface. With all this information, scientists can generate maps that show where earthquake damage is likely to occur. **Figure 22** shows an earthquake hazard map for the United States. Maps are also made showing where landslides, liquefaction, or tsunamis are likely to occur when there is an earthquake.

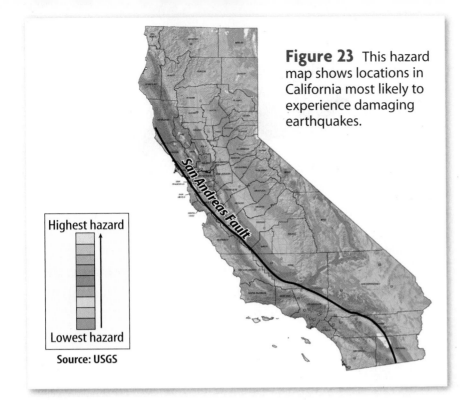

Figure 23 This hazard map shows locations in California most likely to experience damaging earthquakes.

Highest hazard

Lowest hazard

Source: USGS

San Andreas Fault

Long-Term Planning

Figure 23 shows that in California, the risk of earthquake hazards is greatest along the largest fault in the state, the San Andreas Fault. When scientists determine where earthquakes are likely to occur, such as in areas within the San Andreas Fault zone, this information can help city planners develop good strategies for using the land.

Cities can take action to reduce potential damage and loss of life. The frequency and severity of earthquakes must be considered when deciding how to use land. For example, **Figure 24** shows that it may be more appropriate to use land for agricultural purposes or parks, instead of homes, in an area where there is great seismic risk.

Figure 24 The Golden Gate Park at Stow Lake, shown at left, is a good use of land in an area at high risk for earthquakes. At right, liquefaction of sediment beneath homes built on fill resulted from a poor choice for land development.

Earthquakes and Structures

During earthquakes, buildings, bridges, and highways can be damaged or destroyed. Most loss of life during an earthquake occurs when people are trapped in or on these crumbling structures. What can be done to reduce loss of life?

Types of Structures

The damage produced by an earthquake is related to the strength and quality of the buildings, as shown in **Figure 25.** The most severe damage occurs in buildings made of brittle materials, such as non-reinforced concrete, brick, or adobe. Buildings made of flexible materials, such as wood, generally suffer less damage during an earthquake. Taller buildings also are more susceptible to earthquake damage than single-story buildings.

Reading Check Buildings made of what materials, usully sustain the most severe damage?

Earthquake-Resistant Structures

Earthquake-resistant structures tolerate shaking that occurs during an earthquake. Today, some buildings are supported by flexible, circular moorings. The moorings, shown in **Figure 26,** are made of steel plates filled with alternating layers of rubber and steel. The rubber acts like a cushion to absorb seismic waves. Tests have shown that buildings supported in this way should be able to withstand 8.3-magnitude earthquake without major damage.

In older buildings, steel rods are often installed to reinforce building walls. This is an example of retrofitting, or fixing an existing structure, to withstand an earthquake. These measures protect buildings in areas that are likely to experience earthquakes.

Figure 25 The top photo shows that the steel structure of an unfinished building withstood the 1906 San Francisco earthquake. Technological advances have helped make buildings, like the Transamerica building, shown in the lower photo, more earthquake-resistant.

1906

2006

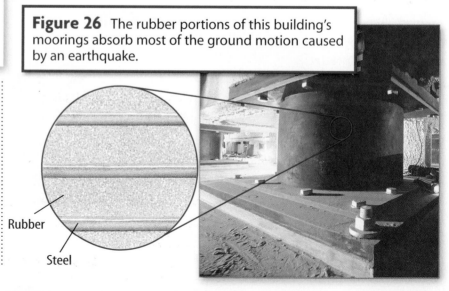

Figure 26 The rubber portions of this building's moorings absorb most of the ground motion caused by an earthquake.

Rubber

Steel

Earthquake Safety

Suppose you are asleep in your bed, when suddenly you awake to your home shaking. Throughout the city, buildings collapse and fires are blazing. What could you do to protect yourself and your family before, during, and after an earthquake?

Before an Earthquake

To protect yourself and your family before an earthquake, you should create, review, and practice an earthquake disaster plan. **Figure 27** shows a family practicing their disaster plan. Your plan should include a safe meeting place after the earthquake. You should also prepare an earthquake supply kit that contains canned food, water, a battery-powered radio, a flashlight, and first aid supplies.

To make your home as earthquake safe as possible, certain steps can be taken. To reduce the danger of injuries from falling objects, move heavy objects to low shelves. With adult supervision learn how to turn off the gas, water, and electricity in your home. To reduce the chance of fire, make sure that water heaters and other gas appliances are held **securely** in place. A newer method that is being used to minimize the danger of fire involves placing sensors on gas lines. These sensors automatically shut off the gas when earthquake shaking is detected.

 **Reading Check** What steps can you take to make your home as earthquake safe as possible?

During an Earthquake

If you are indoors, stay there. Move away from windows and any objects that could fall on you. Seek shelter under a sturdy table or desk. If you are outdoors, stay in the open—away from power lines or anything that might fall.

 Visual Check **Figure 27** Identify and list some of the supplies you would include in your earthquake supply kit.

After an Earthquake

Stay calm and remember your family's earthquake disaster plan. If water and gas lines are damaged and leaking, the valves should be shut off by an adult. If you smell gas, leave the building immediately and call authorities from a phone away from the leak area. Be careful around broken glass and rubble, and wear boots or sturdy shoes to keep from cutting your feet. Finally, stay away from beaches. Tsunamis sometimes occur after the ground has stopped shaking.

ACADEMIC VOCABULARY
securely (seh KYUR lee)
(*adverb*) free from danger
The horse was kept securely in the pasture by a fence.

Figure 27 This family frequently reviews and practices its earthquake disaster plan. They have gathered all of their supplies, including canned food and water.

Surviving an Earthquake

Most of the damage and loss of life is not from the direct shaking from seismic waves, but from the collapse of objects that are shaken. If a structure is not built to withstand shaking from seismic waves, it is more likely to be destroyed by a large earthquake. Hazards of earthquakes, which include fires, landslides, liquefaction, and tsunamis, can also devastate a region.

It is not possible today to predict the precise time and location of an earthquake. Therefore, good land use and development choices, proper design of structures, retrofitting existing structures, and development of emergency plans are the best ways to protect human and wildlife habitats.

LESSON 4 Review

Summarize

Create your own lesson summary as you design a **visual aid.**

1. **Write** the lesson title, number, and page numbers at the top of your poster.

2. **Scan** the lesson to find the **red** main headings. Organize these headings on your poster, leaving space between each.

3. **Design** an information box beneath each **red** heading. In the box, list 2–3 details, key terms, and definitions from each **blue** subheading.

4. **Illustrate** your poster with diagrams of important structures or processes next to each information box.

 ELA6: R 2.4

Standards Check

Using Vocabulary

1. When loose sediment behaves like a liquid, the process is called _____. **1.g**

2. In your own words, write a definition for *tsunami*. **2.d**

Understanding Main Ideas

3. Which building material does not generally withstand shaking from earthquakes? **1.g**

   A. adobe
   B. wood
   C. rubber
   D. steel

4. **Describe** several earthquake hazards. **2.d**

5. **Describe** a safe place to be during an earthquake. **2.d**

6. **Categorize** Draw a diagram like the one below. Categorize factors that increase or decrease earthquake damage. **2.d**

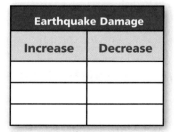

Earthquake Damage	
Increase	**Decrease**

Applying Science

7. **Design** a house that is earthquake safe. Label the features that make the house safe. **2.d**

8. **Evaluate** the seismic safety of your home. **2.d**

Science nline
For more practice, visit **Standards Check** at ca6.msscience.com.

DataLab

Can you locate areas at risk for earthquakes?

Some maps can provide you with information about where earthquakes are likely to occur. The locations of epicenters and faults can be used to infer the risk of earthquakes in a region.

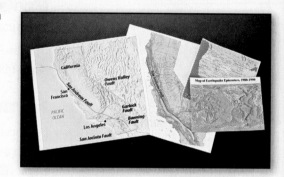

Data Collection

1. Obtain a **political map of California** from your teacher.

2. Plot the location of your school on the map.

3. Obtain **maps with epicenters of significant California earthquakes** from your teacher. Transfer the epicenters of earthquakes that have occurred over the past ten years onto the map. Concentrate on an area that is within about 100 km of your school in any direction.

4. Obtain **maps with faults in California.** Transfer the faults onto the map.

5. Create a map legend that shows the symbols you used for plotting epicenter and fault locations.

Data Analysis

1. **Describe** the general area of your school in terms of the presence of faults and the number of earthquake epicenters.

2. **Infer** whether your school is at moderate or high risk for the occurrence of an earthquake.

3. **Compare** your inference in question 2 with the seismic risk map of California shown in **Figure 23**.

 Science Content Standards

1.g Students know how to determine the epicenter of an earthquake and know that the effects of an earthquake on any region vary, depending on the size of the earthquake, the distance of the region from the epicenter, the local geology, and the type of construction in the region.
7.b Select and use appropriate tools and technology (including calculators, computers, balances, spring scales, microscopes, and binoculars) to perform tests, collect data, and display data.

 **MA6:** MR 2.4

Use the Internet:
Preparing for an Earthquake

Materials

articles
notes
research materials
art materials
computer with
  internet access

Science Content Standards

1.g Students know how to determine the epicenter of an earthquake and know that the effects of an earthquake on any region vary, depending on the size of the earthquake, the distance of the region from the epicenter, the local geology, and the type of construction in the region.
2.d Students know earthquakes, volcanic eruptions, landslides, and floods change human and wildlife habitats.
7.a Develop a hypothesis.
7.d Communicate the steps and results from an investigation in written reports and oral presentations.

Problem

You've read about the hazards to humans and other organisms that are associated with earthquakes. Use your knowledge about earthquakes to educate others on how to prepare for one.

Form a Hypothesis

Understanding certain factors at home, at school, and in rural or urban areas can help with earthquake preparedness. What do you think needs to be included in an educational brochure? Write a concise statement supporting the components you think need to be included in your product to help others be prepared for an earthquake.

Collect Data and Make Observations

1. Gather information you already have about causes, effects, and readiness for earthquakes.
2. Plan with your group who will be responsible for each segment of your project and assess what each of you has available already.
3. Bring photos, maps, and data tables to support your arguments.
4. Visit ca6.msscience.com to research organizations and government programs that assist in disasters.
5. Organize your materials and eliminate anything extraneous.
6. Design a format for an educational brochure to teach the important behaviors related to earthquakes.
7. Construct your brochure.

Analyze and Conclude

1. **Consider** the list of requirements and check your information against the list.

2. **Evaluate** your brochure. Include a discussion about your list of emergency supplies, directions for the safest places to go, a list of things to do and not to do, and a plan for where to meet family afterward.

Communicate

WRITING in Science   **ELA6: W 2.2**

Share your brochures with your classmates. Evaluate those of each other. Make a bulletin board of all the brochures so others can learn from you.

Real World Science

Designing Safer Structures

The engineers shown here are using the world's first outdoor shake table at the University of California, San Diego, to simulate the effects of a 7.3-magnitude earthquake on a 21-m-tall wind turbine. The data will be used by structural engineers to make structures safer and more stable during earthquakes.

Numerous California colleges and universities offer structural engineering programs. Visit **Careers** at **ca6.msscience.com** to investigate what students in these programs study. **List** 10 courses you would be most likely to take if you were a structural engineering student. Explain your choices.

Measuring Major Earthquakes

The California Strong Motion Instrumentation Program is building a network of strong motion accelerographs that will include over 900 locations statewide. This accelerograph is installed in a fire station. This technology could save lives by identifying unstable structures in the aftermath of a major earthquake.

Visit **Technology** at **ca6.msscience.com** to view recent California earthquake data. Plot the approximate location of each event on a state map. **Label** the event magnitude and date. **Analyze** the data for patterns or trends.

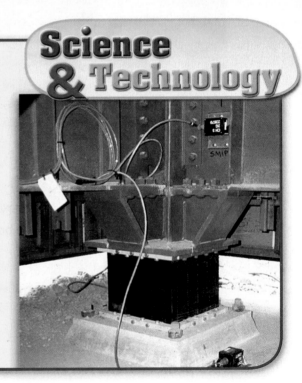

EARTHQUAKE OF 1906

At 5:12 A.M. on April 18, 1906, residents of San Francisco were jolted awake by an earthquake. It was the first of 27 separate earthquakes that would rattle along the San Andreas Fault that day and by far the strongest, with an estimated magnitude of 8.3. Fires broke out all over the city. The fires burned for more than two days, consuming 30 schools, 80 churches, and 250,000 homes. Historians believe more than 3,000 people lost their lives in the aftermath of the earthquake.

Visit **History** at **ca6.msscience.com** to read accounts of events during and after the 1906 earthquake. **Prepare** an oral presentation that describes what you might have seen, heard, and felt had you been in San Francisco on April 18, 1906.

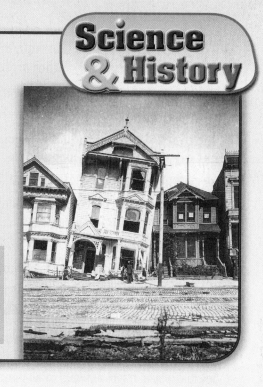

Preparing for the Next Tsunami

On December 26, 2004, an underwater earthquake occurred. This earthquake caused a tsunami to smash into largely unprepared coastal areas. Could an early warning system have prevented some of the estimated 283,000 deaths? Today, scientists are developing the Indian Ocean Tsunami Warning System. The role of coastal communities will be essential in preventing another disaster of this magnitude. Local governments will be required to develop action plans, provide emergency response training, and establish evacuation procedures.

Brainstorm a list of answers to the following question: "What factors should a community consider when helping citizens prepare for a tsunami?" Based on your list, **create** an outline showing key components of an effective disaster plan.

TSUNAMI EARLY WARNING SYSTEM

The BIG Idea Earthquakes cause seismic waves that can be devastating to humans and other organisms.

Lesson 1 Earthquakes and Plate Boundaries

1.d, 1.e, 7.e

Main Idea **Most earthquakes occur at plate boundaries when rocks break and move along faults.**

- An earthquake is a sudden rupture and movement on a fault.
- Energy from motion of lithospheric plates builds up in rocks as elastic strain until it ruptures as an earthquake.
- An earthquake begins at the focus.
- Some earthquakes occur in the middle of plates.

- **earthquake** (p. 246)
- **elastic strain** (p. 247)
- **focus** (p. 248)

Lesson 2 Earthquakes and Seismic Waves

1.g, 7.e

Main Idea **Earthquakes cause seismic waves that provide valuable data.**

- Seismic waves travel outward in all directions from the focus. Primary and secondary waves travel through Earth's interior.
- Surface waves travel around the planet within a thin layer of the crust.
- Earth's internal structure was discovered by analysis of seismic waves.

- **epicenter** (p. 253)
- **primary wave** (p. 254)
- **secondary wave** (p. 254)
- **seismic wave** (p. 252)

Lesson 3 Measuring Earthquakes

1.g, 7.b, 7.g

Main Idea **Data from seismic waves are recorded and interpreted to determine the location and size of earthquakes.**

- Instruments called seismographs produce records of seismic waves called seismograms.
- The difference in P- and S-wave arrival times allows scientists to determine the distance of a seismograph from the epicenter.
- Different scales are used to measure earthquake magnitude and intensity.

- **seismogram** (p. 261)
- **seismograph** (p. 261)

Lesson 4 Earthquake Hazards and Safety

1.g, 2.d, 7.a, 7.b, 7.d

Main Idea **Effects of an earthquake depend on its size and the types of structures and geology in a region.**

- Scientists use knowledge of faults and ground conditions to determine earthquake hazard.
- Land use helps reduce earthquake damage and loss of life.
- Some types of construction are more earthquake-resistant than others.
- Earthquake-resistant structures save lives and property.
- You can protect yourself from future earthquakes by being prepared.

- **liquefaction** (p. 271)
- **tsunami** (p. 272)

STUDY TO GO Download quizzes, key terms, and flash cards from ca6.msscience.com.

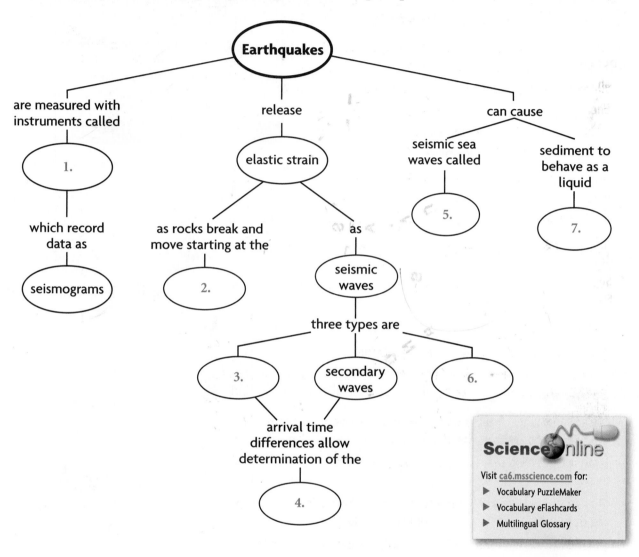

Linking Vocabulary and Main Ideas

Use the vocabulary terms from page 284 to complete this concept map.

Earthquakes

are measured with
instruments called

1.

which record
data as

seismograms

release

elastic strain

as rocks break and
move starting at the

2.

as

seismic
waves

three types are

3.

secondary
waves

6.

arrival time
differences allow
determination of the

4.

can cause

seismic sea
waves called

5.

sediment to
behave as a
liquid

7.

Science Online

Visit <u>ca6.msscience.com</u> for:

▶ Vocabulary PuzzleMaker
▶ Vocabulary eFlashcards
▶ Multilingual Glossary

Using Vocabulary

Fill in the blanks with the correct vocabulary words. Then read the paragraph to a partner.

When an earthquake occurs, _____8._____ is released in the form of rocks breaking and moving and also in the form of _____9._____. To determine the _____10._____, the point on Earth's surface directly above the focus, scientists analyze the arrival times of _____11._____ waves and _____12._____ waves on data records known as _____13._____. _____14._____ are also recorded, which are the seismic waves that generally cause the most damage and loss of life resulting from an earthquake.

Understanding Main Ideas

Choose the word or phrase that best answers the question.

1. Which builds up in rocks before an earthquake?
　A. heat energy
　B. elastic strain
　C. chemical strain
　D. kinetic energy
`1.d`

2. Where do the largest earthquakes occur?
　A. convergent plate boundaries
　B. elastic boundaries
　C. transform plate boundaries
　D. divergent plate boundaries
`1.e`

3. Where does the energy that causes earthquakes come from?
　A. Earth's interior
　B. Earth's surface
　C. the Sun
　D. tsunamis
`1.d`

4. The map below includes data from an earthquake.

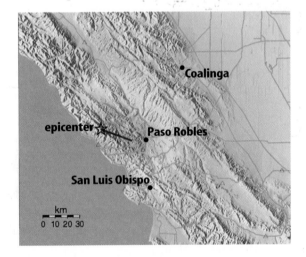

Which measure of an earthquake does this map show?
　A. fault slip
　B. intensity
　C. moment magnitude
　D. Richter magnitude
`1.g`

Use the figure below to answer questions 5 and 6.

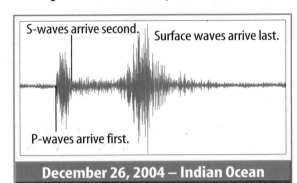

December 26, 2004 — Indian Ocean

5. What is the printed record of ground shaking during an earthquake?
　A. epicenter
　B. seismogram
　C. seismograph
　D. seismologist
`1.g`

6. Which seismic waves travel the fastest?
　A. primary waves
　B. secondary waves
　C. surface waves
　D. tsunami waves
`1.g`

7. Which is a scale used to measure the energy released by an earthquake?
　A. elastic scale
　B. magnitude scale
　C. modified Mercalli scale
　D. shaking scale
`1.g`

8. What is the name of the location where a fault rupture starts?
　A. epicenter
　B. focus
　C. fracture center
　D. hanging wall
`1.d`

9. To locate an epicenter, data are needed from at least how many seismographs?
　A. 1
　B. 2
　C. 3
　D. 100
`1.g`

Applying Science

10. Relate the transfer of kinetic energy from a lithospheric plate to rocks near a fault to what happens when you break a stick. **1.d**

11. Suggest a way to prepare an earthquake supply kit. **2.d**

12. Predict whether earthquakes would be larger or smaller if rocks that rupture along faults were stronger. **1.g**

13. Organize these locations with respect to earthquake safety: an open field, under a desk, outside next to your home, inside a tall building. Defend your organization. **2.d**

14. Suggest a reason that surface waves do more damage than primary waves. **1.g**

15. Identify the type of seismic wave illustrated below. **1.g**

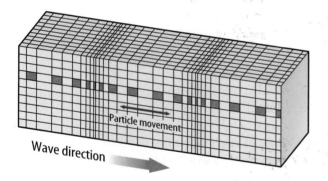

Particle movement

Wave direction

WRITING in Science

16. Write an outline that includes strategies for city planning in an area at high risk for earthquakes. Include design ideas for structures as well as land use strategies. Use the outline to give an oral presentation to the class. **ELA6: W 2.5**

Cumulative Review

17. Suggest a reason there is no continental drift on the Moon. **1.a**

18. Compare and contrast continental and oceanic lithosphere. **1.b**

19. Locate the Cascade and Transverse mountain ranges on a map of California. **1.f**

Applying Math

Use this table to answer questions 20–24.

Seismic Wave Number	Difference in Arrival Time from P-Wave to S-Wave (s)
1	4.9
2	8.7
3	12.3
4	17.8
5	18.0
6	19.0
7	24.4
8	42.7
9	51.9
10	52.9

20. Determine the distance from the origin of the third seismic wave to your current location. **MA6: NS 2.0**

21. If both the second and third seismic waves occur at the same depth and direction from your location, how far apart are their origins? **MA6: NS 2.0**

22. Determine the distance from the origin of the fourth seismic wave to your current location. **MA6: NS 2.0**

23. If both the third and fourth seismic waves occur at the same depth and direction from your location, how far apart are their origins? **MA6: NS 2.0**

24. Determine the distance from the origin of the fifth seismic wave to your current location. **MA6: NS 2.0**

Use the figure below to answer questions 1 and 2.

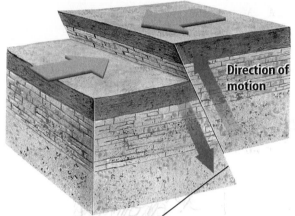

Direction of motion

Reverse fault surface

1 **A reverse fault typically occurs at what type of boundary?**

  A convergent plate boundary

  B divergent plate boundary

  C transform plate boundary

  D strike-slip   `1.d`

2 **Reverse faults at convergent plate boundaries typically produce what kind of earthquakes?**

  A shallow, small in size

  B shallow, large in size

  C deep, devastating

  D deep, no damage   `1.e`

3 **Which best describes the epicenter of an earthquake?**

  A the point on Earth's surface located directly above the earthquake focus

  B the point inside Earth where movement first occurs during an earthquake

  C the surface of a break in a rock along which there is movement

  D the snapping back of a rock that has been strained by force   `1.g`

4 **The figure below shows one type of instrument that records seismic waves.**

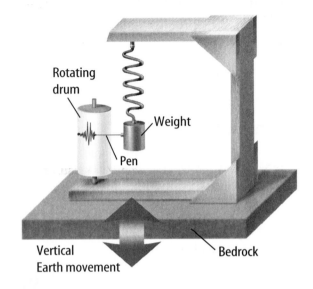

Rotating drum

Weight

Pen

Vertical Earth movement

Bedrock

What is this instrument called?

  A creepmeter

  B seismograph

  C spectrograph

  D tiltmeter   `1.g`

5 **Waves of energy that travel through Earth's interior and along Earth's surface are called?**

  A sound waves

  B energy waves

  C light waves

  D seismic waves   `1.g`

6 **Which is the region where no seismic waves reach Earth's surface?**

  A pressure zone

  B seismic zone

  C shadow zone

  D waveless zone   `1.g`

7 Imagine a tsunami occurs near the Aleutian Islands. The wave reaches the Hawaiian Islands, a distance of 3,800 km, 5 h later. At what speed is the wave traveling?

A   570 km/h

B   670 km/h

C   700 km/h

D   760 km/h   `2.d`

8 What does earthquake intensity measure?

A   damage done

B   earthquake's focus

C   energy released

D   seismic risk   `1.g`

9 The photo below shows the effects of liquefaction on structures during an earthquake.

What is liquefaction?

A   ice melting during an earthquake to cause flooding

B   rivers diverted by flooding caused by earthquakes

C   seismic waves shaking sediment, causing it to become more liquid-like

D   the stopping of S-waves by Earth's liquid outer core   `2.d`

10 Why should you stay away from beaches after an earthquake has occurred?

A   Aftershocks can occur.

B   Fires can occur.

C   Tsunamis can occur.

D   Water becomes polluted.   `2.d`

11 The map below shows three circles drawn around three different seismograph stations.

Which labeled point on the map represents the earthquake's epicenter?

A   Los Angeles

B   Sacramento

C   San Francisco

D   San Diego   `1.g`

12 What is elastic strain?

A   the point on Earth's surface located directly above the earthquake focus

B   the point inside Earth where movement first occurs during an earthquake

C   the surface of a break in a rock along which there is movement

D   the snapping back of a rock that has been strained by force   `1.d`

CHAPTER 7

Volcanoes

The BIG Idea

Volcanoes are locations where magma reaches Earth's surface. They affect human and wildlife habitats.

LESSON 1 1.d, 1.e, 7.b
Volcanoes and Plate Boundaries
Main Idea Most volcanic activity occurs along plate boundaries where plates move relative to one another.

LESSON 2 1.d, 1.f, 7.g, 7.h
Volcanic Eruptions and Features
Main Idea The composition of magma controls volcanic eruptions and determines the different types of lava flow and volcanic features.

LESSON 3 2.d, 7.a, 7.b, 7.d
Hazards of Volcanic Eruptions
Main Idea Volcanic eruptions can change human and wildlife habitats.

Beautiful, but Dangerous

These tourists watch a fountain of on the coast of Hawaii, at the Kilauea volcano. The Hawaiian Islands are forming as a result of volcanic activity. The Hawaiian Islands are in the mi of the Pacific Plate, far from its edges. They sit above a hot spot under the Pacific Plate. Volcanic eruptions create scenic landscapes as the lava is ejected from volcanoes.

Science Journal Imagine you are among the tourists viewing the volc eruption. Make a list of what you might hear, smell, feel, see, and possibly taste while watching the eruption.

How did these rocks form?

Examine three photos of rocks or rock samples from your teacher. Each of these rocks formed from lava ejected during a volcanic eruption.

Procedure

1. Design a chart to record the color, texture, and mass of each **rock.**

2. Estimate the number and size of bubbles and crystals in each rock.

Think About This

- **Infer** which rock cooled the fastest and which one cooled most slowly.

- **Predict** what environmental conditions might have existed when each rock was formed.

 1.d, 7.e

Visit ca6.msscience.com to:

- ▶ view **Concepts in Motion**
- ▶ explore Virtual Labs
- ▶ access content-related Web links
- ▶ take the Standards Check

 **Volcanic Features** Make the following Foldable to identify volcanic features.

▷ **STEP 1** **Fold** a sheet of paper in half lengthwise.

▷ **STEP 2** **Cut** along every third line of the top flap to form tabs.

▷ **STEP 3** **Label** as shown.

Volcanic
Features

Reading Skill ELA6: R 2.4

Reviewing
As you read Lesson 2, identify and list the volcanic features on the tabs. Include a sketch and information under the tabs.

Get Ready to Read

Make Predictions

① Learn It! A prediction is an educated guess based on what you already know. One way to predict while reading is to guess what you believe the author will tell you next. As you are reading, each new topic should make sense because it is related to the previous paragraph or passage.

② Practice It! Read the excerpt below from Lesson 2. Based on what you have read, make predictions about what you will read in the rest of the lesson. After you read Lesson 2, go back to your predictions to see if they were correct.

Think about how the composition of magma could affect a volcanic eruption.

The **composition of magma** is an important characteristic when attempting to predict a volcanic eruption. Scientists can predict the **energy of a volcanic eruption** based on the percentage of silica and oxygen that is present in the magma. **Silica is just one of the chemical characteristics** that controls the eruptive behavior of a body of magma.

—*from page 302*

Predict how the amount of silica present in magma could affect the energy of a volcanic eruption.

Determine what other chemical characteristics control the eruptive behavior of a body of magma.

③ Apply It! Before you read, skim the questions in the Chapter Assessment. Choose three questions and predict the answers.

Target Your Reading

Use this to focus on the main ideas as you read the chapter.

Reading Tip

As you read, check the predictions you made to see if they were correct.

1 **Before you read** the chapter, respond to the statements below on your worksheet or on a numbered sheet of paper.

- Write an **A** if you **agree** with the statement.
- Write a **D** if you **disagree** with the statement.

2 **After you read** the chapter, look back to this page to see if you've changed your mind about any of the statements.

- If any of your answers changed, explain why.
- Change any false statements into true statements.
- Use your revised statements as a study guide.

Science  **nline**

Print a worksheet of this page at ca6.msscience.com.

Before You Read A or D	Statement	After You Read A or D
	1 Most volcanic eruptions occur at plate boundaries.	
	2 Magma rises buoyantly and exerts an upward force on Earth's surface.	
	3 Scientists are able to predict when a volcano will erupt.	
	4 Volcanic ash is dangerous in the air and on the ground.	
	5 Hot spots form where two oceanic plates converge.	
	6 Lava flows are the most dangerous type of volcanic hazard.	
	7 A lava dome is filled with thick, viscous lava.	
	8 Lava that contains high amounts of silica can be extremely explosive.	
	9 Shield volcanoes are the smallest type of volcano.	
	10 Composite volcanoes are composed of alternating layers of lava and tephra.	

Reading Guide

What *You'll Learn*

▶ **Explain** what causes volcanic activity.

▶ **Relate** the location of volcanoes to plate boundaries.

Why *It's Important*

Understanding how volcanoes form and where they occur helps scientists predict volcanic eruptions.

Vocabulary

volcano
hot spot
vent
fissure eruption

Review Vocabulary

lithospheric plate: large, brittle pieces of Earth's outer shell composed of crust and uppermost mantle (p. 183)

Volcanoes and Plate Boundaries

Main Idea Most volcanic activity occurs along plate boundaries where plates move relative to one another.

Real-World Reading Connection Maybe you've been outside when a hot-air balloon flew overhead. When the pilot turned on the flame under the balloon, the balloon rose higher in the sky. As temperature increases, particles move faster, causing the air to expand and become less dense. Particles in molten rock also move faster. The molten rock expands and becomes less dense as the temperature increases.

What is a volcano?

A **volcano** is a land or underwater feature that forms when magma reaches the surface of Earth. Recall from Chapter 2 that magma is molten, liquid rock material found underground and forms igneous rocks. Magma rises up to Earth's surface because its density is less than the rock through which it moves. When magma rises to Earth's surface it is called lava.

Volcanoes can cause the eruption of liquid, gas, or solid materials. Hot gases, magma, and even solid rock particles can be explosively erupted from openings on Earth. Scientists, as shown in **Figure 1,** study the composition of lava to better understand volcanic eruptions.

Figure 1 This volcanologist is monitoring volcanic activity.

How do volcanoes form?

Recall from Chapter 2 that some of Earth's internal heat is left over from when the planet first formed. When rocks become hot enough, and pressure and other conditions are right, they can melt.

 **Reading Check** What causes rock to melt?

In order for a volcano to form, magma must first reach Earth's surface. Magma is less dense then the rock from which it melted. Therefore, magma tends to rise above the denser rock. The density of magma depends upon its composition, the amount of dissolved gas, and its temperature.

Magma also tends to be more buoyant when compared to the rocks that surround it. Magma is forced to rise toward Earth's surface because of the buoyant force pushing up on the magma.

Where do volcanoes occur?

Volcanoes are not common in all regions of Earth's crust. Volcanoes typically occur along divergent and convergent plate boundaries and hot spots, where the flow of heat from Earth's interior is high. **Figure 2** shows the locations of active volcanoes and plate boundaries on Earth. Do you notice a relationship between the locations of active volcanoes and plate boundaries?

Figure 2 This world map shows that most active volcanoes are located near the edges of the plate boundaries. The Ring of Fire is a belt of active volcanoes that circles the Pacific Ocean.

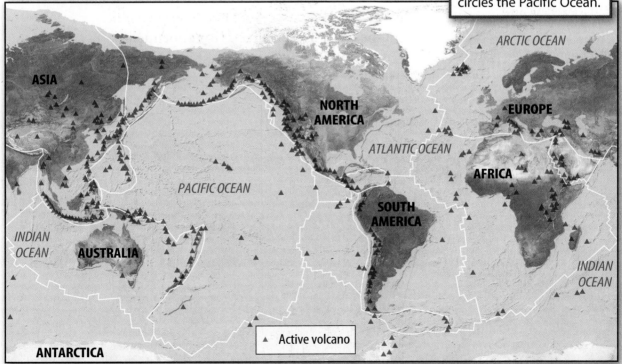

▲ Active volcano

Convergent Plate Boundaries

Recall from Chapter 5 that when two lithospheric plates move toward each other, a convergent plate boundary forms. They include areas where an oceanic plate is subducted below a continental plate or where an oceanic plate is subducted below another oceanic plate. As the plate subducting below the other plate gets deep enough, it becomes hot enough to partially melt and form magma. The magma is then forced up toward Earth's surface and forms volcanoes. Volcanoes that form along convergent plate boundaries tend to erupt more violently than other volcanoes. **Figure 3** shows examples of convergent plate boundary landforms.

 Figure 3 Identify the locations of two convergent plate boundary landforms.

Volcanic Arcs When an oceanic plate subducts beneath a continental plate, a volcanic arc is formed. A volcanic arc is a string of volcanoes that forms on land parallel to the leading edge of the continent. An example of a volcanic arc is shown in **Figure 3.**

Island Arcs When two oceanic plates move toward each other, an island arc is formed. Island arcs are long, curved strings of volcanic islands. **Figure 3** shows an example of an island arc volcano.

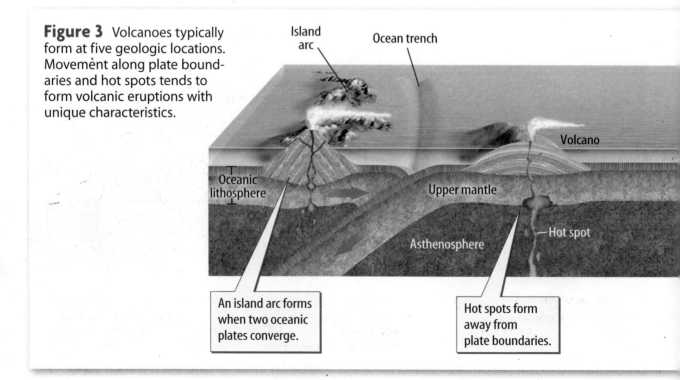

Figure 3 Volcanoes typically form at five geologic locations. Movement along plate boundaries and hot spots tends to form volcanic eruptions with unique characteristics.

Island arc

Ocean trench

Volcano

Oceanic lithosphere

Upper mantle

Asthenosphere

Hot spot

An island arc forms when two oceanic plates converge.

Hot spots form away from plate boundaries.

Divergent Plate Boundaries

Imagine a divergent plate boundary as a huge crack, or rift, in Earth's surface. The boundary can occur in either the ocean, as a mid-ocean ridge, or on a continent, as a continental rift. Tension stresses cause the lithospheric plates to be pulled apart. This pulling action forms rifts that allow magma to reach Earth's surface through fissure eruptions.

In the ocean, lava flows from these rifts are cooled quickly by seawater and form new volcanic rock. As more lava flows and hardens, it builds up on the seafloor. Sometimes, the volcanoes rise above sea level, forming islands such as Iceland.

 **Reading Check** How do volcanoes form at divergent plate boundaries?

Heat Escapes Volcanic eruptions are one of the most noticeable signs that heat is escaping from Earth's interior. After many of thousands or even millions of years, magma reaches Earth's surface. When this magma erupts from the central, circular or oval-shaped opening of a volcano, called a **vent,** a cone-shaped landform develops. This landform develops because lava flows out from one **source** in many directions. As lava flows out, it cools quickly and becomes solid, forming layers of igneous rock around the vent. The steep-walled depression around a volcano's vent is the crater.

SCIENCE USE v. COMMON USE
rift
Science Use fissure, crevasse. *As the lithospheric plates were pulled apart, a rift valley was created.*
Common Use to burst open, divide. *The argument caused a rift between the two friends.*

ACADEMIC VOCABULARY
source (SORS)
(*noun*) point of origin
We found the source of the loud music to be a concert.

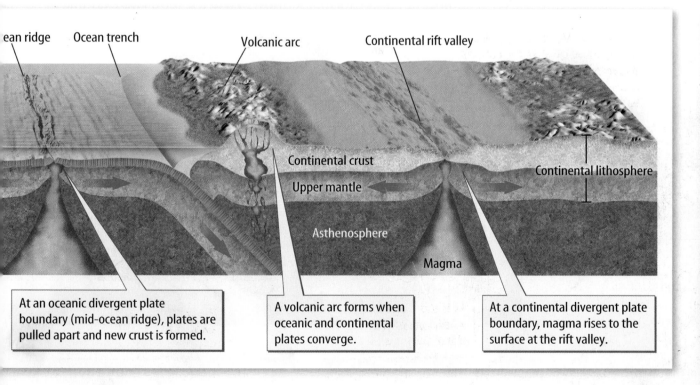

At an oceanic divergent plate boundary (mid-ocean ridge), plates are pulled apart and new crust is formed.

A volcanic arc forms when oceanic and continental plates converge.

At a continental divergent plate boundary, magma rises to the surface at the rift valley.

Figure 4 This crack formed because of a fissure eruption at a mid-ocean ridge. The currents caused by this activity attract the crabs.

Fissure Eruptions When magma escapes from narrow and elongated cracks in Earth's crust, a **fissure eruption** (FIH shur · ih RUP shunz) occurs. The magma pushes out along both sides of the crack and flows smoothly. Long, sheet-shaped landforms develop.

Fissure eruptions can occur in the ocean at a divergent plate boundary. They form mid-ocean ridges and new sea-floor, as shown in **Figure 4.** Fissure eruptions can also occur on a continent at a divergent plate boundary. They form a rift and produce new crust on Earth's surface.

Volcanoes Away from Plate Boundaries

Some volcanoes do not form along plate boundaries. These volcanoes are known as **hot spots,** which are localized areas of high heat in Earth's interior. Scientists continue to study how hot spots form and move. **Figure 5** shows an example of a hot spot.

Figure 5 The Hawaiian Islands are actually volcanoes that formed over a hot spot. They formed over a period of about 5 million years.

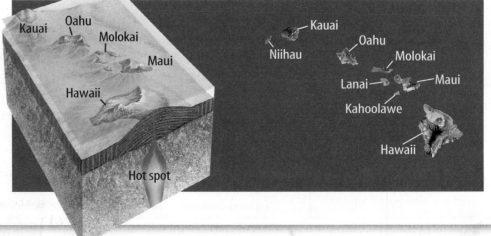

Formation of Volcanoes

Heat and pressure from Earth's interior cause rock to melt and become magma. Because magma is less dense than the surrounding rock, the buoyant force causes it to rise to Earth's surface. Most volcanoes form at divergent or convergent plate boundaries. However, hot spots form away from plate boundaries. Scientists continue to study hot spots to better understand how and why they form.

LESSON 1 Review

Summarize

Create your own lesson summary as you design a **study web.**

1. **Write** the lesson title, number, and page numbers at the top of a sheet of paper.

2. **Scan** the lesson to find the **red** main headings.

3. **Organize** these headings clockwise on branches around the lesson title.

4. **Review** the information under each **red** heading to design a branch for each **blue** subheading.

5. **List** 2–3 details, key terms, and definitions from each **blue** subheading on branches extending from the main heading branches.

 ELA6: R 2.4

Standards Check

Using Vocabulary

Complete the sentences using the correct term.

vent   fissure eruption

1. A cone-shaped landform develops when magma erupts from the _____ forming a volcano. **1.d**

2. Long, sheet-shaped landforms develop when a(n) _____ pushes magma out along both sides of a crack in Earth's crust. **1.d**

Understanding Main Ideas

3. **Illustrate** and label two types of openings from which magma erupts. **1.d**

4. **Organize** Draw a diagram like the one below. List five settings where volcanoes occur. **1.e**

Volcanoes occur
1. _____
2. _____
3. _____
4. _____
5. _____

5. **Sketch** the formation of a volcano. Include the forces at work in Earth's interior that cause a volcano to form and magma to flow to the surface. **1.d**

6. **Compare and contrast** the movement of lithosphere at a mid-ocean ridge and a subduction zone. **1.e**

Applying Science

7. **Hypothesize** why volcanoes are not likely where two continental plates converge. Support your answer with information from this lesson. **1.e**

8. **Relate** the occurrence of volcanoes along the Pacific Ring of Fire to the types of plate boundaries in the region. Consider the direction in which the boundaries are moving and any geologic features that affect the volcanic activity. **1.e**

 **Science Online**

For more practice, visit **Standards Check** at ca6.msscience.com.

MiniLab

How do volcanoes form?

Buoyant forces push magma up and cause it to rise to Earth's surface. This building of pressure causes volcanic eruptions. Volcanic eruptions allow heat to escape from Earth's interior. Can you model how volcanoes form and erupt?

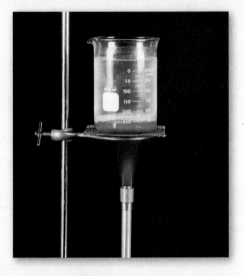

Procedure

1. Read and complete a lab safety form.
2. Obtain a **beaker** with hardened **wax** in the bottom from your teacher.
3. Layer 1 cm of **sand** on top of the wax.
4. Pour 8 cm of **cold water** onto the sand.
5. Set up a **ring stand** with **wire gauze** to hold your beaker over a **Bunsen burner**. Light the burner.
6. Observe the wax as it erupts through the sand and water.
7. Extinguish the burner. Allow the wax to cool.
8. Record your observations of the exposed wax formations.
9. Compare your volcano to your classmates'.

Analysis

1. **Sequence** how your volcano developed, erupted, and formed volcanic features.
2. **Infer** how the density of the wax changed as it was heated.
3. **Compare and contrast** your model to a real-life volcano.

Science Content Standards

1.d Students know that earthquakes are sudden motions along breaks in the crust called faults and that volcanoes and fissures are locations where magma reaches the surface.

7.b Select and use appropriate tools and technology (including calculators, computers, balances, spring scales, microscopes, and binoculars) to perform tests, collect data, and display data.

Science Content Standards

1.d Students know that earthquakes are sudden motions along breaks in the crust called faults and that volcanoes and fissures are locations where magma reaches the surface.

1.f Students know how to explain major features of California geology (including mountains, faults, volcanoes) in terms of plate tectonics.

7.g Interpret events by sequence and time from natural phenomena (e.g., the relative ages of rocks and intrusions).

Also covers: 7.h

Reading Guide

What *You'll Learn*

▶ **Relate** the composition of magma to characteristics of volcanic eruptions.

▶ **Compare** and contrast different types of volcanoes.

▶ **Analyze** California's volcanic activity.

▶ **Differentiate** between volcanic features.

Why *It's Important*

Understanding the composition of magma helps scientists better predict how volcanoes erupt.

Vocabulary

viscosity
shield volcano
cinder cone volcano
tephra
composite volcano

Review Vocabulary

landform: feature sculpted by processes on Earth's surface (p. 79)

Volcanic Eruptions and Features

Main Idea The composition of magma controls volcanic eruptions and determines the different types of lava flow and volcanic features.

Real-World Reading Connection Think about squeezing toothpaste out of the tube. A thick, gooey blob of paste rolls slowly out the end. Imagine the tube filled with water instead of toothpaste. The composition of toothpaste and water are different. How would this difference affect the manner in which these materials flow?

What controls volcanic eruptions?

The composition of magma controls how lava flows and the way a volcano erupts. For example, in 2004, a thick, sticky mass of lava began flowing from Mount St. Helens' vent. Scientists have termed the mass the *whaleback* because of its unique shape. This blob of lava, shown in **Figure 6,** flows the way it does—as a sticky, gooey, lumpy mass—because of its composition.

Figure 6 A thick, sticky mass of lava is flowing from the vent of Mount St. Helens.
Identify the volcanic feature in the center of the volcano.

Figure 7 The composition of magma affects the viscosity. Generally, magma that has a high percentage of silica has high viscosity. Magma that has a low percentage of silica has low viscosity.

Granitic

High viscosity
• Thick, lumpy, sticky
• Erupts explosively
• Flows slowly

Low heat
High silica

Low viscosity
• Thin and runny
• Erupts nonexplosively
• Flows freely

High heat
Low silica

Basaltic

Composition of Magma

The composition of magma is an important characteristic when scientists attempt to predict a volcanic eruption. Scientists can predict the energy of a volcanic eruption based on the percentage of silica and oxygen that is present in the magma. Silica is just one of the chemical characteristics that controls the eruptive behavior of magma.

High Silica In general, magma that contains a high percentage of silica is thicker and slower moving. Magma with high silica content tends to be thick and sticky, like honey or frosting. The **viscosity** (vihs KAH suh tee) of a material is a physical property that describes the material's resistance to flow. In other words, if magma has high viscosity, it does not flow easily.

Low Silica In general, magma that contains a low percentage of silica and more iron and magnesium has low viscosity. In other words, the magma will flow easily. Magma with low silica content tends to be thin and runny, like warm syrup. **Figure 7** shows how the composition of magma affects viscosity.

 **Reading Check** How does the percentage of silica affect the viscosity of magma?

Dissolved Gases and Temperature

The amount of dissolved gases and the temperature of magma also affect how a volcano erupts. In general, the higher the temperature of magma, the more easily it flows.

The most common dissolved gas in magma is water vapor, or H_2O. Other important gases that may be present in magma and released during volcanic eruptions are carbon dioxide, sulfur dioxide, and hydrogen sulfide.

Typically, the more gas that is present in magma, the greater the chance of an explosive eruption, even if the composition of the magma would suggest that it should have a quiet eruption.

Types of Magma and Lava

When you classify something, it's good to keep the main types in your mind as you compare examples. Try this for the types of magma and lava. **Figure 7** also shows two types of magma, basaltic (buh SAWL tihk), and granitic (gra NIH tihk).

Basaltic Magma and Lava

Magma that contains a low percentage of silica is called basaltic magma. Basaltic magma typically has low viscosity, meaning it flows freely. It is also a much thinner, more fluid magma.

Basaltic lava that erupts from a volcano tends to pour from the vent and run down the sides of the volcano. As this *pahoehoe* (pa HOY hoy) lava cools, it develops a smooth skin and forms ropelike patterns. As stiff, slowly moving *aa* (AH ah) lava forms, it flows at a lower temperature. Basaltic lava that erupts underwater at fissure eruptions forms bubble-like pillow lava. Basaltic lava tends to flow easily and produce quiet, nonexplosive eruptions, as shown in **Figure 8.**

Granitic Magma and Lava

Magma that contains a high percentage of silica is called granitic magma. When granitic magma **emerges** on Earth's surface, it typically will have high viscosity, meaning it flows slowly. It will also be more sticky and lumpy. Granitic lava tends to trap gases, which causes pressure to build up and produces explosive eruptions, as shown in **Figure 8.**

Figure 8 Compare the explosiveness of the two volcanic eruptions.

Quiet eruption

Explosive eruption

Figure 8 The percentage of silica, dissolved gases, and temperature are all physical characteristics that affect the eruptive behavior of magma.

Types of Volcanoes

There are three types of volcanoes that are common on Earth's surface—shield volcanoes, cinder cone volcanoes, and composite volcanoes. These volcanoes do not represent all possible types of volcanic eruptions, but they illustrate many common volcanic features and the processes that form them.

 **What are the three types of volcanoes?**

Shield Volcanoes

A **shield volcano** is a huge, gently sloping volcanic landform that is mainly composed of basaltic lava. Basaltic lava moves freely. Shield volcanoes develop as layer upon layer of gently flowing basaltic lava piles up. Shield volcanoes, like the one shown in **Table 1,** were named because they resemble the slightly bent shape of a warrior's shield. The Hawaiian Islands are examples of shield volcanoes. Types of lava that typically flow from a shield volcano are shown in **Figure 9.**

Cinder Cone Volcanoes

A **cinder cone volcano** is mainly composed of solid fragments. These solid fragments are known as **tephra** (TEH fruh) and include fragments of volcanic rock or lava. These fragments range in size from tiny particles to huge boulders. When lava erupts from the vent, cools quickly in the air, and falls to the surface as tephra, it forms a distinctly steep-sided, cone-shaped landform, as shown in **Table 1.**

Cinder cone volcanoes typically produce explosive volcanic eruptions.

Composite Volcanoes

A **composite volcano** is mainly composed of alternating layers of lava and tephra. These layers accumulate from the alternating of quiet and explosive volcanic eruptions. The quiet and explosive eruptions occur because the composition of magma associated with composite volcanoes is somewhere between basaltic and granitic.

 What causes the alternating layers of a composite volcano?

Composite volcanoes often form tall, majestic mountains, as shown in **Table 1.** Some volcanoes have reached altitudes of more than 2,438 meters.

WORD ORIGIN
tephra
from Greek; means *ashes*

Table 1 Types of Volcanoes

Type of Volcano	Examples of Volcanoes
Shield Volcanoes • Huge, gently sloping sides • Composed of basaltic lava • Develop from layers of basaltic lava • Quiet eruptions	 Belknap volcano, Oregon
Cinder Cone Volcanoes • Steep-sided, cone shaped • Composed of basaltic lava • Develop from layers of tephra • Explosive eruptions	 Black Butte volcano, California
Composite Volcanoes • Tall, majestic mountains • Composed of lava between basaltic and granitic • Develop from alternating layers of lava and tephra • Eruptions are both quiet and explosive	 Augustine volcano, Alaska

Visualizing Lava

Figure 9

Lava rarely travels faster than a few kilometers per hour. Therefore, it poses little danger to people. However, homes, property, and crops can be damaged. On land, there are two main types of lava flows—aa and pahoehoe. When lava comes out of cracks in the ocean floor, it forms pillow lava. The lava cooling here came from a volcanic eruption on the island of Hawaii.

Aa lava flows, like this one on Mount Etna in Italy, carry sharp, angular chunks of rock. Aa flows move more slowly than hotter pahoehoe flows.

Pahoehoe lava flows, like this one near Kilauea's Mauna Ulu Crater in Hawaii, are hotter and more fluid than aa flows. They develop a smooth skin and form ropelike patterns as they push forward and then cool.

Pillow lava occurs where lava oozes out of fissure eruptions in the ocean floor. It forms bubble-shaped lumps as it cools. Pillow lava is the most common type of lava on Earth.

Contributed by National Geographic

Volcanoes in California

As you have read in previous chapters, California has a large variety of landscapes and landforms. Examples of volcanic landforms found in California are shown in **Figure 10.**

These volcanoes form at a convergent plate boundary as part of a volcanic arc. They form where the Juan de Fuca Plate subducts beneath the North American Plate at the southern end of the Cascades Range.

Figure 10 This map shows the variety of active volcanic landforms along the western coast of the United States.

Identify a California volcano that is part of the volcanic arc.

Black Butte—Cinder Cone Volcano

Medicine Lake—Shield Volcano

Mount Shasta—Composite Volcano

Lassen Peak—Lava Dome

Canada

▲ Mount Baker
▲ Glacier Peak
● Seattle
Olympia ★ ▲ Mount Rainier
Washington
Mount St. Helens ▲ ▲ Mount Adams
Portland ● ▲ Mount Hood
Mount Jefferson
Belknap
Three Sisters
Diamond Peak ▲ ▲ Lava Butte
Mount Bailey ▲
Crater Lake
Oregon
Black Butte ▲ ▲ Medicine Lake-Lava Beds
Mount Shasta ▲ ▲ Brushy Bush
▲ Big Cave
Eagle Lake ▲ ▲ Lassen Peak
Potato Butte
Nevada
▲ Clear Lake ★ Carson City
Sacramento ★
Mono Lake
San Francisco ● Mono Crater ▲ Red Cones
Inyo Crater ▲ Long Valley Caldera
Golden Trout Creek ▲ ▲
▲ Coso
California
Lavic Lake ▲
Amboy
Los Angeles ●
San Diego ●
Mexico

N
W E
S

▲ Active volcano

Figure 11 The batholith in the top photo and the volcanic neck in the bottom photo are examples of volcanic features.

Intrusive Igneous Features

Most magma never reaches Earth's surface to form volcanoes or flow from fissure eruptions. This magma slowly cools underground and produces intrusive features that later could become exposed by erosion, as shown in **Figure 11.**

Batholiths

The largest intrusive igneous features are batholiths. Batholiths form when magma slowly cools and solidifies before reaching the surface. As shown in **Figure 12,** they can be many hundreds of kilometers in width and length and several kilometers thick.

Dikes and Sills

Magma sometimes squeezes into cracks in rocks below the surface, as shown in **Figure 12.** Magma that cuts across rock layers and hardens is called a dike. Magma that is parallel to rock layers and hardens is called a sill. Most dikes and sills run from a few meters to hundreds of meters long.

Figure 12 This diagram shows intrusive and other features associated with volcanic activity.

Identify which features shown are formed above ground. Which are formed below ground?

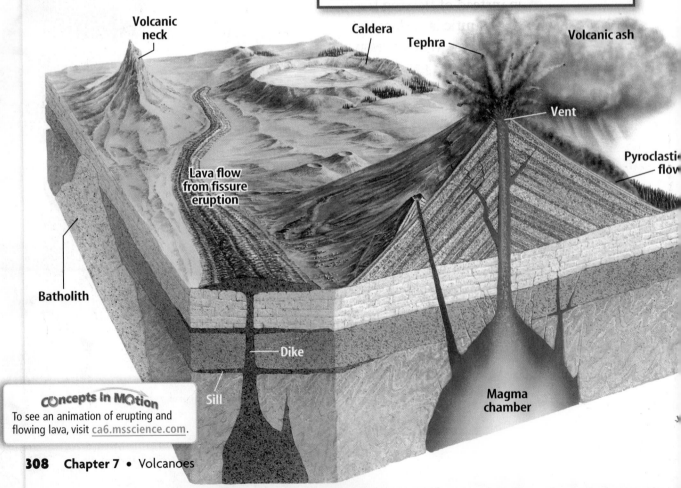

Volcanic neck

Caldera

Tephra

Volcanic ash

Vent

Pyroclastic flow

Lava flow from fissure eruption

Batholith

Dike

Sill

Magma chamber

CONcepts in MOtion
To see an animation of erupting and flowing lava, visit ca6.msscience.com.

Other Volcanic Features

Not all volcanic features form underground from igneous activity. Some volcanic features form from past and current volcanic eruptions and lava flows.

Volcanic Neck

When a volcano stops erupting, magma hardens inside the vent. Erosion, usually by water and wind, begins to wear away the volcano. The cone is much softer than the igneous rock in the vent, and tends to erode first, leaving behind the igneous core as a volcanic neck. Shiprock, located in New Mexico and shown in **Figure 13,** is an example of an eroded volcanic neck.

 Reading Check Which part of a volcano tends to erode first?

Lava Domes

A lava dome is a rounded volcanic feature that forms when a mass of highly viscous lava slowly erupts from the vent. The thick, sticky granitic lava piles up instead of flowing freely.

The viscosity of the granitic lava does not allow it to release gases easily. When pressure builds up within the lava dome, gas, lava, and solid materials can be violently ejected into the air.

Lava domes like the one shown in **Figure 13,** can become dangerous, explosive volcanic eruptions.

Lava Tubes

As swiftly moving magma flows from a magma chamber, its outer surface can cool, harden, and form a lava tube. When the magma stops flowing or changes direction, the lava tube drains and becomes a rocky, hollow tube. A lava tube can be over 8 m in diameter and as long as 48 km in length. Some lava tubes form underground. **Figure 13** shows a lava tube that was formed from pahoehoe.

Figure 13 Examples of other volcanic features include volcanic necks, lava domes, and lava tubes.

Volcanic Neck

Lava Dome

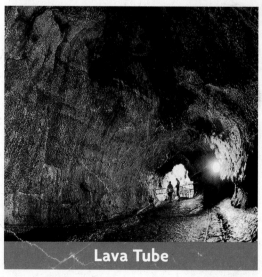

Lava Tube

Figure 14 Calderas form when the top of a volcano collapses.

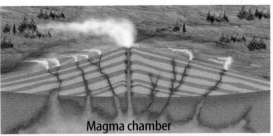

Magma chamber

Volcanic activity occurs when magma is forced upward.

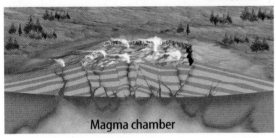

Magma chamber

A caldera forms when the magma chamber partially empties, causing rock to collapse into the emptied chamber below the surface.

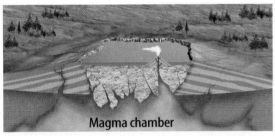

Magma chamber

When surface material collapses, water can collect and form scenic lakes.

Caldera

A caldera (cal DAYR uh) is a large, usually circular depression. A caldera forms when the top of a volcano collapses and becomes wedged into the nearly empty magma chamber as shown in **Figure 14.** The large, sunken crater can become filled with water, forming scenic lakes and landscapes. Calderas can be up to 50 km in diameter. Volcanic eruptions that form calderas are often the largest eruptions on Earth.

 Visual Check

Figure 14 Describe the steps in the formation of a caldera.

Long Valley Caldera, located just east of the Sierra Nevada, is an example of a caldera. It is a large, oblong caldera on top of an active volcano. It is plugged with a large, bubblelike lava dome.

Figure 15 shows another example of a caldera, Crater Lake. Crater Lake Caldera is 597 m deep. The lake partly fills a nearly 1,220 m deep caldera that was formed around 6,900 years ago by the collapse of the volcano Mount Mazama.

Figure 15 Wizard Island in Crater Lake is a cinder cone volcano that erupted after the formation of the caldera.

Characteristics of Volcanic Eruptions

The way a volcano erupts is controlled by the composition of the magma, the amount of dissolved gas, and its temperature. These factors also affect the viscosity and flow of lava, which result in different volcanic features. Shield volcanoes, cinder cone volcanoes, and composite volcanoes all are common types of volcanoes. Fissure eruptions also release heat from Earth's interior through lava flows. The Cascade Range located across California, Oregon, and Washington, contains potentially active volcanoes. Next, you will read about how volcanic eruptions can produce hazards that change human and wildlife habitats.

LESSON 2 Review

Summarize

Create your own lesson summary as you write a script for a **television news report.**

1. **Review** the text after the **red** main headings and write one sentence about each. These are the headlines of your broadcast.

2. **Review** the text and write 2–3 sentences about each **blue** subheading. These sentences should tell *who, what, when, where,* and *why* information about each **red** heading.

3. **Include** descriptive details in your report, such as names of reporters and local places and events.

4. **Present** your news report to other classmates alone or with a team.

ELA6: LS 1.4

 Standards Check

Using Vocabulary

1. In your own words, write a definition for *viscosity*. **1.d**

2. What is the difference between a cinder cone volcano and a composite volcano? **1.d**

Understanding Main Ideas

3. **Identify** three common types of volcanoes. **1.d**

4. **Organize** Draw a diagram like the one below. List three factors that affect the flow of lava. **1.d**

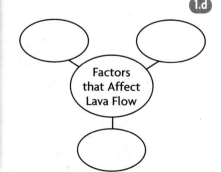

Factors that Affect Lava Flow

5. **Compare and contrast** the shape and composition of a cinder cone volcano with a shield volcano. **1.d**

6. **Predict** how the viscosity of lava affects volcanic features. **1.d**

Applying Science

7. **Infer** why some volcanic eruptions are explosive and some eruptions are quiet. **1.d**

8. **Evaluate** the conditions you would expect to find where the Pacific Plate and the North American Plate converge. Use an example of volcanic activity to support your reasoning. **1.f**

For more practice, visit **Standards Check** at ca6.msscience.com.

DataLab

Model Structures of Volcanoes

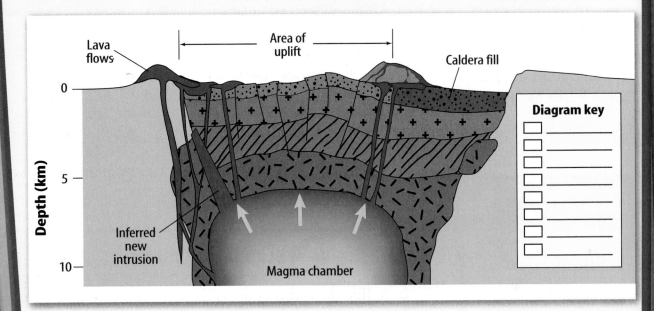

Diagram labels: Lava flows, Area of uplift, Caldera fill, Diagram key, Depth (km), 0, 5, 10, Inferred new intrusion, Magma chamber

Data

1. Redraw the diagram of the volcano in your notebook using colored pencils.

2. Create a diagram key to identify the depth of the three lava flows, magma, magma chamber, caldera, vent, and surrounding topography. Give your key a title.

Data Analysis

1. **Summarize** the event that formed the caldera.

2. **Suggest** three possible processes that might have formed the caldera fill.

3. **Sequence** the processes by which a lava dome might be formed in this volcano.

4. **Predict** how a batholith, dikes, and sills would become evident in this landscape.

 Science Content Standards  MA6: MR 1.1

1.d Students know that earthquakes are sudden motions along breaks in the crust called faults and that volcanoes and fissures are locations where magma reaches the surface.

7.g Interpret events by sequence and time from natural phenomena (e.g., the relative ages of rocks and intrusions).

7.h Identify changes in natural phenomena over time without manipulating the phenomena (e.g., a tree limb, a grove of trees, a stream, a hillslope).

Science Content Standards

2.d Students know earthquakes, volcanic eruptions, landslides, and floods change human and wildlife habitats.
7.a Develop a hypothesis.
7.b Select and use appropriate tools and technology (including calculators, computers, balances, spring scales, microscopes, and binoculars) to perform tests, collect data, and display data.
7.d Communicate the steps and results from an investigation in written reports and oral presentations.

Reading Guide

What *You'll Learn*

▶ **Describe** effects of volcanic eruptions on human and wildlife habitats.

▶ **Discover** geologic events that scientists observe and measure to help predict volcanic eruptions.

Why *It's Important*

As scientists become better able to predict and monitor volcanic eruptions, more lives and property are saved.

Vocabulary

volcanic ash
lahar
pyroclastic flow

Review Vocabulary

seismic waves: wave that travels through Earth generated by earthquake (p. 252)

Hazards of Volcanic Eruptions

Main Idea Volcanic eruptions can change human and wildlife habitats.

Real-World Reading Connection You might remember a day when you had plans to go outside, but couldn't because of poor air quality. Solar radiation can cause a chemical reaction with ozone in the atmosphere and gases emitted from cars and industry. Smog can be produced. Volcanic eruptions can cause changes in air quality too. But, the effects often last longer and can be more devastating than smog.

Effects on Habitats

When a volcano remains quiet for centuries, people tend to forget about the forces at work in Earth's interior, including those that could cause environmental hazards. One such hazard that affects human and wildlife habitats is the release of large amounts of fine-grained tephra, called **volcanic ash. Figure 16** shows volcanic ash that was ejected thousands of meters into the atmosphere from the eruption of Mount St. Helens in 1980.

Figure 16 Volcanic ash released from Mount St. Helens spread through the atmosphere and began falling like snow.

Figure 17 Examples of volcanic hazards include volcanic ash, lahars, gases, pyroclastic flows, and lava flows. They all cause change to human and wildlife habitats.

Volcanic Ash

Lahar

Volcanic Ash

Ash made from burning wood is similar to volcanic ash. They both contain tiny grains of material. But, unlike ash made from wood, volcanic ash is made of tiny, sharp mineral and glasslike particles.

 Reading Check What is the difference between ash made from burning wood and volcanic ash?

Damage to human and wildlife habitats can be caused by volcanic ash. When layers of ash build up on rooftops, the increased weight can cause structural damage. Volcanic ash can bury plants and animals and their food sources and contaminate the water supply. A cloud of volcanic ash from the eruption of Mount Vesuvius is shown in **Figure 17.**

Landslides and Lahars

Scientists refer to a **lahar** (LAH har) as a rapidly flowing mixture of volcanic debris and water. An example is shown in **Figure 17.** Some of the largest lahars begin as landslides. Landslides can occur from volcanic eruptions, earthquakes, precipitation, or gravity. Volcanic ash, tephra, dirt, rocks, and even trees can mix with ground water and precipitation and form a lahar. Rivers of debris can move downhill at rates up to tens of meters per second. It is not possible for humans to outrun a swiftly moving lahar.

When human habitats are built in river valleys near volcanoes, the flow of debris can become directed toward the town. A fast-moving lahar provides little time for warning. The effects can be disastrous.

Gases

Pyrocastic Flow

Lava Flow

Gas Emissions

Gases **released** from volcanoes can also be a silent hazard to humans and wildlife. Sulfur dioxide and hydrogen sulfide gases, when mixed with water, can form acidic precipitation. This precipitation can be harmful to animals and vegetation. **Figure 17** shows trees killed on Mammoth Mountain as a result of carbon dioxide seeping from beneath the volcano.

Pyroclastic Flows

Another hazard from explosive eruptions is a **pyroclastic flow,** which is a fast-moving body of hot gases and solids. Pyroclastic flows tend to follow valleys and can devastate everything in their paths. **Figure 17** shows an example of a pyroclastic flow as it moves quickly down the slopes of the Soufriere Hills volcano on Montserrat Island.

Intensely hot gas that travels within the pyroclastic flow can contaminate the surrounding air, burning and destroying wildlife and human habitats.

 Figure 17 Predict what might happen if a city were in the path of a pyroclastic flow.

Lava Flows

Lava flows can occur from either an explosive or quiet volcanic eruption. **Figure 17** shows an example of a lava flow in Hawaii. Lava flows can destroy human and wildlife habitats by starting fires, destroying property and crops, and releasing smoke, which affects air quality. Most lava flows move slowly enough that humans can be warned of possible dangers.

ACADEMIC VOCABULARY
release (rih LEES)
(verb) to set free from confinement
We released the balloons into the gentle breeze.

WORD ORIGIN
pyroclastic
pyro– from Greek; means *fire*
clastic from Greek *klastos;* means *broken*

Figure 18 Scientists from the United States Geological Survey monitor volcanoes using sensitive seismic equipment.

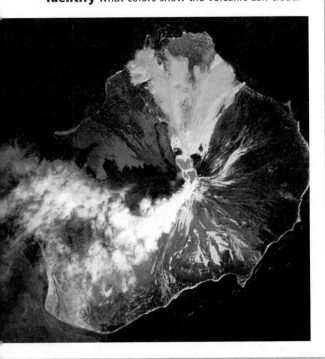

Figure 19 This infrared satellite image shows an eruption of the Augustine volcano in Alaska. The volcanic ash cloud was ejected 11,500 m into the air. The land and the sea around the volcano have been coated with ash.

Identify what colors show the volcanic ash cloud.

Predicting Volcanic Eruptions

Volcanoes usually show signs of activity before they erupt. These signs include small earthquakes, emission of gases, and changes in ground movement and temperature of the surface of the volcano. This allows scientists to issue warnings about possible volcanic activity.

Small Earthquakes

Small earthquakes can occur when magma is forced up through brittle rock. Usually before a volcanic eruption, there are many small earthquakes ranging from about 1.0 to 3.0 in magnitude. **Figure 18** shows an example of a ground-based seismic network designed to detect earthquake activity.

Gas Emissions

Scientists collect gas emissions from volcanoes. Different amounts of gases—carbon dioxide, for example—indicate how deep below the surface magma is located. Changes in the type or amount of gases coming from a volcano might signal that an eruption will occur soon.

Ground Movement and Temperature

As magma moves toward Earth's surface, the ground around the volcano can increase in temperature. As the magma is forced up, the ground might actually begin to rise or bulge slightly. Using remote-sensing devices, scientists are able to detect changes in temperature and ground movement.

Monitoring Volcanic Activity

Even small changes in the surface of a volcano can be detected from space. Scientists can closely monitor volcanic activity with the use of satellite imaging. **Figure 19** shows a satellite image of the movement of volcanic ash clouds as seen from space.

Volcanic Hazards

Volcanic landforms and features range in shape and size. All types of volcanoes can emit gases, solids, lava, and tephra during explosive eruptions. Gas emissions, lahars, and pyroclastic flows are some hazards that result from volcanic eruptions. Scientists are improving their understanding of how volcanic processes work. Through technology, scientists are better able to monitor volcanic activity from space. This helps them more accurately predict dangerous eruptions, which can impact human and wildlife habitats.

LESSON 3 Review

Summarize

Create your own lesson summary as you organize an **outline.**

1. **Scan** the lesson. Find and list the first **red** main heading.

2. **Review** the text after the heading and list 2–3 details about the heading.

3. **Find** and list each **blue** subheading that follows the **red** main heading.

4. **List** 2–3 details, key terms, and definitions under each **blue** subheading.

5. **Review** additional **red** main headings and their supporting **blue** subheadings. List 2–3 details about each.

 ELA6: R 2.4

Standards Check

Using Vocabulary

1. **Define** *volcanic ash* in your own words. **2.d**

2. **Distinguish** between a pyroclastic flow and a lahar. **2.d**

Understanding Main Ideas

3. **Predict** the hazards to a town that is located in a valley at the base of an active volcano. **2.d**

4. **Summarize** how scientists can use small earthquakes to predict a volcanic eruption. **2.d**

5. **Analyze** the advantages of monitoring volcanic activity with technology in space as compared to ground-based sensoring. **2.d**

Applying Science

6. **Assess** the potential volcanic hazards from a glacier-topped volcano over 4 km high. Support your reasoning with specific examples. **2.d**

7. **Design** Draw a chart like the one below. List the effects of volcanic hazards on human and wildlife habitats. **2.d**

Effects of Volcanic Hazards	
Name of Hazard	**Effects of Hazard**
Volcanic ash	
Lahar	
Gas emission	
Pyroclastic flow	
Lava flow	

 **Science Online**

For more practice, visit **Standards Check** at ca6.msscience.com.

MiniLab

minutes

How does lava affect habitats?

Volcanic eruptions can change human and wildlife habitats. Scientists make models of volcanoes and their landscapes to test how lava might flow during an eruption.

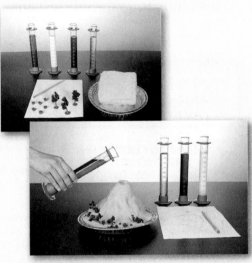

Procedure

1. Read and complete a lab safety form.

2. Make a volcano on an **aluminum pie pan** with **modeling clay.** Include trees and buildings on the landscape.

3. Draw a topographic map of your landscape to record your results.

4. Design a procedure to test different types of materials to imitate lava, such as **frosting, molasses, syrup, honey,** and **water.** Measure the same amount of each material.

5. Pour the lava from the top of the volcano. Record the path, area, and distance of the lava flow on your topographic map.

6. Wash and dry the landscape. Repeat with the same amount of each lava sample.

Analysis

1. **Describe** the viscosity of each lava sample.

2. **Sequence** the lava flows from fastest to slowest.

3. **Compare** the rate of lava flow to the viscosity for each sample.

4. **Infer** how each type of lava would affect the environment.

Science Content Standards

2.d Students know earthquakes, volcanic eruptions, landslides, and floods change human and wildlife habitats.

7.b Select and use appropriate tools and technology (including calculators, computers, balances, spring scales, microscopes, and binoculars) to perform tests, collect data, and display data.

Applying Math

Finding Range

1.d, 1.e

MA6: NS 2.0

Some volcanoes show evidence of past eruptions. But, an exact date of the past eruption is unknown. The dates of past volcanic eruptions are sometimes given in ranges as shown in the table below. The ± (plus or minus) symbol indicates the date is approximate, but is probably a certain number of years before or after the date given.

Example

Find the range of years for the last eruption of Medicine Lake.

Volcano	Approximate Date of Last Eruption
Golden Trout Creek	5550 B.C. ± 1,000 years
Inyo Craters	A.D. 1380 ± 50 years
Mammoth Mountain	A.D. 1260 ± 40 years
Medicine Lake	A.D. 1080 ± 25 years
Mono Craters	A.D. 1350 ± 20 years
Mount Shasta	A.D. 1786

1 **This is what you know:** The range of years that Medicine Lake volcano might have last erupted is A.D. 1080 ± 25 years.

2 **This is what you need to find:** The latest possible eruption year and the earliest possible eruption year.

3 **Use these equations:** A.D. 1080 + 25 = A.D. 1105 and A.D. 1080 − 25 = A.D. 1055

Answer: The last volcanic eruption occurred some time between the years A.D. 1055 and A.D. 1105.

Practice Problems

1. Find the range in years for the last eruption of Mono Craters.

2. Find the range in years for the last eruption of Mammoth Mountain.

Science Online

For more math practice, visit **Math Practice** at ca6.msscience.com.

Use the Internet:
The Ring of Fire

Materials

world tectonic
    plate map
3-in × 5-in cards
colorful yarn
push pins
computer with
    internet access

Science Content Standards

1.d Students know that earthquakes are sudden motions along breaks in the crust called faults and that volcanoes and fissures are locations where magma reaches the surface.

1.e Students know major geologic events, such as earthquakes, volcanic eruptions, and mountain building, result from plate motions.

7.a Develop a hypothesis.

7.d Communicate the steps and results from an investigation in written reports and oral presentations.

Problem

As Earth's lithospheric plates move in relation to one another, magma rises to the surface, and erupts as lava containing gases, solids, and tephra. How can scientists predict catastrophic volcanic events and their hazards in an attempt to avoid dangers to humans and wildlife habitats?

Form a Hypothesis

Make a prediction about how the location of active volcanoes relates to the movements of lithospheric plates to create volcanic eruptions.

Collect Data and Make Observations

1. Make a data table similar to the one shown below.
2. Visit ca6.msscience.com to research the three volcanoes your teacher has assigned to you.
3. For each volcano, record the location, type of volcano, and date of most recent eruption in your data table.
4. For each volcano, copy this information on one card.
5. Plot the location of your volcanoes on a world tectonic plate map using push pins, yarn, and your card.
6. Compare your data to the results of your classmates.
7. Refer to the location and distribution of the active volcanoes on the world tectonic plate map as you respond to the Analyze and Conclude section of this lab.

Locations of Active Volcanoes			
Name of Volcano	**Location**	**Type of Volcano**	**Date of Most Recent Eruption**

Analyze and Conclude

1. **Identify** the location of the three most recent volcanic eruptions. Record this information in your data table.
2. **Infer** the movement of the lithospheric plates at the location of the three most recent volcanic eruptions.
3. **Relate** the locations of volcanoes and earthquakes to plate boundaries and their movement.
4. **Predict** the locations and frequency of volcanoes to earthquakes and plate movements.

Communicate

Create a Travel Brochure Design a travel brochure for a tour group. Include a variety of types of transportation to visit the different volcanic landforms. Present your brochure to the class as you inform them of your educational and entertaining volcanic tour.

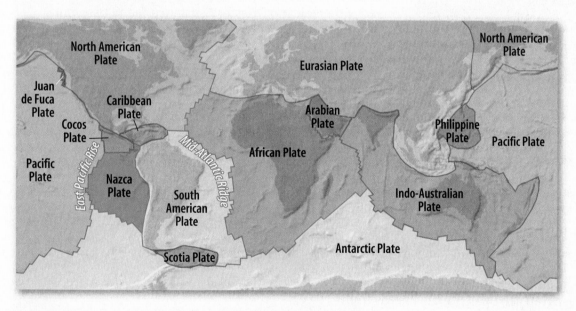

Real World Science

Science & Career

Predicting Volcanic Eruptions

As part of the Volcano Disaster Assistance Team based in Menlo Park, California, seismologist Randy White has visited many volcanoes. Once alerted by increased earthquake activity in an area, White's team can arrive at a volcano site within 24 hours, where they interpret seismic data, ground bulging measurements, and atmospheric CO_2 and SO_2 levels. The team's goal? Predicting if and when a "sleeping volcano" might erupt, and how violent the eruption could be. Their recent prediction of eruptions in the Philippines and Latin America saved countless lives.

Visit **Careers** at ca6.msscience.com to learn more about the Volcano Disaster Assistance Team. Use your research to **write** a newspaper article describing the work of the Volcano Disaster Assistance Team.

Science & Technology

From Blast to Blast

To contain the force of an exploding bomb, engineers are using rock blasted out of a volcano. Perlite, a volcanic glass, is part of a material called BlastWrap, which is being built into walls and ceilings, concrete barriers, vehicle protection systems, and even trash cans. BlastWrap can dissipate the energy from an exploding bomb the size of a backpack in less than one one-thousandth of a second.

Visit **Technology** at ca6.msscience.com to research the properties of BlastWrap. **Create** a brochure or advertisement describing the benefits of this technology.

The Year Without a Summer

When the Indonesian volcano Tambora erupted in 1815, ash was ejected far into the upper atmosphere. Circulating through the atmosphere, the ash reduced sunlight, affecting temperatures worldwide. In 1816, northeastern America was particularly hard hit. Frosts in May and July killed most crops. In June, two large snowstorms caused many deaths. In August, ice was seen in lakes and rivers as far south as Pennsylvania.

Visit **History** at ca6.msscience.com to learn more about the summer of 1816. **Write** a poem or song lyrics that describe living in northeastern America during this period. **Share** your work aloud with your class.

Making the Most of Volcanoes

Icelanders use their country's unique geography astride the Mid-Atlantic Ridge to their advantage. While one of Iceland's 200 volcanoes erupts every few years, these same volcanoes provide geothermal energy power produced by heat from within Earth, such as steam from geysers. There's little land for farming, but residents grow nearly everything they need in greenhouses heated with geothermal energy. Eighty-five percent of Iceland's 300,000 residents use geothermal energy to heat their homes. By using geothermal energy rather than imported oil, Iceland's economy saves an estimated $100 million annually.

Visit **Society** at ca6.msscience.com to learn how Iceland uses geothermal resources. **Locate** and **quantify** specific categories of use. **Create** a bar graph or pie chart illustrating the percentage of each category.

The BIG Idea Volcanoes are locations where magma reaches Earth's surface. They affect human and wildlife habitats.

Lesson 1 Volcanoes and Plate Boundaries

 1.d, 1.e, 7.b

Main Idea **Most volcanic activity occurs along plate boundaries where plates move relative to one another.**

- A volcano is a land or underwater feature that forms when magma reaches the surface of Earth.

- Buoyant force pushes magma toward Earth's surface.

- Volcanoes occur in areas of high heat flow.

- Most volcanic eruptions occur at convergent and divergent plate boundaries.

- Some volcanoes form at hot spots not associated with plate boundaries.

- **fissure eruption** (p. 298)
- **hot spot** (p. 298)
- **vent** (p. 297)
- **volcano** (p. 294)

Lesson 2 Volcanic Eruptions and Features

1.d, 1.f, 7.g, 7.h

Main Idea **The composition of magma controls volcanic eruptions and determines the different types of lava flow and volcanic features.**

- Chemical composition, dissolved gases, and the temperature of magma are important properties that control volcanic eruptions.

- Two types of magma are low-silica, low-viscosity basaltic magma and high-silica, high-viscosity granitic magma.

- Three common volcanic landforms on Earth's surface are shield volcanoes, cinder cone volcanoes, and composite volcanoes.

- Volcanoes are forming in the Cascade Range as the Juan de Fuca Plate subducts beneath the North American Plate.

- **cinder cone volcano** (p. 304)
- **composite volcano** (p. 304)
- **shield volcano** (p. 304)
- **tephra** (p. 304)
- **viscosity** (p. 302)

Lesson 3 Hazards of Volcanic Eruptions

2.d, 7.a, 7.b, 7.d

Main Idea **Volcanic eruptions can change human and wildlife habitats.**

- Some volcanic hazards include volcanic ash, lahars, gas emissions, pyroclastic flows, and lava flows.

- Warning signs for an eruption include numerous small earthquakes, changes in gases emitted, and changes in the temperature and movements of the ground surface.

- Eruptions are detected using ground-based and space-based methods.

- **lahar** (p. 314)
- **pyroclastic flow** (p. 315)
- **volcanic ash** (p. 313)

STUDY TO GO Download quizzes, key terms, and flash cards from **ca6.msscience.com**.

Linking Vocabulary and Main Ideas

Use vocabulary terms from page 324 to complete this concept map.

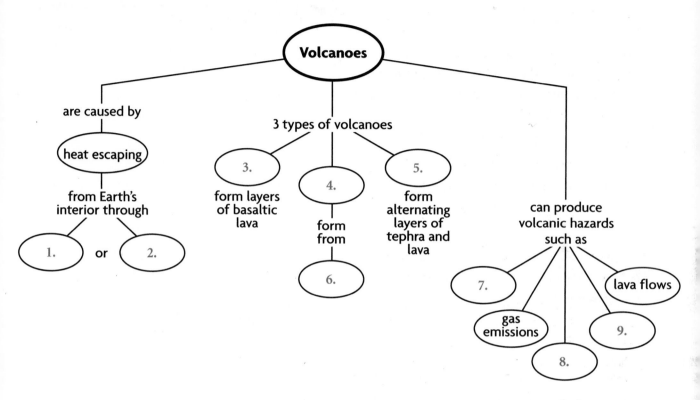

Using Vocabulary

Fill in each blank with the correct vocabulary term.

10. Lava is ejected from a volcano at the _____.

11. Scientists refer to how easily lava flows as its _____.

12. Volcanoes that do not form near plate boundaries are called _____.

13. Pillow lava forms as lava flows from _____ on the seafloor.

14. When volcanic debris mixes with water, it can become a fast-moving _____ destroying everything in its path.

15. The _____ of basaltic lava is low because the percentage of silica is low.

Understanding Main Ideas

Choose the word or phrase that best answers the question.

1. What is a central opening from which material is erupted from a volcano?
   A. dike
   B. fissure
   C. lava dome
   D. vent `1.d`

2. Which of these is a volcanic eruption from an elongated crack in Earth's surface?
   A. central eruption
   B. fissure eruption
   C. gas eruption
   D. magma eruption `1.d`

3. Which type of erupted material is particularly hazardous to aircraft?
   A. ash
   B. dike
   C. lava
   D. magma `2.d`

4. The illustration below shows a type of volcano.

Which type of volcano, shown above, is made entirely of tephra?
   A. shield
   B. caldera
   C. cinder cone
   D. composite `1.d`

Use the map below to answer questions 5–6.

5. What do the Aleutian Islands shown on this map form?
   A. island arc
   B. caldera
   C. dike
   D. oceanic trench `1.d`

6. Which type of plate boundary forms a volcanic arc?
   A. continent-to-continent convergent
   B. divergent
   C. ocean-to-ocean convergent
   D. ocean-to-continent convergent `1.e`

7. Where does magma form before it moves toward Earth's surface?
   A. within the slab
   B. at the inner core
   C. in the mantle above the slab
   D. along the center of the ridge `1.d`

8. What is the most common dissolved gas in magma?
   A. carbon dioxide
   B. carbon monoxide
   C. sulfur dioxide
   D. water vapor `1.d`

Applying Science

9. Predict which volcanic hazards you would expect from Mount Shasta if it were to erupt in your lifetime. `2.d`

10. Explain why scientists would want to monitor glacial ice located near the vent of an erupting volcano. `2.d`

11. Hypothesize Imagine you are hiking and encounter a dense, dark volcanic rock that is spread out in layers for many kilometers. Write a statement about the type of eruption you think formed this rock and the composition of the magma that supplied it. `1.d`

12. Infer what type of volcano Popocatepetl is, which is located in Mexico at the boundary region between the North American and Cocos Plates. `1.d`

13. Explain why subduction zones form. `1.e`

14. Describe the difference between basaltic and granitic magma. `1.d`

15. Describe the effects pyroclastic flows have on humans. `2.d`

16. Illustrate how a sill forms. How is it different from a dike? `1.d`

17. Explain why active volcanoes do not occur along the east coast of the United States. `1.f`

18. Identify the plate tectonic setting that formed Mount Shasta. `1.e`

19. Defend the following statement: Most volcanoes that occur at the leading edges of continents along the Pacific Ocean are composite volcanoes. `1.e`

WRITING in Science

20. Write an informative pamphlet that lists examples of volcanic hazards associated with volcanoes in the Cascade Range. `ELA6: W 1.5`

Cumulative Review

21. Explain how seafloor spreading works. `1.a`

22. Relate the three major types of plate boundaries to main types of stress expected at these boundaries. `1.e`

23. Describe what happens to the energy of seismic waves as the distance from the focus increases. `1.g`

Applying Math

Use the chart below to answer questions 24–26.

Volcano	Approximate Date of Last Eruption
Golden Trout Creek	5550 B.C. $\pm$ 1,000 years
Inyo Craters	A.D. 1380 $\pm$ 50 years
Mammoth Mountain	A.D. 1260 $\pm$ 40 years
Medicine Lake	A.D. 1080 $\pm$ 25 years
Mono Craters	A.D. 1350 $\pm$ 20 years
Mount Shasta	A.D. 1786

24. Find the range in years for the last eruption of Golden Trout Creek. `MA6: NS 2.0`

25. Find the range in years for the last eruption of Medicine Lake. `MA6: NS 2.0`

26. Find the range in years for the last eruption of Mammoth Mountain. `MA6: NS 2.0`

Use the table below to answer questions 1 and 2.

Plate Boundaries		
Plate	Number of Convergent Boundaries	Number of Divergent Boundaries
African	1	4
Antarctic	1	2
Indo-Australian	4	2
Eurasian	4	1
North American	2	1
Pacific	6	2
South American	2	1

1 Determine which plate has the most boundaries that are being pulled apart.

A   African   `1.e`

B   Antarctic

C   Indo-Australian

D   Pacific

2 Predict which plate will be surrounded by the most volcanoes.

A   Antarctic   `1.f`

B   Eurasian

C   Indo-Australian

D   Pacific

3 Identify the characteristics associated with a shield volcano.

A   layers of tephra and lava flow   `1.d`

B   explosive eruptions

C   large, broad volcanoes

D   found at convergent plate boundaries

4 The diagram below shows Earth's lithospheric plates.

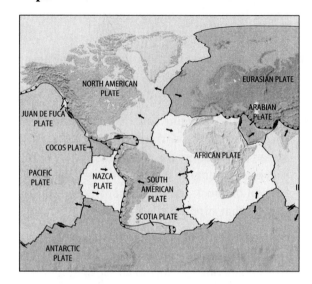

Determine movement of the Pacific Plate.

A   north-northwest   `1.f`

B   north-northeast

C   south-southwest

D   south-southeast

5 The lava of a particular volcano is high in silica, water vapor, and other gases. What kind of eruption will likely result?

A   explosive   `1.d`

B   quiet

C   thick, bubble-like lava dome

D   flood basalt

6 Volcanoes are not usually associated with

A   oceanic divergent   `1.d`

B   continental divergent

C   ocean-to-continent convergent

D   continent-to-continent convergent

Science online   Standards Assessment ca6.msscience.com

7 What causes magma to be forced upward toward Earth's surface?

A The magma has a higher density than the surrounding rock. `1.d`

B The magma has a lower density than the surrounding rock.

C Magma has a different composition than the surrounding rock.

D Magma has the same composition as the surrounding rock.

8 The photo below shows a volcanic eruption.

Infer what volcanic hazard buried cities, towns, and fields during this eruption.

A lahar `2.d`

B lava flow

C mud flow

D pyroclastic flow

9 What is made of layers of tephra and lava?

A lava dome `1.d`

B cinder cone volcano

C composite volcano

D shield volcano

10 What is the relationship between silica content and viscosity?

A Magma that has less silica is more viscous.

B Magma that has medium viscosity has high silica. `1.d`

C Magma that has medium viscosity has low silica.

D Magma that has more silica is more viscous.

11 What is the relationship between temperature and viscosity?

A The higher the temperature, the more viscous the magma. `1.d`

B The lower the temperature, the more viscous the magma.

C The lower the temperature, the less viscous the magma.

D Temperature has no affect on viscosity.

12 Which feature forms when the top of a volcano collapses into a partially emptied magma chamber?

A caldera `1.d`

B crater

C fissure

D sill

Weathering and Erosion

The Mighty Restless Sea

The endless crashing of waves again these rocks wore away the softest parts, leaving this arch between the land and a sea stack.

Science *Journal* Make a list of five things you know about the ocean. Select two of them and write a paragraph about each topic. Then, write a third paragraph that compares the two.

Launch Lab

00:20 minutes

Set in Stone?

Have you ever gone to a cemetery and noticed that the writings on some headstones are clear after hundreds of years, while others are so worn that the names can hardly be read? Different types of stone react differently with the environment.

Procedure

1. Use the **marble** and **chalk** provided by your teacher.

2. Fill **two clear containers** with 100 mL of **water** each. Fill two more clear containers with 100 mL of **white vinegar** each. Label the jars *Water/Marble, Water/Chalk, Vinegar/Marble, and Vinegar/Chalk.*

3. Add the marble and chalk to the correctly labeled containers and observe for ten min.

Think About This

Analyze Why do you think the marble and the chalk reacted differently?

 2.a, 7.a

Science Online

Visit ca6.msscience.com to:

► view **Concepts in Motion**
► explore Virtual Labs
► access content-related Web links
► take the Standards Check

Weathering Make the following Foldable to compare and contrast chemical and physical weathering.

► **STEP 1 Fold** a sheet of paper in half from top to bottom and then in half from side to side.

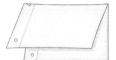

► **STEP 2 Unfold** the paper once. **Cut** along the fold of the top flap to make two flaps.

► **STEP 3 Label** the flaps as shown.

Chemical Weathering	Physical Weathering

Reading Skill  ELA6: R 2.2

Compare and Contrast

As you read this chapter, compare and contrast the actions and effects of chemical weathering to those of physical weathering. Include examples of weathering.

Identify Cause and Effect

1 Learn It! A *cause* is the reason something happens. The result of what happens is called an *effect*. Learning to identify causes and effects helps you understand why things happen. By using graphic organizers, you can sort and analyze causes and effects as you read.

2 Practice It! Read the following paragraph. Then use the graphic organizer below to show what happens when water freezes in the cracks of rocks.

Water has the unique property of expanding when it freezes. In climates where the temperature drops below the freezing point of water, water that has seeped into the cracks of rocks will freeze and expand. The expanding ice causes pressure to increase. This forces the crack in the rock to open slightly. After many cycles of freezing and thawing, the crack is forced completely to open. The rock breaks into pieces.

—*from page 337*

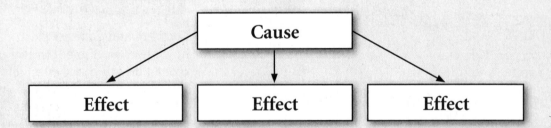

Cause → Effect, Effect, Effect

3 Apply It! As you read the chapter, be aware of causes and effects of gravity and water. Find five causes and their effects.

Target Your Reading

Use this to focus on the main ideas as you read the chapter.

Reading Tip

Graphic organizers such as the Cause-Effect organizer help you organize what you are reading so you can remember it later.

① **Before you read** the chapter, respond to the statements below on your worksheet or on a numbered sheet of paper.

- Write an **A** if you **agree** with the statement.
- Write a **D** if you **disagree** with the statement.

② **After you read** the chapter, look back to this page to see if you've changed your mind about any of the statements.

- If any of your answers changed, explain why.
- Change any false statements into true statements.
- Use your revised statements as a study guide.

Science Online

Print a worksheet of this page at ca6.msscience.com.

Before You Read A or D	Statement	After You Read A or D
	1 Water, wind, gravity, and ice are weathering agents.	
	2 Exposure to atmospheric water and gases causes rocks to change chemically.	
	3 Frost wedging is a common method of breaking rocks into fragments in all climates.	
	4 Mass wasting is the slow process of changing rock into soil.	
	5 Movement of water in streams causes them to constantly change their path.	
	6 Sand on beaches comes from rock weathered by the ocean.	
	7 Flooding is a common feature of all rivers and streams.	
	8 Sand dunes are deposits from wind, water, and ice agents.	
	9 The mountains around the Central Valley are uplifted landforms.	

Reading Guide

What *You'll Learn*

▶ **Compare and contrast** chemical and physical weathering.

▶ **Describe** weathering actions.

▶ **Explain** the effects of weathering.

▶ **Determine** the roles of humans and living things in weathering.

Why *It's Important*

Everything around us is affected by weathering—the roads, our homes, and the land we live on.

Vocabulary

weathering
chemical weathering
physical weathering
frost wedging
soil

Review Vocabulary

mineral: naturally occuring, inorganic solid that has a definite chemical composition and an orderly atomic structure (p. 87)

Weathering

Main Idea Rocks exposed at Earth's surface are broken down into sediment and soils by the action of weathering.

Real-World Reading Connection Think about your favorite pair of jeans. Perhaps when they were new they were dark blue, but now they are faded. The knees, pockets, and cuffs might be worn with holes and have large threads hanging from them. These are the effects of wear and aging. Earth shows similar signs of wear. Rocks get worn smooth and cracked open, gullies form as soil gets moved, and minerals corrode and change color. They are all caused by weathering.

What is weathering?

Weathering is the destructive process that breaks down and changes rocks that are exposed at Earth's surface. Weathering is caused by the action of water, wind, ice, and gravity. They are referred to as agents of weathering. These agents create two different weathering processes that can change rocks. The processes are chemical weathering and physical weathering. An example of weathering is shown in **Figure 1.** Weathering has slowly destroyed the features of the Sphinx, which was carved out of limestone in Egypt 7,000 to 9,000 years ago.

 Figure 1 Predict how precipitation might contribute to the weathering of the Sphinx.

Figure 1 Weathering has almost completely destroyed the face of the Sphinx. An artist's reconstruction shows what the original face might have looked like.

Figure 2 The breakdown of iron, shown in these old cars, creates rust. This is an example of chemical weathering called oxidation.

Chemical Weathering

Chemical weathering results when minerals and rocks at Earth's surface are weakened and broken down from exposure to water and gases in the atmosphere. This exposure causes the composition of the minerals of a rock to change. The result is the formation of new minerals such as the iron oxide, or rust, on the cars shown in **Figure 2.**

 What causes rock surfaces to break down?

Water

The most common agent of chemical weathering is water. Rocks and minerals that dissolve in water are said to be soluble. When water mixes with carbon dioxide from the air, carbonic acid is formed. This is the same weak acid found in carbonated soft drinks. Most rainwater contains some dissolved carbon dioxide from the air. This makes rainwater slightly acidic. Carbonic acid in water is a more destructive weathering agent than pure water.

 Figure 2 What does oxidation of iron or steel look like?

Figure 3 Chemical reactions with atmospheric elements cause metals to corrode.

Figure 4 Some rock types are better suited for outdoor use than others.

Identify What agents of weathering have affected these headstones?

Acid

What happens when slightly acidic rainwater comes in **contact** with rock? It reacts with the minerals in the rock, such as in feldspar. Feldspar weathers rapidly, changing into clay minerals. The formation of clay is one of the most common results of chemical weathering.

Human-made pollution, like that produced from burning coal, can cause chemical weathering to occur even more rapidly. When coal is burned, sulfur dioxide is released into the atmosphere. The sulfur dioxide combines with water vapor in the air, creating sulfuric acid. This ultimately becomes acid rain. When acid rain reaches the ground, it damages rocks and buildings. Plants, soil, and lake habitats also are affected by the increase in the acidity of the soil and water.

 What forms in the atmosphere when coal is burned?

Oxygen

When oxygen that is dissolved in water comes in contact with compounds of some metals, a chemical reaction occurs, forming a new substance. The greenish color on the statue in **Figure 3** is a substance that formed from a reaction of water and oxygen with copper compounds on this bronze statue. Other metals may get a white or gray powder on their surface.

Rock Type and Weathering

The type of rock also determines how quickly its surface is chemically weathered. Compare the two old headstones shown in **Figure 4.** They are about the same age and have been exposed to the same climate. However, the carved details of the top headstone are still clear after 100 years. This headstone is made of a rock that resists chemical weathering. The headstone on the bottom has lost most of the carved detail because of chemical weathering.

Physical Weathering

Physical weathering is the breaking of rock into smaller pieces without changing its mineral composition. Processes of physical weathering include frost wedging and the work of plants and animals. These are described below.

Frost Wedging

Frost wedging occurs when water freezes, expands, and melts in the cracks of rocks. Water has the unique property of expanding when it freezes. In climates where the temperature drops below the freezing point of water, water that has seeped into the cracks of rocks will freeze and expand. The expanding ice causes pressure to increase. This forces the crack in the rock to open slightly. After many cycles of freezing and thawing, the crack is forced completely open. The rock breaks into pieces. An example of frost wedging is shown in **Figure 5.**

 What happens to water when it freezes?

Plants and Animals

The breaking down of rock into smaller pieces also can be caused by plants and animals. Have you ever noticed a sidewalk that is broken and buckled upward? An example is shown in **Figure 5.** This occurs because as the tree grows, the roots also grow bigger. Over time, the increase in the size of the root forces the concrete to crack. Plant roots in search of water can also grow into cracks within rocks. As the plant roots grow in size, they eventually wedge the rocks apart.

Burrowing animals can move loose rocks and dirt to the surface. The material is exposed to wind and water. This causes the weathering process to increase.

 Figure 5 What caused the rock to break in the left photo?

ACADEMIC VOCABULARY ⋯
contact (KON takt)
(noun) a union or junction of surfaces
The foul occurred when the two players made contact.

Frost Wedging

Root Pressure

Figure 5 Like frost wedging, plants can break rocks into fragments with root pressure.

Water and Weathering

Water has an effect on the world around you. Water erodes and transports sediment to new locations. Physical weathering can be caused by the tumbling action of running water.

Procedure

1. Complete a lab safety form.
2. Your teacher will provide you with some **pieces of broken rock.**
3. Rinse the rocks and drain off the water. Pat the rocks dry and weigh them. Record the mass in grams.
4. Put your rocks in a **plastic bottle** and add **water** to cover the rocks. Seal the bottle with a lid.
5. Shake the bottle for 5 min.
6. Drain the water, pat dry, and reweigh your rocks.
7. Record the mass of the rocks before and after shaking.

Analysis

1. **Compare and contrast** the appearance of the rocks before and after shaking. Explain the difference.
2. **Compare** your weight difference with the rest of the class. Did everyone's rocks weather at the same rate? Explain any differences.

 2.a, 7.e

Soil Formation

The weathering of rock on Earth's surface produces soil. **Soil** is a mixture of weathered rock, minerals and organic matter, such as decaying plants and animals. Water and air fill the spaces between soil particles. **Figure 6** shows how soil forms through physical and chemical weathering. The formation of soil is affected by several factors, such as the type of rock, the climate, the length of time a rock has been weathering, and the interaction of plants and animals with the soil.

Remember that most rocks contain the mineral feldspar, which typically breaks down through chemical weathering to form clay minerals. This explains why clay is one of the most abundant ingredients in soils. Soil is important because it contains the nutrients necessary for plant growth and food crop production. In addition to being anchored in the soil, plants growing in soil help to keep it from eroding away.

 What is soil made of?

Composition of Soil

If soil remains in the same location where it formed, it is called a residual soil. The composition of the soil matches the composition of the rock from which it formed. For example, granite contains quartz. Quartz is resistant to weathering. A soil that develops from granite will be sandy because of the sand-sized grains of quartz it contains. But a soil developed from basalt, which contains large amounts of feldspar, will have sticky clay particles instead. Transported soils develop from weathered material that has been moved to a new location by wind, water, or glaciers. The composition of the soil does not match the composition of the rock beneath it.

 Figure 6 How do plants help create soil?

Visualizing Soil Formation

Figure 6

Thousands of years of weathering solid rock results in the formation of soil. Soil is made up of mineral fragments, bits of rock, and the remains of dead plants and animals. Water and air fill the spaces between the particles.

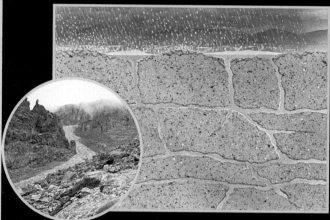

A Natural acids in rainwater weather the surface of exposed bedrock. Frost-wedging can enlarge a crack, causing rocks to fracture and break apart. The inset photo shows weathered rock in the Tien Shan Mountains of Central Asia.

B Plants take root in the cracks and among the bits of weathered rock—shown in the inset photo above. As they grow, plants absorb mineral from the rock, weakening it. Along with other natural forces, the process of breaking down rocks continues and a thin layer of soil begins to form.

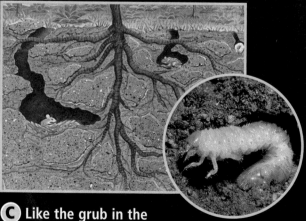

C Like the grub in the inset photo, insects, worms, and other living things take up residence among plant roots. Their wastes, along with dead plant material, add organic matter to the soil.

D As organic matter increases and the underlying bedrock continues to break down, the soil layer thickens. Rich topsoil supports trees and other plants with large root systems.

Soil Layers

If you have ever dug a deep hole, you might have noticed layers with different colors and appearances. These layers are called soil horizons. They can take thousands of years to develop. Three soil horizons make up a complete soil profile, as shown in **Figure 7.**

 Reading Check What are layers in a soil profile called?

A Horizon The topmost soil horizon is called the A horizon. It contains small rocks, minerals, and different amounts of decomposed plant material called humus. This horizon is usually a dark color because it contains organic matter. Water seeping through this horizon dissolves minerals from it, resulting in the bottom of the A horizon being light in color.

B Horizon The dissolved minerals are deposited in the next soil horizon, called the B horizon. This layer contains large amounts of clay and commonly is stained red or brown.

C Horizon Below the B horizon is the C horizon. The C horizon consists of partly weathered parent material or bedrock. Below this horizon is unweathered parent material, solid rock.

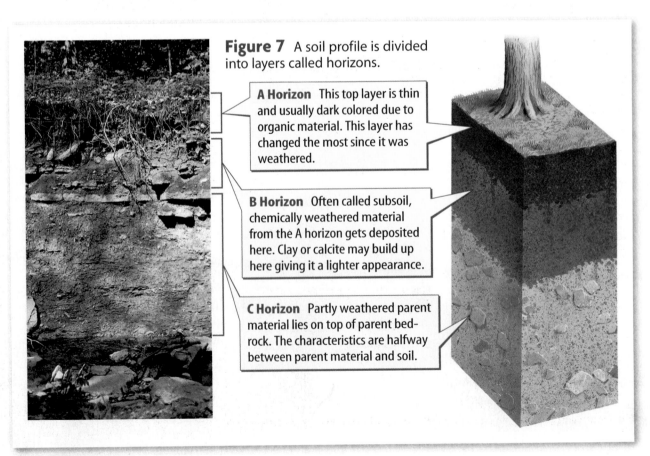

Figure 7 A soil profile is divided into layers called horizons.

A Horizon This top layer is thin and usually dark colored due to organic material. This layer has changed the most since it was weathered.

B Horizon Often called subsoil, chemically weathered material from the A horizon gets deposited here. Clay or calcite may build up here giving it a lighter appearance.

C Horizon Partly weathered parent material lies on top of parent bedrock. The characteristics are halfway between parent material and soil.

Weathering and Landforms

The processes of chemical and physical weathering work together to break down rocks. Chemical weathering changes the composition of rocks. Physical weathering breaks down rocks without changing the composition. Rocks that have been broken into smaller and smaller pieces by physical weathering have more surface area that can be exposed to chemical weathering. The process of weathering helps to form soil. Soils develop in layers called horizons. Weathering and other factors influence the character of the soil that forms.

LESSON 1 Review

Summarize

Create your own lesson summary as you write a **newsletter.**

1. **Write** this lesson title, number, and page numbers at the top of a sheet of paper.

2. **Review** the text after the **red** main headings and write one sentence about each. These will be the headlines of your newsletter.

3. **Review** the text and write 2–3 sentences about each **blue** subheading. These sentences should tell *who, what, when, where,* and *why* information about each headline.

4. **Illustrate** your newsletter with diagrams of important structures and processes next to each headline.

 ELA6: W 1.2

Standards Check

Using Vocabulary

1. Breaking rocks into pieces by physical and chemical means is called _____. `2.a`

2. _____ occurs when water in a cracked rock freezes and expands. `2.a`

Understanding Main Ideas

3. **Compare and contrast** chemical and physical weathering processes. `2.a`

Similarities	Differences

4. **Draw** and **label** a complete soil profile. `2.a`

5. **Construct** a diagram with arrows showing three weathering processes that could act on a rock that is exposed on Earth's surface. `2.a`

6. **Differentiate** between a soil profile from the rain forest and a soil profile from the Great Plains. `2.a`

7. Which of the following is not a factor in the effect of weathering? `2.a`

   A. rock type
   B. acidity of rain water
   C. climate
   D. soil type

Applying Science

8. **Hypothesize** how a soil profile from a warm, humid, tropical environment would differ from a soil profile from a dry, desert environment. `2.a`

9. **Infer** how a river could cause both physical and chemical weathering. `2.a`

 Science Online

For more practice, visit **Standards Check** at <u>ca6.msscience.com</u>.

 Weathering ca6.msscience.com

Lesson 1 • Weathering **341**

Science Content Standards

2.a Students know water running downhill is the dominant process in shaping the landscape, including California's landscape.
2.b Students know rivers and streams are dynamic systems that erode, transport sediment, change course, and flood their banks in natural and recurring patterns.
2.c Students know beaches are dynamic systems in which sand is supplied by rivers and moved along the coast by the action of waves.
2.d Students know earthquakes, volcanic eruptions, landslides, and floods change human and wildlife habitats.
Also covers: 7.c, 7.g

Reading Guide

What *You'll Learn*

▶ **Tell** how the land surface is changed by water action.

▶ **Describe** stream formation.

▶ **Discuss** mass wasting and how it relates to land use in California.

▶ **Explain** erosion and deposition.

Why *It's Important*
Landscapes are the result of erosion and deposition.

Vocabulary

erosion	flood
deposition	flood plain
mass wasting	beach
landslide	glacier
meander	

Review Vocabulary

sediment: rock that is broken down into smaller pieces or is dissolved in water (p. 99)

Erosion and Deposition

Main Idea Movement of rock and soil are natural occurrences caused by specific geological conditions.

Real-World Reading Connection The city of LaConchita, California, experienced a mudslide in January 2005. This area has a history of mudslides dating back to the 1800s. Why might some areas be more prone to geological events such as these?

What are erosion and deposition?

If you ever have seen a river or stream, you may have noticed that the flowing water can move pieces of rock and soil downstream. Recall that the process of moving weathered material from one location to another is called **erosion.** Erosion can be caused by running water, rain, waves, glaciers, wind, and in the case of landslides, gravity. When sediments are laid down in a new location by one of these processes, it is called **deposition. Figure 8** shows Laguna Beach in southern California, where a landslide occurred on June 1, 2005. Erosion of this type is very rapid, but normal erosion may take years to move this much soil and rock.

Figure 8 Landslides commonly occur during rainy periods in southern California.
Consider What factors might have contributed to the occurrence of this landslide?

Mass wasting

Mass wasting is a form of erosion caused mainly by gravity. It involves the downhill movement of rocks and/or soil in one large mass. Mass wasting commonly occurs when the ground becomes soaked with rainwater. This weakens the forces that hold the various material on the hillside together. The steeper the slope of hillside, the more likely or frequently mass wasting will occur. When the weight of the soil and water becomes too great, the mass of soil will begin to slide. As the soil and water mix more evenly it may then begin to flow like a liquid.

Mass wasting also can occur from vibrations, such as shaking from earthquakes, heavy machinery, blasting, or even thunder. Several types of mass wasting are described below.

 **Reading Check** What can cause mass wasting to occur?

Fast Mass Wasting

Landslides are rapid, gravity-caused events that move soil, loose rock, and boulders. Mudslides, like the one in La Conchita, contain mixtures of soaked soil and rock material. Rock falls involve loosened rock falling from steep cliffs. The result of a rock fall in Utah is shown in **Figure 9.** Finally, slumps occur when a block of rock and the overlying soil slide down a slope as one large mass. Slumping can also involve soil movement only.

Figure 9 These episodes of mass wasting each occurred under different geological conditions.
Infer Which of the mass wasting events might have been triggered by wet ground?

Landslide

Mudslide

Rock Fall

Figure 10 Over time, creep has caused these tree trunks to lean downhill.

ACADEMIC VOCABULARY ····

ultimate (UHL tih mut)
(adjective) farthest, last, final, in the end
In most sporting events the ultimate goal is to win.

Creep

Sometimes mass wasting does not occur quickly. It occurs over long periods of time. Sediment moves slowly downhill, pulled by the force of gravity. This is called creep and is the slowest form of mass wasting. As shown in **Figure 10,** signs of creep include the tilting of telephone poles, trees, or fences in the downhill direction. Creep often results from freezing and thawing and burrowing animals.

 Figure 10 Describe a physical weathering process that may have caused the trees to tilt.

Climate and Erosion

The climate of an area determines the amount of water that a region receives. Regions that receive large amounts of rainfall are more likely to experience mass wasting than areas with dry climates. Climate also influences the type and abundance of vegetation. The presence of thick vegetation on slopes tends to prevent landslides because the root systems of the plants help to hold sediment in place. Vegetation also acts as a cushion for falling raindrops which reduces their erosive effect.

 How does vegetation help to prevent mass wasting?

Water and Erosion

What happens to water that does not soak into the ground or evaporate into the air? It flows over Earth's surface into lakes, streams, and rivers, and **ultimately** into the oceans. Streams and rivers are active systems that erode the land, transport sediment, and deposit sediment in new locations.

Stages of Stream Development

Have you ever noticed that waterfalls and rapids occur in steep mountain regions rather than in flat valleys? This is because the characteristics of rivers change as water moves from high in the mountains down to lakes or oceans at sea level. Rainfall and melting snow feed streams that originate in hills and mountains. The steep slopes allow the water to flow downhill rapidly. This produces a high level of energy that erodes the bottom of the stream more than the sides. These streams cut steep, V-shaped valleys and have white-water rapids and waterfalls.

Development of Meanders

When a stream has eroded the steep valleys to gentler slopes, the stream flows more slowly. Now water in the stream erodes along the sides of the stream bed rather than along the stream bottom. This causes the stream to develop meanders. **Meanders** (mee AN durs) are the curves in the stream, as shown in **Figure 11.** Once a stream develops meanders, the curves tend to become wider and wider. This is because the speed of the water is greatest at the outside of a bend. **Figure 11** also illustrates the erosion that occurs at the outside of the meanders. On the other hand, the water flows more slowly on the inside of the meanders. Deposition, the dropping of sediment being carried by the stream, occurs on the inside of a meander.

Reading Check Where does erosion in a meander occur?

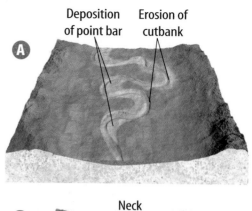
Deposition of point bar · Erosion of cutbank

A

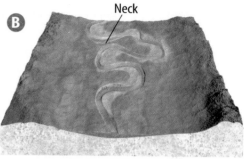
Neck

B

C

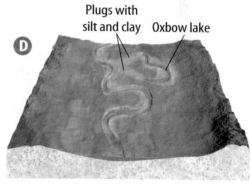
Plugs with silt and clay · Oxbow lake

D

Figure 11 A. Erosion occurs on the outside of a bend and deposition occurs on the inside of a bend. B. When the erosion of the river brings the outside bends close together, it leaves a narrow piece of land called a neck. C and D. When the neck is eroded away the river deposits silt and mud to create an oxbow lake.

Deposition and Water

The sediment and rock that are eroded and carried by river systems are transported, or moved. Eventually they are deposited at a new location. Deposition might occur anywhere along a stream where the water slows down. Slowing reduces the amount of energy that the stream has to carry sediment.

Deposited sediments can form distinct features. Deposition on the inside of a meander can cut off a large U-shaped meander from a river, producing a small lake called an oxbow lake. This is illustrated in **Figure 11** on the previous page. When a stream or river reaches a large body of water, such as a lake or ocean, it slows down. Most of the sediment drops out, forming a triangular-shaped deposit called a delta. **Figure 12** shows an example of a delta. When rivers empty from steep narrow canyons out onto flat plains at the foot of mountains, they form a similar triangular deposit called an alluvial fan. The alluvial fan in **Figure 12** is in the Mojave Desert. **Table 1** shows how the steepness of the slope affects the river as it develops.

Figure 12 In the top photo, the Mississippi delta forms as the river enters the Gulf of Mexico. The bottom photo shows how the Sheep Creek alluvial fan sediment is deposited on land in the Mojave Desert.

Explain What causes sediment to drop out when a river reaches the ocean?

Delta

Alluvial Fan

Stream Development

Streams develop as water falls on Earth's surface and runs off. Steep slopes increase the erosion power of water. V-shaped valleys result. As the land flattens out, the water slows down into S-shaped meanders.

Table 1 Stream Development	
What Happens	**What It Looks Like**
Mountain streams flow in steep valleys, have V-shaped stream beds, and are often rocky and filled with rapids.	
Farther downstream, rivers become wider and less steep. Their stream beds are wider with fewer rocks and rapids in them.	
On the plains, river beds flatten and the rivers develop wide floodplains, meanders, and oxbow lakes. There are no rapids.	

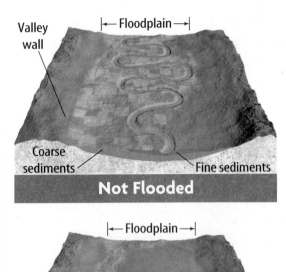

Figure 13 Floodplains are often mapped and given names such as the 50-year floodplain, the 100-year floodplain, and so on.

Conclude What do you think these names mean?

Valley wall

Floodplain

Coarse sediments

Fine sediments

Not Flooded

Floodplain

Natural levees

Flooded

Flooding

A **flood** occurs when the water level in a river rises above the usual height and overflows the sides of its banks. Floods are caused by major rain storms or rapid melting of winter snow. During times of flooding, water spills onto the floodplain. The **floodplain** is a wide, flat valley that is located along the sides of some rivers and streams. Floodplains form from the side-to-side erosional action of a meandering stream, as shown in **Figure 13.** Yearly flooding supplies these areas with mineral-rich, fertile soils that are ideal for farming.

Preventing Flooding Sediment carried by floodwaters is deposited along both sides of the river into long, low ridges on the floodplain. These natural levees protect the area from flooding. Artificial levees sometimes are built along the banks of a river to help control floodwaters. In New Orleans, a break in the artificial levee, shown in **Figure 14,** allowed the river to flood the nearby region. In urban areas such as Los Angeles, lining small streams and rivers with concrete has reduced flood hazards. Such a structure is called an aqueduct. Because floods are unpredictable, building on floodplains or near dams and levees is not a good idea. All geological factors need to be considered before any construction begins.

 What function do levees perform?

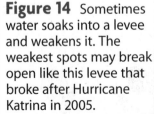

Figure 14 Sometimes water soaks into a levee and weakens it. The weakest spots may break open like this levee that broke after Hurricane Katrina in 2005.

Shorelines and Erosion

California has 1,100 miles of shoreline along the Pacific Ocean. If you ever have had a chance to swim in the ocean, you know that waves are incredibly powerful forces. The energy produced by the constant action of waves continuously changes the shape of the shore. This change occurs because of rocks breaking into smaller pieces, transporting and grinding sediment, and depositing material farther along the shore.

Beaches and Wave Erosion

A **beach** is a landform consisting of loose sand and gravel. It is located along a shore. Beaches are dynamic, actively changing systems. Most of California's steep shores have been formed by beach erosion. Sand is also supplied by the continuous flow of rivers to the oceans. Sediment carried by the rivers gets deposited on the beach. Wave action then moves it along the shore.

 Where does beach sand come from?

Erosion Features Cliffs are formed by the cutting action of waves at the base of rocks that are exposed along the coasts. **Figure 15** shows what happens when a cliff is eroded. It moves back from the shoreline, leaving behind a flat area called a wave-cut platform. Sometimes these platforms can be lifted above the water level by upward movement along faults. The platform then is called a marine terrace. Erosional features with unusual shapes such as sea caves, sea stacks, and sea arches can form when waves erode the softer or more fractured portions of rocks.

Figure 15 Wave-cut platforms and uplifted marine terraces are common erosional features along California's shore, such as these at Bolinas Point near San Francisco.

Decide What would happen to these marine terraces if sea level rose?

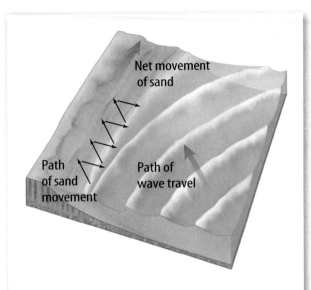

Figure 16 Longshore transport is a process that moves beach sediment parallel to the shore in the direction the wind is blowing.

Figure 17 These groins are helping to reduce sediment transport along the Marina del Ray beach.

Infer In which direction is the longshore current moving?

The Longshore Current

Figure 16 shows that waves approach the shore in a three-step process. They usually come in at an angle to the shore. The friction of hitting the beach at an angle causes the waves to bend until they are approximately parallel to the coast. Finally, they retreat from the beach perpendicular to the shore. This process is called longshore transport. The movement of the water is called the longshore current. It is this current that moves vast amounts of sediment along coasts.

 How do waves change direction as they approach the shore?

Preventing Erosion

There are several ways that beaches can be protected from erosion. Shoreline armoring is the name applied to the building of structures to help reduce erosion. These structures include retaining walls, harbor channels, and groins like those shown in **Figure 17.** Groins are positioned at right angles and placed at certain intervals along the shore. As the longshore current moves sediment along the shore, the groins trap the sediment. Shoreline armoring changes natural shoreline processes. But in some cases, it is absolutely necessary to prevent the collapse of cliffs or the complete destruction of a beach.

What are glaciers?

You have read of water's weathering power and its effect on erosion. Ice is also a strong eroding agent. **Glaciers** are large masses of ice and snow. They form on land in areas where the amount of winter snowfall is greater than the amount of summer melting. It takes hundreds to thousands of years to form a glacier. Although glaciers may appear to be motionless, they actually move very slowly, at a rate of about 2.5 cm per day. Glacial ice makes up about 2 percent of all the water on Earth. That is roughly 66 percent of the freshwater.

Types of Glaciers

There are two types of glaciers. Valley glaciers, or alpine glaciers, form in existing stream valleys high in the mountains. They flow from high to low elevations. There are more than 100,000 of this type of glacier on Earth today. Continental glaciers, or ice sheets, are several kilometers thick and cover entire land areas. The only continental glaciers on Earth today are in Antarctica and Greenland. Geological evidence indicates that these were the types of glaciers that covered portions of Earth during past ice ages.

 Reading Check Where do valley glaciers form?

Glaciers Eroding Land

Glaciers erode the surface as they pass over it. Rocks and boulders that are trapped at the bottom of the ice create grooves and scratches. This is similar to how sandpaper leaves scratches on wood. These grooves can be used to determine the direction the glacier was moving. Erosion by valley glaciers produces the distinct features shown in **Figure 18.** The presence of these features are evidence that valley glaciers once covered an area.

Depositing Sediment

As glaciers melt, they deposit sediment that had been frozen in the ice. Till and outwash are two types of sediment deposited by glaciers. **Figure 18** shows that till often builds up along the sides and fronts of glaciers into long, high ridges called moraines. It also shows that till can be molded beneath the glacier into a variety of landforms. Outwash consists mostly of sand and gravel. Many of these deposits have been quarried for use as construction materials.

Figure 18 This diagram shows how a glacier might change the features in a narrow, V-shaped river valley. Eroded rock material and water flow out from beneath a melting glacier. Distinct landforms form as this sediment builds up.

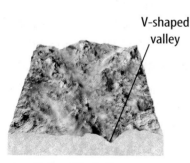

Before Glaciation

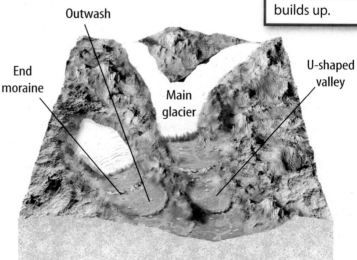

Outwash

End moraine

Main glacier

U-shaped valley

V-shaped valley

After Glaciation

Figure 19 Unsecured soil in the Great Plains was at the mercy of the wind and caused the Great Dust Bowl in the 1930s. It is believed that years of drought and poor agricultural practices contributed to the soil conditions.

Wind

Have you read about the Great Dust Bowl of the 1930s? This event occurred when the southern Great Plains of the United States were devastated for an entire decade by a drought. Deeply plowed fields and overgrazed pastures left soil unprotected and exposed to the elements. As shown in **Figure 19,** strong winds removed this soil and carried it into the air. Skies were blackened by great wind-generated dust storms.

 Why was the soil so easily eroded during the Great Dust Bowl?

Wind Erosion and Deposition

Wind lifts and redeposits loose material. There are two common types of wind-blown deposits. Sand dunes are shown in **Figure 20.** These mounds and ridges form from heavier sediment that blows along the ground surface. Eventually it is pushed into piles and dunes form. Loess (LUHS) is the second type of wind-blown deposit. It consists of wind-blown silt that was carried in the air. Loess is the smallest grain size produced by glacial erosion. Strong winds that blow across glacial outwash pick up the loess and redeposit it elsewhere. As wind-blown sediment is carried along, it cuts and polishes exposed rock surfaces.

 Figure 20 What is the basic shape of a dune?

Figure 20 Sand dunes, such as these in Death Valley, California, are formed as wind-blown sand moves over the ground surface.

Shaping by Erosion and Deposition

Several geologic processes are involved in erosion and deposition. Mass wasting causes landslides, rock falls, mudslides, and more. Climate and the amount of rainfall an area receives are directly related to mass wasting. Rivers erode streambeds and transport sediment to new locations. Their erosive power changes the shape of their streambeds. Wave action on ocean shores breaks up rocks and creates distinct features along beaches. Erosion and deposition by glaciers create familiar mountain scenery. Wind can be strong enough to cause erosion and to form dunes. The results of all these processes are seen in the landscapes present in California today.

LESSON 2 Review

Summarize

Create your own lesson summary as you design a **visual aid.**

1. **Write** the lesson title, number, and page numbers at the top of your poster.

2. **Scan** the lesson to find the **red** main headings. Organize these headings on your poster, leaving space between each.

3. **Design** an information box beneath each **red** heading. In the box, list 2–3 details, key terms, and definitions from each **blue** subheading.

4. **Illustrate** your poster with diagrams of important structures or processes next to each information box.

 **ELA6:** R 2.4

Standards Check

Using Vocabulary

1. Use the following terms together in a sentence: *erosion, deposition, mass wasting.* **2.b**

2. **Explain** the relationship between the following two terms: *flood, floodplain.* **2.d**

Understanding Main Ideas

3. **Create** a list of items to consider when deciding to build a house in an area that is on a hillside and has not been previously used for construction. **2.d**

4. **Explain** the three-step process of longshore transport along ocean beaches and what can be done to slow down or stop erosion. **2.c**

5. **Distinguish** three characteristics of valleys that had previously contained a glacier from characteristics of valleys formed by flowing water. **2.a**

6. **Sequence** Draw a graphic organizer like the one below to describe a possible history for a mineral grain of quartz that begins in the mountains within a piece of granite and ends as a piece of beach sand on the coast. Include at least three erosional processes. **2.b**

Applying Science

7. **Design** an experiment to see how vegetation planted in soil affects wind erosion. **2.d**

8. **Explain** how the main channel of the Mississippi River has moved from side to side. **2.d**

For more practice, visit **Standards Check** at ca6.msscience.com.

DataLab

00:25 minutes

Try at Home

Sorting It Out

When rivers flood they carry a lot of sediment and debris with them because they move with high energy. Along the shore the water loses energy, slows down, and drops its load of sediment.

Data Collection

1. Read and complete a lab safety form.

2. Create a data table to record your predicted time and your measured time.

3. Measure 50 g of each of the **sediments** provided.

4. Use a paper cone to pour the sediments into a **2-L plastic bottle.**

5. Fill the bottle with **water,** leaving an air space at the top.

6. Predict and record how long you think the sediments will remain in suspension after shaking them.

7. After shaking the bottle for 30 s, start a **timer.**

8. **Record** how long it takes for each of the sediments to settle down to the bottom.

Sample Data Table		
Type of Sediment	**Predicted Time (s)**	**Measured Time (s)**
White sand		
Beach sand		
Gravel		

Data Analysis

Create a class data table. Average the data and make a bar graph. How close were your predictions?

 Science Content Standards    MA6: MR 2.4

2.c Students know beaches are dynamic systems in which the sand is supplied by rivers and moved along the coast by the action of waves.
7.g Interpret events by sequence and time from natural phenomena (e.g., the relative ages of rocks and intrusions).

Science Content Standards

1.f Students know how to explain major features of California geology (including mountains, faults, volcanoes) in terms of plate tectonics.

2.a Students know water running downhill is the dominant process in shaping the landscape, including California's landscape.

2.b Students know rivers and streams are dynamic systems that erode, transport sediment, change course, and flood their banks in natural and recurring patterns.

2.c Students know beaches are dynamic systems in which sand is supplied by rivers and moved along the coast by the action of waves.

7.d Communicate the steps and results from an investigation in written reports and oral presentations.

Reading Guide

What *You'll Learn*

▶ **Describe** the effects of weathering and erosion on California's landscape.

▶ **Relate** California's uniqueness to the land use.

Why *It's Important*

Understanding weathering and erosion help you understand how California's landscape was formed.

Vocabulary

basin and range
arroyo

Review Vocabulary

uplift: any process that moves Earth's surface to a higher elevation (p. 79)

Reshaping the California Landscape

Main Idea The geology of California is expressed as mountains, deserts, valleys, and shorelines. These are natural physical features of Earth's surface.

Real-World Reading Connection California's early farmers and gold miners headed west following the California Trail to get to their final destinations. In this lesson, you will read about some of the landscapes they had to cross as they searched for the best place to settle.

Mountain Landscapes

The landscapes of California can be divided into a number of different regions. **Figure 21** shows how California can be divided into four major types of landscapes: mountains, deserts, the Central Valley, and the coast. Mountain ranges cover most of California. Many of the rocks within the ranges formed below Earth's surface hundreds of millions of years ago. They have been uplifted and exposed at the surface by tectonic processes.

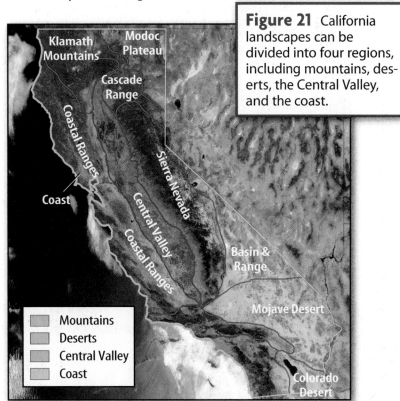

Figure 21 California landscapes can be divided into four regions, including mountains, deserts, the Central Valley, and the coast.

Klamath Mountains
Modoc Plateau
Cascade Range
Coastal Ranges
Coast
Sierra Nevada
Central Valley
Coastal Ranges
Basin & Range
Mojave Desert
Colorado Desert

Mountains
Deserts
Central Valley
Coast

Figure 22 U-shaped valleys have been scoured out by glacial ice.

Steep Slope

Gentle Slope

Figure 23 A change in slope can have dramatic effects along a river's path.

Glaciated Mountains

During the last 2.5 million years, glaciers have carved many features in the Sierra Nevada and Klamath Mountains. The scenery of the Yosemite valley in the Sierra Nevada, shown in **Figure 22,** is the result of glacial erosion. U-shaped valleys and hanging valleys, such as those shown in **Figure 22,** are common in Yosemite National Park. The largest present-day glacier in California is the Palisade Glacier in the Sierra Nevada. It is estimated to be about 3 km long and several hundred meters thick. Recall that glaciers deposit material as well as erode and move rock. Depositional features in California's mountains include moraines that were produced when the glaciers melted and deposited their till.

 Reading Check What types of glacial features are present in the Sierra Nevada?

Other Erosional Features

California mountain ranges have other types of erosional features that are not related to glaciers. Remember that streams and rivers change as water moves from the mountains to the oceans. Steep white-water streams in V-shaped valleys, as shown in **Figure 23,** are common in the high, steep parts of mountains. Wider and more meandering rivers, like that shown in **Figure 23,** are more common in the plains. Landslides and rockfalls also are common in California's mountains.

 Visual Check **Figure 23** What features are present along this river?

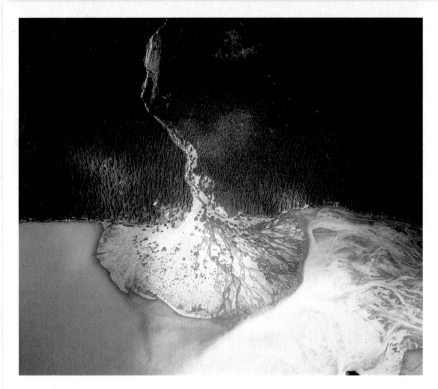

Figure 24 An alluvial fan formed by a mountain stream empties onto a desert valley floor.

Compare How is this alluvial fan similar to a delta?

Desert Landscapes

Refer to **Figure 21,** which shows that the deserts of California are primarily located in the southeastern corner of the state. The deserts consist of flat, sandy valleys and dry lake-beds called playas. The Mojave Desert, located in California, is an example of a desert landscape. It is sometimes referred to as the high desert because of its high elevation and little vegetation. In contrast is the Colorado Desert, which lies as much as 75 m below sea level and is referred to as the low desert. The Colorado Desert has become an agricultural area thanks to irrigation from the nearby Colorado River.

Features of the Desert

Alluvial fans, like the one shown in **Figure 24,** are common depositional features in the deserts. Wind-blown sand dunes are common in the desert as well. Some regions have such consistent, strong winds that hundreds of windmills are used to generate electricity for nearby towns. The windmills shown in **Figure 25** provide electricity for the popular vacation community of Palm Springs.

 Why is wind-generated power so successful in Palm Springs?

Figure 25 This field of windmills, located near Palm Springs, California, is generating a nearly constant supply of electricity.

Figure 26 The Amargosa River is an arroyo in Death Valley National Monument. It contains flowing water only after a rare rainstorm.

SCIENCE USE V. COMMON USE

basin (BAY sihn)

Science Use a tract of land drained by a river and its tributaries. *The Death Valley basin has no outlet.*

Common Use an open vessel with sloping or curving sides for holding water. *She bathed the baby in a small basin.*

WORD ORIGIN

arroyo

from Latin *ad–* (means *toward*) and *ruga* (means *wrinkle*)

The Basin and Range

The **Basin and Range** is a large area of north-south trending mountain ranges and valleys. It is primarily located in Nevada and Utah. You can refer to **Figure 21** to determine its location. Most of this area has a desert climate. At the western edge of the Basin and Range in California is Death Valley.

 What geologic features make up the basin and range?

Death Valley Gold-seekers named Death Valley in 1849 because of the valley's harsh conditions. The part of Death Valley known as Badwater is the lowest point in the western hemisphere at 86 m below sea level. Death Valley contains little vegetation to stop erosion during storms. The area contains many **arroyos,** which are streambeds that only contain water during heavy rains or floods. During these times, rock and sediment are transported downstream. The Amargosa River, shown in **Figure 26,** is an arroyo. In addition to present-day alluvial fans, ancient alluvial fan deposits are also present in Golden Canyon in Death Valley. The alluvial fan deposits shown in **Figure 27,** formed about six million years ago and have turned into rock.

Figure 27 These ancient alluvial fan deposits were uplifted and tilted from their original horizontal orientation by the same tectonic forces that produced the Basin and Range.

Infer What type of rock do you think these alluvial fans are made of?

Figure 28 The Sacramento River originates in the northern part of the Great Valley and flows to the Pacific Ocean through San Francisco Bay. Notice the lush agricultural region along the river's flood-plains.

Consider Why is this area so good for farming?

The Central Valley

Refer to **Figure 21** to locate the Central Valley of California. It is also called the Great Valley. It is about 800 km long and 50 km wide. It is a fault-bounded valley. This means that the mountains around it have been uplifted along faults while the Central Valley has dropped to lower elevations.

Rivers in the Central Valley

Rivers in the Central Valley have meanders and flow slowly along shallow slopes. There are two major rivers that flow through the Central Valley. The south-flowing Sacramento River shown in **Figure 28** is in the north part of the valley. The north-flowing San Joaquin River is in the south part of the valley. The rivers meet and form a delta into the Pacific Ocean through San Francisco Bay, shown in **Figure 29.**

 What features do the rivers of the Central Valley have?

Figure 29 The Sacramento and San Joaquin Rivers of the Central Valley join and flow into the Pacific Ocean where they deposit sediment to form the Sacramento River Delta.

October 1997

April 1998

Figure 30 The amount of beach sand along the coast at the mouth of Tomales Bay at Point Reyes National Seashore changed significantly between October 1997 and April 1998.

Explain Would you recommend building a beach house on the shore at Tomales Bay?

Deposition in the Central Valley

The Central Valley receives abundant sediment from the rivers flowing into it from the surrounding mountains. This sediment provides the valley with a thick fertile soil and has made it the most productive agricultural area in California. In fact, the Central Valley provides half the produce in the United States.

 Reading Check Why is this region so good for farming and agriculture?

Coastal Landscapes

California is known for its sandy beaches, which can change appearance **significantly** during different seasonal conditions. For example, **Figure 30** shows how the beach changed at Point Reyes National Seashore over a six-month period. Stormy El Niño conditions during the 1997–1998 winter caused many landslides and general erosion near the shore. This provided more sediment to the shore for waves to transport and deposit along beaches. California's rocky coasts have many erosional features described earlier in this chapter. Can you identify the sea stack, sea arch, and cliffs shown in **Figure 31?** Although this scene is in France, these types of landforms can be found along California's coast.

Figure 31 The sea arch and sea stack were formed by waves eroding the cliff. The sea stack at the left has been eroded to a sharp point.

California's Landscapes

California has a variety of landscapes. Mountains formed either by tectonic uplift or by the formation of volcanoes. Glaciers, wind, streams, and mass wasting have carved the mountains into the shapes they are today. Deserts experience strong winds and erosion that form dunes. The Basin and Range formed from tectonic activity. Fertile soils and wide rivers cover the Central Valley. The coast of California has beaches and rocky shorelines.

ACADEMIC VOCABULARY ···

significant (sig NIH fuh kent)

(adjective) to have influence or effect

Wind direction has a significant effect on sand dune shape.

LESSON 3 Review

Summarize

Create your own lesson summary as you design a **study web.**

1. **Write** the lesson title, number, and page numbers at the top of a sheet of paper.

2. **Scan** the lesson to find the **red** main headings.

3. **Organize** these headings clockwise on branches around the lesson title.

4. **Review** the information under each **red** heading to design a branch for each **blue** subheading.

5. **List** 2–3 details, key terms, and definitions from each **blue** subheading on branches extending from the main heading branches.

 **ELA6: R 2.4**

Standards Check

Using Vocabulary

1. Use the term *arroyo* in a sentence. **2.b**

2. The region in California that consists of long mountain ranges and valleys is called the _____. **1.f**

Understanding Main Ideas

3. **Name** one erosional feature and one depositional feature you might expect to find (a) along the coast, (b) in a desert, and (c) in the mountains. **2.a**

4. **Interview** three people to find out if they have visited each of the various types of California landscapes. Construct a table summarizing which erosional/depositional features (if any) they each noticed during their visits. **2.b**

Site	Erosional	Depositional

5. **Determine** which type of landscape you live in. Include three pieces of evidence for your choice. **1.f**

6. **Distinguish** two erosional and two depositional features in Death Valley from those in the Central Valley. **1.f**

Applying Science

7. **Design** the shortest driving trip possible that would take you from where you live through each of the different types of California landscapes. **2.a**

8. **Recommend** the best locations closest to where you live to see the following features: (a) sea arches, (b) alluvial fans, (c) sand dunes, and (d) cirque. Be specific about location names; you may need to check a map. **2.c**

 **Science Online**

For more practice, visit **Standards Check** at ca6.msscience.com.

MiniLab

Will it slump, or will it creep?

Thin soils, steep slopes, and rainfall lead to mass movements of the land. What causes a hillside to slump or creep?

Procedure

1. Read and complete a lab safety form.
2. Obtain a **stream table** set up from the teacher.
3. Cover one-third of one end with a layer of packed **sand**.
4. Place a piece of **foil or plastic wrap** over the sand.
5. Loosely cover the foil with sand to a depth of 3–5 cm.
6. Raise the sand end of the stream table using **three books**.
7. Predict what will happen when the upper layer of sand becomes wet.
8. Gently pour **water** on the sand until it moves.
9. Repeat steps 5–8 with a steeper angle **(six books)**.

Water source

Incline block

Sand

Stream table

Hose

Clamp

Analysis

1. **Compare and contrast** movements of the sand when the slopes are different.
2. **Infer** how you could keep the sand from moving.
3. **Form a hypothesis** about the cause of the wet sand movement. Was it slump or creep?

 Science Content Standards

2.a Students know water running downhill is the dominant process in shaping the landscape, including California's landscape.

2.d Students know earthquakes, volcanic eruptions, landslides, and floods change human and wildlife habitats.

7.a Develop a hypothesis.

7.e Recognize whether evidence is consistent with a proposed explanation.

Applying Math

Erosion Rates

California's Big Sur coast is affected by erosion due to landslides that cause the cliffs to retreat. The graphs show loss rate in inches per year for three areas of Big Sur—Big Slide-Pitkins Curve, Grayslip, and Hurricane Point.

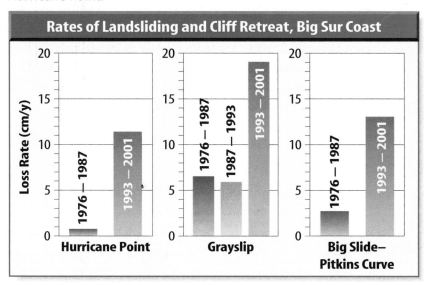

Rates of Landsliding and Cliff Retreat, Big Sur Coast

How much more erosion loss is there in Grayslip than Hurricane Point in the years 1976–1987?

What you know:
- Erosion rate for Grayslip: 6.5 cm/yr
- Erosion rate for Hurricane Point: 1.0 cm/yr

What you need to find:
- The difference in erosion rates

Subtract: $6.5 - 1.0 = 5.5$ cm/yr.

Answer: The difference in erosion rates is 5.5 cm/yr.

Practice Problems

1. In the years 1993–2001, how much more erosion occurred at Grayslip than Hurricane Point?
2. Which span of years and location has the most loss due to erosion?

Science Online
For more math practice, visit **Math Practice** at ca6.msscience.com.

Stream Sediment

Materials

stream table
plastic bottles
(1-L and 2-L)
pieces of rubber
tubing (2)
sand, gravel, and rock
water
bucket
clamps (2, optional)
incline materials

Safety Precautions

 Science Content Standards

7.b Select and use appropriate tools and technology (including calculators, computers, balances, spring scales, microscopes, and binoculars) to perform tests, collect data, and display data.

Problem

Erosion is an important agent of change in the landscape. As you learned in the previous lesson, water can carry sediments, and often those sediments are from the surrounding land. The movement of sediment is part of erosion. How does water move sediment? What affects the movement of sediments? What formations are created by the movement?

Collect Data and Make Observations

1. Complete a lab safety form.
2. Your teacher will supply you with materials and instructions to assemble a stream table.
3. Record the size of the bottle you will be using.
4. Shake the sand and gravel together in the box. Slide the mixture to one end of the box and form a "mountain slope."
5. What do you think will happen when you add water to your mountain?
6. Fill the bottle with water and stretch the rubber tubing over the mouth. Place the other end of your tubing on top of your mountain and turn the bottle upside down. Let the water run until the bottle is empty.
7. Draw or photograph the results.
8. In part of the sand that is undisturbed, trace out a streambed. Make a new hypothesis about the results of adding more water.
9. Repeat step 6 introducing water to the streambed.
10. Draw or photograph the results.
11. Add stones to the streambed; vary the slope of the mountain, or change another variable you might want to test.
12. Repeat steps 5–6.
13. Draw or photograph the results of each test.

Analyze and Conclude

1. **Describe** the differences between your first and other trials. Did you correctly predict what would happen in your stream table? Did any of the trials form recognizable formations?
2. **Compare and contrast** your drawings with other students. What bottle size moved the most sediment?

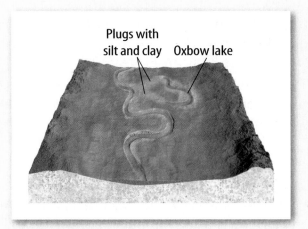

Plugs with silt and clay   Oxbow lake

Communicate

WRITING in Science

Erosion occurs in areas that humans have settled. Coastal real estate disappears every year in some areas, while harbors get filled in with sediment in other areas. Private land and parkland bordering rivers experience trouble with flooding and deposition of silt.

How do communities deal with erosion? Research a community or engineering strategy that is used in California to combat erosion. Make a report explaining how land owners or park custodians could use the strategy to prevent erosion.

Real World Science

You can build dams!

Environmental engineers work to prevent air and water pollution. Some environmental engineers design, build, or repair structures to prevent soil erosion. Many of these projects require you to work outdoors and travel to new places. To become an environmental engineer, you will need to take science, math, and computer courses in high school to prepare for a four-year college degree.

Visit **Careers** at **ca6.msscience.com** to learn more about the work of environmental engineers. Think about a river or stream near your school. **Research** three different dam types and decide which would be the most economical and practical for the stream. **Decide** what purpose the dam would have.

PREVENTING BEACH EROSION

The sand on beaches naturally moves over time—sometimes parallel to the beach with the longshore current and sometimes perpendicular to the beach. This is only a problem for human-made structures such as houses, roads, boardwalks, and lighthouses. Oceanfront home owners and ocean-side towns risk loss of property and tourist revenue with each winter storm. Many different ways to prevent beach erosion have been developed to reduce these losses.

Visit **Technology** at **ca6.msscience.com** to explore different ways to prevent beach erosion. Evaluate some methods of preventing beach erosion according to cost and time constraints.

Fossils at Lyme Regis

This spiral-shaped ammonite fossil is located on Monmouth Beach at Lyme Regis. The cliffs at Lyme Regis on the coast of Great Britain contain millions of fossils from the Jurassic period. The cliffs have been etched by wind and tides for millions of years, exposing layers of sediment. In the nineteenth century, fossil collector Mary Anning was born and raised in Lyme Regis. She spent her childhood collecting fossils, and when she was about 11, she uncovered the first ichthyosaur known to the British scientific community. She took over the family fossil business, and became an expert paleontologist.

Visit **History** at **ca6.msscience.com** to find out more about Mary Anning and her fossil business. Imagine you are Mary Anning. Write a journal entry about discovering a fossil and deciding what to do with it.

Acid Rain

Acid rain is caused when air pollution mixes with rain. The acid that results damages ecosystems, humanmade structures, and farmable soil. Most acid rain is produced by power plant and road vehicle emissions. Because air does not stop at state or national boundaries, acid rain may fall hundreds of miles from the pollution source. Certain solutions have been proposed to diminish the creation of acid rain, but most are expensive or require a change in our habits.

Visit **Society** at **ca6.msscience.com** to research acid rain. Work in small groups to generate a list of ideas that individuals, businesses, utilities, and the government can implement to prevent acid rain. Select your group's three best solutions considering cost, effectiveness, and ease of implementation. Present your solutions to the class.

The BIG Idea — Weathering, erosion, and deposition shape Earth's surface.

Lesson 1 Weathering
2.a, 7.e

Main Idea Rocks exposed at Earth's surface are broken down into sediment and soils by the action of weathering.

- Chemical and physical weathering break down rocks and minerals.
- Exposure to water and gases in the atmosphere causes chemical weathering.
- Frost wedging, burrowing animals, and the growth of tree and plant roots cause physical weathering.
- Soil contains weathered rock and minerals, organic matter, water, and air.
- Soil profiles have distinct layers called soil horizons.

- **chemical weathering** (p. 335)
- **frost wedging** (p. 337)
- **physical weathering** (p. 337)
- **soil** (p. 338)
- **weathering** (p. 334)

Lesson 2 Erosion and Deposition
2.a, 2.b, 2.c, 2.d, 7.g

Main Idea Movement of rock and soils are natural occurrences caused by specific geological conditions.

- Mass wasting is the downhill movement of sediments. It includes landslides, mudslides, creep, and slump. Steepness of a slope, the presence or absence of vegetation, and the climate affect mass wasting.
- Rivers begin in V-shaped valleys and end in flatter, wider plains. Rivers form meanders, floodplains, oxbow lakes, and deltas.
- Ocean waves erode shorelines and create sediment and unusual landforms. Waves constantly move and redeposit sediment along coastlines.
- Glaciers are large masses of ice and snow that form on land. They are strong agents of erosion. They deposit two types of sediment called till and outwash.
- The two most common types of wind deposits are dunes and loess.

- **beach** (p. 349)
- **deposition** (p. 342)
- **erosion** (p. 342)
- **flood** (p. 348)
- **floodplain** (p. 348)
- **glacier** (p. 350)
- **landslide** (p. 343)
- **mass wasting** (p. 343)
- **meander** (p. 345)

Lesson 3 Reshaping the California Landscape
1.f, 2.a, 2.b, 2.c, 2.d, 7.d

Main Idea The geology of California is expressed as mountains, deserts, valleys, and shorelines. These are natural physical features of Earth's surface.

- California's landscape is made up of mountains, coasts, deserts, and a central valley. The mountains have features made from both rivers and glaciers.
- Wind erosion, sand dunes, arroyos, and alluvial fans are desert features.
- The Basin and Range consists of mountains and valleys.
- The Central Valley has thick, fertile soil.
- Erosion along rocky coasts produces sea stacks, sea arches, and sea cliffs.

- **arroyo** (p. 358)
- **basin and range** (p. 358)

STUDY TO GO — Download quizzes, key terms, and flash cards from ca6.msscience.com.

Linking Vocabulary and Main Ideas

Use vocabulary terms from page 368 to complete this concept map.

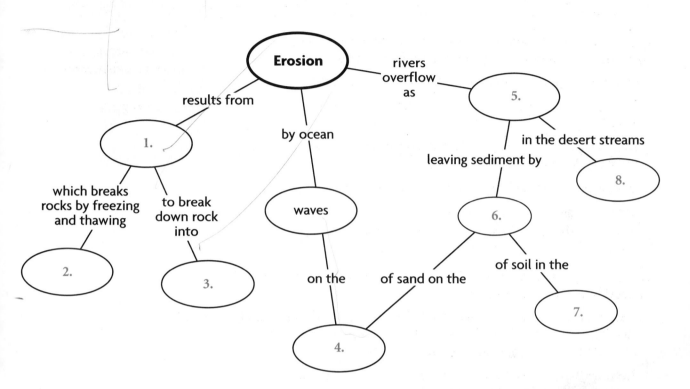

Using Vocabulary

Fill in each blank with the correct vocabulary term.

9. _____ is the total effects of sun, rain, snow, and humidity on the erosion of rocks into tiny bits and pieces.

10. The weathering process during which water freezes in the cracks of rocks, causing the cracks to expand, is called _____.

11. An example of mass wasting is _____.

12. A river delta is formed by the process of _____.

13. Longshore transport is a process that moves sediment along a(n) _____.

14. A desert stream channel that usually contains water only during flash floods is known as a(n) _____.

15. The accumulation of grains and particles of rock mixed with dead organic material is called _____.

Understanding Main Ideas

Choose the word or phrase that best answers the question.

1. What is the cycle of freezing and thawing that breaks rocks called?
   A. rock cycle   2.a
   B. frost wedging
   C. exfoliation
   D. root pressure

2. Which causes most chemical weathering?
   A. frost wedging   2.a
   B. air pollution
   C. acidic water
   D. biological activity

3. What is another term for decayed organic matter?
   A. humus   2.a
   B. soil
   C. worms
   D. sediment

4. The photo below shows a soil profile.

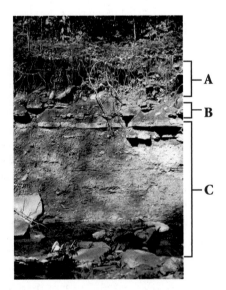

   Which soil horizons contains mostly unweathered rock with no organic matter?
   A. A horizon   2.a
   B. B horizon and C horizon
   C. C horizon
   D. A horizon and B horizon

5. One type of mass wasting is shown below.

   The tilted tree is a sign of which type of mass wasting?
   A. landslide   2.b
   B. creep
   C. slump
   D. mudslide

6. Which valleys have been eroded by glaciers?
   A. V-shaped valleys   2.a
   B. S-shaped valleys
   C. U-shaped valleys
   D. C-shaped valleys

7. In which type of landscape do more than half of all Californians live?
   A. coast   1.f
   B. mountains
   C. deserts
   D. Central Valley

8. In which type of landscape does most agricultural activity occur in California?
   A. coast   1.f
   B. mountains
   C. deserts
   D. Central Valley

9. Which feature is erosional?
   A. alluvial fan in Mojave Desert   2.b
   B. floodplain along the Sacramento River
   C. glacial moraines in the Sierra Nevada
   D. sea arch at Point Lobos State Reserve

Science Online Standards Review ca6.msscience.com

Applying Science

10. Prioritize four concerns you might have related to chemical and physical weathering processes if you were involved in plans to build a new school on a sea cliff overlooking the ocean. **1.f**

11. Hypothesize why loess deposits travel farther than sand dune deposits. **2.b**

12. Design an experiment to test the effectiveness of different kinds of retaining walls (such as solid versus wire mesh) to prevent mass wasting. **1.f**

The diagram below shows longshore transport. Use it to answer questions 13 and 14.

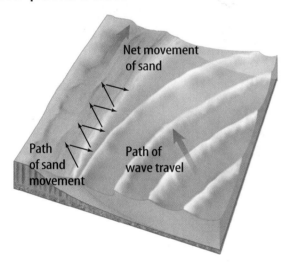

Net movement of sand

Path of sand movement

Path of wave travel

13. Identify the directions of water and sand as the longshore current moves the sediment. **2.c**

14. Predict how the placement of a groin will affect the sand movement. Illustrate the movement of the sand before placement, after placing one groin, and after placing two groins. **2.d**

15. Design a class field trip to the beach to investigate erosional and depositional features. What would you look for and how would you have the class record their observations? **2.c**

16. Compile a list of at least five hazardous erosional or depositional conditions that would be worse during a particularly stormy, rainy season. **2.a**

WRITING in ▶ Science

17. Predict how the landscapes of California might change if another ice age occurred that brought large ice sheets from the north into the state. Write a paragraph describing where you think the ice would go and the kinds of features it would produce.

Cumulative Review

18. Compare and contrast a volcanic lahar and a mudslide. **2.d**

19. Predict During an earthquake, sand can be thrown up from the ground in a sand boil. Why might it be difficult to locate a sand boil a year after the earthquake? **2.d**

Applying Math

Use the graphs below to answer questions 20–22.

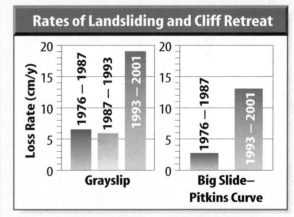

Rates of Landsliding and Cliff Retreat

Loss Rate (cm/y)

Grayslip: 1976 – 1987, 1987 – 1993, 1993 – 2001

Big Slide–Pitkins Curve: 1976 – 1987, 1993 – 2001

20. In the years 1976-1987, how much more erosion occurred at Grayslip than Big Slide-Pitkins Curve? **MA6: NS 2.0, MR 2.4**

21. In the years 1993-2001, how much more erosion occurred at Grayslip than Big Slide-Pitkins Curve? **MA6: NS 2.0, MR 2.4**

22. Which span of years and location has the least loss due to erosion? **MA6: NS 2.0, MR 2.4**

1 Which is an example of physical weathering?

A creep

B frost wedging

C oxidation

D slump
　　2.b

2 Which forms as a glacier moves into a stream valley?

A cirque

B outwash

C U-shaped valley

D V-shaped valley
　　2.b

3 Which factor in soil formation deals with the slope of the land?

A climate

B parent rock

C time

D topography
　　2.a

4 Which is a mixture of weathered rock, organic matter, water, and air?

A humus

B organisms

C parent rock

D soil
　　2.a

5 Which type of erosion occurs when a thin sheet of water flows downhill?

A creep

B gulley erosion

C runoff

D sheet erosion
　　2.a

6 What causes potholes to form in roadways?

A creep

B frost wedging

C oxidation

D slump
　　2.a

Use the illustration below to answer questions 7 and 8.

7 Which form of mass movement is shown in this picture?

A creep

B mudslide

C rockslide

D slump
　　2.d

8 Which agent of erosion causes this effect?

A gravity

B ice

C water

D wind
　　2.d

9 What form of mass movement occurs when a pasty mix of water and sediment moves downhill?

A creep

B mudflow

C rockslide

D slump   2.a

10 Which climate conditions produce the most rapid chemical weathering of rock?

A cold and dry

B cold and wet

C hot and dry

D hot and wet   2.a

11 What type of erosion can make pits in rocks and produce smooth, polished surfaces?

A abrasion

B deflation

C glaciation

D sedimentation   2.a

12 If an iron-containing mineral is exposed to rain, a rustlike material forms on its surface. Which best explains this?

A chemical weathering involving carbon dioxide and water

B chemical weathering involving oxygen and water

C physical weathering caused by strong winds

D physical weathering caused by rain   2.a

13 Which area is likely to be most affected by soil erosion?

A a steep slope after a fire burned the vegetation

B a section of low-elevation tropical rain forest

C a meadow with several kinds of grasses

D a hillside that has been terrace-farmed   2.b

The illustration below shows a statue damaged by acid rain.

14 Which statue is likely to be affected the most by chemical weathering?

A a marble statue in a cool, dry climate

B a granite statue in a cool, dry climate

C a marble statue in a warm, wet climate

D a granite statue in a warm, wet climate   2.a

15 What is the main reason that plants benefit from the presence of decaying organic material in the soil?

A It adds nutrients to the soil.   2.a

B It encourages mechanical weathering.

C It speeds the rate of evaporation from soil.

D It protects the plants from harmful insects.

From the Recommended Literature for Science and Math

Are you interested in learning more about how Earth's surface is shaped and reshaped by geologic events, weathering, and erosion? If so, check out these great books.

Narrative Nonfiction

The Buried City of Pompeii: What It Was Like When Vesuvius Exploded, by Shelly Tanaka, makes the events surrounding the volcanic eruption of Mount Vesuvius in A.D. 79 come alive. This book describes what might have occurred on that day. ***The content of this book is related to*** *Science Standard 6.1.*

Fiction

The Big Wave, by Pearl Buck, describes life and the events following a tsunami. Kino lives in the shadow of a volcano and his friend Jiya lives by the sea. When the big wave destroys both the village and Jiya's family, the boys learn important lessons about sorrow and acceptance. ***The content of this book is related to*** *Science Standard 6.1.*

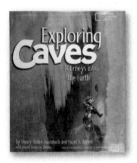

Nonfiction

Exploring Caves: Journeys into the Earth, by Nancy Aulenback, describes adventures exploring different types of caves including those in Greenland's ice cap, Mexico's Yucatan Peninsula, and the Grand Canyon. This book includes views of rock formations and the animals that live there. Information on cave formation and mapping are included. ***The content of this book is related to*** *Science Standard 6.2.*

Nonfiction

Earthquakes, by Mark Maslin, is an informative overview of the causes and the detection of earthquakes. This book contains photographs, colorful diagrams, quotes from eyewitnesses, and excerpts from news accounts of earthquakes. Details about well-known earthquakes and their devastation are included throughout the book to illustrate specific concepts. ***The content of this book is related to*** *Science Standard 6.2.*

Choose the word or phrase that best answers the question.

1. The photo below shows a fault in which the rock above the fault surface is moving down relative to the rock below the fault surface.

Tension forces pull rocks apart.

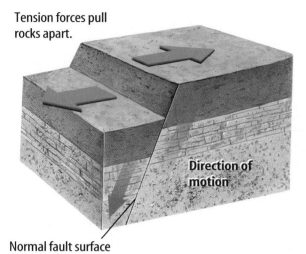

Direction of motion

Normal fault surface

What kind of fault is shown above?
A. normal
B. strike-slip
C. reverse
D. shear   `1.d`

2. What type of boundary is associated with composite volcanoes?
A. plates moving apart
B. plates sticking and slipping
C. plates moving together
D. plates sliding past each other   `1.e`

3. What type of magma produces violent volcanic eruptions?
A. those rich in silica
B. those with low viscosity
C. those forming shield volcanoes
D. those rich in iron   `1.d`

4. Which is the area of land from which a stream collects runoff?
A. drainage basin
B. gully
C. runoff
D. underground   `2.b`

Write your responses on a sheet of paper.

5. **Analyze** how soil erosion reduces the quality of soil.   `2.a`

6. **Sequence** the process through which earthquakes occur. Include a description of how energy builds up in rocks and is later released.   `1.d`

7. **Explain** how some volcanoes occur where one plate sinks beneath another plate. Support your answer with a labeled diagram.   `1.e`

8. **Discuss** how volcanoes affect humans and wildlife. List four safety precautions for people living in volcanic areas.   `2.d`

9. The map below shows three circles drawn around three different seismograph stations. The circles have radii equal to the distance between the seismograph station and the earthquake's epicenter.

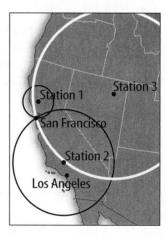

Station 1
Station 3
San Francisco
Station 2
Los Angeles

Identify which labeled point on the map represents the earthquake's epicenter? Explain.   `1.g`

10. **Discuss** what humans can do to control flood waters.   `2.b`

11. **Explain** how waves and longshore currents affect sand and sediments.   `2.c`

Energy in the Earth System

Energy in the Air Lightning bolts have temperatures hotter than the surface of the Sun. This energy is still mysterious, but scientists know it is generated in electrically charged storm systems.

West-Coast EVENTS

40,000–10,000 Years Ago
Sharp drop in temperature causes much of Earth's water to freeze, forming huge sheets of ice (glaciers) that cover northern areas of Earth, including the Pacific Northwest.

1865
Coastal spot that is present-day La Conchita, California, experiences its first recorded landslide.

1860 A.D.   1880   1900

2560 B.C.
Egyptians build the Great Sphinx and Great Pyramid at Giza; yearly flooding of the Nile River made the soil rich for farming.

WORLD EVENTS

March 1888
Blizzard leaves 1.5 m of snow and 400 people dead on the East Coast of the United States.

Science Online

To learn more about meteorologists and their work, visit ca6.msscience.com.

Concepts In Motion

Interactive Time Line To learn more about these events and others, visit ca6.msscience.com.

March 1910
Avalanche of snow in Wellington, Washington, pushes 2 trains into canyon.

March 1928
Flood in Santa Paula, California, kills 450 people.

January 1969
Heavy rains in Southern California cause floods and mudslides, leaving wide-spread property damage.

January 1995
Heavy rain caused by El Niño leads to mudslide at La Conchita, California, destroying homes and roads; another mudslide occurs there in January 2005.

1920   **1940**   **1960**   **1980**   **2000**   **2020**

July–August 1931
Flood along Yangtze River in China causes death of 3.7 million people from drowning, disease, or starvation.

March 1952
28 tornadoes touch down in Arkansas and Tennessee.

August 2005
Hurricane Katrina hits the coasts of Louisiana, Mississippi, and Alabama, flooding about 350,000 homes in New Orleans alone.

Earth's Atmosphere

The BIG Idea

The Sun's energy and Earth's atmosphere are critical for creating the conditions needed for life on Earth.

LESSON 1  4.a, 4.b
Energy from the Sun

(Main Idea) The Sun is the major source of energy for Earth.

LESSON 2 3.c, 3.d, 4.d
Energy Transfer in the Atmosphere

(Main Idea) Earth's atmosphere distributes thermal energy.

LESSON 3

3.d, 4.a, 4.b, 4.d, 4.e, 7.c

Air Currents

(Main Idea) Solar energy is responsible for the continuous movement of air in the troposphere, which transports and distributes thermal energy around Earth.

Up, Up, and Away!

Because the air inside this hot-air balloon is dense than the surrounding air, the balloon rises. Sometimes moisture tha carried by warm air rising above Earth's surface condenses to form clouds the ones shown above Mt. Shasta. Air can also be deflected by the geogra of the land, such as mountains.

(Science *Journal*) Write a hypothesis that explains how you think thes clouds formed above Mt. Shasta.

Launch Lab

00:10 minutes

Does temperature affect air density?

Air density can be affected by something as simple as temperature.

Procedure

1. Read and complete a lab safety form.
2. Loosen up the opening of a **balloon.**
3. Stretch the opening of the balloon over the opening of a **bottle.**
4. Hold the bottle in a **bucket of hot water.** Observe what happens to the balloon.
5. Place the bottle in a **bucket of cold water.** Observe what happens to the balloon.

Think About This

- **Describe** What happened when the bottle was placed in the hot water? In the cold water?

- **Explain** Why do you think these things happened? Is it what you expected?

4.d

FOLDABLES™
Study Organizer

Heat Transfer Make the following Foldable to identify the types of heat transfer that occur in Earth's atmosphere.

STEP 1 Fold a sheet of paper into thirds lengthwise and fold the top down about 3 cm.

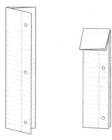

STEP 2 Unfold and **draw** lines along all folds. **Label** as shown.

Reading *Skill*   **ELA6:** R 2.4

Analyzing As you read Lesson 2, identify the important concepts about the ways thermal energy is transferred in the atmosphere.

Science Online

Visit ca6.msscience.com to:

▶ view **Concepts in Motion**
▶ explore Virtual Labs
▶ access content-related Web links
▶ take the Standards Check

Get Ready to Read

Make Connections

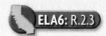 ELA6: R.2.3

1 **Learn It!** Make connections between what you read and what you already know. Connections can be based on personal experiences (text-to-self), what you have read before (text-to-text), or events in other places (text-to-world).

As you read, ask connecting questions. Are you reminded of a personal experience? Have you read about the topic before? Did you think of a person, a place, or an event in another part of the world?

2 **Practice It!** Read the excerpt below and make connections to your own knowledge and experience.

Text-to-self:
What happens to your skin when you are in the Sun? Have you ever gotten a sunburn on a cloudy day?

Text-to-text:
What have you read about radiation in other chapters? Which layer of the atmosphere absorbs UV radiation?

Text-to-world:
How do people avoid getting a sunburn?

Ultraviolet (uhl truh VI uh luht) (UV) waves have shorter wavelengths than visible light. Humans do not see or feel ultraviolet radiation. However, you might have felt the effects of ultraviolet radiation. Ultraviolet light is the radiation that is responsible for causing skin to tan or sunburn. Some animals, such as bees, butterflies, and birds, can detect ultraviolet light with their eyes.

—*from page 385*

3 **Apply It!** As you read this chapter, choose five words or phrases that make a connection to something you already know.

Target Your Reading

Use this to focus on the main ideas as you read the chapter.

Reading Tip

Make connections with memorable events, places, or people in your life. The better the connection, the more likely you will remember.

1 **Before you read** the chapter, respond to the statements below on your worksheet or on a numbered sheet of paper.

- Write an **A** if you **agree** with the statement.
- Write a **D** if you **disagree** with the statement.

2 **After you read** the chapter, look back to this page to see if you've changed your mind about any of the statements.

- If any of your answers changed, explain why.
- Change any false statements into true statements.
- Use your revised statements as a study guide.

Before You Read A or D	Statement	After You Read A or D
	1 Earth's atmosphere is made of gases.	
	2 The Sun's energy heats Earth's surface.	
	3 The sky looks blue because light is absorbed by the atmosphere.	
	4 Only about half the Sun's energy reaches Earth's surface.	
	5 Hot air rises and cold air sinks.	
	6 Carbon dioxide is an example of a greenhouse gas.	
	7 Earth's surface is heated evenly by the Sun.	
	8 Air currents can move vertically.	
	9 Air moves from areas of low pressure to areas of high pressure.	
	10 Earth's rotation affects the direction in which air and water move.	

Science Online

Print a worksheet of this page at ca6.msscience.com.

Reading Guide

What *You'll Learn*

▶ **Identify** some of the differences between layers of the atmosphere.

▶ **Describe** how solar radiation reaches Earth's surface.

▶ **Understand** that solar radiation has its maximum in the range of visible light.

▶ **Explain** why the sky looks blue.

▶ **Identify** the Sun as a constant and almost uniform source of energy for Earth.

Why *It's Important*

The heat from the Sun helps keep Earth's surface warm.

Vocabulary

atmosphere
troposphere
stratosphere
electromagnetic spectrum
infrared wave
ultraviolet wave

Review Vocabulary

radiation: energy transfer by electromagnetic waves (p. 150)

Energy from the Sun

Main Idea The Sun is the major source of energy for Earth.

Real-World Reading Connection When you sit outside on a sunny day, you can feel the Sun's energy warming you. Have you ever known anyone who tanned or sunburned on a hazy day? Even on cloudy days, the Sun's energy reaches Earth.

Earth's Atmosphere

The **atmosphere** (AT muh sfihr) is a mixture of gases that surrounds Earth. This mixture is often referred to as *air*. The atmosphere is made up of several different layers. Each layer has distinct properties.

The Composition of Air

The main gases that make up Earth's atmosphere are nitrogen and oxygen, as shown in **Figure 1.** Oxygen gas (O_2) makes up about 21 percent of the atmosphere. Humans and other animals need to breathe oxygen to live. Nitrogen gas (N_2) makes up about 78 percent of the atmosphere. Particles and gases such as water vapor (H_2O), argon (Ar), carbon dioxide (CO_2), and ozone (O_3) make up about 1 percent of the atmosphere. Even though these substances are present in small concentrations, they are still important. Some of these substances affect weather and climate and protect living things from harmful solar radiation. Others can have damaging effects on the atmosphere and the organisms that breathe them in the air.

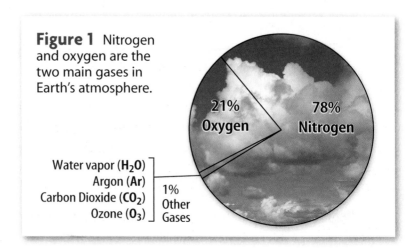

Figure 1 Nitrogen and oxygen are the two main gases in Earth's atmosphere.

21% Oxygen

78% Nitrogen

Water vapor (**H₂O**)
Argon (**Ar**)
Carbon Dioxide (**CO₂**)
Ozone (**O₃**)
1% Other Gases

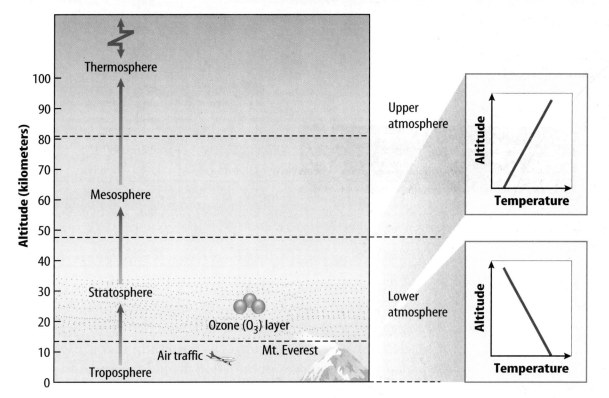

Figure 2 Earth's atmosphere can be divided into layers based on the different characteristics of each layer.
Identify In which layer of the atmosphere do planes fly?

Layers in the Atmosphere

Figure 2 shows the altitudes, or height above sea level, of atmospheric layers. The lowest layer of the atmosphere is the troposphere. The **troposphere** (TRO puh sfihr) is the region of the atmosphere that extends from Earth's surface to a height of about 8–15 km. It holds the majority of Earth's air and has weather. In the troposphere, as altitude increases, air temperature decreases, as shown in the bottom graph.

The stratosphere is above the troposphere. The **stratosphere** (STRA tuh sfihr) is the region of the atmosphere that extends from about 15 km to 50 km. In the stratosphere, as altitude increases, air temperature increases, as shown in the top graph. This occurs because the concentration of ozone is much higher in the stratosphere than in the troposphere. The layer of ozone in the stratosphere absorbs some of the Sun's harmful ultraviolet radiation, causing air temperature to rise.

The top two layers of the atmosphere are the mesosphere and the thermosphere. The mesosphere extends to about 80 km above Earth's surface. The thermosphere does not have a defined upper limit. Beyond the thermosphere is space.

Figure 2 Does temperature increase or decrease with height in the stratosphere?

WORD ORIGIN· · · · · · · · · · · ·
troposphere
from Greek *tropos,* means *a turn, change;* and *spharia,* means *sphere*

SCIENCE USE V. COMMON USE· ·
concentration
Science Use the amount of a substance in a given area or volume. *There is a higher concentration of nitrogen in the atmosphere than oxygen.*
Common Use the direction of attention to a single object or task. *I used all my concentration to finish the test in time.*

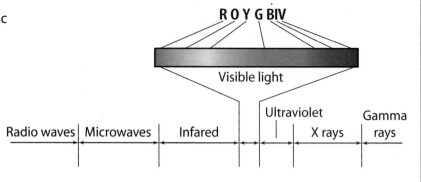

Figure 3 The electromagnetic spectrum shows the different types of radiation from short waves to long waves.

Identify Which colors make up visible light?

The Sun's Continuous Spectrum

The **electromagnetic spectrum** (ih lek troh mag NEH tik • SPEK trum) includes the entire range of wavelengths or frequencies of electromagnetic radiation. Shown in **Figure 3,** the electromagnetic spectrum is a continuum that is used to describe differences in radiation, from long waves to short waves. Ninety-nine percent of solar radiation consists of ultraviolet light, visible light, and infrared radiation.

Visible Radiation

Sometimes sunlight is referred to as **visible** light or white light. Recall from Chapter 2 that wavelengths in the visible range are those you can see. Have you ever used a prism to separate white light into different colors? White light can be divided into red, orange, yellow, green, blue, indigo, and violet, as shown in **Figure 3.** Visible light, including all of the colors of a rainbow, is actually visible radiation. The energy coming from the Sun peaks in the range of visible light, as you can see in **Figure 4.**

ACADEMIC VOCABULARY
visible (VIH zuh bul)
(adj) able to be seen
The moon was visible on the clear, cloudless night.

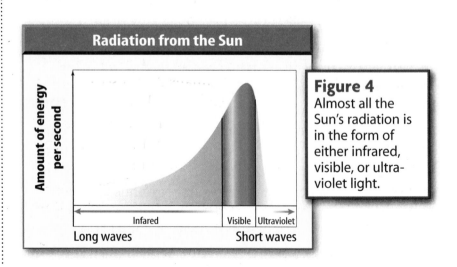

Figure 4
Almost all the Sun's radiation is in the form of either infrared, visible, or ultraviolet light.

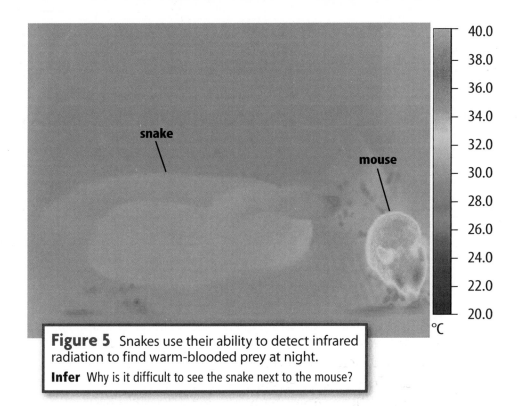

Figure 5 Snakes use their ability to detect infrared radiation to find warm-blooded prey at night.

Infer Why is it difficult to see the snake next to the mouse?

Near-Visible Radiation

In addition to visible light, **Figure 4** shows that we also receive infrared and ultraviolet radiation from the Sun. The wavelengths of these two forms of radiation are just beyond the range of visibility to human eyes. However, these forms of radiation can be detected by some organisms.

Infrared (IR) waves have longer wavelengths than visible light and sometimes are felt as heat. If you have ever felt the warmth from a fire, you have felt infrared radiation. You also can feel infrared radiation when you are being warmed by the Sun as you lie on the beach. Some snakes, such as rattlesnakes, have special sensors near their eyes that can detect infrared radiation. **Figure 5** shows how a mouse looks to a snake with infrared sensors.

Ultraviolet (ul truh VI uh luht) **(UV) waves** have shorter wavelengths than visible light. Humans do not see or feel ultraviolet radiation. However, you might have felt the effects of ultraviolet radiation. Ultraviolet light is the radiation that is responsible for causing skin to tan or sunburn. Some animals, such as bees, butterflies, and birds, can detect ultraviolet light with their eyes. The ability to sense ultraviolet light helps bees find flower nectar. **Figure 6** compares how a flower looks to the human eye to how it looks to a honeybee.

Normal

UV

Figure 6 Honeybees can detect ultraviolet light. The bottom photo shows how a flower would look through the eyes of a honeybee.

 **Reading Check** What are some differences between infrared waves and ultraviolet waves?

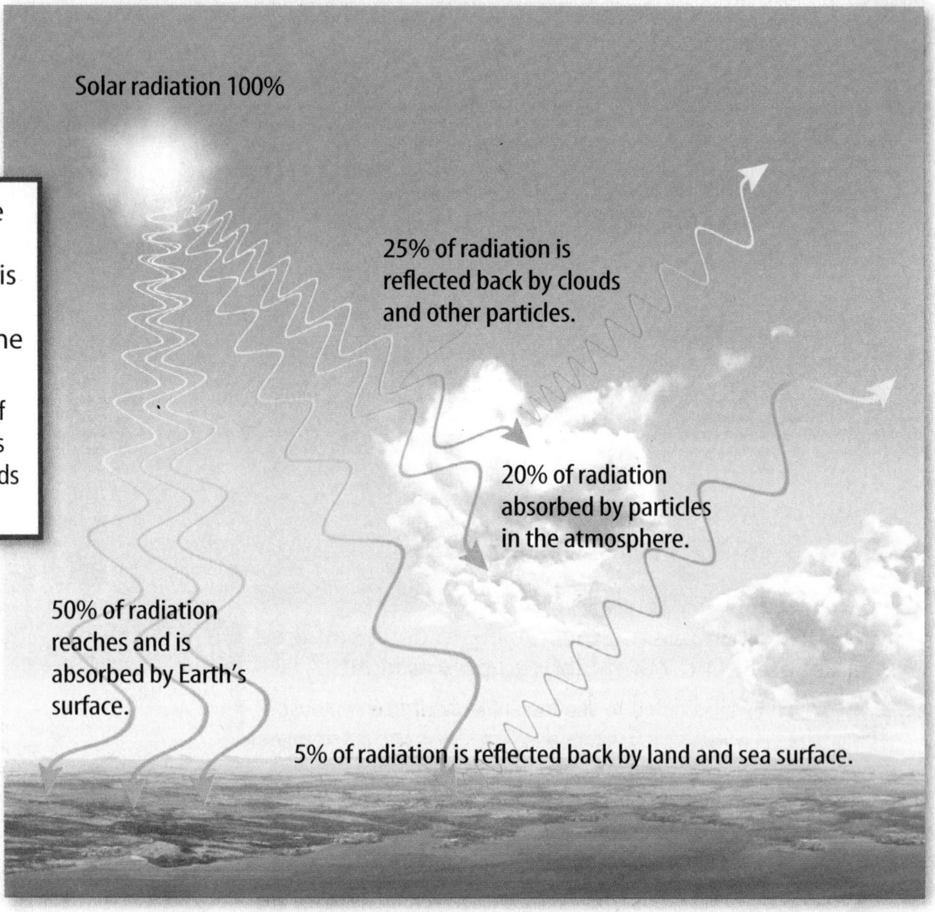

Solar radiation 100%

Figure 7 Not all the Sun's energy reaches Earth's surface. Some is reflected or absorbed as it passes through the atmosphere.

Identify what percent of incoming solar radiation is reflected to space by clouds and other particles.

25% of radiation is reflected back by clouds and other particles.

20% of radiation absorbed by particles in the atmosphere.

50% of radiation reaches and is absorbed by Earth's surface.

5% of radiation is reflected back by land and sea surface.

Sunlight Penetrating the Atmosphere

As the Sun's radiation passes through the atmosphere, some of it is absorbed by gases and particles, and some of it is reflected back into space. As a result, not all the radiation coming from the Sun reaches Earth's surface. Study **Figure 7.** About 20 percent of incoming solar radiation is absorbed by gases and particles in the atmosphere. Oxygen, ozone, and water vapor all absorb incoming ultraviolet radiation. Some of the infrared radiation from the Sun is absorbed by water and carbon dioxide in the troposphere. However, the wavelengths of visible light are not greatly absorbed by Earth's atmosphere.

About 25 percent of incoming solar radiation is reflected to space by clouds and tiny particles in the air. Another 5 percent is reflected into space by land and sea surfaces. So, a total of 30 percent of the incoming solar radiation is reflected to space. This means that, along with the 20 percent that is absorbed by gases and particles, only about 50 percent of incoming solar radiation reaches and is absorbed by Earth's surface.

Why does the sky look blue?

Do you ever wonder why the sky is blue? The answer to this question lies in the interaction between incoming solar radiation and the gases and particles present in our atmosphere. As visible light passes through the atmosphere, it is absorbed, reflected, and scattered by particles and gas molecules in the atmosphere. Light with a shorter wavelength, including violet and blue, is absorbed and then reflected first as light passes through the atmosphere. As shown in **Figure 8,** when blue light scatters through the atmosphere and reaches our eyes, the sky appears blue.

A Yellow Sun For the same reason the sky appears blue, the Sun looks yellow. As the violet and blue light are scattered when they pass through the atmosphere, the remaining colors of light—green, yellow, orange, and red—together appear yellow.

A Red Sunset As the Sun sets low in the sky, light must travel a longer distance through Earth's atmosphere. As the light travels, not only is much of the blue light scattered, but green light is also scattered. At first, the setting Sun looks orange. As it sinks lower in the sky, light has to travel even farther to reach Earth's surface. Even the longer yellow and orange wavelengths are scattered and reflected, leaving only the longest wavelengths of red to reach our eyes, as shown in **Figure 8.**

A Black Sky If you were to view the sky from space, it would appear black. In space, there is no atmosphere to reflect or scatter any light. When no scattered light reaches your eyes, the sky appears dark and black, as shown in **Figure 8.**

 **Reading Check** What happens when blue light is scattered by particles in the atmosphere?

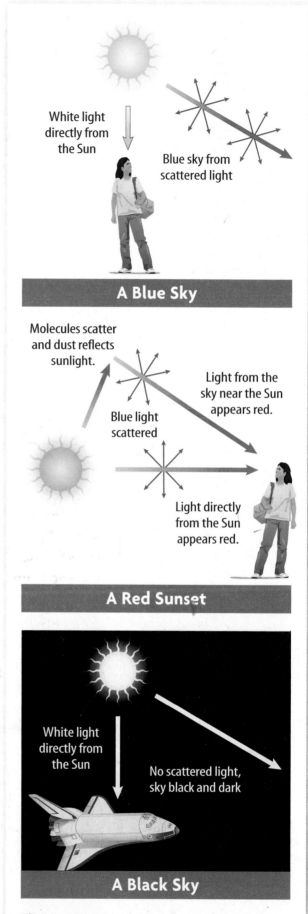

A Blue Sky

White light directly from the Sun

Blue sky from scattered light

Molecules scatter and dust reflects sunlight.

Light from the sky near the Sun appears red.

Blue light scattered

Light directly from the Sun appears red.

A Red Sunset

White light directly from the Sun

No scattered light, sky black and dark

A Black Sky

Figure 8 The colors of the sky and the Sun are affected by the type of light that is scattered as it passes through the atmosphere.

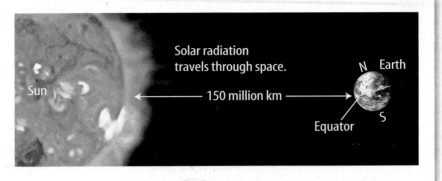

Figure 9 Although it is far from Earth, the Sun is the source of heat for Earth.

Identify How far is the Sun from Earth?

Solar radiation travels through space.

150 million km

N   Earth

S

Equator

Sun

The Sun's Power

The Sun emits an enormous amount of radiation. **Figure 9** shows that radiation from the Sun has to travel a long distance through space before reaching Earth. Even though this is true, solar heating creates the climate conditions on Earth. Due to the Sun's radiation and the atmosphere, the conditions at Earth's surface can sustain life. Without the heat from solar radiation, this would not be possible.

A Constant and Uniform Source of Energy

The Sun will continue to produce energy for a long time—billions of years. For this reason, scientists consider the Sun a constant source of energy. There are some changes in the amount of solar radiation reaching Earth but, in general, the energy coming from the Sun is nearly uniform.

The Angle of Sunlight

Even though radiation coming from the Sun is constant and uniform, it is not evenly distributed on Earth. **Figure 10** shows how sunlight is distributed over Earth's curved surface. Notice that a beam of sunlight near the equator is almost perpendicular to Earth's surface. The beam of sunlight is concentrated into a small area.

Since there is more sunlight for the amount of surface area near the equator, the land, water, and air become warm. However, the same-size beam of sunlight also strikes Earth's surface near the poles. But this time, it strikes at a low angle. Now the beam of sunlight is spread out over a larger area. Since there is less sunlight for the amount of surface area near the poles, the land, water, and air do not warm as much. When the Sun stays below the horizon during the winter months, the poles become very cold.

 Figure 10 At which areas on Earth does the Sun strike at a low angle?

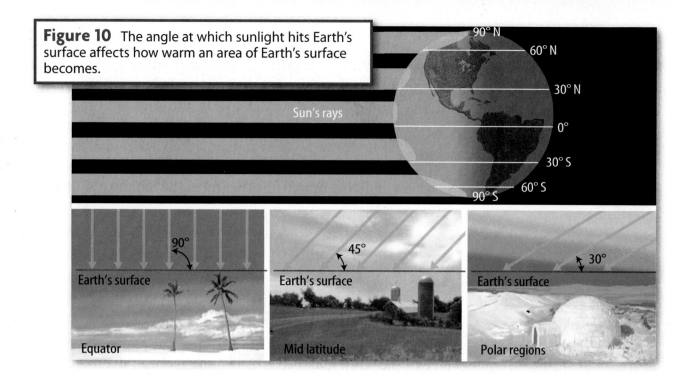

Figure 10 The angle at which sunlight hits Earth's surface affects how warm an area of Earth's surface becomes.

Sun's rays

90° N
60° N
30° N
0°
30° S
60° S
90° S

90°
Earth's surface
Equator

45°
Earth's surface
Mid latitude

30°
Earth's surface
Polar regions

Solar Energy on Earth

The Sun's energy heats the air, the oceans, and the land on Earth. The Sun's energy is responsible for climate and weather. Not only does the Sun's energy make climate conditions on Earth suitable for life; the Sun's energy serves as the power for other cycles on Earth. Air currents, or wind, are generated as the Sun's energy heats Earth's surface. Wind leads to the formation of waves on the surface of the ocean. Powerful weather systems, including hurricanes and tornadoes, ultimately get their energy from the Sun.

The energy that drives the water cycle comes from the Sun. The water cycle is the cycle in which water at Earth's surface continually evaporates and returns to Earth's surface as precipitation. The Sun is also necessary for photosynthesis. Plants undergo photosynthesis, which produces energy-rich molecules that release energy when broken down.

Humans use energy from the Sun in many ways, either directly or indirectly. Solar energy can be collected directly with devices that capture the rays coming from the Sun. Solar energy can be changed to electricity and used to power lights and other electrical devices in homes and businesses. The Sun powers the water cycle, which includes fast-flowing rivers. Energy from rivers can be transformed into electrical power by using dams. Energy from the Sun powers wind. Windmills, like the ones shown in **Figure 11,** can be used to convert energy from wind into electrical energy.

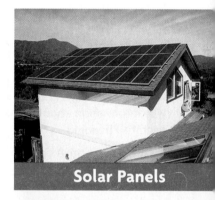

Solar Panels

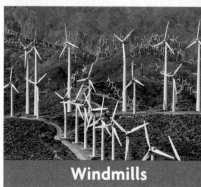

Windmills

Figure 11 Windmills and solar panels are used to generate electrical energy.

The Sun's Energy

Energy from the Sun, along with the composition of the atmosphere, makes life as we know it on Earth possible. Energy from the Sun reaches Earth in the form of visible light, infrared radiation, and ultraviolet light and is constant and nearly uniform. Solar radiation warms water, air, and land at Earth's surface. It powers the water cycle and photosynthesis in living organisms that form the base of the food chain.

LESSON 1 Review

Summarize

Create your own lesson summary as you write a script for a **television news report.**

1. **Review** the text after the **red** main headings and write one sentence about each. These are the headlines of your broadcast.

2. **Review** the text and write 2–3 sentences about each **blue** subheading. These sentences should tell *who, what, when, where,* and *why* information about each **red** heading.

3. **Include** descriptive details in your report, such as names of reporters and local places and events.

4. **Present** your news report to other classmates alone or with a team.

 ELA6: LS 1.4

Using Vocabulary

1. Distinguish between *atmosphere* and *troposphere*. **4.a**

2. In your own words, write the definition for *electromagnetic spectrum*. **4.a**

Understanding Main Ideas

3. **Compare and Contrast** Copy and fill in the graphic organizer below to compare and contrast the features of the troposphere and the stratosphere. **4.a**

Layer of Atmosphere	Similarities	Differences
Troposphere		
Stratosphere		

4. **Analyze** what happens to solar radiation from the moment it leaves the Sun until it reaches Earth's surface. **4.b**

5. Approximately what percentage of incoming solar radiation actually reaches Earth's surface? **4.b**

   A. 20 percent   **C.** 50 percent
   B. 35 percent   **D.** 75 percent

 Standards Check

6. **Explain** why most of the solar radiation reaching Earth's surface is in the range of visible light. **4.b**

7. **Explain** why the sky looks blue and the Sun looks yellow in the afternoon. **4.b**

8. **Explain** why the Sun is considered to be a constant source of energy for Earth. **4.a**

Applying Science

9. **Predict** what would happen if all the ozone in the stratosphere were to disappear. Would more or less ultraviolet radiation from the Sun reach Earth's surface? **4.b**

10. **Infer** Suppose Earth's atmosphere were not able to reflect incoming solar radiation. How would that change conditions at Earth's surface? **4.b**

For more practice, visit **Standards Check** at ca6.msscience.com.

MiniLab

Why is the sky blue?

00:20 minutes

You may have asked this question when you were younger. You might have received an answer that didn't make sense to you at the time. Soon, you will be able to answer the question and demonstrate it at home.

Procedure

1. Read and complete a lab safety form.

2. Pour water into a **jar** until it is two-thirds full.

3. Measure the amount of **milk** assigned by your teacher and add it to the jar.

4. Record the color of the liquid.

5. Hold a **flashlight** above the jar so the light shines down into it. Look into the glass from the side. Record the color the liquid appears to be.

6. Shine the flashlight from one side of the jar and look into it from the other side. Record the color the liquid appears to be.

7. Shine the flashlight from the bottom of the jar and look in at the top. Record the color the liquid appears to be.

8. Compare the colors you recorded with those of other lab groups.

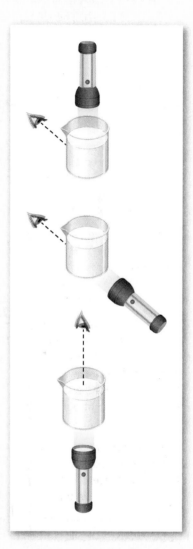

Analysis

1. **Compare and contrast** the colors you recorded with the colors of students who used less milk and more milk than you did. Did everyone see the same colors?

2. **Explain** Did the liquid actually change color? Why did it appear to be different colors?

Science Content Standards

4.b Students know solar energy reaches Earth through radiation, mostly in the form of visible light.

Applying Math

Making a Scale Model of Earth's Atmosphere

Earth's atmosphere is composed of gaseous layers that surround the surface of Earth. The distance from Earth's surface to the top of each layer is shown in the table below.

Example

Find the measures needed to make a scale model of these distances using 5 km of the true distance equal to 0.5 cm in the scale model. How many centimeters would be needed in the scale model to represent the true distance to the top of the ozone layer?

Distances of Atmospheric Layers Above Earth's Surface	
Name of Layer	**Distance (km)**
Troposphere	15
Tropopause	20
Stratosphere	50
Ozone layer	45
Mesosphere	90
Ionosphere	350
Thermosphere	600
Exosphere	10,000

What you know:
- Distance from Earth to the top of the ozone layer: 45 km
- Scale: 5 km = 0.5 cm

What you need to find:
- How many 5-km pieces are in the given distance of the ozone layer
- Distance of the ozone layer on the model

Divide to find how many 5-km pieces there are in 45 km:
$\frac{45}{5} = 9$ pieces that are 5 km in length

Multiply:
9 pieces × 0.5 cm = 4.5 cm

Answer: The scale model distance of ozone would be 4.5 cm.

Practice Problems

1. How many centimeters long would the scale model be for the exosphere?

2. How much longer is the scale model length for the exosphere than the ozone layer?

Science nline

For more math practice, visit **Math Practice** at ca6.msscience.com.

Science Content Standards

3.c Students know heat flows in solids by conduction (which involves no flow of matter) and in fluids by conduction and convection (which involves flow of matter).
3.d Students know heat energy is also transferred between objects by radiation (radiation can travel through space).
4.d Students know convection currents distribute heat in the atmosphere and oceans.

Reading Guide

What *You'll Learn*

▶ **Describe** how the air is heated from the lower layers of the atmosphere.

▶ **Explain** why hot air rises and cold air sinks.

▶ **Distinguish** the properties of the radiation emitted by the Sun from those of the radiation emitted by Earth.

▶ **Identify** the effects of greenhouse gases on Earth's climate.

Why *It's Important*

Heat energy from the Sun that is distributed through the atmosphere helps keep Earth warm.

Vocabulary
inversion
greenhouse gas
global warming

Review Vocabulary
convection: heat transfer by the movement of matter from one place to another (p. 147)

Energy Transfer in the Atmosphere

Main Idea Earth's atmosphere distributes thermal energy.

Real-World Reading Connection When you place a spoon into a cup of hot chocolate and leave it there for a few minutes, the spoon handle gets warm. Air that is touching or very close to Earth's surface is heated in a similar way.

Conduction in Air

Recall from Chapter 2 that there are three types of heat transfer—conduction, convection, and radiation, as shown in **Figure 12.** Conduction takes place when molecules transfer their kinetic energy by collisions. Think about the molecules in air and how much space is between them. Do you think air is a good conductor of thermal energy?

Heating the Air from Below

Conduction is the process that heats air close to Earth's surface. Radiant energy from the Sun warms the land and the oceans. However, air is a poor heat conductor. As a result, the hot ground only warms a shallow layer of air above it. Even in calm weather, conduction only warms a layer of air that is no more than a few centimeters thick. So, how is thermal energy transferred from one region to another?

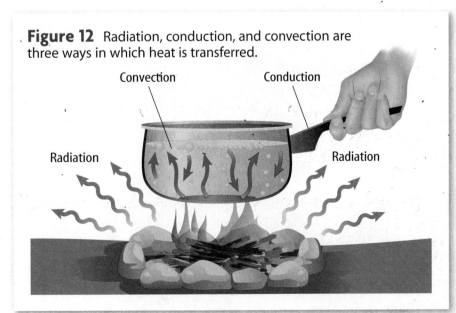

Figure 12 Radiation, conduction, and convection are three ways in which heat is transferred.

Convection

Conduction

Radiation

Radiation

Figure 13 As its temperature changes, air rises and sinks, forming convection currents.

A Air inside the balloon is heated.

B As the air is heated, the molecules become more distant from each other and the air expands.

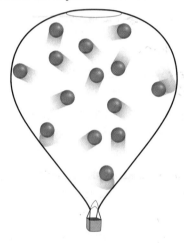

C As the density of the air inside the balloon lessens, it will rise above cooler, denser air.

Convection in Air

Hot air that is close to Earth's surface moves to higher altitudes by convection. As a result, air in the troposphere is almost always moving. In the troposphere, the movement of air by convection is mostly in a vertical direction. Why are the people shown in **Figure 13 A** shooting flames into their hot-air balloon?

Expanding Air

As the temperature of air increases, the kinetic energy of the molecules increases. This means the air molecules are moving faster and becoming more distant from one another, as shown in **B**. When this happens, the air expands. When the air expands, its density decreases.

 Figure 13 What happens to the air molecules inside the balloon as the temperature increases?

Rising and Sinking Air

When air that is close to the surface warms up, it rises in a similar way as a hot-air balloon rises. Why does a balloon need hot air to rise? At first, the balloon rises because the air inside the balloon is hotter and, therefore, less dense than the air surrounding it. The balloon will rise, as shown in **C**, while colder air with higher density moves beneath it forcing the balloon upward. When the temperature inside the balloon is equal to the temperature outside the balloon, it will stop rising.

What would happen if the air inside the balloon were colder than the air around it? In this case, the air inside the balloon will have a higher density than the air around it. The denser air will move down while air with lower density remains above it. The balloon with colder air in it sinks.

 How does the density of air affect whether it will rise or fall?

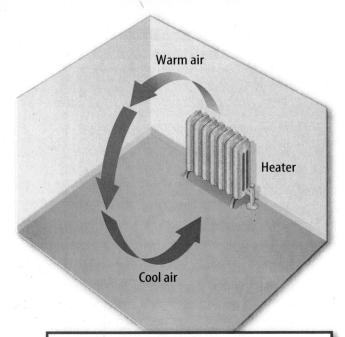

Figure 14 Convection currents created by the heated air circulate throughout the room.

Air Circulation Patterns

As you have just learned through the balloon example, hot air rises and cold air sinks due to differences in density compared to surrounding air. This principle also works when heating a room. As shown in **Figure 14,** warm air near a heater on the floor rises as its density decreases. When this happens, colder air near the ceiling flows down to replace the warmer air that is rising. When the colder air moves closer to the heater, it warms up, so eventually it will move up again. When air rises, it cools down, so eventually it sinks again. This leads to a continuous pattern of circulating air.

Since hot air moves up and cold air moves down, there is a continuous movement of air. The continuous vertical movement of air that occurs in a circular pattern is called a convection current. The current shown in **Figure 14** is a convection current. Convection currents distribute thermal energy within the troposphere.

 How do convection currents work?

How do clouds form from convection currents?

Clouds are condensed water droplets. As air rises due to convection currents, it expands and cools. As it cools, the water vapor in the air condenses and small water droplets form. All the water droplets close together form a cloud.

Procedure

1. Complete a lab safety form.
2. Barely cover the bottom of clear **3.5-L jar** with **water.**
3. Stretch a **rubber glove** over the top of the jar with the fingers pointing into the jar.
4. Put your **hand** in the glove and pull it quickly up, but do not pull the glove off of the jar.
5. When you are comfortable performing the action in step 4, remove the glove from the jar. Your teacher will drop a lighted match into it. Quickly put the glove back on the jar as in step 3.
6. Repeat step 4.

Analysis

1. **Describe** What happened when you pulled up the glove in step 4? What happened in step 6?
2. **Compare and contrast** the conditions in the jar with conditions in the atmosphere when clouds form.

 4.d

WORD ORIGIN

inversion
from Latin *invertere;* means *turn upside down*

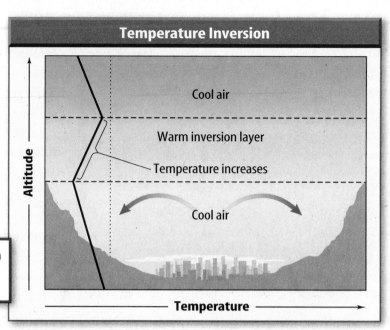

Temperature Inversion

Altitude

Cool air

Warm inversion layer

Temperature increases

Cool air

Temperature

Figure 15 Inversions occur when a layer of cool air is trapped beneath a layer of warm air.

Normal

Inversion

Figure 16 The haze in the lower photo was caused by an inversion. A layer of air is trapped by a warm layer of air above it.

When Air Is Stable

Recall that within the troposphere, the temperature normally decreases as altitude increases. But, what happens when this situation is reversed? When something is turned upside down, it is inverted. Sometimes, air temperature in the troposphere increases as altitude increases. This is called an **inversion.** As shown in **Figure 15,** an inversion occurs when warm air sits on top of cold air. This means that air that is rising from Earth's surface can only reach a certain altitude, and then it becomes trapped by the warm layer of air above it.

An inversion can have serious consequences. Imagine air that contains a harmful substance, such as a pollutant. It would be best for the air to rise as high as possible so that it is far from where it can be breathed by humans and other animals. In the case of an inversion, the air cannot move upward. Harmful substances remain trapped close to Earth's surface.

Inversions can happen many times during a year depending on location, weather conditions, and other factors. For example, frequent inversions contribute to increased levels of air pollution and decreased visibility in Los Angeles. **Figure 16** compares two photos of downtown Los Angeles. The photo above was taken on a day when no inversion was present. Compare that to the photo below, which was taken during an inversion.

Reading Check How can an inversion be harmful to human health?

Radiation Traveling Through Space

Recall from Chapter 2 that radiation is the transfer of energy in the form of electromagnetic waves. Unlike convection and conduction, which need a medium, or material, such as air or water through which to travel, radiation can travel through empty space, or a vacuum. In this way, solar radiation can travel through space and reach our planet.

Heating with Sunlight

People used to start fires by rapidly spinning a wooden stick. Friction caused the wood particles to vibrate, producing heat. Today, people have learned that concentrated sunlight can also be used as a source of heat.

Notice the shaded area within the blue dashed line in **Figure 17.** If the magnifying lens weren't between the Sun and the concrete, the shaded area would be as bright as the surrounding concrete area. However, now the Sun's rays are bent by the magnifying lens toward the center of the shaded area, as shown by the red circle. Notice how the area within the red circle is much brighter than the surrounding area. The concentrated solar rays make the molecules within the red circle vibrate rapidly and become very hot.

The example described above can be compared to the situation at Earth's equator and poles. But, there is a slight difference. Recall from Lesson 1 that a beam of sunlight covers a small surface area at the equator. All the Sun's rays heat the land within a small surface area. However, at Earth's poles the same size beam of sunlight is spread out over a larger area. Earth's poles are cold because there is less energy available for the amount of area that needs to be heated.

SCIENCE USE v. COMMON USE

vacuum

Science Use the emptiness of space, without air or matter. *A feather will drop as fast as a rock in a complete vacuum.*

Common Use a household appliance for cleaning floors, carpets, upholstery, etc. *He cleaned his car with a vacuum.*

Figure 17 When the Sun's radiation is concentrated in a small area, the area becomes very warm.

Absorbing Different Wavelengths

Different molecules absorb different types of radiation. Although ozone is a harmful pollutant in the troposphere, it absorbs harmful ultraviolet radiation in the stratosphere. Water and carbon dioxide absorb infrared radiation.

Only about 7 percent of the Sun's radiation consists of ultraviolet radiation. Ultraviolet radiation can be harmful to animals and plants. Fortunately, as shown in **Figure 18,** much of this damaging radiation is absorbed by gas molecules in the atmosphere before it reaches Earth's surface.

Emitting Radiation

Is the Sun the only object in the universe to emit radiation? No, radiation is emitted by all objects in the universe as long as they have a temperature above absolute zero. For example, your own body emits infrared radiation known as heat.

In the same way as the Sun emits radiation, Earth also emits radiation. The main difference is that the radiation emitted by the Sun is mostly in the visible range, while the radiation emitted by Earth is mostly in the infrared region.

Maintaining Radiation Balance

Why doesn't Earth get hotter and hotter with time as it continues to receive radiation from the Sun? It is because there is a balance between the amount of incoming radiation from the Sun and the amount of outgoing radiation from Earth. The total amount of energy reaching Earth from the Sun is equal to the amount of energy leaving Earth. This is shown in **Figure 19.** Solar radiation that reaches Earth's surface is absorbed by land, trees, asphalt, sand, soil, and oceans. After the solar radiation is absorbed, it is reemitted from Earth, mostly as infrared radiation.

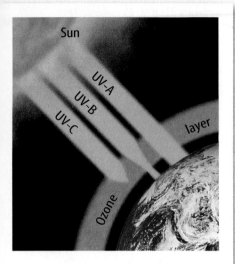

Figure 18 Much of the harmful ultraviolet radiation from the Sun is filtered by the atmosphere before reaching Earth's surface.

Infer In which layer of the atmosphere is most of ultraviolet radiation absorbed?

Figure 19 The amount of solar radiation reaching Earth's surface is equal to the amount of outgoing radiation form Earth's surface.

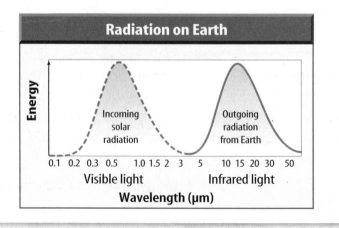

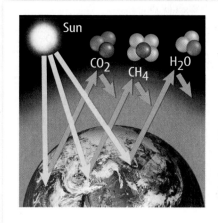

Figure 20 Some outgoing radiation is directed back toward Earth's surface by greenhouse gases.

List What are some examples of greenhouse gases?

Greenhouse Gases

Not all the radiation emitted by Earth escapes the planet. There is a fraction of the outgoing infrared radiation emitted by Earth that is absorbed by gases present in the atmosphere, as shown in **Figure 20.** These gas molecules absorb infrared radiation and reemit it in all directions. Some of the radiation that is reemitted is directed back to Earth. This is **similar** to the way a greenhouse works. The glass of the greenhouse in **Figure 20** traps warm air, and the greenhouse stays warm as a result. When gases in Earth's atmosphere direct radiation back toward Earth's surface, this warms Earth's atmosphere more than normal. Gases that strongly absorb a portion of Earth's outgoing radiation are called **greenhouse gases.** Water vapor (H_2O), methane gas (CH_4), and carbon dioxide(CO_2) are examples of greenhouse gases. Water molecules in the atmosphere are responsible for a large percentage of the additional warming of Earth's surface.

Many scientists are studying data that represents the average surface temperature of Earth over time. In recent years, some scientists have presented data that show that Earth's average surface temperature is increasing every year, a condition called **global warming.** One possible explanation for this temperature increase is an increased amount of greenhouse gases being released into the atmosphere. For example, carbon dioxide is given off when gas, oil, and coal are burned. An increase in the concentration of greenhouse gases could lead to Earth's surface getting hotter.

ACADEMIC VOCABULARY
similar (SIH muh lar)
(adj) having characteristics in common
The two paintings were similar in color.

 Explain the analogy between a greenhouse and Earth's atmosphere.

Energy in the Troposphere

Energy from the Sun is distributed through the troposphere by the processes of conduction, convection, and radiation. Air that is close to Earth's surface is heated by conduction. Convection currents, caused by differences in the density of air masses, circulate vertically and distribute thermal energy within the troposphere. Radiation from the Sun heats Earth's surface unevenly, with more energy concentrated in equatorial regions than polar regions.

The balance between incoming solar radiation and outgoing radiation that is reemitted by Earth helps keep Earth's surface temperatures stable. Greenhouses gases, which trap some of the radiation reemitted by Earth, are also involved in the regulation of Earth's surface temperature.

LESSON 2 Review

Summarize

Create your own lesson summary as you organize an **outline**.

1. **Scan** the lesson. Find and list the first **red** main heading.

2. **Review** the text after the heading and list 2–3 details about the heading.

3. **Find** and list each **blue** subheading that follows the **red** main heading.

4. **List** 2–3 details, key terms, and definitions under each **blue** subheading.

5. **Review** additional **red** main headings and their supporting **blue** subheadings. List 2–3 details about each.

 ELA6: R 2.4

Standards Check

Using Vocabulary

Complete the sentences using the correct term.

1. Gases that direct radiation back toward Earth's surface are called _____. **3.d**

2. A(n) _____ occurs when air temperature in the troposphere increases as altitude increases. **4.d**

Understanding Main Ideas

3. Air at Earth's surface is heated by which process? **3.c**
   A. conduction
   B. convection
   C. evaporation
   D. inversion

4. **Explain** how convection currents transfer heat. **3.c**

5. **Identify** the effects of greenhouse gases on Earth's climate. **3.d**

6. **Compare** the radiation emitted by the Sun with the radiation emitted by Earth. **3.d**

7. **Determine Cause and Effect** Copy and fill in the graphic organizer below to list the possible causes and effects of global warming. **3.d**

Applying Science

8. **Compare and contrast** the transfer of heat in the atmosphere by conduction, convection, and radiation. **3.c**

9. **Evaluate** the effect of inversions on air pollution levels. **4.d**

For more practice, visit **Standards Check** at ca6.msscience.com.

Science Content Standards

4.a Students know the sun is the major source of energy for phenomena on Earth's surface; it powers winds, ocean currents, and the water cycle.

4.d Students know convection currents distribute heat in the atmosphere and oceans.

4.e Students know differences in pressure, heat, air movement, and humidity result in changes of weather.

Also covers: 3.d, 4.b, 7.c

Reading Guide

What *You'll Learn*

▶ **Describe** how solar energy gives rise to winds.

▶ **Explain** why Earth's surface is heated unevenly.

▶ **Understand** how pressure differences affect winds.

▶ **Explain** how great air currents circle Earth, transporting heat and water vapor.

Why *It's Important*

Air currents transport and distribute heat throughout Earth's atmosphere.

Vocabulary
wind
updraft
downdraft
Coriolis effect
jet stream

Review Vocabulary
density: amount of matter in a substance per unit volume (p. 91)

Air Currents

Main Idea Solar energy is responsible for the continuous movement of air in the troposphere, which transports and distributes thermal energy around Earth.

Real-World Reading Connection Have you ever been outside on a windy day? Where does wind come from? How does it start? A cool breeze on a hot day feels good. High-speed winds, on the other hand, can cause damage to buildings and trees.

Local Winds and Eddies

Wind is air that is in motion relative to Earth's surface. Differences in air pressure over Earth's surface cause winds. The Sun is the major source of energy that powers winds.

Many communities are familiar with their own local winds—and sometimes give them names. Examples include ocean breeze, mountain breeze, valley breeze, monsoon, El Norte, and Santa Ana wind. Some of these local winds will be described in Chapter 11.

A current of water or air that runs counter to the main current—especially a circular current—is called an eddy. Sometimes eddies are visible as whirlwinds or dust devils when they carry leaves or dust through the air. A few small eddies are shown in **Figure 21.** What causes local winds and eddies?

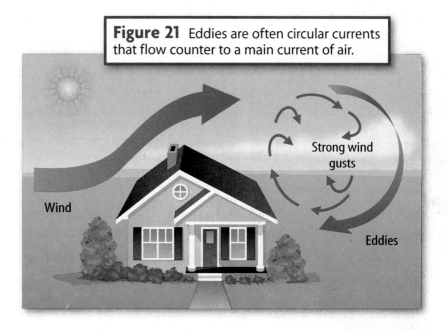

Figure 21 Eddies are often circular currents that flow counter to a main current of air.

Wind

Strong wind gusts

Eddies

Table 1 Percent of Reflection of Solar Radiation

Material	Example	Percent Reflected
Forest		5–10
Asphalt or concrete		5–10
Water		8
Sand		30–60
Snow and ice		80–90

Figure 22 Convection currents are created when Earth's surface is heated unevenly.

Uneven Heating of Earth's Surface

Uneven heating of Earth's surface causes air to be heated unevenly. Aside from differences in the angle of sunlight, what else can cause Earth's surface to be heated unevenly?

The materials found at Earth's surface absorb or reflect different amounts of sunlight. **Table 1** shows the percentage of solar radiation reflected by some materials at Earth's surface. For example, black asphalt in a parking lot can absorb so much sunlight that it will get hot enough to burn your feet. Many land surfaces, such as soil, rock, and sand, absorb sunlight, heat rapidly, and then heat the air above them.

Some materials remain cool even in direct sunlight. For example, snow reflects solar radiation and remains cool. Plants absorb a lot of sunlight, but much of the solar radiation is converted to stored energy through photosynthesis. Water must absorb a lot of energy to produce a small change in its temperature. As a result, Earth's surface is heated unevenly. Because Earth's surface is heated unevenly, the air above its surface is also heated unevenly. **Figure 22** illustrates how air circulates above surfaces that are heated unevenly.

 **Reading Check** What happens when a material at Earth's surface reflects a large percentage of the Sun's radiation?

Figure 23 Birds and hang-gliders use thermals to glide through the air.

Updrafts

Sometimes a large area of land absorbs more solar radiation than the land nearby. When the land becomes warm, it heats the air above it. As the air is heated, it expands and becomes less dense than the surrounding air. Eventually, the heated air rises as part of a convection current. A rising column of air is called an **updraft.**

Sometimes updrafts are called thermals. Many birds, such as the California condor shown in **Figure 23,** use thermals to soar high in the air. These birds have very large wings to catch the rising air. They can remain in the rising column of air for long periods of time without flapping their wings. People also use thermals for hang gliding and parasailing.

When air rises in an updraft or a thermal, the air at Earth's surface **temporarily** decreases in density and pressure. The surrounding air will move in to fill the area of low pressure, as shown in **Figure 24.** This is similar to sucking liquid through a straw. When liquid is pulled through the straw, the surrounding liquid moves in the cup to replace it.

ACADEMIC VOCABULARY
temporarily (tehm puh RAYR uh lee)
(adv) lasting for a limited time
The office was moved to a new location temporarily while the building was repaired.

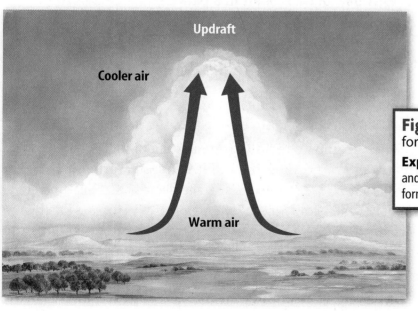

Updraft

Cooler air

Warm air

Figure 24 Updrafts are formed when heated air rises.
Explain how radiation, conduction, and convection are all involved in the formation of an updraft.

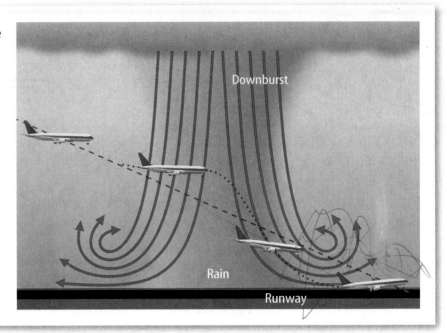

Figure 25 Downbursts are strong, downward-flowing air currents that can cause problems for airplanes during landing.

Downburst

Rain

Runway

Downdrafts

When dense air sinks toward Earth's surface, a **downdraft,** or a sinking column of air, occurs. Sometimes updrafts and downdrafts can occur within a single cloud, such as the thunderhead. A downdraft that is rapid and forceful is called a downburst. This is similar to dropping a water balloon on the pavement. At Earth's surface, the air from a downburst pushes outward in all directions like the spokes on a wheel. Aircraft pilots need to be aware of hazardous conditions created by downdrafts and downbursts, as shown in **Figure 25.**

The dense air that descends from a downdraft or a downburst temporarily creates an area of high pressure at Earth's surface. This occurs because the dense air is concentrated on a small area. This situation can be contrasted with the area of low pressure created by an updraft. Recall that the troposphere contains weather. On a very small scale, these areas of high and low pressure are similar to the high- and low-pressure systems that are described in weather reports.

High Pressure to Low Pressure

When air moves rapidly from an area of high pressure to an area of low pressure, we call it wind. As warm air rises, the air pressure close to Earth's surface decreases. The opposite happens when cold air sinks—air pressure close to Earth's surface increases. Air moves away from areas of high pressure and toward areas of low pressure.

 **Reading Check** Describe how air flows in terms of pressure.

Air Currents Around Earth

Winds are important because they transfer heat and water vapor from one location to another, which, in turn, affects weather and climate. However, dense air sinking as less-dense air rises does not explain everything about wind.

The Coriolis Effect

Figure 26 shows the path of winds in the northern and southern hemispheres. What is the difference between the direction wind flows in the two hemispheres? In the northern hemisphere, winds are deflected to the right as they move across Earth's surface. In the southern hemisphere, winds are deflected to the left. This deflection of air and water currents, known as the **Coriolis effect,** is caused by Earth's rotation.

Think about what would happen if you threw a ball to someone sitting across from you on a moving merry-go-round. Would the ball reach that person? By the time the ball got to the opposite side, the person would have moved, and the ball would have appeared to have curved. Like the merry-go-round, the rotation of Earth causes moving air and water to appear to turn to the right in the northern hemisphere and to the left in the southern hemisphere.

The flow of air caused by differences in the amount of solar radiation received on Earth's surface and by the Coriolis effect creates distinct wind patterns on Earth's surface. These global winds distribute heat and water vapor around Earth's surface and influence changes in weather.

WORD ORIGIN

hemisphere
from Greek *hemisphairion;*
hemi– means *half;*
–sphaira means *sphere*

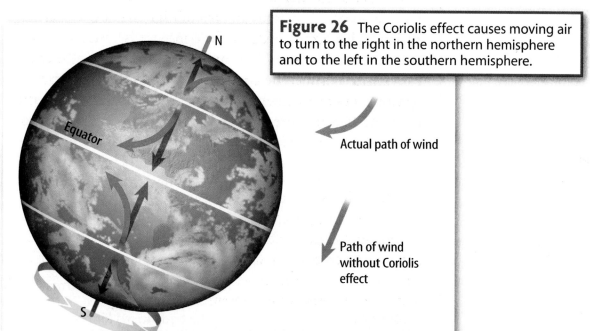

Figure 26 The Coriolis effect causes moving air to turn to the right in the northern hemisphere and to the left in the southern hemisphere.

Actual path of wind

Path of wind without Coriolis effect

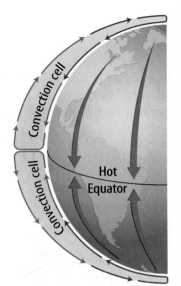

Figure 27 The old model of atmospheric circulation consisted of a single large cell in each hemisphere.

Contrast How does this model differ from the model currently used by scientists, shown in **Figure 28?**

Global Convection Currents

Scientists used to think that Earth's atmosphere circulated in a single, large convection cell—one cell in each hemisphere—like the model shown in **Figure 27.** Air would be heated at the equator, rise to the top of the troposphere, and then sink at the poles. They called this a one-cell model. Later, they discovered that this model did not work because of Earth's rotation. Today, scientists use a three-cell model to describe the circulation of Earth's atmosphere. The three convection cells look like three large donuts wrapped around Earth, like the model shown in **Figure 28.**

The Three-Cell Model Examine the three circulation cells in the northern hemisphere shown in **Figure 28.** In the first cell, hot air rises at the equator and moves to the top of the troposphere. Then, the air moves toward the poles until it cools and sinks back to Earth's surface near 30° latitude. The first cell is completed when most of the air returns toward the equator near Earth's surface. This convection cell is called the Hadley cell.

The third cell, or polar cell, is also a convection cell. Cold, dense air from the poles moves toward the equator along Earth's surface. The air becomes warmer and eventually rises near 60° latitude. The second cell, or Ferrel cell, between 30° and 60° latitude, is not a convection cell. Its motion is partially driven by the other two cells, like rolling cookie dough between your two hands. These three cells exist in the southern hemisphere as well.

 Describe the three types of cells in the model.

Prevailing Winds The three global convection cells in each hemisphere create northerly and southerly winds. When the Coriolis effect acts on the winds, they turn and blow to the east or the west, creating relatively steady, predictable winds, often referred to as prevailing winds. Sailors have known about and used the trade winds and the westerlies, both shown in **Figure 28,** for centuries to move sailboats across the ocean. Sometimes sailors found little or no wind to power their boats near the equator. As the Sun heats the air and water near the equator, the air rises, creating low pressure and little wind. This area is referred to as the doldrums.

 Figure 28 How are the trade winds formed?

Visualizing Global Winds

Figure 28

The Sun's uneven heating of Earth's surface forms giant loops, or cells, of moving air. The Coriolis effect deflects the surface winds to the west or the east, setting up belts of prevailing winds that distribute heat and moisture around the globe.

A WESTERLIES Near 30° north and south latitude, Earth's rotation deflects air from west to east as air moves toward the polar regions. In the United States, the westerlies move weather systems, such as this one along the Oklahoma-Texas border, from west to east.

B DOLDRUMS Along the equator, heating causes air to expand, creating a zone of low pressure. Cloudy, rainy weather, as shown here, develops almost every afternoon.

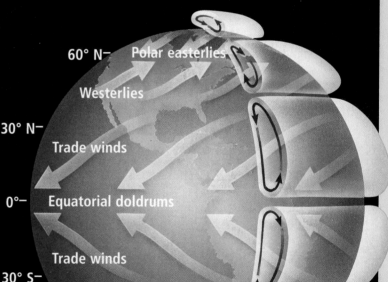

60° N— Polar easterlies

Westerlies

30° N— Trade winds

0°— Equatorial doldrums

Trade winds

30° S—

Westerlies

60° S— Polar easterlies

C TRADE WINDS Air warmed near the equator travels toward the poles but gradually cools and sinks. As the air flows back toward the low pressure of the doldrums, the Coriolis effect deflects the surface wind to the west. Early sailors, in ships like the one above, relied on these winds to navigate global trade routes.

D POLAR EASTERLIES In the polar regions, cold, dense air sinks and moves away from the poles. Earth's rotation deflects this wind from east to west.

Is it windy here?

How do winds differ between two locations in California?

Data

These data show the monthly average winds speeds in Santa Barbara, California, and Barstow, California, respectively.

Average Wind Speeds (km/h)		
Month	**Santa Barbara**	**Barstow**
January	8	13
February	10	14
March	11	21
April	13	23
May	11	23
June	11	23
July	11	21
August	10	19
September	10	16
October	10	16
November	8	14
December	8	13

Data Analysis

1. **Graph** the monthly wind speeds for each city on the same graph.

2. **Analyze** your graph. What differences exist in the wind speeds between the two cities?

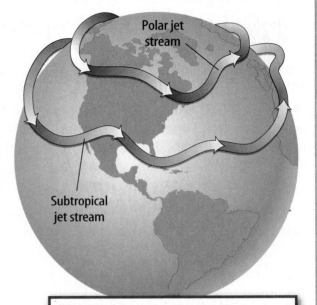

Figure 29 The jet streams move at high altitudes in both the northern and southern hemispheres.

Jet Streams

Jet streams are strong, continuous winds that range from 200 km/h to 250 km/h and are found at the top of the troposphere. There are two jet streams in each hemisphere. The polar jet stream and the subtropical jet stream are both shown in the northern hemisphere in **Figure 29.**

Jet streams are high altitude winds, located between 6 and 10 km above Earth's surface. They travel in a snakelike fashion around Earth from west to east, like rivers of strong, fast wind. Jet streams affect weather as well as the flight patterns and speeds of airplanes.

 What is a jet stream?

Americans didn't discover jet streams until World War II. Near the end of the war, high-altitude bombers experienced what they called "freak" winds that caused problems. Pilots flying from Saipan in the Pacific Ocean to Japan found headwinds so strong that they couldn't reach Tokyo and still have enough fuel to return to Saipan.

Air Currents at Earth's Surface

The Sun is the major source of energy that powers winds. Differences in the amount of solar radiation absorbed or reflected at Earth's surface, as well as the angle at which the Sun's rays strike Earth's surface, result in the uneven heating of Earth's surface. This causes the air above Earth's surface to be heated unevenly, creating pressure differences between locations. Wind, which flows from high pressure to low pressure, results from these pressure differences. The direction in which wind flows is affected by the Coriolis effect. Distinct wind patterns can be found around the globe. Three circulation cells exist in each hemisphere—Hadley cells, polar cells, and Ferrel cells—that distribute and transport heat and water vapor throughout the atmosphere.

LESSON 3 Review

Summarize

Create your own lesson summary as you write a **newsletter.**

1. **Write** this lesson title, number, and page numbers at the top of a sheet of paper.

2. **Review** the text after the **red** main headings and write one sentence about each. These will be the headlines of your newsletter.

3. **Review** the text and write 2–3 sentences about each **blue** subheading. These sentences should tell *who, what, when, where,* and *why* information about each headline.

4. **Illustrate** your newsletter with diagrams of important structures and processes next to each headline.

ELA6: W 1.2

Using Vocabulary

1. Distinguish between *updraft* and *downdraft.* **4.d**

2. In your own words, write the definition for *Coriolis effect.* **4.d**

Understanding Main Ideas

3. Which is shown in the figure below? **4.d**

Wind

A. downdraft
B. eddy
C. Ferrel cell
D. updraft

4. **Explain** how solar energy causes wind. **4.a**

Standards Check

5. **Describe** the main factors that contribute to the uneven heating of Earth's surface. **4.d**

6. **Explain** how differences in air pressure influence wind. **4.e**

Applying Science

7. **Compare** Copy and fill in the graphic organizer below and compare the three main convection currents in each hemisphere. Include an explanation of how each current transports heat around Earth. **4.e**

Hadley cell	
Ferrel cell	
Polar cell	

Science Online

For more practice, visit **Standards Check** at ca6.msscience.com.

Experiment:
Water and Sand Temperatures

Materials

sand
tap water
salt water
beakers (3)
light source
thermometers

Safety Precautions

Science Content Standards

3.d Students know heat energy is also transferred between objects by radiation (radiation can travel through space).
4.b Students know solar energy reaches Earth through radiation, mostly in the form of visible light.
7.c Construct appropriate graphs from data and develop qualitative statements about the relationships between variables.

Problem

If you have ever strolled barefoot on a beach on a very sunny day, you know that sand can be very hot. But the ocean water is always much cooler than the sand. If sand and water both absorb energy from the Sun, why do they feel so different?

Form a Hypothesis

➤ **Predict** whether sand and water will heat up equally fast.

➤ **Explain** why you think they will heat up at the same speed or at different speeds.

Collect Data and Make Observations

1. Read and complete a lab safety form.
2. Obtain three beakers. Measure 100 g each of tap water, sand, and salt water, and place each material in a separate beaker.
3. Place the beakers beneath the light source so that all three beakers receive the same amount of light.
4. Measure the temperature in each beaker before turning on the light source.
5. Turn on the light source.
6. Record a temperature from each beaker once every minute for 15 min.
7. Record your data in a table like the one shown below.

Temperature Changes in Materials			
Time (min)	Temperature of Material (°C)		
	Tap Water	Sand	Salt Water
0			
1			
2			
3			

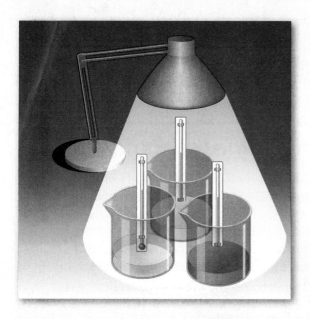

Analyze and Conclude

1. **Graph** the temperature changes for each material on the same graph.

2. **Compare and Contrast** What differences occurred in temperature changes among the tap water, the sand, and the salt water? Which material had the highest temperature? Which had the least temperature change?

3. **Describe** what is happening to the air over the different materials. In which direction would the air move?

4. **Think Critically** Review your lab setup. Besides moving the lightbulb farther from the sand, what could you have done to keep the sand from heating up? How would your results be different if you had shined the light through a jar that contained milk before it reached the materials?

5. **Explain** why the ocean is cooler than the sand.

Communicate

WRITING in > Science

Write an Advertisement Design a way or a product to keep visitors to the beach from overheating their feet as they walk to the water. Write an advertisement that explains and shares your idea with other visitors to the beach.

Real World Science

Managing Atmospheric Expeditions

As a field project coordinator with the National Center for Atmospheric Research (NCAR), Jose Meitin plans and leads expeditions to study the atmosphere. Speaking Spanish, French, and Italian helps Meitin during international expeditions, as research teams rely on him to choose study sites, acquire supplies and equipment, and coordinate travel. Meitin is also responsible for providing scientists with accurate, organized data for analysis, a task that can last for a year or more after field operations end.

Writers Needed! Visit **Careers** at **ca6.msscience.com** to learn about research projects sponsored by the NCAR. **Research** a current project and **write** a magazine article describing the project's purpose and the scientists involved.  ELA6: W 1.2

Wind Machines

Wind machines, or windmills, can convert kinetic energy from wind to electrical energy. Two types of wind machines are used today—horizontal-axis wind machines and vertical-axis wind machines. Ninety-five percent of the wind machines today are horizontal-axis machines. These machines have three blades shaped like airplane propellers, can be as tall as a 20-story building, and catch more wind than vertical-axis machines. Vertical-axis machines have blades that look like giant eggbeaters and are about as tall as a 10-story building.

WARP New technology has lead to the development of the Wind Amplified Rotor Platform (WARP). Visit **Technology** at **ca6.msscience.com** to **research** information about WARP. **Write** a report that describes how it works and why it's better than current technology.

ELA6: W 2.3

Sailing in the Jet Stream

In late winter, 1999, balloonists Bertrand Piccard and Brian Jones caught a ride on a jet stream and made history. They became the first aviators to complete a nonstop balloon flight around the world on March 20, 1999, traveling a distance of 40,814 km. Had the balloonists relied on the low-altitude winds which drive conventional hot-air balloons, the trip would have required many weeks of flight. Instead, the team climbed to altitudes as high as 11,373 m to take advantage of jet stream winds that sometimes reached 298 km per hour, propelling them around the world in just under 20 days.

Map it! Visit **History** at <u>ca6.msscience.com</u> to **analyze** the flight plan of the Breitling Orbiter 3. **Map** the route and transfer details of the plan to a **chart** on a world map. Note the balloon's maximum altitude and the countries over which it passed. Which jet stream did the Breitling Orbiter 3 appear to follow?

THE SANTA ANA WINDS

In autumn, southern Californians uneasily anticipate the arrival of warm, dry, north-easterly winds that howl through narrow canyons and mountain passes at speeds of at least 46 km/h. Santa Ana winds occur when a high pressure system forms in the Great Basin between the Sierra Nevada and Rocky Mountains. Air flowing down from the high plateau warms and speeds up dramatically, sometimes moving fast enough to carry dust over the Pacific Ocean. Santa Ana winds increase fire danger by removing moisture from vegetation and soil, cause air turbulence for planes approaching Los Angeles International Airport, and can topple trees and power lines in their path.

Legends and Lore Visit **Society** at <u>ca6.msscience.com</u> to learn about the legends and lore associated with the Santa Ana winds. Write a poem using these winds as its central theme. Consider sharing your poem with other students in your class or school.

ELA6: W 1.3

The BIG Idea The Sun's energy and Earth's atmosphere are critical for creating the conditions needed for life on Earth.

Lesson 1 Energy from the Sun
4.a, 4.b

Main Idea **The Sun is the major source of energy for Earth.**

- Earth's atmosphere is composed of a mixture of gases.

- The troposphere is the region of the atmosphere closest to Earth's surface where weather and climate take place.

- A portion of the incoming solar radiation is absorbed by the atmosphere. Another portion is reflected back to space by clouds, surfaces, and tiny particles in the atmosphere.

- The Sun is the main source of energy for Earth. It provides a constant source of energy and makes life on Earth possible.

- **atmosphere** (p. 382)
- **electromagnetic spectrum** (p. 384)
- **infrared wave** (p. 385)
- **stratosphere** (p. 383)
- **troposphere** (p. 383)
- **ultraviolet wave** (p. 385)

Lesson 2 Energy Transfer in the Atmosphere
3.c, 3.d, 4.d

Main Idea **Earth's atmosphere distributes thermal energy.**

- Air is a poor heat conductor.

- Continuous vertical motion of air occurs in convection currents.

- An inversion is an increase in air temperature as altitude increases.

- Radiation is the transfer of energy in the form of electromagnetic waves.

- All objects emit radiation.

- **global warming** (p. 399)
- **greenhouse gas** (p. 399)
- **inversion** (p. 396)

Lesson 3 Air Currents
3.d, 4.a, 4.b, 4.d, 4.e, 7.c

Main Idea **Solar energy is responsible for the continuous movement of air in the troposphere, which transports and distributes thermal energy around Earth.**

- The Sun is the major source of energy that powers winds.

- Earth's surface is heated unevenly.

- Wind blows from a region of high pressure to a region of low pressure.

- Great air currents transport thermal energy around Earth.

- Coriolis force acts on air currents as they move within Earth's atmosphere.

- **Coriolis effect** (p. 405)
- **downdraft** (p. 404)
- **jet stream** (p. 408)
- **updraft** (p. 403)
- **wind** (p. 401)

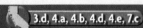 **STUDY TO GO** Download quizzes, key terms, and flash cards from ca6.msscience.com.

Linking Vocabulary and Main Ideas

Use vocabulary terms from page 414 to complete this concept map.

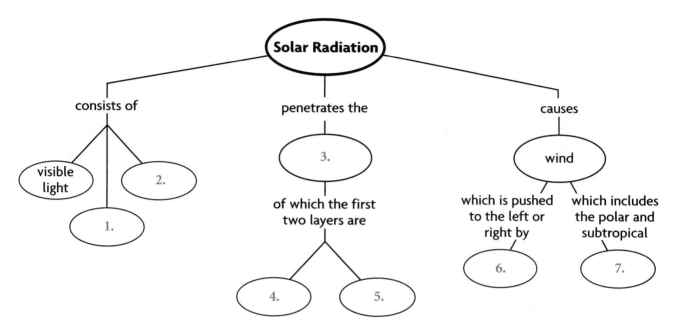

Using Vocabulary

Complete each statement using a word from the vocabulary list.

8. Carbon dioxide, methane, and water vapor are examples of _____.

9. The mixture of gases that surrounds Earth is the _____.

10. Air that is in motion relative to Earth's surface is called _____.

11. The _____ is the region of the atmosphere in which life exists.

12. During a(n) _____, the temperature in the troposphere increases with height.

13. In the _____, temperature increases with height.

14. A(n) _____ is a rising column of air.

15. Waves that have longer wavelengths than visible light and are felt as heat are _____. *Infrared wave*

Science Online

Visit ca6.msscience.com for:
▶ Vocabulary PuzzleMaker
▶ Vocabulary eFlashcards
▶ Multilingual Glossary

Understanding Main Ideas

Choose the word or phrase that best answers the question.

1. Which is the main source of energy for Earth?
   A. the Sun
   B. the Moon
   C. water
   D. wind   `4.a`

2. Energy from the Sun has its maximum in which region of the electromagnetic spectrum?
   A. ultraviolet
   B. visible
   C. infrared
   D. X-ray   `4.b`

3. Use the illustration below to answer the question.

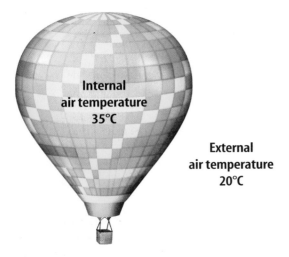

Internal air temperature 35°C

External air temperature 20°C

   Which best describes what will happen to the balloon?
   A. It will move up.
   B. It will move down.
   C. It will move to the left.
   D. It will move to the right.   `4.d`

4. Ozone in the stratosphere absorbs which type of incoming solar radiation?
   A. ultraviolet
   B. visible
   C. infrared
   D. X-ray   `4.b`

5. Use the graph below to answer the question.

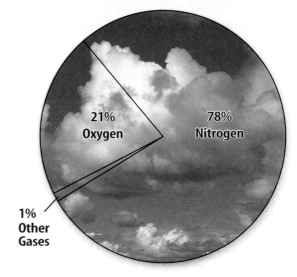

21% Oxygen

78% Nitrogen

1% Other Gases

   Oxygen and nitrogen make up which percentage of the components of air?
   A. 21.0 percent
   B. 78.9 percent
   C. 99.0 percent
   D. 99.9 percent   `4.b`

6. Which best describes how air moves?
   A. It moves from an area of high pressure to an area of low pressure.
   B. It moves from an area of low pressure to an area of high pressure.
   C. Warm air sinks and cool air rises.
   D. Warm air sinks and cool air sinks.   `4.e`

7. Which best describes a Hadley cell?
   A. circulation cell in which hot air rises at the equator, cools, and sinks back toward Earth's surface near 30° latitude   `4.e`
   B. circulation cell in which cold air rises at the equator, warms, and sinks back toward Earth's surface near 60° latitude
   C. circulation cell in which cold air from the poles moves toward the equator along Earth's surface
   D. circulation cell in which the motion of the air is driven by two other convection cells

Science🌐nline   Standards Review ca6.msscience.com

Applying Science

8. **Explain** why the Sun is considered a constant and almost-uniform source of energy.   **4.a**

9. **Compare** the roles of ozone in the troposphere and the stratosphere.   **4.b**

10. **Infer** Airplane A travels west to east from Redding, California, to New York, New York. It leaves at the same time as airplane B, which is traveling from New York to Redding. The planes travel at the same speed and do not stop. Airplane A arrives takes 1 h less to reach New York than airplane B takes to reach Redding. Why is there a difference in the travel times?   **4.e**

11. **Analyze** the differences between incoming solar radiation and outgoing terrestrial radiation.   **3.d**

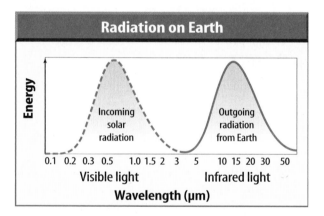

12. **Assess** the reasons why air is heated from below.   **3.c**

13. **Predict** if the temperature of Earth without an atmosphere would be higher or lower than it is today.   **4.b**

14. **Analyze** the relationship between latitude and the angle of sunlight striking Earth's surface. How can this relationship affect temperature at different locations on Earth's surface?   **4.d**

15. **Assess** the importance of global circulation cells for the distribution of thermal energy around our planet.   **4.e**

WRITING in ▶ Science

16. **Write** one paragraph explaining how the distribution of thermal energy over Earth would change if Earth were flat.

Applying Math

Use the table below to answer questions 17 through 19.

Distances of Atmospheric Layers Above Earth's Surface	
Name of Layer	**Distance (km)**
Troposphere	15
Tropopause	20
Stratosphere	50
Ozone layer	45
Mesosphere	90
Ionosphere	350
Thermosphere	600
Exosphere	10,000

17. The table above provides distances from Earth's surface to the top of the designated atmospheric layers. Use a scale of 5 km (true distance) = 0.5 cm (scale distance) for questions 18–22. How many centimeters would the scale distance be for the ionosphere?   **MA6: NS 1.2, AF 2.1**

18. How much longer is the scale distance for the ionosphere than the ozone layer?   **MA6: NS 1.2, AF 2.1**

19. How many centimeters would the scale distance be for the troposphere?   **MA6: NS 1.2, AF 2.1**

1 Which is the most abundant gas in Earth's atmosphere?

  A carbon dioxide

  B nitrogen

  C oxygen

  D water vapor   **4.b**

Use the illustration below to answer questions 2–4.

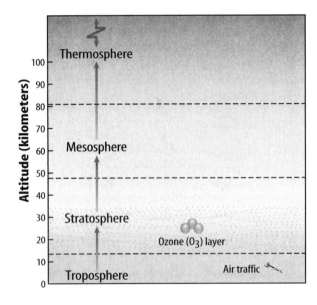

2 Which layer of the atmosphere contains the ozone layer?

  A mesosphere

  B stratosphere

  C thermosphere

  D troposphere   **4.b**

3 Which atmospheric layer contains weather?

  A mesosphere

  B stratosphere

  C thermosphere

  D troposphere   **4.e**

4 Which deflects winds to the right in the northern hemisphere?

  A convection

  B Coriolis effect

  C eddy

  D jet stream   **4.d**

5 Which process transfers thermal energy by contact?

  A conduction

  B convection

  C inversion

  D radiation   **3.c**

6 Which process transfers thermal energy through a vacuum?

  A conduction

  B convection

  C inversion

  D radiation   **3.d**

7 The California condor glides through the air.

Which is used by the California condor to soar high above the ground?

  A Coriolis effect

  B downdraft

  C jet stream

  D updraft   **4.d**

8 In which layer does Earth's atmosphere absorb most of the Sun's ultraviolet radiation?

   A mesosphere

   B stratosphere

   C thermosphere

   D troposphere   `4.b`

9 The photo below shows the movement of air around a shoreline.

Warm air

Cool air

Which best describes the process of air moving around the shoreline in the photograph?

   A conduction

   B contraction

   C convection

   D expansion   `4.d`

10 By what process does thermal energy transfer from a hot cup of hot chocolate to your hand?

   A conduction *collision*

   B convection

   C inversion

   D radiation   `3.d`

11 Which global wind helps move weather across the United States?

   A doldrums

   B easterlies

   C trade winds

   D westerlies   `4.e`

12 The dark arrows show changes of wind direction.

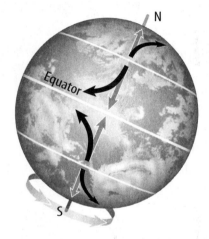

N

Equator

S

What causes this change in wind direction?

   A convection

   B Coriolis effect

   C jet stream

   D radiation   `4.d`

13 Why does the sky look blue and the Sun look yellow during the day?

   A. Only blue light can penetrate Earth's atmosphere.

   B. Only yellow and blue light can penetrate Earth's atmosphere.

   C. Blue light is scattered by the atmosphere more than yellow light.

   D. Yellow light is scattered by the atmosphere more than blue light.   `4.b`

The BIG Idea

Oceans are a major feature of Earth.

Heating the Far Shores

Energy from the Sun is absorbed and stored in the oceans. Near Earth's equator, stored thermal energy makes the water warm. Then, ocean currents transfer this thermal energy to distant shores around the world.

(Science *Journal*) Near Earth's poles, where the angle of sunlight is low, the water is cold. Write a hypothesis that explains how warm ocean currents reach higher latitudes and cold ocean currents reach lower latitudes.

Will hot water sink?

What happens when hot water mixes with cold water?

Procedure

1. Complete a lab safety form.

2. Fill a **fish tank** with **cold water.**

3. Get a **jar with a lid** and record how many holes are in the lid of your jar.

4. Fill the jar with **hot water.**

5. Place a few drops of **food coloring** in the jar and screw on the lid.

6. Cover the hole or holes with your fingers and lower the jar to the bottom of the fish tank.

7. Use a **stopwatch** to time how long it takes for all the hot water to escape.

Think About This

- **Observe** What happened to the hot water?

- **Compare** Did other students have faster or slower times? How many holes did their lids have?

Science Online

Visit ca6.msscience.com to:

▶ view **Concepts in Motion**

▶ explore Virtual Labs

▶ access content-related Web links

▶ take the Standards Check

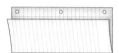

Ocean Currents Make the following Foldable to compare and contrast ocean currents.

▷ **STEP 1** **Fold** a sheet of paper in half lengthwise. Make the back edge about 3 cm longer than the front edge.

▷ **STEP 2** **Fold** into thirds.

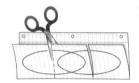

▷ **STEP 3** **Unfold** and **cut** along the folds of the top flap to make three flaps.

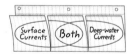

▷ **STEP 4** **Label** the flaps as shown.

Reading *Skill*   **ELA6: R 2.2**

Comparing and Contrasting
As you read Lesson 2, list under the appropriate flap information about surface currents and deep-water currents. Be sure to include causes, direction, and deflection by land.

Get Ready to Read

Summarize

① Learn It! Summarizing helps you organize information, focus on main ideas, and reduce the amount of information to remember. To summarize, restate the important facts in a short sentence or paragraph. Be brief and do not include too many details.

② Practice It! Read the text on page 425 labeled *Bathymetric Maps.* Then read the summary below and look at the important facts from that passage.

Important Facts

The ocean floor has mountains, trenches, and flat areas.

The depth of water is measured from sea level to the ocean floor.

Summary

Bathymetric maps show the contours of the ocean floor and its geologic features. The ocean floor has the same types of geologic shapes seen on land.

Sea level is the level of the sea's surface halfway between high and low tides. The ocean floor is Earth's surface underneath ocean water.

Before modern technology, sailors would make soundings to record water depth and make bathymetric maps.

③ Apply It! Practice summarizing as you read this chapter. Stop after each lesson and write a brief summary.

Target Your Reading

Use this to focus on the main ideas as you read the chapter.

1 **Before you read** the chapter, respond to the statements below on your worksheet or on a numbered sheet of paper.

- Write an **A** if you **agree** with the statement.
- Write a **D** if you **disagree** with the statement.

2 **After you read** the chapter, look back to this page to see if you've changed your mind about any of the statements.

- If any of your answers changed, explain why.
- Change any false statements into true statements.
- Use your revised statements as a study guide.

Reading Tip

Reread your summary to make sure you didn't change the author's original meaning or ideas.

Before You Read A or D	Statement	After You Read A or D
	1 The ocean floor is completely flat.	
	2 A map of the ocean floor can be made using sound waves.	
	3 A continuous chain of underwater volcanoes extends through all oceans.	
	4 Surface currents in the ocean are caused by wind.	
	5 Deep currents in the ocean are caused by wind.	
	6 Waves cause erosion along the shoreline.	
	7 Sand is transported by currents along the beach.	
	8 Beaches can be made of different types of sand.	
	9 Hurricanes do not occur in California.	
	10 The rocky shore has a high diversity of organisms.	

Science Online

Print a worksheet of this page at ca6.msscience.com.

Science Content Standards

7.c Construct appropriate graphs from data and develop qualitative statements about the relationships between variables.
7.f Read a topographic map and a geologic map for evidence provided on the maps and construct and interpret a simple scale map.

Reading Guide

What *You'll Learn*

▶ **Identify** the different oceans on Earth.

▶ **Understand** how bathymetric maps of the oceans are made.

▶ **Describe** the features of the ocean floor.

Why *It's Important*

Oceans cover more than 70 percent of Earth's surface.

Vocabulary

sea level
ocean floor
bathymetric map
echo sounding
continental shelf

Review Vocabulary

topographic map: a map that uses lines of equal elevation to show the shape of Earth's surface (p. 54)

Earth's Oceans

Main Idea Mapping the ocean floor is important to understanding Earth's global features.

Real-World Reading Connection In photographs taken from space, Earth's surface is covered with blue. The blue comes from the oceans, which make up about 71 percent of Earth's surface. This is why Earth is sometimes called the Water Planet.

Mapping Earth's Oceans

As shown in **Figure 1,** Earth contains five major oceans—the Pacific Ocean, the Atlantic Ocean, the Indian Ocean, the Arctic Ocean, and the Southern Ocean. The Pacific Ocean is the largest ocean. However, it is slowly decreasing in size because of the subduction zones that surround it. The Atlantic Ocean is the second largest ocean. The Atlantic Ocean is slowly growing larger because lava continually rises to the surface from deep within Earth. New ocean floor is continually created in the middle of the Atlantic Ocean. The Indian Ocean is the shallowest ocean. The Arctic Ocean is at the most northern part of Earth and much of it is often covered in ice. The Southern Ocean surrounds the continent of Antarctica and extends north to latitude of 60°S. It connects the Pacific, Indian, and Atlantic Oceans.

Figure 1 Modern maps usually include five major oceans.

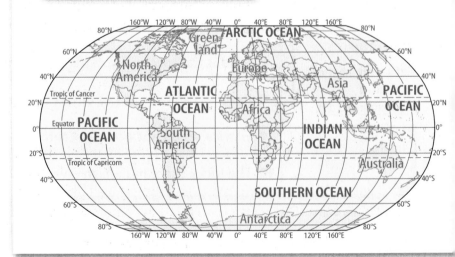

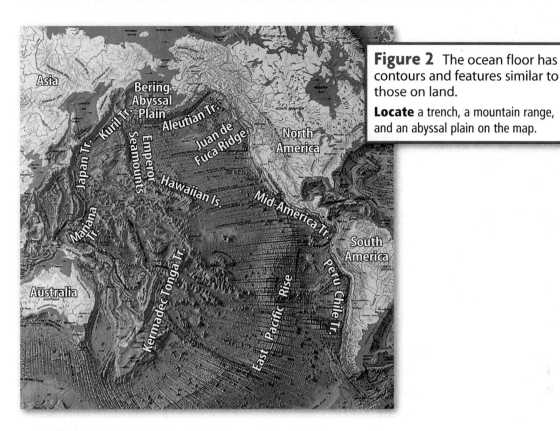

Figure 2 The ocean floor has contours and features similar to those on land.

Locate a trench, a mountain range, and an abyssal plain on the map.

Bathymetric Maps

Figure 2 shows the contours and features of the ocean floor in the Pacific Ocean. On the ocean floor hidden beneath the water, there are the same kinds of geological shapes that we see on land. There are underwater mountain ranges, trenches, and flat areas in each ocean.

One of the most important things you would want to know if you were sailing around the world is the depth of the water and the location of underwater obstacles. The depth of water is measured from sea level to the ocean floor. **Sea level** is the level of the ocean's surface halfway between high and low tides. The **ocean floor** is Earth's surface underneath the ocean water.

Before modern technology, sailors would drop a rope from their ship until it hit the bottom of the ocean. Then they would measure the length of rope they let out and record the water depth. This **method** of measuring water depth is called sounding. By making a large number of soundings and compiling them, a map of the ocean floor, or a bathymetric map, can be created. A **bathymetric** (BATH ih meh trihk) **map,** like the one shown in **Figure 2,** is a map of the bottom of the ocean showing the contours of the ocean floor and its geologic features. Bathymetric maps are like topographic maps except they show land formations that are underwater.

 Reading Check What is the purpose of a bathymetric map?

ACADEMIC VOCABULARY
method (MEH thud)
(noun) a way or process for doing something
Even though John and Sara used different methods to carry out the experiment, they both got the same results.

WORD ORIGIN
bathymetric
from Greek *bathys*; means *deep*

Echo Sounding

Today oceanographers map the ocean floor using sound and radio waves. Sonar **echo sounding** is a determination of the depth of water using sound waves. Scientists attach an instrument to the bottom of a ship that emits a sound wave. They then measure the amount of time that it takes for the sound wave to bounce off the ocean floor and return to the ship, as shown in **Figure 3.** If the sound bounces back quickly, the depth of the ocean is shallow. If it takes a long time, the depth of the ocean is deep.

Sound waves, radio waves, and light waves can be used to map the locations of coastlines, the geological features on the bottom of the oceans, and the location and direction of currents. Satellites use radio waves to detect small bumps and dips in the ocean surface. These bumps and dips reflect the locations of mountains and trenches on the ocean floor. **Figure 4** shows a bathymetric map created by echo sounding of an area of the Pacific Ocean just off the coast of Los Angeles, California.

Figure 3 Oceanographers use sound waves to create bathymetric maps of the ocean floor.

Reading Check What methods are used to map ocean floors?

Figure 4 These canyons are called submarine canyons because they are below sea level.

Newport Canyon

LOS ANGELES

Santa Monica Canyon

Redondo Canyon

San Pedro Sea Valley

San Gabriel Canyon

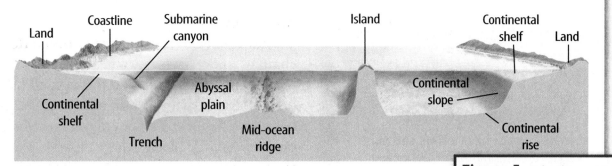

Labels on figure: Land, Coastline, Submarine canyon, Island, Continental shelf, Land, Continental shelf, Abyssal plain, Continental slope, Trench, Mid-ocean ridge, Continental rise

Figure 5 A bathymetric profile of a typical ocean floor shows different geological features.

Describe What are some differences between the continental shelf and the continental slope?

The Ocean Floor

Imagine taking a slice through the ocean floor and looking at it from the side. This is called a bathymetric profile, or a cross-section of the ocean. **Figure 5** shows some typical geologic features you will see in a bathymetric profile of the ocean floor.

Continental Shelf

The **continental shelf** is an underwater portion of continental crust that extends from the continental shoreline and gently slopes toward the deeper parts of the ocean. Along the east coast of the United States, the continental shelf is wide. California has a narrow continental shelf.

Continental Slope

The continental slope is the steep slope between the continent and the deep ocean. Some of these slopes represent locations where the supercontinent, Pangaea, split apart. Many of these slopes contain deep canyons called submarine canyons. Sediments flow down the canyons, sometimes in huge avalanches. The sediments are deposited on the continental rise, between the continental slope and the ocean floor.

Abyssal Plain

Beyond the continental slope and rise, the ocean floor is extremely flat. This region is called the abyssal (uh BIH sul) plain. The abyssal plain is made of blocks of basalt that are thought to have originated along mid-ocean ridges. Later, the blocks were covered with thick layers of sediment.

Ocean Trenches

Deep ocean trenches are extremely deep underwater valleys. The deepest point in the ocean is 11,033 m in the Mariana Trench in the Pacific Ocean. Ocean trenches are subduction zones, places where the tectonic plates are recycled into Earth's interior.

How do you read a bathymetric map?

Bathymetric maps have letters and numbers that represent areas of the ocean floor. Depth is represented by different colors. Each color has a matching depth given below.

Data
Examine the bathymetric map and table.

	A	B	C	D	E	F	G	H	I	J
1										
2										
3										
4										
5										
6										

Example of a Simple Bathymetric Map

Ocean Depth (meters)	Color Code
500	pink
1,000	red
1,500	orange
2,000	yellow
2,500	green
3,000	light blue
3,500	dark blue
4,000	purple
4,500	black

US Geological Survey Color Scheme

Data Analysis

1. **Describe** the features of the ocean floor in the map. In which grid is the water the deepest? Where is it the shallowest?

2. **Graph** Choose one row of the map. Draw a profile of what the ocean floor would look like from point A to point J.

Mid-Ocean Ridges

Mid-ocean ridges are a continuous chain of underwater volcanoes more than 65,000 km long that extend through all the ocean basins. The mid-ocean ridges rise 2 km above the ocean floor on average. Mid-ocean ridges are places where tectonic plates are moving away from each other and new sea floor is being created. **Figure 6** shows how the Juan de Fuca Ridge, in the Northeast Pacific, is formed.

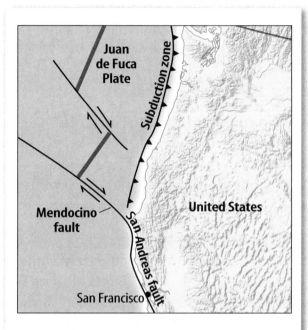

Figure 6 The Juan de Fuca Ridge is formed as the Pacific plate and the Juan de Fuca plate move away from each other.

Features of the Ocean Floor

The five oceans—the Pacific Ocean, the Atlantic Ocean, the Indian Ocean, the Arctic Ocean, and the Southern Ocean—cover more than 70 percent of Earth's surface. The ocean floor has contours and features similar to those found on land, including mid-ocean ridges, trenches, and flat abyssal plains. Bathymetric maps show the contours of the ocean floor and its geologic features. Information about the features of the ocean floor can be collected through echo sounding, using sound waves bounced off the bottom of the ocean; and through satellites, using radio waves bounced off the surface of the ocean. Bathymetric profiles of the ocean floor show the continental shelf, the continental slope and rise, and ocean trenches, ridges, and abyssal plains.

LESSON 1 Review

Summarize

Create your own lesson summary as you design a **visual aid.**

1. **Write** the lesson title, number, and page numbers at the top of your poster.

2. **Scan** the lesson to find the **red** main headings. Organize these headings on your poster, leaving space between each.

3. **Design** an information box beneath each **red** heading. In the box, list 2–3 details, key terms, and definitions from each **blue** subheading.

4. **Illustrate** your poster with diagrams of important structures or processes next to each information box.

 ELA6: R 2.4

Standards Check

Using Vocabulary

1. Distinguish between *sea level* and *ocean floor*. **7.f**

2. In your own words, write the definition for *echo sounding*. **7.f**

Understanding Main Ideas

3. **Identify** the five major oceans and their locations. **7.f**

4. Which describes extremely flat regions of the ocean floor?
   A. continental shelf
   B. trench
   C. abyssal plain
   D. mid-ocean ridge **7.f**

5. **Explain** how bathymetric maps are made. **7.f**

6. **Describe** mid-ocean ridges and how they are formed. **1.a**

Applying Science

7. **Draw** a bathymetric profile through an ocean with a narrow continental shelf, a steep continental slope, two mid-ocean ridges, and at least one abyssal plain. **Label** the different features. **7.f**

8. **Determine Cause and Effect** Copy and fill in the graphic organizer below to explain how features on the ocean floor are the result of the movements of Earth's plates. **1.a**

 Causes → Ocean Floor Features → Effects

 **Science** nline

For more practice, visit **Standards Check** at ca6.msscience.com.

Science Content Standards

4.a Students know the sun is the major source of energy for phenomena on Earth's surface; it powers winds, ocean currents, and the water cycle.

4.d Students know convection currents distribute heat in the atmosphere and oceans.

Reading Guide

What *You'll Learn*

▶ **Explain** how ocean currents are formed.

▶ **Explain** how ocean currents distribute thermal energy around Earth.

▶ **Describe** the major global ocean currents and gyres.

Why *It's Important*

Ocean currents transfer heat and influence weather and climate.

Vocabulary
ocean current
salinity
gyre

Review Vocabulary
latitude: the distance in degrees north or south of the equator (p. 49)

Ocean Currents

Main Idea Ocean currents help distribute heat around Earth.

Real-World Reading Connection You may have felt a current carrying you downstream when swimming in a river. Similarly, you may have felt an ocean current when swimming at the beach. If you have ever tried to swim against the current, you know it can be strong and fast-moving.

Influences on Ocean Currents

Earth is covered with a network of rivers that are important in many ways. They carry water, redistribute nutrients, and move sediments from place to place. Because it is made up of water, you might not think that the ocean also contains a network of moving water. Ocean water moves from place to place in **ocean currents,** which are like rivers in the ocean. Ocean currents, like the one shown in **Figure 7,** transport water, heat, nutrients, animals and plants, and even ships from place to place in the oceans.

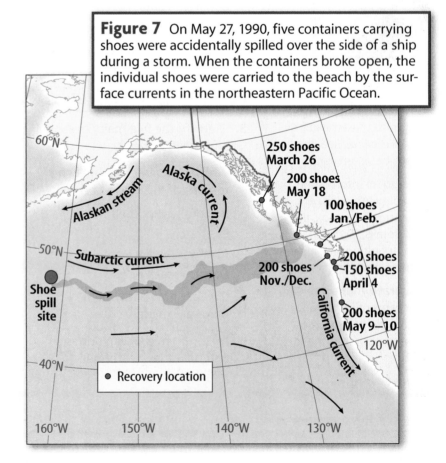

Figure 7 On May 27, 1990, five containers carrying shoes were accidentally spilled over the side of a ship during a storm. When the containers broke open, the individual shoes were carried to the beach by the surface currents in the northeastern Pacific Ocean.

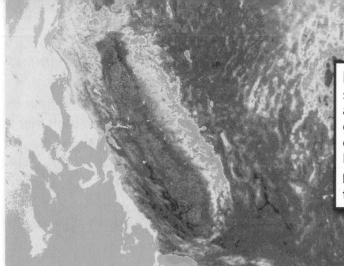

Figure 8 This satellite image shows surface temperatures in California during a heat wave. Blue is coldest, followed by green and yellow. Red is hottest. The cool green-blue area at the left is the Pacific Ocean.

Infer Why do you think the ocean is cooler than the land?

A Huge Reservoir of Energy

Figure 8 shows the difference in temperature between ocean water and land in California on a hot summer day. Have you ever been on the beach on a really hot day? When you step on the sand, it is so hot it feels like it could burn your feet. But when you step in the ocean, the water feels cool and refreshing. The Sun beats down on both the water and the sand with the same energy. What causes the difference?

One of the properties of water is that a large amount of heat can be added to or removed from it before it changes temperature. It takes five times more heat to change the temperature of an area of water than it does to change the same area of sand. As a result, sand changes temperature much more quickly than water does on a hot day. The oceans, because they are a huge reservoir of water, hold an enormous amount of heat.

Heat Transfer by the Oceans

Recall that the amount of energy received from the Sun varies greatly depending on latitude. In general, heat is gained by oceans in areas between 30°N and 30°S latitudes. Heat is lost by oceans at latitudes of above 40°. Even though heat is gained at the equator, the oceans do not boil there. At the same time, the amount of ice in polar regions remains about the same. What keeps heat balanced throughout the planet?

Water's ability to absorb and lose large amounts of heat energy without changing temperature makes it perfect for moving heat around the planet. In general, ocean currents carry heat from the tropics to the poles. This helps equalize the amount of heat throughout the planet.

How are oceans involved in balancing heat throughout Earth?

Figure 9 Faster wind produces more whitecaps on waves.

Calm Day

Windy Day

Surface Currents

Have you ever stood on a beach on a calm day? You probably noticed that the waves rolling toward shore are smooth until they break on the sand. But have you watched the ocean on a windy day? The waves not only crash on the beach, but they also form whitecaps out at sea, as shown in **Figure 9.**

As the wind blows over the ocean, it tugs on the surface of the ocean, moving the ocean surface water. On windy days, the wind moves the surface water faster than the wave is moving, causing it to crash in front of the wave. This produces whitecaps. The winds are the most important force driving the movement of surface water in the ocean. They have the strongest effect on the location and movement of the global ocean surface currents.

The Coriolis Effect in the Oceans

Recall that the Coriolis (kor ee OH lihs) effect is caused by Earth spinning on its axis. As a result, winds in the northern hemisphere are deflected to the right and winds in the southern hemisphere are deflected to the left. The spinning of Earth affects liquids in the same way it affects the gases that make up air. As shown in **Figure 10,** in the northern hemisphere, ocean currents are deflected to the right and in the southern hemisphere, they are deflected to the left. The overall effect is that currents tend to move in a clockwise pattern in the northern hemisphere and in a counter-clockwise pattern in the southern hemisphere.

 Figure 10 How does the Coriolis effect affect surface currents in the ocean?

Figure 10 The Coriolis effect deflects ocean currents in the same way that it affects winds.

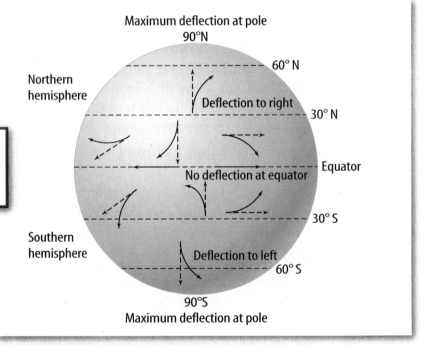

Maximum deflection at pole
90°N

60° N

Northern hemisphere

Deflection to right

30° N

No deflection at equator — Equator

30° S

Southern hemisphere

Deflection to left

60° S

90°S
Maximum deflection at pole

Density and Deep Ocean Currents

Not all currents in the ocean are driven by wind across the surface. Some currents are found deep in the ocean where there is no effect of the wind. What drives these currents? The answer has to do with the density of water. The density of water depends on both its temperature and the amount of salt it contains. Recall from the Launch Lab that cooler water has a higher density than warmer water. How does salinity affect the density of water?

The amount of salt that is dissolved in a quantity of water is called **salinity** (say LIH nuh tee). As the salinity of water increases, its density increases. Areas of water in different parts of the ocean have different densities. These differences in density form deep ocean currents.

When surface water becomes denser than the water below it, the surface water sinks. For example, surface water in Antarctica is cooled by air temperatures. It becomes saltier when salt is left behind as ocean water freezes. This makes the surface water dense. As a result, it sinks. This water mass then flows across the ocean floor, producing deep ocean currents, as shown in **Figure 11**.

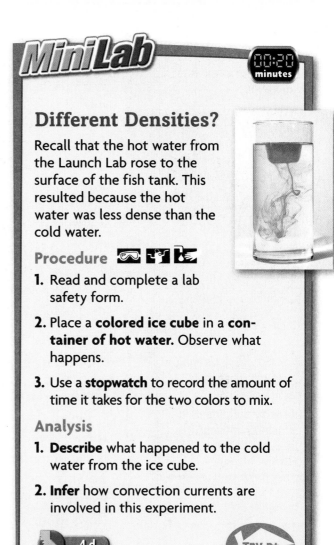

MiniLab

00:20 minutes

Different Densities?

Recall that the hot water from the Launch Lab rose to the surface of the fish tank. This resulted because the hot water was less dense than the cold water.

Procedure 🥽 👔 🔥

1. Read and complete a lab safety form.

2. Place a **colored ice cube** in a **container of hot water**. Observe what happens.

3. Use a **stopwatch** to record the amount of time it takes for the two colors to mix.

Analysis

1. **Describe** what happened to the cold water from the ice cube.

2. **Infer** how convection currents are involved in this experiment.

4.d

Try at Home

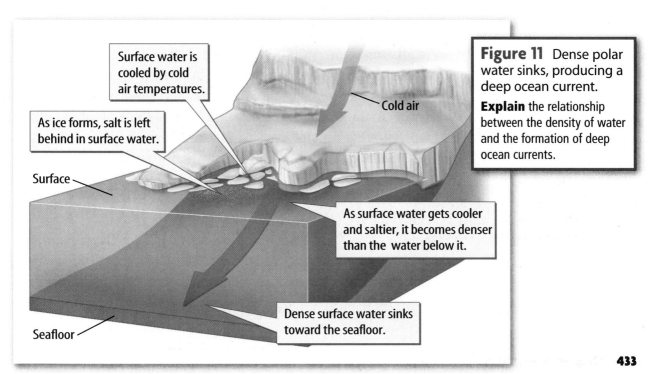

Surface water is cooled by cold air temperatures.

As ice forms, salt is left behind in surface water.

Cold air

Surface

As surface water gets cooler and saltier, it becomes denser than the water below it.

Dense surface water sinks toward the seafloor.

Seafloor

Figure 11 Dense polar water sinks, producing a deep ocean current.

Explain the relationship between the density of water and the formation of deep ocean currents.

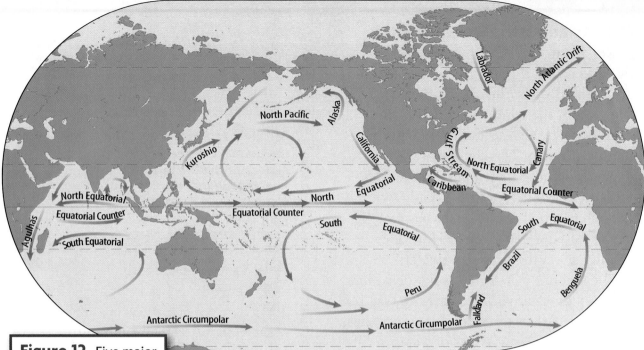

Figure 12 Five major gyres circulate in Earth's oceans.

Identify the currents in the southern Atlantic Ocean and describe how the gyre circulates.

ACADEMIC VOCABULARY

cycle (SI kul)

(noun) a series of events that occur regularly and usually lead back to the starting point; a circular or spiral arrangement

The cycle of four seasons takes one year to complete.

Gyres—Great Ocean Surface Currents

Recall that certain winds, including the trade winds and the westerlies, are concentrated in bands around Earth. Wind is the major factor that influences the movement of surface water in the ocean. The major surface currents are shown in **Figure 12.** Notice the location of the North Pacific Current. It flows along the same path as the westerlies in the northern hemisphere. But what happens when the North Pacific Current reaches the continent of North America? The presence of the land mass, as well as the Coriolis effect, deflects the current to the right, and it becomes the California Current, flowing south along the coast of California.

When the California Current reaches the tropics, the trade winds tug on it, moving it westward. Its name then changes to the North Equatorial Current. When the North Equatorial Current reaches Asia, the land mass and the Coriolis effect again turn it to the right, and it becomes the Kuroshio Current, moving northward past Japan. When the Kuroshio Current reaches the westerlies, it is pushed toward the east into the North Pacific Current again.

A **cycle** of ocean currents, like the ones in **Figure 12,** is called a **gyre** (JI ur). There are five major gyres in Earth's oceans. The North Atlantic and North Pacific Gyres rotate in a clockwise direction. The South Atlantic, South Pacific, and Indian Ocean Gyres rotate in a counterclockwise direction.

 Reading Check Which factors influence the direction in which a major ocean surface current will flow?

Figure 13 The Gulf Stream is a warm-water current that transports heat from the equator toward the poles. Warm water is shown in red and orange. Cooler water appears in blue and green.

Special Currents and Their Effects

The strongest and deepest currents are found on the western sides of the gyres. These currents are called western boundary currents because they are on the western side of the ocean basins. Eastern boundary currents are on the eastern side of the ocean basins.

The biggest western boundary current is the Gulf Stream, shown in **Figure 13,** which is part of the North Atlantic Gyre. It transports enough water to fill the entire Rose Bowl Stadium about 25 times per second. This water rushes north from the tropics toward the poles. The Gulf Stream and all the other western boundary currents are important to the redistribution of heat throughout the oceans. The Gulf Stream causes the climate in Europe to be milder than you might expect given its high latitude.

 Reading Check Why are western boundary currents important to Earth's heat balance?

Surrounding the continent of Antarctica is the Antarctic Circumpolar Current, shown in **Figure 14.** It is a continuous flow of water, but it is not a gyre because it surrounds land rather than water. It is the largest current in the oceans, with twice as much flow as the Gulf Stream. The Antarctic Circumpolar Current is driven in an eastward direction around the southern part of Earth by the strong westerlies.

Figure 14 The Antarctic Circumpolar Current moves around Antarctica in a clockwise direction.

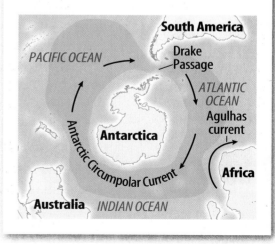

Figure 15 These drawings illustrate the differences in the trade winds and the currents during normal years and during El Niño and La Niña events. Green represents warm water and blue represents cold water.

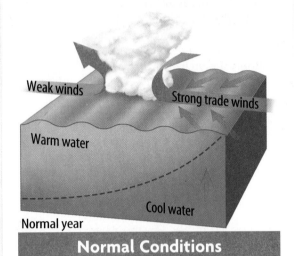

Weak winds
Strong trade winds
Warm water
Cool water
Normal year

Normal Conditions

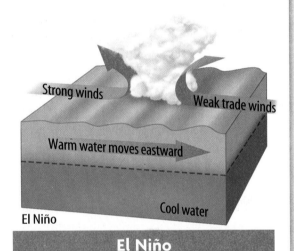

Strong winds
Weak trade winds
Warm water moves eastward
Cool water
El Niño

El Niño

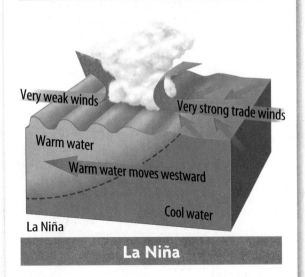

Very weak winds
Very strong trade winds
Warm water
Warm water moves westward
Cool water
La Niña

La Niña

Effects of El Niño and La Niña on Currents

In the Southern Pacific Ocean, the trade winds push tropical water westward from Central and South America toward Australia. As shown in **Figure 15,** cool, deep water normally rises to the surface near South America. However, sometimes the trade winds weaken or even reverse direction.

When the trade winds stop driving the flow of water across the Pacific, the South Equatorial Current slows down. Warm water from the western side of the Pacific sloshes back across the ocean, as shown in **Figure 15.** This phenomenon is known as an El Niño event. Because ocean currents and winds are connected throughout the planet, El Niño conditions have effects all over Earth. Effects of El Niño include droughts in the western Pacific areas of Australia and Indonesia and increased rain and flooding in the eastern Pacific including Peru and California. During an extremely strong El Niño in 1997 and 1998, the rainfall in California was twice the normal amount. Landslides and avalanches occurred more frequently than usual.

 **Reading Check** What changes occur in winds and the ocean during an El Niño event?

When the trade winds begin to blow again, they usually do with great strength, as shown in **Figure 15.** Warm tropical water is pulled across the Pacific toward Australia. The coast of South America becomes unusually cold and chilly. These conditions are called La Niña. El Niño and La Niña events occur about every three to eight years. Researchers are still trying to determine what drives these global-scale changes to the world's weather and ocean currents.

 Visual Check Figure 15 How do wind and ocean conditions change during a La Niña event?

Water Movement in the Ocean

Surface currents are driven by wind and their direction is influenced by the Coriolis effect and land formations. Large gyres circulate in each major ocean basin. Deep ocean currents are driven by differences in water density. Ocean currents transfer and distribute heat throughout Earth and help keep Earth's heat balanced. Ocean currents also affect weather and climate. Western boundary currents, such as the Gulf Stream, are warm-water currents that can influence regional climates by making them milder. Weather all over the world can be strongly influenced by El Niño and La Niña events, which act on winds and ocean currents.

LESSON 2 Review

Summarize

Create your own lesson summary as you design a **study web.**

1. **Write** the lesson title, number, and page numbers at the top of a sheet of paper.

2. **Scan** the lesson to find the **red** main headings.

3. **Organize** these headings clockwise on branches around the lesson title.

4. **Review** the information under each **red** heading to design a branch for each **blue** subheading.

5. **List** 2–3 details, key terms, and definitions from each **blue** subheading on branches extending from the main heading branches.

 ELA6: R 2.4

Standards Check

Using Vocabulary

Complete the sentences using the correct vocabulary term.

1. _____ transfer thermal energy from the Sun around the globe. **4.a**

2. A cycle of currents is called a(n) _____. **4.d**

Understanding Main Ideas

3. Which can cause an increase in the density of water?

   A. a decrease in temperature

   B. a decrease in salinity

   C. an increase in volume

   D. an increase in wind   **4.d**

4. **Identify** Copy and fill in the graphic organizer below to identify the three influences on ocean surface currents.

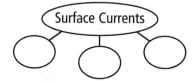

5. **Describe** the major global ocean currents and how they form gyres. **4.d**

6. **Examine Figure 12.** Determine which current is the western boundary current in the South Atlantic Gyre. **Explain** how this current transfers heat. **4.d**

Applying Science

7. **Design an Experiment** You are given two samples of ocean water. One is from the North Pacific Gyre and the other is from the Mediterranean Sea. Design an experiment to determine which is denser. **4.d**

8. **Hypothesize** What would happen to the North Pacific Current if the westerlies stopped blowing? **4.d**

Science Online

For more practice, visit **Standards Check** at ca6.msscience.com .

Reading Guide

What *You'll Learn*

▶ **Understand** how waves shape the shore.

▶ **Distinguish** between different types of sand.

Why *It's Important*

Beaches are always changing shape.

Vocabulary

shore
shoreline
longshore current
longshore drift
rip current
sand

Review Vocabulary

sediment: rock that is broken down into smaller pieces or that is dissolved in water (p. 99)

The Ocean Shore

(Main Idea) The shore is shaped by the movement of water and sand.

Real-World Reading Connection Have you ever watched a stream of water as it runs through dirt? You might have noticed that the water carries pieces of the dirt from one place to another, redistributing them. In the same way, the movement of water shapes the ocean's shore.

Shoreline Processes

As shown in **Figure 16,** the **shore** is the area of land found between the lowest water level at low tide and highest area of land that is affected by storm waves. The **shoreline** is the place where the ocean meets the land. The location of the shoreline constantly changes as the tide moves in and out. Tides are the alternate rising and falling of the surface level of the ocean. A beach is the area in which sediment is deposited along the shore. Beaches can be made of fine sand, tiny pebbles, or larger stones. The size and composition of the sediment that makes up a beach depends on where the sediment comes from. Sometimes the shore is rocky and there is little sand. Waves and currents close to shore influence the shape of a shore as they erode or deposit sediment.

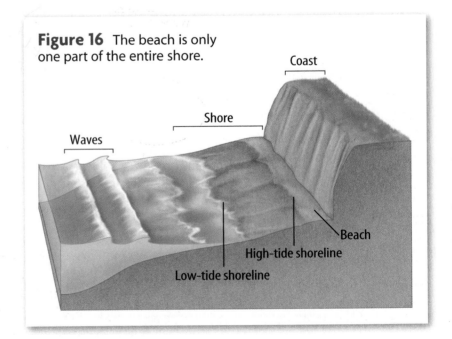

Figure 16 The beach is only one part of the entire shore.

Coast

Shore

Waves

Beach

High-tide shoreline

Low-tide shoreline

Effects of Wind and Waves

Wind and waves constantly beat the shoreline, causing erosion. Wind picks up tiny pieces of sediment, called grit, and then smashes it against rocks. The grit acts like sandpaper, rubbing large rocks into smaller ones. Crashing waves force air and water into cracks in rocks, breaking them into pieces. Waves also hurl sand and gravel at the shoreline, wearing larger rocks down into smaller pieces. Finally, water itself can dissolve many minerals in rocks, causing them to break apart.

Erosion Shoreline erosion by wind and waves depends on two factors—the type of rock found in the area and the intensity of the wind and waves. Hard rocks, like granite and basalt, erode very slowly. Soft rocks, like limestone and sandstone, may wear away quickly. As waves erode the rocks, shoreline features such as the sea arch and the sea stack shown in **Figure 17** are created. A sea arch is a tunnel that has been carved out of rock by erosion due to wind and waves. A sea stack is formed when a sea arch collapses and one side becomes separated from the main land formation.

 What type of shoreline features can be created from erosion?

Deposition Sediment that is eroded from one area of the shoreline eventually is deposited in another area. The deposition occurs where the energy of waves is low. The sediment falls out of the water and settles on the seafloor. Shoreline features such as baymouth bars and tombolos (TOHM boh loh), shown in **Figure 17,** are formed by the deposition of sediment. A baymouth bar is an accumulation of sediment that completely crosses the opening of a bay, sealing it off from the open ocean. A tombolo is a ridge of sediment that connects an island to the mainland or another island.

Sea Arch and Sea Stack

Baymouth Bar

Tombolo

Figure 17 The effect of wind and waves shapes the shoreline. Sea arches and sea stacks are the result of erosion. Baymouth bars and tombolos are the result of deposition.

Contrast What is the difference between a baymouth bar and a tombolo?

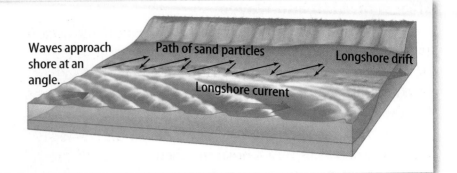

Figure 18 Because waves hit the shoreline at an angle, sediments are moved down the shore.

Describe How does the longshore current transport sand?

Waves approach shore at an angle.

Path of sand particles

Longshore drift

Longshore current

Longshore Drift

Once sediments are eroded from rocks, they usually do not stay in one place very long. The water from a breaking wave pushes sand up the beach at an angle. However, when the water from the wave runs back toward the ocean, it goes straight downhill because of gravity. As shown in **Figure 18,** this process moves sand along the beach. Part of the energy from the waves coming into the beach at an angle moves parallel to the shoreline. This energy drives a narrow current parallel to the shore called the **longshore current.** Sometimes the longshore current can move up to 4 km/h. Longshore currents transport sand that is **suspended** in the surf along the shoreline. The combination of the movement of sand on the beach by breaking waves and the movement of sand in the longshore current is called **longshore drift.**

Rip Currents

Sometimes a lot of waves hit the shore at once and pile up a lot of water. Usually the longshore current moves excess water along the beach. But when too much water piles up, the current cannot move it fast enough. The water breaks through the surf in a few places and rushes back out to the ocean. These swift currents that flow away from the beach are called **rip currents,** and are shown in **Figure 19.**

ACADEMIC VOCABULARY

suspend (suh SPEND)
(verb) to keep from falling or sinking by some invisible support, e.g., dust in the air
The silt that was stirred up from John walking in the river remained suspended in the water for several minutes.

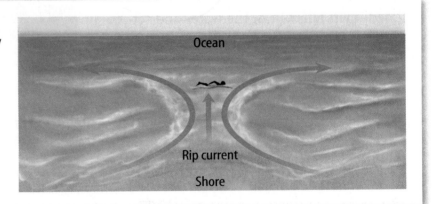

Figure 19 Rip currents are narrow currents of water moving offshore.

Ocean

Rip current

Shore

Human Activity and Beaches

Have you ever tried to build a sand castle near the ocean? If one wave comes farther up the beach, it can cause the whole thing to collapse. Since beaches are always changing, building on the beach is a difficult task, both for sand-castle builders and people who build real buildings.

To try to stabilize the beach, artificial structures often are put in place, as shown in **Figure 20.** Jetties, groins, and break-waters all are structures that extend from the beach out into the water. Seawalls are built on land and usually are parallel to the shore. Sometimes building structures to protect beaches has unintended results. Breakwaters, jetties, and groins trap sand, which stops the normal flow of sand along the shoreline. Farther down the shoreline, the beaches may become smaller. The sand that would usually be deposited by longshore drift is trapped by the structures. Seawalls also can cause erosion. The wave energy that is deflected by the seawall can be redirected on either side of it and below it. This can erode sand from around the seawall, causing it to collapse.

 Reading Check How do jetties and groins affect the longshore current?

SCIENCE USE V. COMMON USE
deposit
Science Use to let fall, as in sediment. *The sediment was deposited as the speed of the current slowed down.*
Common Use to place something, such as money, for safekeeping. *Laura deposited her paycheck into her bank account.*

Figure 20 Jetties and groins stop erosion because they slow longshore drift. Seawalls deflect the energy of the waves on the beach.

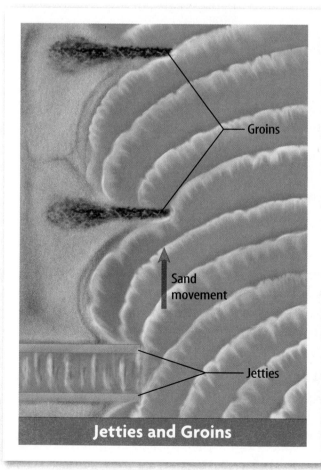

Jetties and Groins

Groins

Sand movement

Jetties

Seawall

Table 1 Range of Sediment Sizes	
Name	**Size**
Boulder	> 256 mm
Cobble	64–256 mm
Very coarse gravel	32–64 mm
Coarse gravel	16–32 mm
Medium gravel	8–16 mm
Fine gravel	4–8 mm
Very fine gravel	2–4 mm
Very coarse sand	1–2 mm
Coarse sand	$\frac{1}{2}$–1 mm
Medium sand	$\frac{1}{4}$ – $\frac{1}{2}$ mm
Fine sand	125–250 μm
Very fine sand	62.5–125 μm
Silt	3.90625–62.5 μm
Clay	< 3.90625 μm

Boulder

Medium Gravel

Fine Sand

Concepts In Motion

Interactive Table To explore more about sediment size, visit ca6.msscience.com.

Sand and Weathered Material

Sand is a term that is used to describe rocks that are between 0.0625 mm and 2.0 mm in diameter. Within this range, sand is categorized as very coarse, coarse, medium, fine, and very fine, as shown in **Table 1**. Rocks that are larger and smaller than sand are also shown in **Table 1.**

Sand Origins

Weathering breaks large boulders into smaller rocks. Rain then washes small rocks into rivers. Rivers transport these rocks to the ocean. Along the way, the rocks continually break into smaller pieces. These small pieces of rock then are transported in currents along the shoreline. The broken up rocks are eventually deposited as sand on sandy beaches. The sand that ends up on a beach in San Francisco Bay may have originated hundreds of kilometers away in the Sierra Nevada. Sand also may be from local rocks that have been weathered away just meters from the beach.

Sand Composition

Sand is made up of different minerals and rocks, depending on where the sand originated. White sand might be quartz or calcium carbonate from ground-up shells and coral skeletons. Black sand can be basalt, mica, or magnetite. Magnetite is a mineral that is attracted to magnets. Green sand can be a mineral called olivine that originates in lava. Sand that is pink or white originates from the mineral feldspar. Red sand can come from coral and from iron in volcanic cinders.

 **Reading Check** What types of minerals can be found in sand?

Sand Deposition

Even after sand reaches the ocean and then rests on a sandy beach, it does not stay there for long. Sand is continuously eroded, transported, and deposited along the shoreline. In the process, sand is sorted according to its size. The smaller grains of sand end up on low-energy beaches, and the larger grains remain on high-energy beaches. Most sand usually ends up on the ocean floor. As shown in **Figure 21,** the ocean floor near California contains several deposition beds, where sand and sediments accumulate.

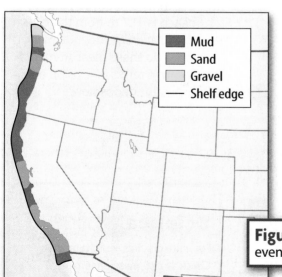

Figure 21 Sand and other sediments eventually are deposited in the ocean.

Isn't all sand the same?

Different beaches have different types of rocks and minerals in their sand. An examination of a sample of beach sand can reveal what types of rocks and minerals are in the area.

Procedure

1. Read and complete a lab safety form.

2. Examine the **sand** your teacher has given you. Make a note of the various sizes, shapes, and colors of the grains.

3. Study the grains with a **magnifying lens.** Try to identify the types of minerals and rocks that are present in your sample—quartz, feldspar, shell fragments, or volcanic rock.

4. Wrap a large **magnet** in a **plastic bag** and pass it through your sample. Are any sand grains attracted to the magnet?

Analysis

1. **Summarize** the composition of your sand sample. What rocks or minerals could you identify? What colors and shapes were the different grains? Where do you think the sand came from?

2. **Compare** your summary to other students' summaries. Did everyone have the same type of sand?

3. **Infer** What type of mineral would you find using a magnet?

 2.c

Mud
Sand
Gravel
— Shelf edge

Shaping the Shoreline

Beaches are areas of constant change. Wind and waves erode the rocks along the shoreline, creating features such as sea arches and sea stacks. Deposition of sediment occurs in areas of low energy, where sediment falls out of the water and settles on the seafloor. Baymouth bars and tombolos are shoreline features created by sediment deposition. Human activities can affect the shape of beaches as well. Artificial structures such as breakwaters, jetties, and groins can interfere with the longshore current, resulting in abnormal erosion and deposition of sediment.

LESSON 3 Review

Summarize

Create your own lesson summary as you write a script for a **television news report.**

1. **Review** the text after the **red** main headings and write one sentence about each. These are the headlines of your broadcast.

2. **Review** the text and write 2–3 sentences about each **blue** subheading. These sentences should tell *who, what, when, where,* and *why* information about each **red** heading.

3. **Include** descriptive details in your report, such as names of reporters and local places and events.

4. **Present** your news report to other classmates alone or with a team.

 **ELA6:** LS 1.4

Standards Check

Using Vocabulary

1. Distinguish between *longshore current* and *longshore drift.* **2.c**

2. Use each term in a separate sentence: *shoreline* and *shore.* **2.c**

Understanding Main Ideas

3. Which is a swift current that flows away from the beach?

   A. longshore current **2.c**
   B. rip current
   C. Bengula Current
   D. California Current

4. **Explain** the major processes that affect the shape of shorelines. **2.c**

5. **Sequence** Draw a graphic organizer like the one below and arrange the following in order from largest to smallest: coarse gravel, coarse sand, cobble, and silt. **2.c**

6. **Explain** how sand gets to sandy beaches. **2.c**

Applying Science

7. **Design** You are given a bag of sand from a beach. Design a method to separate it into fine sand, medium sand, and coarse sand. **2.c**

8. **Interpret** The first year a beach is measured, it is 13.1 m from the average low-tide mark to the cliffs behind the beach. The next year, the beach is 11.7 m from the cliffs to the average low-tide mark. What do these measurements mean in terms of erosion and sediment deposition? **2.c**

Science nline

For more practice, visit **Standards Check** at ca6.msscience.com.

Applying Math

Sediment Deposition Rates

Sediment deposition rates were measured in the areas below lake level for Searsville Lake in Santa Clara and San Mateo counties.

Example

Find the mean annual deposition rate of the data from 1913 to 1946.

What you know: From the graph, the datum from 1913 to 1929 is 3 ac-ft/yr.
From the graph, the datum from 1929 to 1946 is 7 ac-ft/yr.

What you need to find: The mean of the data

The mean of the data is the sum of all the numbers in the data set divided by the number of values. Rather than list all the data and then add, you can find how many years have a value of 3 ac-ft/yr and how many years have a value of 7 ac-ft/yr.

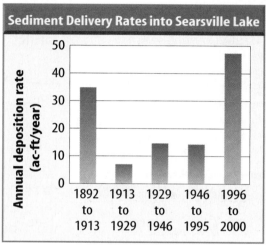

Sediment Delivery Rates into Searsville Lake

Source: Balance Hydrologics, Inc, Searsville Lake Sediment Impact Study

① Find how many years are included in the data beginning with 1913 to the beginning of 1929, but not including 1929, to find how many years have a value of 3 ac-ft/yr.

16 years × 3 ac-ft/yr = 48 ac-ft

② Now find how many years are included in the data beginning with 1929 to the beginning of 1946, but not including 1946, to find how many years have a value of 7 ac-ft/yr.

17 years × 7 ac-ft/yr = 119 ac-ft

③ Now find the mean.

$\frac{48 + 119}{33}$ or about 5.06 ac-ft/yr

Answer: For the 33-year time span, there is an annual deposition rate of about 5.06 ac-ft/yr.

Practice Problems

1. Find the mean annual deposition rate from 1892 to 1929.
2. Find the median annual deposition rate from 1913 to 1995.

Science nline
For more math practice, visit **Math Practice** at ca6.msscience.com.

Science Content Standards

1.e Students know major geologic events, such as earthquakes, volcanic eruptions, and mountain building, result from plate motions.

4.d Students know convection currents distribute heat in the atmosphere and oceans.

7.c Construct appropriate graphs from data and develop qualitative statements about the relationships between variables.

7.f Read a topographic map and a geologic map for evidence provided on the maps and construct and interpret a simple scale map.

Reading Guide

What *You'll Learn*

▶ **Understand** the geology of the California coastline.

▶ **Explain** how ocean currents affect California.

Why *It's Important*

The entire western border of California meets the Pacific Ocean.

Vocabulary

California Current
Davidson Current
habitat
marine

Review Vocabulary

transform plate boundary: boundary formed where two lithospheric plates move sideways past one another (p. 220)

Living on the California Coast

Main Idea Geology and ocean currents influence life in California.

Real-World Reading Connection How far away from the coast do you live? Have you ever been to the beach in California? Was the water cold or warm?

Geology of the California Coast

The geology of California is based on the movement of tectonic plates. Most of California lies on the North American plate, while the Pacific Ocean rests on the Pacific plate. Until about 30 million years ago, these two plates smashed directly into each other. The force of this collision created the coastal mountains in Northern and Southern California. About 30 million years ago, the plates changed direction and started slipping past each other. This created a transform boundary. This slipping lifted up and crushed the sea floor into mountains in central California. As a result of this tectonic activity, coastal mountain ranges stretch along the entire state of California, as shown in **Figure 22.**

Figure 22 Mountain ranges dominate the geology of the California coastline.
Describe how the mountains in central California were formed.

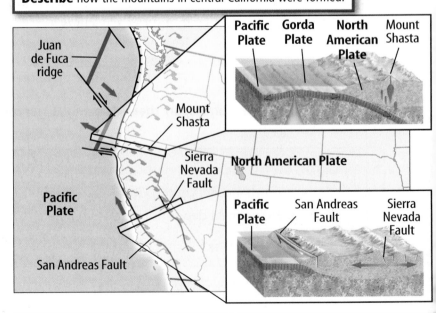

Why does California have so many rocky beaches?

Very few offshore islands protect the coast of California. The waves that hit the shore carry a lot of energy with them. Because California rests on a transform boundary where plates are compressed, the shoreline is elevated. The high-energy of the waves erodes the cliffs along the shoreline, leaving large boulders and cobbles.

What about tsunamis?

Tsunamis are large sea waves. They are caused by anything that displaces a large amount of water, such as landslides, icebergs falling off glaciers, volcanic explosions, and undersea earthquakes. Undersea earthquakes create the largest tsunamis.

When an earthquake occurs under the ocean, the movement of Earth displaces a large amount of water. This creates waves. The waves then move away from the location of the earthquake in all directions. When a tsunami reaches the shore, the bottom of the wave drags on the seafloor. This causes the enormous amount of water carried by the wave to pile up on itself. The excess water runs up on shore, similar to a fast and strong high tide. **Figure 23** shows the difference between a tsunami and waves caused by wind.

 What causes a tsunami?

Because there is tectonic activity throughout the Pacific Ocean, the California coast is at risk from tsunamis. Since 1812, 14 tsunamis have hit California, and 12 have caused damage. A sudden rise or drop in sea level is a warning sign of an approaching tsunami. To stay safe, people should to move to higher ground immediately. When tsunamis are generated far from California, the Tsunami Warning Center alerts local officials who will make decisions about evacuations.

Figure 23 Tsunamis are created when underwater earthquakes displace ocean water.

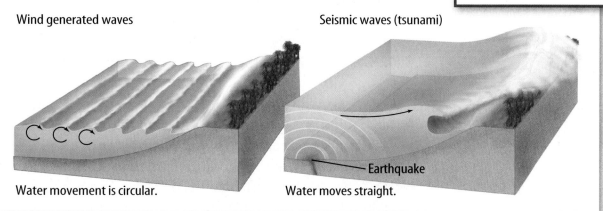

Wind generated waves

Seismic waves (tsunami)

Water movement is circular.

Water moves straight.

Earthquake

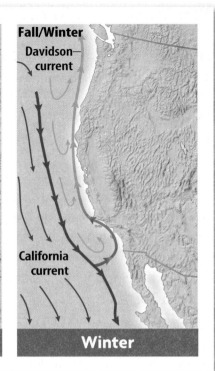

Spring/Summer **Fall/Winter**
Davidson—
current

California
current

California
current

Spring **Winter**

Figure 24 The cold California Current and the warmer Davidson Current are the major ocean currents along the California coast.

Compare and contrast the Davidson Current in the spring and in the winter.

Currents Along the Coast

Figure 24 shows the major ocean currents near California. The **California Current** is a wide, slow-moving, cold-water current along the California coast. It flows southward, bringing cool water from northern latitudes. The **Davidson Current** is a narrow, warm water current that flows northward, close to the California coast, from lower latitudes to higher latitudes. The Davidson Current is a seasonal current. It is stronger in the winter than the summer. California is cooled in the summer by the California Current, and warmed in the winter by the Davidson Current. The two currents often collide near Point Conception, just north of Los Angeles. At this point the California Current moves offshore.

The California Current

If you have ever jumped in the ocean in northern or central California, you might have been shocked by how cold the water was. The chill in the ocean off California is caused by the California Current.

The California Current is an eastern boundary current up to 1,000 km wide. It comes from the Gulf of Alaska, where most of the easterly flowing North Pacific Current is deflected to the south by the Coriolis effect and by hitting North America. Also, offshore winds and deflection by the Coriolis effect draw nutrient-rich deep waters to the surface.

Why are there no hurricanes in California?

Hurricanes are large storms that form in warm, tropical **regions.** As shown in **Figure 25,** they usually move from the east to the west, pushed by the trade winds. Every year, about 18 tropical storms form near Central America in the Pacific Ocean. About half become hurricanes. These storms curve northward and either hit land or lose energy at higher latitudes.

But no hurricanes have been recorded hitting California. The largest storm to hit California was September 25, 1939, in Long Beach, just south of Los Angeles. It had winds of 80 km/h, about two-thirds as strong as the winds of a hurricane. Why are there no hurricanes in California?

The power of a hurricane comes from the warm water in tropical areas. Hurricanes need approximately 27°C water temperature to survive. When the storm system moves into cold water, it loses energy. Recall that the California Current brings cold water along the California coast from northern latitudes. The California Current's surface temperature can be 15°C or even lower. The cold water around California acts as a hurricane shield.

 **Reading Check** Explain why there are no hurricanes in California.

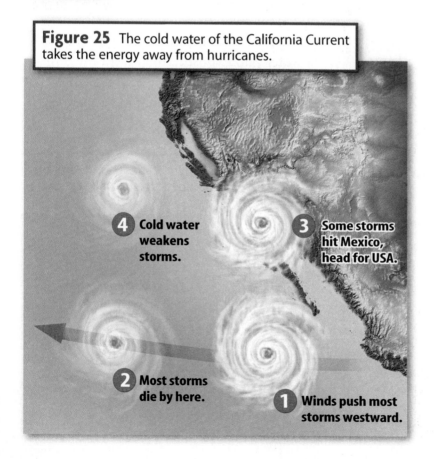

Figure 25 The cold water of the California Current takes the energy away from hurricanes.

4 **Cold water weakens storms.**

3 **Some storms hit Mexico, head for USA.**

2 **Most storms die by here.**

1 **Winds push most storms westward.**

How many whales are in the ocean?

The Channel Islands are at a latitude of about 34°N with an average water temperature of about 55–60°F. The Hawaiian Islands lie further south in the Pacific at latitudes between 19–22° N. The average water temperature in Hawaii is 73–78°F.

Data Collection

These data show the number of species recorded in the Channel Islands and the Hawaiian Islands.

Species Counts		
Species	**Channel Islands**	**Hawaiian Islands**
Whales and dolphins	30	4
Seals and Sea lions	7	1
Fishes	at least 40	at least 59
Birds	60	11
Invertebrates	at least 35	at least 33
Marine plants	10	10
Reptiles	0	4

Data Analysis

1. **Graph** the data in the table above.

2. **Analyze** Which groups differ the most between the two locations? Which location has the highest total possible number of species?

3. **Draw Conclusions** How would you explain the differences between the two locations?

Sea Life

The place in which an organism lives is called its **habitat.** Critical elements that affect habitat include the types of food available, the shelter, the moisture, and the temperature ranges needed for survival. In places where warm and cold water come together, the ocean usually is full of life. This is exactly what happens in the Channel Islands where the warm Davidson Current meets the cold California Current. The Channel Islands, along the coast of southern California, are home to a wide variety of different marine animals and plants, as shown in **Figure 26. Marine** refers to anything that is related to the ocean.

 Why is the diversity of organisms high in the Channel Islands?

The rocky shore also provides a variety of habitats for the organisms that live there. The intertidal zone is the area of shore that is between the lowest low-tide line and the highest high-tide line. On rocky shores, the intertidal zone provides many different habitats for marine creatures. The amount of water and exposure to air and the Sun varies in the intertidal zone. **Figure 27** shows some organisms that live in the intertidal zone.

Figure 26 The Channel Islands, where the cold California Current meets the warm Davidson Current, are home to a great diversity of marine life.

Visualizing the Rocky Shore Habitat

Figure 27

Life is tough in the intertidal zone—the coastal area between the highest high tide and the lowest low tide. Organisms here are pounded by waves and alternately covered and uncovered by water as tides rise and fall. These organisms tend to cluster into three general zones along the shore. Where they live depends on how well they tolerate being washed by waves, submerged at high tide, or exposed to air and sunlight when the tide is low.

Upper intertidal zone

Mid-intertidal zone

Lower intertidal zone

UPPER INTERTIDAL ZONE This part of the intertidal zone is splashed by high waves and is usually covered by water only during the highest tides each month. It is home to crabs that scuttle among periwinkle snails, limpets, and a few kinds of algae that can withstand long periods of dryness.

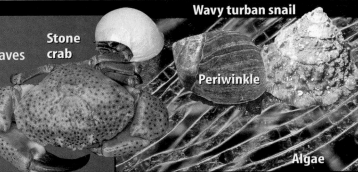

Wavy turban snail

Stone crab

Periwinkle

Algae

MID-INTERTIDAL ZONE Submerged at most high tides and exposed at most low tides, this zone is populated by brown algae, sponges, barnacles, mussels, chitons, snails, and sea stars. These creatures are resistant to drying out and good at clinging to slippery surfaces.

Gooseneck barnacles

Lined chiton

Blue mussels

LOWER INTERTIDAL ZONE This section of the intertidal zone is exposed only during the lowest tides each month. It contains the most diverse collection of living things. Here you find sea urchins, large sea stars, brittle stars, nudibranchs, sea cucumbers, anemones, and many kinds of fish.

Sea lemon nudibranch

African sea star

Sea urchins

Human Impact on the Coast

Humans use California's coastal waters for a variety of activities including swimming, boating, fishing, and even generating energy from the changing water level of the tides.

Fisheries

Many people fish in the ocean for business. They remove large amounts of herring, sea bass, sea urchins, and squid to sell to stores or restaurants. If too many organisms are removed, the organism may become threatened or endangered. Laws regulate the amount of organisms that can be removed, the time they can be removed, and the sizes of organisms that are removed. Fisheries are regulated by government agencies such as the California Department of Fish and Game.

Each fishery has a specific management plan. For example, the squid fishery is California's largest fishery with more than 160 million pounds of squid caught in 2002. The Market Squid Fishery Management Plan has policies to keep the fishery healthy and the population of squid out of danger.

Habitat Changes

Some human activities can alter the habitats of organisms. Wetlands and other coastal marsh areas are habitats for birds, fish, insects, plants, and marine invertebrates. **Figure 28** shows the difference in the amount of marsh area in the San Francisco Bay area between 1858 and 1983. Between these dates, more than 80 percent of the marsh area was developed for agriculture and buildings. This habitat loss has led to the decline in numbers of some species. Some native organisms, including the California clapper rail and the salt marsh harvest mouse, are endangered partly due to loss of habitat.

Figure 28 Marshes and wetlands often provide food, shelter, and nursery areas for many land and marine species.

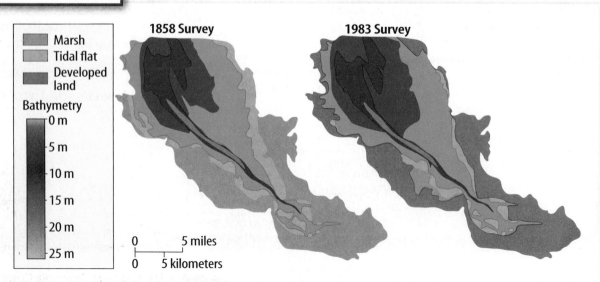

1858 Survey   1983 Survey

Marsh
Tidal flat
Developed land

Bathymetry
0 m
5 m
10 m
15 m
20 m
25 m

0   5 miles
0   5 kilometers

Coastal California

The coast of California has been shaped by the tectonic activity of the North American Plate and the Pacific Plate, resulting in coastal mountains and rocky shorelines, as well as the threat of tsunamis. The California Current and the seasonal Davidson Current affect water temperatures, sea life, and weather along the coast of California. Due to the cold temperature of the water carried by the California Current, hurricanes do not occur in California. Marine life in the waters off of California is diverse. Organisms living on beaches, rocky shores, and in coastal waters can be affected by habitat change.

LESSON 4 Review

Summarize

Create your own lesson summary as you organize an **outline.**

1. **Scan** the lesson. Find and list the first **red** main heading.

2. **Review** the text after the heading and list 2–3 details about the heading.

3. **Find** and list each **blue** subheading that follows the **red** main heading.

4. **List** 2–3 details, key terms, and definitions under each **blue** subheading.

5. **Review** additional **red** main headings and their supporting **blue** subheadings. List 2–3 details about each.

 **ELA6:** R 2.4

Standards Check

Using Vocabulary

1. Distinguish between *California Current* and *Davidson Current*. **4.d**

2. Use each term in a separate sentence: *marine* and *habitat*. **5.e**

Understanding Main Ideas

3. Which describes a sea wave caused by an earthquake? **1.e**

    A. current

    B. transform wave

    C. tsunami

    D. wind wave

4. **Organize Information** Copy and fill in the graphic organizer below to describe the two major currents that affect California. **4.d**

Current	Description
California Current	
Davidson Current	

5. **Explain** the connection between the geology of California and how many rocky beaches California has. **1.e**

6. **Explain** why you can find both fish that live in warm water and fish that live in cold water on San Clemente Island, one of the Channel Islands. **5.e**

Applying Science

7. **Construct** a set of guidelines that could be handed out to people who go to the beach that provides information about tsunamis. **1.e**

8. **Infer** The east coast of the United States is hit by hurricanes every year. What can you infer about the current that flows by this part of the country? **4.d**

Science online

For more practice, visit **Standards Check** at ca6.msscience.com.

Mapping the Ocean Floor

Materials

shoe box with lid
dowel rod marked
  in centimeters
modeling clay
colored markers

Safety Precautions

Science Content Standards

7.f Read a topographic map and a geologic map for evidence provided on the maps and construct and interpret a simple scale map.

Problem

Mapping the ocean floor is a challenging activity involving data collection and analysis. In this lab, you will make a model of the ocean floor and then collect data to make a map.

Form a Hypothesis

How can you make a bathymetric map?

Collect Data and Make Observations

Build the Ocean Floor

1. Read and complete a lab safety form.

2. Obtain a shoe box. Use the modeling clay to make a model of the ocean floor at the bottom of the shoe box.

3. Your model should include some of the features of the ocean floor such as ridges, abyssal plains, trenches, the continental shelf, and the continental slope. Do not allow other groups to see your ocean floor.

4. Place the lid on the box and exchange your box with another group.

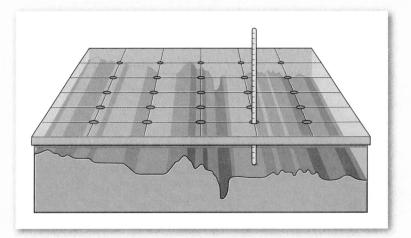

Collect Data

5. Examine the grid on top of the lid.
6. Design a data grid like the one shown to the right that has the same number of squares as the box has holes.
7. Starting at square A1, put the dowel rod into the hole in the box.
8. Find the square on your data grid that matches the hole on the box lid. Record the depth to the nearest centimeter. Continue until you have recorded a depth for each square.

	A	B	C	D	E
1					
2					
3					
4					
5					

Analyze and Conclude

1. **Interpret** the data you have collected. Use the Depth and Color table to color code your data grid. Color each square of your data grid the color that corresponds most closely to a measurement found in the table below.
2. **Classify** the different areas of the bottom of the box. Are there any ridges, trenches, or abyssal plains?
3. **Describe** what the bottom of the box looks like. Did you find anything unexpected?
4. **Think critically** about the method you used to make your map. How does it differ from how scientists make their maps? Is it similar to any other method you have learned about?
5. **Analyze** the bottom of the box once your teacher opens the lid. How closely did your map mirror the bottom of the box?

Depth and Color	
Ocean Depth (cm)	Color Code
1	Pink
2	Red
3	Orange
4	Yellow
5	Green
7	Light blue
8	Dark blue
10	Purple
12	Black

Communicate

 **WRITING in Science**   ELA6: W 2.5

Write an Advertisement Suppose you make bathymetric maps of near-shore areas based on data collected by scientists. You want to sell the maps to boat captains and other navigators. Write an ad to be placed in a boating magazine to sell your maps.

Real World Science

Shipboard Instructor

Would you like to sail on a research vessel, teaching students about the physics, chemistry, biology, and geology of the ocean? You can spend 3–12 months on voyages as an instructor. Many colleges and universities around the world offer such programs. To qualify for such a position, you will need to take science classes in high school and college. Most instructors have master's degrees or significant outside experience.

Visit **Careers** at <u>ca6.msscience.com</u> to learn more about becoming a shipboard instructor. Visit at least three Web sites of oceangoing instructional ships and **make a chart** to compare what qualifications you would need to work at each. **Decide** upon which ship you would prefer to work and give three reasons why.

OTEC

Ocean Thermal Energy Conversion (OTEC) relies upon a difference in the temperature of the ocean at the surface and at depth. Offshore OTEC power plants use this temperature difference to produce electricity, freshwater, and air conditioning. Such a plant is expensive at this time, but may be an excellent choice in the future for island communities with no fossil fuel resources and insufficient fresh water.

Visit **Technology** at <u>ca6.msscience.com</u> to learn more about OTEC technology. Pretend you are living in American Samoa, an island in the South Pacific. **Write** a letter to the editor of the local newspaper in support of using this technology to provide power and drinking water for the island. Give at least five reasons to support your position.

ELA6: W 1.2

Marie Tharp

During the Cold War, there was a great demand for precise knowledge of the seafloor. Having studied geology at the University of Michigan while the men were away at war, Marie Tharp joined a research group at Columbia University in New York in 1948. She spent several decades with Dr. Bruce Heezen analyzing and piecing together the data that the team had compiled of the Atlantic seafloor. She patiently hand-inked version after version of contour maps that showed the mid-oceanic ridge that solidified the theory of plate tectonics.

How did Marie Tharp make her contour maps? Visit **History** at **ca6.msscience.com** to learn more about where she got her data. Work through the tutorial to create your own contour maps in the same way that Marie Tharp did.

OVERFISHING THE OCEANS

One in five people of the world rely upon fish as their primary source of protein. However, since the early 1800s, certain cultures have been capable of depleting and sometimes have depleted the local population of certain fish. According to the Food and Agriculture Organization of the United Nations, over 70 percent of fish species are fully exploited or depleted. Although a greater tonnage of fish is caught with each passing year, the fishing industry is catching smaller and smaller fish each year.

Visit **Society** at **ca6.msscience.com** to learn more about overfishing. Pretend that you are a subsistence fisher working to support and feed your family. How do you feel about new restrictions on the size of your catch to keep the fish population from collapsing? How effective do you feel these measures will be?

The BIG Idea Oceans are a major feature of Earth.

Lesson 1 Earth's Oceans
`7.c, 7.f`

Main Idea Mapping the ocean floor is important to understanding Earth's global features.

- Bathymetric maps of the oceans are made by using soundings, and they show geologic features under water.

- The ocean floor typically includes the continental shelf, the continental slope and rise, mid-ocean ridges, trenches, and abyssal plains.

- **bathymetric map** (p. 425)
- **continental shelf** (p. 427)
- **echo sounding** (p. 426)
- **ocean floor** (p. 425)
- **sea level** (p. 425)

Lesson 2 Ocean Currents
`4.a, 4.d`

Main Idea Ocean currents help distribute heat around Earth.

- Wind is the major driver of currents in the oceans.

- Deep ocean currents are driven by differences in the density of ocean water.

- Gyres are cycles of currents that move water and distribute heat around ocean basins.

- **gyre** (p. 434)
- **ocean current** (p. 430)
- **salinity** (p. 433)

Lesson 3 The Ocean Shore
`2.c`

Main Idea The shore is shaped by the movement of water and sand.

- Wind and waves continually erode beaches.

- Longshore drift is the movement of sand along the shoreline.

- Sand is composed of different rocks and minerals.

- **longshore current** (p. 440)
- **longshore drift** (p. 440)
- **rip current** (p. 440)
- **sand** (p. 442)
- **shore** (p. 438)
- **shoreline** (p. 438)

Lesson 4 Living on the California Coast
`1.e, 4.d, 7.c, 7.f`

Main Idea Geology and ocean currents influence life in California.

- The major currents along the California Coast are the California Current and the Davidson Current.

- Where the California Current and the Davidson Current meet, marine life is abundant and diverse.

- Hurricanes do not reach the shore in California.

- **California Current** (p. 448)
- **Davidson Current** (p. 448)
- **habitat** (p. 450)
- **marine** (p. 450)

STUDY TO GO Download quizzes, key terms, and flash cards from ca6.msscience.com.

Linking Vocabulary and Main Ideas

Use the vocabulary terms from page 458 to complete this concept map.

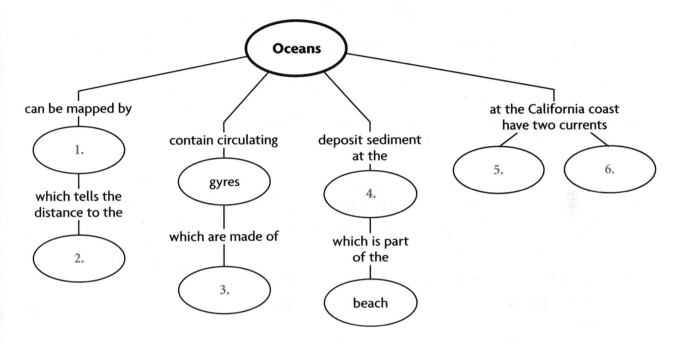

Using Vocabulary

Fill in the blanks with the correct vocabulary terms.

7. _____ is the amount of salt dissolved in water.

8. A(n) _____ is like a river in the ocean.

9. _____ is quicker and less costly than dropping a rope to the bottom of the ocean to determine the depth of the ocean floor.

10. When sand moves along the shoreline as a result of waves hitting the shore at an angle it is _____.

11. _____ is any rock between 2 mm and 62.5 μm.

12. The _____ brings cool water along the California coast.

13. A(n) _____ is a fast-moving current that moves water offshore.

14. Animals that are found in the ocean are called _____.

Science nline

Visit ca6.msscience.com for:
- ▶ Vocabulary PuzzleMaker
- ▶ Vocabulary eFlashcards
- ▶ Multilingual Glossary

Understanding Main Ideas

1. Where is new ocean floor formed?
   A. mid-ocean ridge
   B. abyssal plain
   C. deep ocean trench
   D. continental rise   **1.a**

2. The map below shows currents in the northern Pacific Ocean.

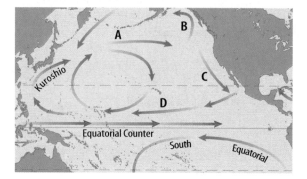

   Which current is labeled A?
   A. Alaska Current
   B. California Current
   C. North Equatorial Current
   D. North Pacific Current   **4.d**

3. Which is most important to the flow of deep ocean currents?
   A. the density of water
   B. the wind
   C. the specific heat capacity of water
   D. the convection in the atmosphere **4.d**

4. During El Niño, which occurs?
   A. The westerlies blow with increased force.
   B. The trade winds intensify.
   C. The California Current reverses directions.
   D. The North Equatorial Current weakens. **4.d**

5. Which is true in the northern hemisphere?
   A. The Coriolis effect deflects the winds to the left.
   B. There are four major gyres.
   C. The strongest currents are on the eastern side of the oceans.
   D. The gyres rotate clockwise.   **4.d**

6. Which statement is correct?
   A. The sand on beaches originates from rocks within a 5-km radius.
   B. The sand on beaches usually is made up of a single type of rock or mineral.
   C. The sand on beaches may have originated as boulders hundreds of miles from the beach.
   D. The sand on beaches always remains in the same location. **2.c**

7. Examine the figure below.

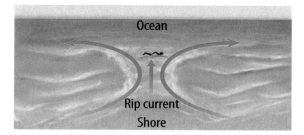

   Which best describes what is happening in this figure?
   A. The water from a tsunami is piling up on itself and surging up the shore.
   B. A rip current is collecting on the beach.
   C. A hurricane is making landfall on the central California coast.
   D. The longshore drift is pulling water above the high-tide mark. **2.c**

8. Why is California relatively protected from hurricanes?
   A. Hurricanes are only found at tropical latitudes.
   B. Hurricanes only move from east to west.
   C. Hurricanes require warm water for power.
   D. Hurricanes develop when underwater earthquakes occur. **4.d**

Applying Science

9. **Compare and contrast** a mid-ocean ridge and a deep ocean trench.   `1.a`

10. **Describe** how you might construct a bathymetric profile across San Francisco Bay.   `7.f`

11. **Hypothesize** What would happen to the North Pacific Gyre if Earth spun the opposite direction on its axis?   `4.d`

12. **Explain** what influences the flow of the major currents in the North Atlantic Gyre. Include the North Equatorial Current, the Gulf Stream, the North Atlantic Current, and the Canary Current in your answer.   `4.d`

13. **Hypothesize** what the effects of longshore drift on a beach would be where the waves always hit exactly parallel to the shoreline.   `2.c`

14. **Interpret** You examine two samples of white sand under the microscope and decide they must be from different beaches. Explain what you might have seen that led you to this.   `2.c`

15. **Analyze** What event is shown in the diagram below? What happens to the winds during this event? How is water affected during this event?   `4.d`

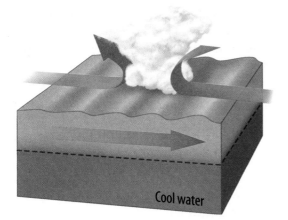

Cool water

WRITING in Science

16. **Suppose** there is a massive earthquake in the seafloor near Puerto Rico. Write a 500-word story from the point of view of someone in Miami, Florida.   `ELA6: W 2.1`

Cumulative Review

17. **Explain** how the Sun and Earth's atmosphere make conditions on Earth suitable for life.   `4.a`

18. **Identify** and describe the three types of radiation that make up solar radiation.   `4.b`

Applying Math

Use the graph below to answer questions 19–23.

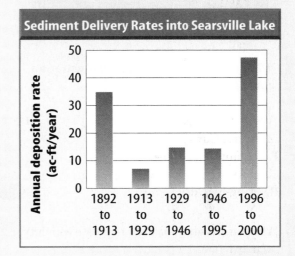

Sediment Delivery Rates into Searsville Lake

19. Find the mean annual deposition rate from 1946 to 2000.   `MA6: SP 1.1, SP 1.2`

20. Find the median annual deposition rate from 1946 to 2000.   `MA6: SP 1.1, SP 1.2`

21. Find the mean annual deposition rate from 1892 to 1946.   `MA6: SP 1.1, SP 1.2`

22. Find the median annual deposition rate from 1892 to 1946.   `MA6: SP 1.1, SP 1.2`

23. Find the mean annual deposition rate from 1929 to 1995.   `MA6: SP 1.1, SP 1.2`

Use the illustration below to answer questions 1–2.

1 Which process is shown in the illustration above?

A conduction

B convection

C echo sounding

D longshore drift   7.f

2 The process above is used to create which type of map?

A bathymetric map

B geologic map

C geologic cross section

D satellite map   7.f

3 Why aren't all ocean surface currents the same temperature?

A Different currents move at different speeds.

B The surface water is deeper in some areas than others.

C Different latitudes receive different amounts of solar energy.

D Surface currents start out warm but transfer all of their thermal energy.   4.d

4 Which drives ocean surface currents?

A conduction

B evaporation

C salinity

D wind   4.d

Use the table below to answer questions 5–6.

Partial Sediment Size Range	
Name	**Size**
Boulder	>256 mm
Coarse gravel	16–32 mm
Fine gravel	4–8 mm
Coarse sand	0.5–1 mm
Fine sand	125–250 μm
Silt	3.90625–62.5 μm
Clay	<3.90625 μm

5 What size sediment would you expect medium sand to have?

A 250 μm—0.5 mm

B 300 μm—1 mm

C 8 mm—9mm

D 8mm—15mm   2.c

6 Which type of sediment is more likely to be found on a low-energy beach?

A boulder

B gravel

C coarse sand

D fine sand   2.c

Science Online   Standards Assessment ca6.msscience.com

7 Which geological feature is the result of plate movements?

A abyssal plain

B continental shelf

C continental slope

D deep ocean trench   `1.e`

8 The photo shows a tombolo behind Morro Rock, California.

Which formed the tombolo shown in the photo?

A convection

B deposition

C erosion

D rip currents   `2.c`

9 Which is an effect of La Niña?

A cold and chilly South American coast   `4.d`

B droughts in the western Pacific region

C global increase in temperature

D increased flooding in California and Peru

Use the illustration below to answer questions 10–11.

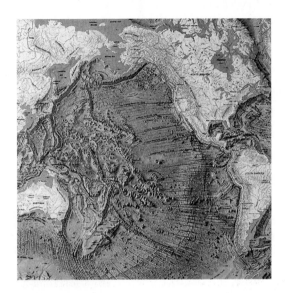

10 The figure above is an example of

A bathymetric map.

B geologic map.

C geologic cross section.

D satellite map.   `7.f`

11 Which feature is not visible on the map above?

A abyssal plain

B asthenosphere

C mountain range

D ocean trench   `7.f`

12 How are the North Pacific Current and the westerlies related?

A They both flow along the coast of California.

B They flow along the same path in the northern hemisphere.

C They flow in opposite paths in the northern hemisphere.

D They both flow in concentrated bands around the Earth.   `4.d`

Weather and Climate

The Golden State

During winter—California's rainy season—the hills turn green and valleys fill with thick, misty fog. Fields of golden poppies—California's state flower—can be seen each spring throughout the state. During summer—California's dry season—the hills turn golden brown. How does California's weather and climate contribute to its nickname—"the golden state?"

Science Journal Describe your observations of California's weather, climate, and seasons. Analyze the importance of water in your descriptions.

Launch Lab

minutes

How does water move in the atmosphere?

Water moves from oceans to clouds to lakes and rivers. How does this happen?

Procedure

1. Use **colored pencils** to draw the following scene on a **sheet of paper:** The Sun shining on Earth's surface, which includes ocean water, land, lakes, rivers, and snow-capped mountains with some clouds in the sky.

2. Draw arrows on your diagram that show how water moves from Earth's surface into the atmosphere and back.

Think About This

- **Infer** Streams flow from melting snow, run down the mountains, join other streams, and become rivers. Where do the rivers end up?

- **Explain** The ocean does not continue to fill until it spills over the land. Where does the ocean water go? Explain.

4.a

Science Online

Visit ca6.msscience.com to:

▶ view **Concepts in Motion**

▶ explore **Virtual Labs**

▶ access content-related Web links

▶ take the Standards Check

FOLDABLES™ **Study Organizer**

Climate Make the following Foldable to define climate and the factors that influence it.

▷ **STEP 1** **Collect** three sheets of paper and layer them about 2 cm apart vertically. Keep the left edges even.

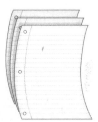

▷ **STEP 2** **Fold** up the bottom edges of the paper to form 5 equal tabs. Crease the fold to hold the tabs in place.

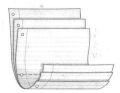

▷ **STEP 3** **Staple** along the fold. **Label** as shown.

| Humans |
| Prevailing Winds |
| Ocean Currents |
| Land and water distribution |
| Latitude |
| **Climate** |

Reading *Skill*    **ELA6: R 2.2**

Recognizing Cause and Effect
As you read Lesson 3, use your Foldable to record information about climates and the factors that influence climates.

Get Ready to Read

Compare and Contrast  ELA6: R 2.2

1 Learn It! Good readers compare and contrast information as they read. This means they look for similarities and differences to help them to remember important ideas. Look for signal words in the text to let you know when the author is comparing or contrasting.

Compare and Contrast Signal Words	
Compare	**Contrast**
as	but
like	or
likewise	unlike
similarly	however
at the same time	although
in a similar way	on the other hand

2 Practice It! Read the excerpt below and notice how the author uses contrast signal words to describe the temperature differences between Newport Beach and Big Bear.

On a sunny winter day, a person can be in Newport Beach, California, and observe the peaks of nearby mountains covered in snow, such as Big Bear. **Although** the latitude of the two locations, Newport Beach and Big Bear, are nearly the **same, differences** in altitude cause a large difference in temperature.

3 Apply It! Compare and contrast mediterranean and highland climates on page 484.

Target Your Reading

Use this to focus on the main ideas as you read the chapter.

As you read, use other skills, such as summarizing and connecting, to help you understand comparisons and contrasts.

Reading Tip

① **Before you read** the chapter, respond to the statements below on your worksheet or on a numbered sheet of paper.
- Write an **A** if you **agree** with the statement.
- Write a **D** if you **disagree** with the statement.

② **After you read** the chapter, look back to this page to see if you've changed your mind about any of the statements.
- If any of your answers changed, explain why.
- Change any false statements into true statements.
- Use your revised statements as a study guide.

Before You Read A or D	Statement	After You Read A or D
	1 Air temperature is a weather factor.	
	2 Air pressure does not affect weather.	
	3 Clouds are made of water droplets or ice crystals.	
	4 Weather conditions can change quickly.	
	5 Excessive rainfall can lead to flooding.	
	6 There is only one climate region in North America.	
	7 Latitude affects the climate of an area.	
	8 Ocean currents do not affect weather and climate.	
	9 The area on the lee side (downwind slope) of a mountain experiences high rainfall.	
	10 During the dry summers in California, the risk of fire increases.	

Science Online

Print a worksheet of this page at ca6.msscience.com.

LESSON 1

Science Content Standards

4.a Students know the sun is the major source of energy for phenomena on Earth's surface; it powers winds, ocean currents, and the water cycle.

4.e Students know differences in pressure, heat, air movement, and humidity result in changes of weather.

Reading Guide

What *You'll Learn*

- **Identify** some of the factors used to describe weather.
- **Differentiate** between the terms *humidity, relative humidity,* and *dew point.*
- **Describe** the processes that move water within the water cycle.

Why *It's Important*

Weather affects our lives in many ways.

Vocabulary

weather
humidity
relative humidity
dew point
precipitation
water cycle

Review Vocabulary

wind: air that is in motion relative to Earth's surface (p. 401)

Weather

Main Idea Weather describes the atmospheric conditions of a place at a certain time.

Real-World Reading Connection Weather affects our lives in many ways—from the type of houses we build, to the way we dress, to the plans we make for activities each day. How would you describe the weather where you live?

Weather Factors

Weather is the atmospheric conditions, along with short term changes, of a certain place at a certain time. If you have ever been caught in a rainstorm on what started out as a sunny day, you know that weather conditions can change quickly—sometimes over just a few short hours. On the other hand, your area may have the same sunny weather for three days in a row before the weather changes.

Perhaps the first things that come to mind when you think about weather are temperature and rainfall. As you dress in the morning, knowing what the temperature will be throughout the day helps you decide what to wear. How would rain affect your plans?

Temperature and rainfall are just two of the factors used to describe weather. Barometric pressure, humidity, cloud coverage, visibility, and wind are other factors used to describe weather. Examine the weather forecast shown in **Figure 1** for Los Angeles, California.

High: 81° F
Low: 62° F
Partly Cloudy
Chance of rain: 10%
Humidity: 42%
Wind Speed: 0–5 mph/calm
Barometric Pressure: 29.97 in
Visibility: 10.00 miles/good
Forecast for Friday September 9, 2005, Los Angeles.

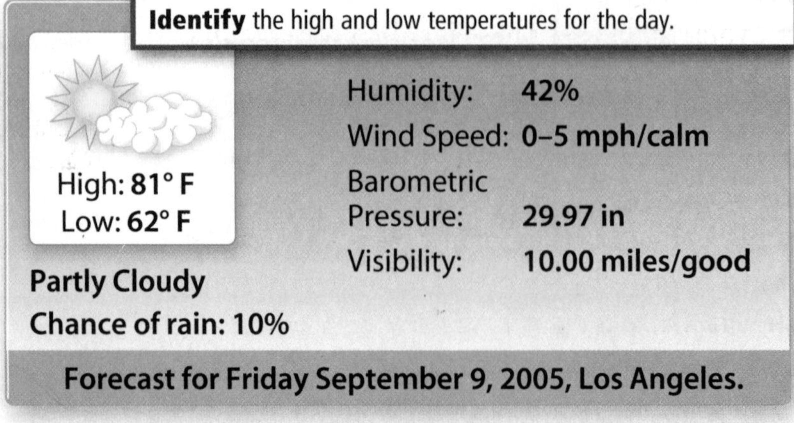

Figure 1 A weather report gives information about the weather factors for the day.
Identify the high and low temperatures for the day.

Air Temperature

The temperature of the air is a measure of the average kinetic energy of air molecules. When the temperature is high, molecules have a high kinetic energy. Therefore, molecules move faster than when temperatures are cold. Air temperatures change with time of the day, season, location, and altitude.

Air Pressure

Air pressure is the pressure that a column of air exerts on the air below it. Atmospheric pressure decreases with height. Therefore, air pressure is higher close to Earth's surface than at higher altitudes. Typical barometers, like the one shown in **Figure 2,** measure air pressure in millibars (mb). This pressure is referred to as barometric pressure in a weather forecast. Knowing the barometric pressure of different areas helps meteorologists predict the weather.

 What happens to air pressure as altitude increases?

Wind

Winds can change direction quickly. However, many winds—such as the westerlies and the trade winds—typically blow from the same direction. **Traditionally,** wind direction is given as the direction from which the wind is coming. For example, the westerlies blow from west to east. The polar easterlies blow from east to west. Some local winds—called *northers*—blow from the north.

Humidity

Water in the gas phase is called water vapor. The amount of water vapor present in air is used to describe weather. **Humidity** (hyew MIH duh tee) is the amount of water vapor per volume of air and is measured in grams of water per cubic meter of air (g/m^3). When the humidity is high, there is more water vapor in the air. On a day with high humidity, your skin might feel sticky and sweat might not evaporate from your skin as quickly.

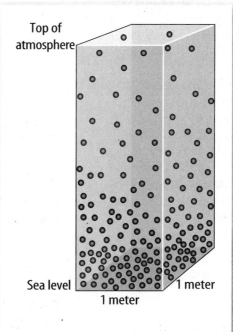

Figure 2 Air pressure is the pressure that a column of air puts on the air below it. A barometer measures air pressure.

ACADEMIC VOCABULARY

traditionally
(tra DIH shuh nuh lee)
(*adv*) an established or customary pattern or way of doing something
Traditionally, final exams are held during the last week of school.

DataLab

00:20 minutes

When will dew form?

Use **Figure 3** to determine whether dew will form as the temperature changes.

Data Collection

1. The relative humidity on a summer day is 80 percent. The temperature is 35°C. You want to find out if the dew point will be reached if the temperature falls to 25°C later that evening. Use **Figure 3** to find the amount of water vapor needed for saturation at each temperature.

2. Calculate the amount of water vapor in air that is 35°C at 80 percent relative humidity. (Hint: Multiply the amount of water vapor air can hold at 35°C by the percent of relative humidity.)

3. At 25°C, air can hold 22 g/m³ of water vapor. If your answer from step 2 is less than 22 g/m³, the dew point is not reached and dew will not form. If the number is greater, then dew will form.

Data Analysis

1. **Draw Conclusions** Will dew form when the temperature drops from 35°C to 25°C at 80 percent humidity?

2. **Predict** During the day, the relative humidity is 75 percent and the air temperature is 5°C. At night, the temperature falls to 0°C. What will happen?

 4.e    MA6: MR 2.4

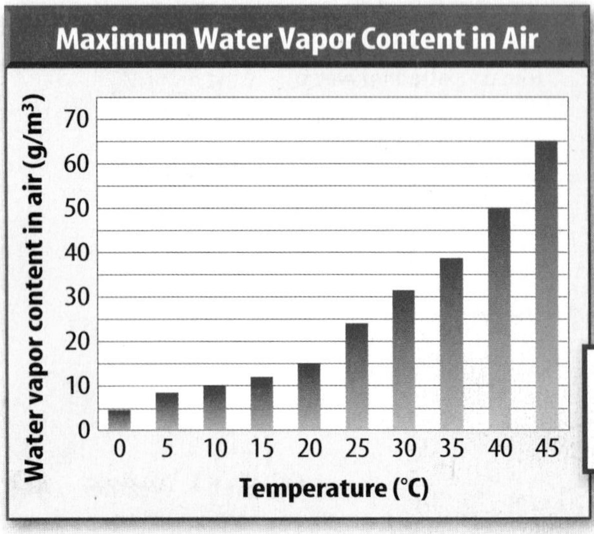

Relative Humidity

When air is saturated, it cannot hold any more water vapor. Think about how a sponge holds water. At some point, the sponge becomes full with the maximum amount of water it can hold. After that maximum amount is reached, water starts to drip from the sponge. The amount of water vapor present in the air relative to, or compared to, the maximum amount of water vapor the air can hold at that temperature before becoming saturated is the **relative humidity.**

Relative humidity is given in percentages. For example, a relative humidity of 50 percent means that the amount of water vapor in the air is one-half of the maximum the air can hold at that temperature. When weather forecasters give information about the humidity levels, they are usually referring to relative humidity.

Dew Point

Think about what happens when you leave a glass containing cold water on a table on a hot summer day. Soon, water droplets begin to form on the outside of the glass. The temperature at which air becomes fully saturated with water vapor and condensation forms is the **dew point.**

When the air temperature drops, the air can hold less water vapor. The water vapor in air will condense to a liquid—dew—if the temperature is above freezing, or form ice crystals—frost—if the temperature is below 0°C. The graph in **Figure 3** shows the total amount of water vapor that can be held by air at different temperatures. When the dew point is reached, the relative humidity is 100 percent.

Maximum Water Vapor Content in Air

Water vapor content in air (g/m³) vs. Temperature (°C)

Figure 3 The graph shows that as air temperature increases, more water vapor can be present in the air.

Stratus

Cumulus

Cirrus

Figure 4 Stratus clouds are flat, wide, "layered" clouds and can be found at all altitudes. Cumulus clouds are fluffy, "heaped," or piled-up clouds and can be found at all altitudes. Cirrus clouds are "wispy," high-altitude clouds.

Clouds and Fog

When air reaches its dew point, water vapor condenses to form droplets. These small droplets can come together to form larger droplets that form clouds. Clouds are water droplets or ice crystals suspended in the atmosphere. Clouds can have different shapes and be present at different altitudes within the atmosphere. Pictures of different types of clouds are shown in **Figure 4.** Since clouds move, they can transport water and heat from one location to another. Recall that clouds are also important in reflecting some of the Sun's incoming radiation.

When clouds form close to Earth's surface, it is called fog. Fog is a suspension of water droplets or ice crystals close to Earth's surface. Dense fog surrounding the Golden Gate Bridge in San Francisco is shown in **Figure 5.** Fog reduces visibility, which is the distance a person can see into the atmosphere.

 How does fog form?

When you exhale air that is warmer than surrounding outside air on a cold winter day, the warm air cools down. If the dew point is reached, you can see the condensed water vapor in a foggy cloud in front of your face.

Figure 5 Fog is a surface cloud. When you are in fog, you're inside a cloud.

Figure 6 Rain, snow, sleet, and hail are forms of precipitation.

Cloud droplets

Warm

Raindrops

Warm

Rain

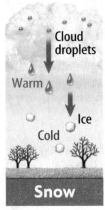

Cloud droplets

Warm

Ice

Cold

Snow

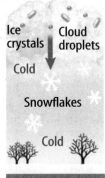

Ice crystals | Cloud droplets

Cold

Snowflakes

Cold

Sleet

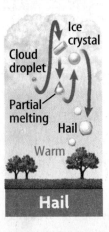

Ice crystal

Cloud droplet

Partial melting

Hail

Warm

Hail

Precipitation

When water, in liquid or solid form, falls from the atmosphere it is called **precipitation** (prih sih puh TAY shun). Examples of precipitation—rain, snow, sleet, and hail—are shown in **Figure 6.** Rain is precipitation that reaches Earth's surface as droplets of water. Snow is precipitation that reaches Earth's surface as solid, frozen crystals of water. Sleet reaches Earth's surface as small ice particles that began as rain in clouds, then froze as they passed through a layer of below-freezing air. Hail reaches Earth's surface as large pellets of ice. Hail is formed when layers of ice are formed around a small piece of ice that is repeatedly caught in an updraft within a cloud.

 What is the difference between snow and sleet?

The Water Cycle

Water is essential for all living organisms. Approximately 96 percent of Earth's water is stored in the oceans. Fresh water, present in glaciers, polar ice, lakes, rivers, and under the ground, represents only 4 percent of the water on Earth. The *hydrosphere* is the term used to describe all the water at Earth's surface. Water constantly moves between the hydrosphere and the atmosphere through the **water cycle,** which is shown in **Figure 7.**

 Figure 7 What happens to water that reaches Earth's surface as rain?

The Sun's Energy

Ultimately, it is the Sun's energy that drives the water cycle. Solar radiation that reaches Earth's surface causes water in the hydrosphere to change from a liquid to a gas, a process called evaporation. Water that evaporates from lakes, streams, and oceans enters Earth's atmosphere as water vapor. As land and water are heated by the Sun, the air masses over them become warm and rise. As the air masses rise, the air expands and cools down. When the air cools down, the water vapor changes from a gas back into a liquid, a process called condensation. As the water vapor condenses, water droplets form. These water droplets then form clouds. When the droplets become larger and heavier, precipitation falls from the clouds to Earth's surface, returning water to the hydrosphere.

 What happens as the Sun's heat warms Earth's surface?

Visualizing the Water Cycle

Figure 7

As the diagram below shows, energy for the water cycle is provided by the Sun. Water continuously cycles between oceans, land, and the atmosphere through the processes of evaporation, transpiration, condensation, and precipitation.

▲ Droplets inside clouds join to form bigger drops. When the droplets become heavy enough, they fall as rain, snow, or some other form of precipitation.

▲ As it rises into the air, water vapor cools and condenses into water again. Millions of tiny water droplets form a cloud.

▶ Rain runs off the land into streams and rivers. Water flows into lakes and oceans. Some water is taken up by plants.

▲ Water evaporates from oceans, lakes, and rivers. Plants release water vapor through transpiration.

Contributed by National Geographic

Describing Weather and the Water Cycle

Temperature, precipitation, air pressure, and wind are some of the factors that are used to describe weather. The amount of water vapor in the air, or humidity, is also an important factor that determines the weather.

Weather, along with the water cycle, is ultimately driven by the Sun's energy. As water moves between the hydrosphere and the atmosphere, water evaporates at Earth's surface, clouds form, precipitation falls, and water returns to Earth's surface.

LESSON 1 Review

Summarize

Create your own lesson summary as you write a **newsletter.**

1. **Write** this lesson title, number, and page numbers at the top of a sheet of paper.

2. **Review** the text after the red main headings and write one sentence about each. These will be the headlines of your newsletter.

3. **Review** the text and write 2–3 sentences about each **blue** subheading. These sentences should tell *who, what, when, where,* and *why* information about each headline.

4. **Illustrate** your newsletter with diagrams of important structures and processes next to each headline.

 ELA6: W 1.2

Standards Check

Using Vocabulary

1. Distinguish between *humidity* and *relative humidity.* `4.e`

2. In your own words, write the definition for *weather.* `4.e`

Understanding Main Ideas

3. Which describes the process by which water vapor changes from a gas into a liquid?

   A. evaporation `4.e`
   B. condensation
   C. run off
   D. precipitation

4. **List** some of the most important factors that are used to describe the weather. `4.e`

5. **Explain** the difference between humidity and relative humidity. `4.e`

6. **Explain** why this statement is true: "The Sun drives the water cycle." `4.a`

7. **Illustrate** the differences between evaporation and condensation. `4.e`

Applying Science

8. **Differentiate** between cloud formation, fog formation, and dew point. `4.e`

9. **Assess** Explain how wind direction is used when naming winds. What are some advantages to using this system to name winds? `4.e`

10. **Organize** Copy and fill in the graphic organizer below. In each oval, list the different types of precipitation. `4.e`

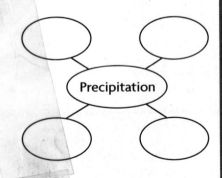

Precipitation

Science Online

For more practice, visit **Standards Check** at ca6.msscience.com.

Clouds ca6.msscience.com

Brain POP

Science Content Standards

2.d Students know earthquakes, volcanic eruptions, landslides, and floods change human and wildlife habitats.

4.e Students know differences in pressure, heat, air movement, and humidity result in changes in weather.

Reading Guide

What *You'll Learn*

▶ **Identify** some of the factors involved in weather variations.

▶ **Define** air masses and weather fronts.

▶ **Differentiate** between high- and low-pressure systems.

▶ **Describe** some severe weather events and their effects.

Why *It's Important*

Weather can change quickly and can be severe.

Vocabulary

air mass
cold front
warm front
season
drought
flash flood

Review Vocabulary

atmosphere: mixture of gases that surrounds Earth (p. 382)

Weather Patterns

Main Idea Several factors drive changes in weather.

Real-World Reading Connection Weather can change quickly. You might be enjoying a nice summer day when suddenly a cold breeze starts. The temperature starts decreasing, and maybe it starts to rain.

The Changing Weather

Weather conditions can change rapidly. Why does this happen? How is weather affected by the movements of air masses and pressure changes?

Air Masses

An **air mass** is a body of air that has consistent features, such as temperature and relative humidity. Air masses get their characteristics from the surface over which they develop. For example, an air mass that forms over a warm dry area will have warm dry conditions. **Figure 8** shows six major air masses that affect weather in the United States.

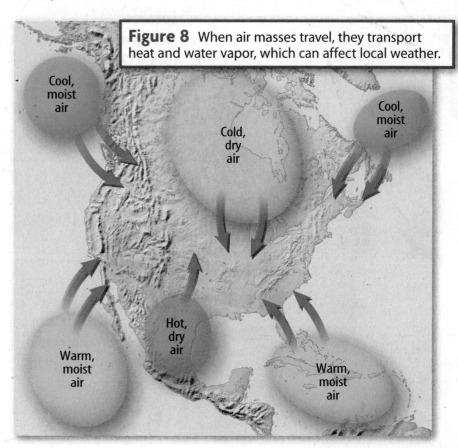

Figure 8 When air masses travel, they transport heat and water vapor, which can affect local weather.

Cool, moist air

Cold, dry air

Cool, moist air

Hot, dry air

Warm, moist air

Warm, moist air

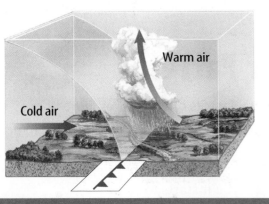

Cold Front

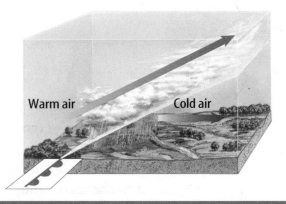

Warm Front

Figure 9 A cold front can move quickly. Thunderstorms often form as warm air is suddenly lifted up over cold air. Warm air slides over colder air along a warm front. This can lead to hours, if not days, of wet weather.

Describe the differences between a cold front and a warm front.

Weather Fronts

The boundary between two air masses of different density, moisture, and temperature is called a front. As air masses move from one location to another, they eventually run into each other. Study the air masses in **Figure 9.** A **cold front** occurs when a colder air mass moves toward warmer air. The cold air pushes the warm air up into the atmosphere. The warm air cools as it rises and water vapor condenses. Clouds form and precipitation begins to fall. In many cases, cold fronts give rise to severe storms.

A **warm front** forms when lighter, warmer air moves over heavier, colder air, as shown in **Figure 9.** Clouds form as the water in warm air condenses. A warm front usually results in steady rain for several days.

Highs and Lows

Have you ever heard a meteorologist using the terms *low-pressure system* and *high-pressure system?* Recall that when warm air rises, it creates a decrease in pressure close to Earth's surface. Therefore an area of low pressure is created. Areas of low pressure are associated with cloudy, stormy weather.

What happens when cold air sinks? When cold air sinks, it moves closer to Earth's surface. The surface air pressure increases as air moves down, and an area of high pressure is created. This is referred to as a high-pressure system. The sinking motion in high-pressure systems makes it hard for air to rise and for clouds to form. High-pressure systems are associated with fair weather.

 **Reading Check** Why are high pressure systems associated with fair weather?

Weather Maps

Information on weather factors, high- and low-pressure systems, and weather fronts is usually represented in maps. These maps, called weather maps, provide useful information on the atmospheric conditions over areas of interest. Weather maps contain a lot of information that is in the form of symbols. A key for each symbol is usually provided next to the map. Study the weather map shown in **Figure 10.** The map contains information about pressure systems and weather fronts for the western United States. Notice the symbols for high- and low-pressure systems, cold fronts, and warm fronts shown in the map legend.

Figure 10 Highs, lows, and fronts help meteorologists forecast the weather.

Predict What will the weather be like in San Diego, California, for the next few days?

Ⓛ	Low pressure
Ⓗ	High pressure
▲▲▲	Cold front
⌒⌒⌒	Warm front

How do you interpret a satellite image?

Satellite images show the pattern of clouds and weather systems across a large region. What can you learn about the weather from satellite images?

Procedure

1. Examine the satellite photo shown above.

2. Identify the colors that represent clouds, ocean water, and land.

3. Identify the United States and the Great Lakes on the map.

4. Identify the regions of the United States that have cloud cover. Identify regions that have clear skies.

Analysis

1. **Describe** In which regions of the United States is there most likely a high-pressure system, according to the image? How do you know?

2. **Infer** Which region of the map is most likely experiencing a low-pressure system? Explain how you know this.

3. **Explain** why satellite images are helpful to weather forecasters.

Cycles that Affect Weather

Some of the cycles that regularly affect the weather include the day and night cycle, the seasons, and El Niño.

Day and Night Cycles

Air goes through a daily cycle of warming and cooling. As the Sun rises in the morning, sunlight warms the ground. The ground warms the air by conduction until a few hours past noon. As the Sun lowers in the afternoon, its energy is spread over a larger area. Sometime in late afternoon or early evening, the ground and air above begin to lose energy and start to cool. By late night or early morning, the coldest air is found next to the ground.

Seasons

WORD ORIGIN············

season
from Latin *sationem*;
means *a sowing*

The regular changes in temperature and length of day that result from the tilt of Earth's axis are **seasons.** Changes in the amount of solar radiation received at different latitudes during different times of the year give rise to the seasons.

Earth revolves around the Sun, as shown in **Figure 11.** As a result of Earth's tilt on its axis, the amount of solar radiation reaching different areas of Earth changes as Earth completes its yearly revolution around the Sun. More solar radiation reaches the northern hemisphere in June, when summer begins. However, less solar radiation reaches the southern hemisphere during this same month, when winter begins. During January, less solar radiation reaches the northern hemisphere, resulting in winter. In the southern hemisphere, summer begins in January.

 Reading Check What causes seasons on Earth?

Figure 11 The amount of solar radiation reaching most of Earth's surface changes as Earth completes its revolution around the Sun.

Identify What season is it in the southern hemisphere when it is summer in the northern hemisphere?

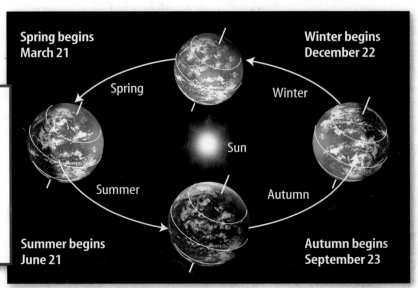

Spring begins
March 21

Winter begins
December 22

Spring

Winter

Sun

Summer

Autumn

Summer begins
June 21

Autumn begins
September 23

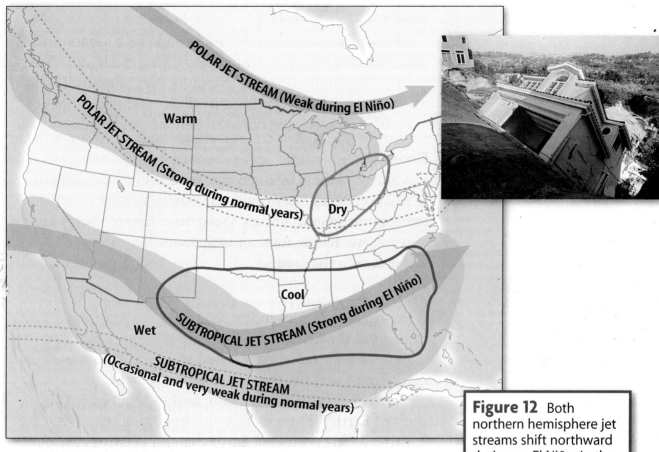

POLAR JET STREAM (Weak during El Niño)

POLAR JET STREAM (Strong during normal years)

Warm

Dry

SUBTROPICAL JET STREAM (Strong during El Niño)

Cool

Wet

SUBTROPICAL JET STREAM (Occasional and very weak during normal years)

Figure 12 Both northern hemisphere jet streams shift northward during an El Niño. In the United States, this often leads to warmer-than-usual temperatures in the north and increased rain and flooding in the south. The home shown above slipped down a hillside eroded by heavy El Niño rains.

El Niño and La Niña

El Niño and La Niña can affect weather worldwide. During an El Niño period, warmer surface water in areas of the eastern equatorial Pacific Ocean leads to more water vapor being in the air above the water. This can result in increased rainfall across the southeastern United States, and the South American countries of Peru and Ecuador. Heavy rainfall can lead to flooding and landslides like the one shown in **Figure 12.** On the other side of the globe, severe droughts can occur in Australia, Indonesia, and southeast Africa. In many cases, the droughts lead to forest fires as well. During El Niño years, the north-central United States sometimes experiences a milder-than-normal winter. The Carribean and southeastern United States experience a decrease in the number of hurricanes.

A La Niña event occurs when sea surface temperatures in the eastern equatorial Pacific are colder than normal. This can produce colder-than-normal winter temperatures in the north-western United States. It also leads to warmer-than-normal winter temperatures in the southeastern United States. **Figure 12** shows how El Niño affects the location of the polar and subpolar jetstreams over the United States.

Severe Weather

Sometimes weather can be severe and cause hazardous conditions. For example, extensive rainfall can lead to floods. Other times, rainfall can be absent for long periods of time. These events can pose a threat to all living things in the affected area, including humans, animals, and plants.

Droughts

A period of time when precipitation is much lower than normal or absent is a **drought.** Droughts can last months or years. Droughts can bring about several hazardous conditions for ecosystems and human populations. If an extended drought occurs in a region where precipitation is normally low, there can be a major decrease in water supply for the population of that region. A decrease in water supply also affects agriculture and, in extreme cases, can result in famine.

California experienced significant droughts from 1987 until the beginning of 1993. These droughts lasted a relatively long time and resulted in a decrease in the region's water reservoirs. Although the droughts in California were significant, they did not have the same severe **consequences** as droughts that occurred in areas of Africa, where thousands of lives were lost. The photos in **Figure 13** show lowered water levels and fires caused by the extreme dryness of a drought.

 **Reading Check** Why is a drought an example of severe weather?

Figure 13 Don Pedro Reservoir, east of Modesto, is the sixth-largest body of water in California. In 2003, drought lowered its water level, as shown in the photo on the left. Also in 2003, drought intensified a fire storm, shown on the right, as it descended from the foothills at night toward East Highlands, near San Bernardino, California.

Don Pedro Reservoir

East Highlands

Figure 14 These utility workers are patrolling a California suburb temporarily deserted after flooding caused by El Niño.

Floods

Sometimes, when severe weather occurs, it leads to flooding. Floods occur when water enters an area faster than it can be taken away by rivers, absorbed by the ground, or contained in lakes. Floods can occur when periods of extended heavy rain or melting snow increase the amount of water in rivers, producing an overflow. Flooding also can occur when snow melts rapidly, producing large amounts of runoff. When soils cannot retain water well, flooding can also happen. As shown in **Figure 14,** floods can damage, wash away, or even bury the living areas of both humans and wildlife.

Floods are common natural disasters and can occur in many locations. For example, in 1993, an enormous flood took place in the Midwest region of the United States. This flood has been one of the most damaging floods in the United States. In 1997, several floods took place in different locations of the United States, including California. More recently, New Orleans was almost completely destroyed by the flood waters that poured into the city from Hurricane Katrina in August 2005.

A **flash flood** is a flood that takes place suddenly. Flash floods, like the one shown in **Figure 15,** are the most dangerous type of floods. With deaths at over 200 people per year, flash floods are the number-one reason for weather-related deaths in the United States. Unfortunately, the damaging effect of flash floods is increasing due to human activities. This is because the construction of buildings, parking lots, and other structures decreases the amount of vegetation and soil that can potentially absorb runoffs of water.

Figure 15 Heavy rain in northern California caused the Russian River to rise 3 meters above flood stage in early January, 2006. Flash floods and mudslides killed at least 14 people and forced thousands to flee their homes.

Changes in Weather

Changes in air pressure and moving air masses can lead to changes in the weather. Cold fronts and warm fronts lead to changes in temperature and precipitation. Low-pressure systems and high-pressure systems can bring rain or fair weather respectively. Weather is affected by daily cycles of warming and cooling. It is also affected by longer cycles, including seasonal changes and El Niño and La Niña. El Niño and La Niña can lead to severe weather such as droughts or floods. Both of these can be dangerous to humans and wildlife.

LESSON 2 Review

Summarize

Create your own lesson summary as you design a **visual aid.**

1. **Write** the lesson title, number, and page numbers at the top of your poster.

2. **Scan** the lesson to find the **red** main headings. Organize these headings on your poster, leaving space between each.

3. **Design** an information box beneath each **red** heading. In the box, list 2–3 details, key terms, and definitions from each **blue** subheading.

4. **Illustrate** your poster with diagrams of important structures or processes next to each information box.

 ELA6: R 2.4

Standards Check

Using Vocabulary

Complete the sentences using the correct term.

seasons   drought

1. The regular changes in temperature and length of day that result from the tilt of Earth's axis are _____. **4.e**

2. A _____ is a period of time when precipitation is much lower than normal. **4.e**

Understanding Main Ideas

3. What kind of weather usually results from a warm front? **4.e**

   A. steady rain for several days
   B. a severe storm
   C. clear, sunny sky
   D. clouds but no rain

4. **List** some of the factors that can cause changes in weather. **4.e**

5. **Explain** how a cold front can affect the weather of an area. **4.e**

6. **Describe** how floods can occur and what affects they can have. **2.d**

Applying Science

7. **Apply** How might a high-pressure system affect the level of air pollution in an area? **4.e**

8. **Compare and Contrast** Copy and fill in the graphic organizer below to compare and contrast a high-pressure system and a low-pressure system. **4.e**

Pressure System	Similarities	Differences
Low		
High		

 **Science** nline

For more practice, visit **Standards Check** at ca6.msscience.com.

Humidity ca6.msscience.com   Brain POP

Science Content Standards

4.d Students know convection currents distribute heat in the atmosphere and oceans.

4.e Students know differences in pressure, heat, air movement, and humidity result in changes in weather.

Reading Guide

What *You'll Learn*

▶ **Compare** the characteristics of a mediterranean climate to those of a highland climate.

▶ **Identify** the factors that influence the climate of a region.

▶ **Describe** some of the ways in which human activities can affect the climate.

Why *It's Important*

Like weather, climate influences all living things in an area.

Vocabulary

climate
mediterranean climate
highland climate

Review Vocabulary

habitat: place in which an organisms lives (p. 450)

Climate

Main Idea Climate is often defined by annual temperatures and precipitation amounts.

Real-World Reading Connection Think about the climate of your area. How does your winter differ from that of Minneapolis, Minnesota, where the average temperature in January is −11°C?

A World of Many Climates

We live in a world of different climates—from hot and dry desert areas, to warm and wet rainforest regions, to frigid cold tundra at the poles. The climate present in different regions affects all of the organisms living there.

What is climate?

The long-term average of the weather patterns of an area is **climate.** This includes temperature, winds, and precipitation over a long period of time. Often, climate data is presented in graphs like the ones in **Figure 16.**

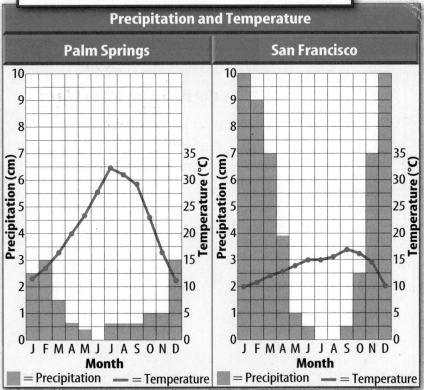

Figure 16 These graphs show the temperature and precipitation data for San Francisco and Palm Springs.

Climate Regions

The different climates of Earth are classified into climate regions. One classification system uses temperature, precipitation, and vegetation as characteristics to describe the climate of a region. Some of the main climate regions include cold polar tundra, dry desert, mediterranean, humid subtropical, highland, humid continental, and marine. **Figure 17** shows the climate regions of North America. As shown in **Figure 17,** most of California has mediterranean and highland climates. A **mediterranean** (me dih tur RAY nee en) **climate** is characterized by mild, wet winters and hot, dry summers. Mediterranean climates usually occur on the western side of a continent. Summer fires, due to dry conditions, often occur in a mediterranean climate.

A **highland climate** is characterized by cool-to-cold temperatures and occurs in the mountains and on high plateaus. Recall that as altitude increases, the temperature decreases. In California, the influence of altitude on climate can be easily observed. On a sunny winter day, a person can be in Newport Beach, California, and observe the peaks of the nearby mountains covered in snow, such as Big Bear. Although the latitude of the two locations, Newport Beach and Big Bear, are nearly the same, differences in altitude cause a large difference in temperature.

 Figure 17 Which regions of North America have a dry climate?

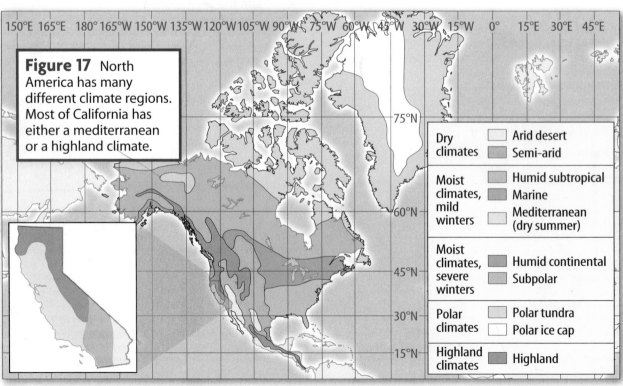

Figure 17 North America has many different climate regions. Most of California has either a mediterranean or a highland climate.

Dry climates	Arid desert
	Semi-arid
Moist climates, mild winters	Humid subtropical
	Marine
	Mediterranean (dry summer)
Moist climates, severe winters	Humid continental
	Subpolar
Polar climates	Polar tundra
	Polar ice cap
Highland climates	Highland

Climate Controls

Earth's climates are shaped by energy received from the Sun. Uneven heating of Earth's surface results in air and ocean currents that influence the different climate regions. Other factors, such as latitude and altitude, the distance from a large body of water, and mountain barriers, also affect climate. These factors are called climate controls. Some of these factors, such as latitude and mountain barriers, are unchanging. Other factors, such as winds and ocean currents, can vary over the course of a year.

Latitude

Recall that areas close to the equator receive more solar radiation per unit of surface than areas located further north or south. Since more solar radiation is received in areas near the equator, these areas have warmer climates than regions at higher latitudes. The farther a region is from the equator, the colder its climate. The graph in **Figure 18** shows how air temperature changes with latitude.

Figure 18 Higher latitudes receive less solar radiation and have lower average temperatures. As latitude decreases, average temperatures increase.

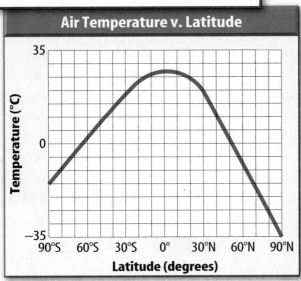

Air Temperature v. Latitude

How does latitude affect the angle of sunlight?

Investigate to see how different latitudes affect the angle at which the Sun's rays strike Earth.

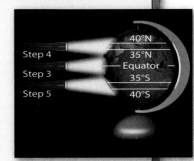

Procedure

1. Complete a **lab safety form.**

2. Place a **globe** on a table or desktop.

3. Hold a **flashlight** parallel to the floor and aim the beam of light directly at the equator on the globe. Record your observations about how the beam hits the globe.

4. Keep the flashlight parallel to the floor. Aim the beam between 35–40° N latitude. Record your observations about how the beam hits the globe.

5. With the flashlight parallel to the floor, aim the beam between 35–40° S latitude. Record your observations about how the beam hits the globe.

6. Find the state of California on the globe.

Analysis

1. **Contrast** What differences did you observe as the light beams hit the globe at the different latitudes?

2. **Infer** How would the differences you observed affect the climate of the different regions?

3. **Describe** Where is California compared to the latitudes at which light was aimed?

4.d

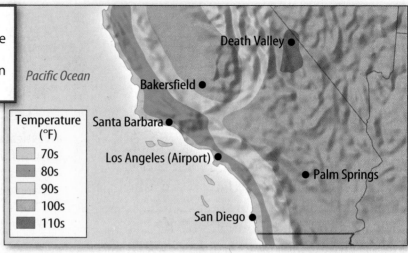

Figure 19 Large bodies of water, such as oceans, tend to modify climate by preventing the more extreme temperature changes that can occur in inland areas.

Pacific Ocean

Temperature (°F)	
	70s
	80s
	90s
	100s
	110s

Death Valley ●

Bakersfield ●

Santa Barbara ●

Los Angeles (Airport) ●

● Palm Springs

San Diego ●

ACADEMIC VOCABULARY

affect (ah FEKT)

(verb) to produce an effect
The weather along the coast of California is affected by the California current.

Distribution of Land and Water

The distribution of land and water has an important influence on climate. Recall that water can absorb or lose large amounts of thermal energy with little change in temperature. Land does not have this characteristic. Land surfaces heat and cool rapidly. Ocean surfaces heat and cool slowly. As a result, the climate of locations near an ocean is **affected.** As shown in **Figure 19,** the average daytime temperature in southern California generally increases as one moves east from coastal areas to nearby inland areas. At night, temperatures along the coast are usually higher than inland.

Ocean Currents

Ocean currents help to redistribute the Sun's energy on Earth in the form of heat. Ocean currents that move water away from the equator, such as the Gulf Stream, carry heat to higher latitudes. Currents moving toward lower latitudes, such as the California Current, replace warm water from the lower latitudes with cold water from higher latitudes.

This redistribution of heat can be seen when comparing the average winter temperatures of Great Britain to those of Labrador, Canada. These two locations are found at the same latitude. However, due to the warm-water influence of the Gulf Stream, the average winter temperature in Great Britain is 15–20°C warmer than that of Labrador.

The California Current is a cold-water current that flows past the coast of California. The water temperature of the California current remains stable year-round. This helps keep temperatures along the coast from rising too high in the summer and from dipping too low in the winter.

 Reading Check How do ocean currents influence climate?

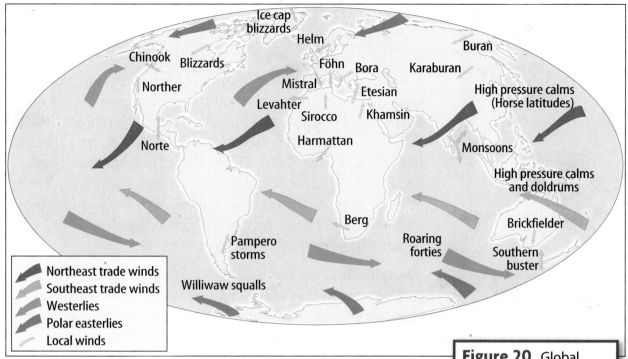

Figure 20 Global winds include the trade winds, the westerlies, and the polar easterlies.

Describe In which direction do the westerlies blow?

Prevailing Winds

A prevailing wind is a wind that blows from one direction most of the time. **Figure 20** shows Earth's major prevailing winds, such as the trade winds and polar easterlies. Prevailing winds significantly influence climate by moving air, heat, and water vapor around Earth. Climate in many parts of the United States is affected by the prevailing westerlies.

In some places, wind direction varies seasonally or is deflected by local topography. For example, the prevailing wind in Sacramento is from the south—due to the orientation of the Sacramento valley. Sometimes a local seasonal wind, called a *California norther,* blows from the north.

Human Influences on Climate

Some scientists believe that human activities can affect the climate of our planet and therefore have important consequences in our lives. Recall that the burning of fossil fuels releases greenhouse gases. An increase in the concentration of greenhouse gases could lead to global warming. If the average surface temperature of Earth increases, scientists hypothesize that changes in global climate could occur. This means that some regions might experience large increases in temperature, while other regions could experience only small increases, or even slight decreases. This could result in more droughts and floods in various regions.

 How might human activities influence climate?

Describing Climate

Climate is usually described in terms of the average weather conditions of an area over a long period of time. Conditions such as temperature, precipitation, and winds are taken into account. Of the many climate regions found on Earth, most of California has either a mediterranean climate or a highland climate. Factors that influence climate include latitude, altitude, distance from a large body of water, and presence or absence of mountain barriers. The climate of the coastal areas of California is influenced by the California Current. Inland areas may be influenced by altitude or mountain barriers.

LESSON 3 Review

Summarize

Create your own lesson summary as you design a **study web**.

1. **Write** the lesson title, number, and page numbers at the top of a sheet of paper.

2. **Scan** the lesson to find the **red** main headings.

3. **Organize** these headings clockwise on branches around the lesson title.

4. **Review** the information under each **red** heading to design a branch for each **blue** subheading.

5. **List** 2–3 details, key terms, and definitions from each **blue** subheading on branches extending from the main heading branches.

 ELA6: R 2.4

Standards Check

Using Vocabulary

1. In your own words, write a definition for *climate*. Then write a sentence using the word. **4.e**

Complete the sentence using the correct term.

   highland   mediterranean

2. A climate characterized by mild, wet winters and hot, dry summers is a _____ climate. **4.e**

Understanding Main Ideas

3. Which statement is true?

   A. As altitude increases, temperature decreases. **4.d**

   B. As latitude increases, temperature increases.

   C. The polar easterlies blow from west to east.

   D. The westerlies blow from east to west.

4. **List** some of the factors that influence the climate of a region. **4.e**

5. **Compare and Contrast** Copy and fill in the graphic organizer below to compare and contrast mediterranean climates and highland climates. **4.e**

Climate	Similarities	Differences
Mediterranean		
Highland		

6. **Explain** why the latitude of a location is an important factor in determining its climate. **4.e**

Applying Science

7. **Analyze** ways in which human activities can alter the climate. **4.e**

 **Science** nline

For more practice, visit **Standards Check** at ca6.msscience.com.

Applying Math

Finding Precipitation Averages

Meteorologists collect precipitation data daily. The daily amounts are then averaged into monthly and annual amounts. The table shows the daily precipitation for the city of Orick, California, on the northern California coast, for the month of November 2005.

Example

Find the average, or mean, amount of precipitation for this area for the entire month of November.

What you know: daily precipitation for 30 days

What you need to find: average amount of precipitation for the data set

1 **First find the sum of the daily precipitation amounts.**

$0.9 + 0.0 + 2.3 + 0.8 + 1.0 + 9.2 + 1.2 + 0.0 + 0.0 + 0.0 + 0.5 + 0.0 + 2.2 + 0.5 + 0.0 + 0.0 + 0.0 + 0.0 + 0.0 + 0.0 + 0.0 + 0.0 + 0.0 + 1.0 + 3.7 + 0.1 + 0.0 + 2.4 + 0.6 + 0.8 = 27.2$ cm

2 **Then divide by the number of data points in the data set.** There are 30 days in November or 30 data points, so divide 27.2 by 30 to find the average.

$27.2 \div 30 = 0.906666$ or about 0.9 cm

Answer: The average precipitation in Orick, California for November is 0.9 cm.

Date	Precipitation (cm)
11/01	0.9
11/02	0.0
11/03	2.3
11/04	0.8
11/05	1.0
11/06	9.2
11/07	1.2
11/08	0.0
11/09	0.0
11/10	0.0
11/11	0.5
11/12	0.0
11/13	2.2
11/14	0.5
11/15	0.0
11/16	0.0
11/17	0.0
11/18	0.0
11/19	0.0
11/20	0.0
11/21	0.0
11/22	0.0
11/23	0.0
11/24	1.0
11/25	3.7
11/26	0.1
11/27	0.0
11/28	2.4
11/29	0.6
11/30	0.8

Practice Problems

1. Find the average amount of precipitation in Orick, California from November 1 to November 15.

2. For the month of July 2005 in the same location, each day there was 0.0 cm of precipitation. What is the average amount of precipitation for this month?

ScienceOnline
For more math practice, visit **Math Practice** at ca6.msscience.com.

Reading Guide

What *You'll Learn*

▶ **Describe** the climate of California, its seasonal changes, and the effect of local wind systems.

▶ **Explain** how rain shadows are developed.

▶ **Compare** sea breezes and land breezes.

▶ **Compare** mountain breezes and valley breezes.

Why *It's Important*

California has several climate regions and varying weather as a result.

Vocabulary

rain shadow
sea breeze
land breeze
valley breeze
mountain breeze
Santa Ana wind

Review Vocabulary

California Current: wide, slow-moving, cold-water current that flows southward along the California coast (p. 448)

California Climate and Local Weather Patterns

Main Idea California's climate is primarily mediterranean and highland.

Real-World Reading Connection You have probably heard comments such as "it never rains in southern California," and "sunny California enjoys great climate." Think about the climate where you live. Do you agree with these statements?

Mediterranean and Highland Climates

Areas of California with a mediterranean climate typically experience mild, rainy winters and hot, dry summers. Many plant species, called chaparral, shown in **Figure 21,** are so well adapted to this type of climate that it is sometimes referred to as a chaparral climate.

Locations in California that are at high altitudes have highland climates, shown in **Figure 21.** The temperatures in these regions are lower than other regions at lower altitudes and the same latitudes. Precipitation is generally greater in highland areas than in locations at lower altitudes.

Figure 21 A mediterranean climate is drier and warmer than a highland climate region.

Mediterranean Climate

Highland Climate

Coastal Fog

Valley Fog

Figure 22 Valley fog, also called ground fog, is shown in the Great Central Valley. It forms during the late fall and winter when moisture in the air is cooled to the dew point by the cold ground below.

Seasonal Changes in California

Although winters in California are mild, there is still a contrast between summer and winter. In California, this contrast is best described in terms of precipitation. Instead of referring to seasons as hot and cold, the terms *rainy* and *dry* are often used. Recall that dry summers and wet winters are characteristic of mediterranean climates. California's dry season is also referred to as the fire season. Less precipitation occurs in summer due to the presence of an offshore high-pressure system referred to as the Pacific High. California's rainy season is typically from November to March.

 What changes in weather occur between summer and winter in a mediterranean climate?

California's Fog Belts

Much of the California coast is known for the presence of fog. During summer, warm air masses accumulate in offshore locations far from the coast. These warm air masses contain a lot of water vapor, or moisture. The westerlies transport the air masses and the moisture toward California's coast. When warm, moist air crosses over cold water in the California Current, the moisture cools to the dew point and condenses to coastal fog, as shown in **Figure 22.**

This fog is essential to the survival of the redwoods on the northern coast of California. Coastal redwoods, shown in **Figure 23,** are able to survive California's dry summers by taking in up to 40 percent of the water they need from the fog that rolls in every day during the summer months. The fog moisture collects on the needles and branches, drips to the ground, and is absorbed by the tree's shallow root system.

Figure 23 Fog drip is moisture that condenses on plants, then drips onto the ground.

Explain Why is fog so important to the survival of coastal redwoods?

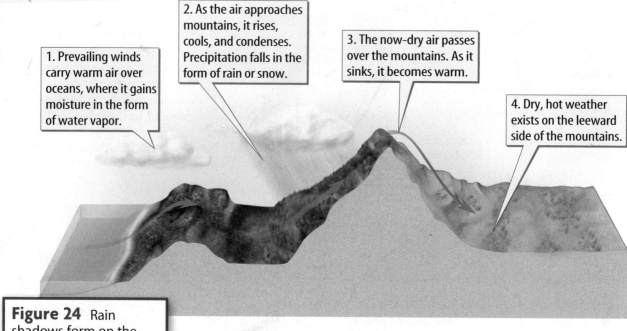

1. Prevailing winds carry warm air over oceans, where it gains moisture in the form of water vapor.

2. As the air approaches mountains, it rises, cools, and condenses. Precipitation falls in the form of rain or snow.

3. The now-dry air passes over the mountains. As it sinks, it becomes warm.

4. Dry, hot weather exists on the leeward side of the mountains.

Figure 24 Rain shadows form on the downwind slope of a mountain.

Infer Why don't rain shadows form on the upwind slope of mountains?

ACADEMIC VOCABULARY
accumulate
(ah KYEW myuh layt)
(*verb*) to increase slowly in quantity or number
Jennie accumulated many awards over the years she ran for the track team.

California's Rain Shadows

There are areas in California where rainfall is low. One reason for this is the presence of rain shadows. An area of low rainfall on the lee side, or downwind slope, of a mountain is called a **rain shadow**. **Figure 24** shows how a rain shadow forms. As the westerlies bring warm air over the ocean, the air **accumulates** water vapor as it passes over the ocean. As the warm, moist air travels inland across California, it can run into mountains, such as the Coast, Cascade, and Sierra Nevada mountain ranges. When the air mass runs into the mountains, it rises. Recall that when air rises, it cools down. As the air cools, the water vapor it holds condenses, and eventually precipitation takes place. This precipitation usually occurs before the air mass gets to the other side of the mountains, as shown in **Figure 24.** After the air mass crosses over the mountains, it will start to sink. As the air mass sinks, it warms up. The relative humidity of the air has decreased due to the precipitation and because it is warmer. The air that reaches the other side, or downwind slope, of the mountain is now dry. As a result, the area on the downwind slope of the mountain is a rain shadow that experiences low rainfall.

Local Winds

Recall that winds blow from regions of high pressure to regions of low pressure. Local changes in pressure can give rise to local winds such as sea breezes, land breezes, valley breezes, and mountain breezes.

Sea Breezes and Land Breezes

A wind that blows from the sea to the land is called a **sea breeze.** A wind that blows from the land to the sea is called a **land breeze.** Sea breezes and land breezes are caused by atmospheric convection currents. Solar radiation heats the land and the ocean. However, the temperature of the land changes more rapidly than the temperature of the ocean. When the air above land is heated, it rises and creates an area of low pressure, as shown in **Figure 25.** Sea breezes occur when the air pressure over the ocean is higher than the air pressure over the land. At night, the land cools faster than the ocean. In this case, air sinks over the land producing an area of high pressure. As a result, a land breeze blows from the land to the ocean, as shown in **Figure 25.**

Valley Breezes and Mountain Breezes

As shown in **Figure 26, valley breezes** blow upward from the valley along the mountain slopes. Usually valley breezes occur during daytime hours when the land is hot. **Mountain breezes** flow downward from mountains, as shown in **Figure 26.** They usually occur at night in wide valleys that were exposed to the Sun during the day. After sunset, air along the exposed mountain slopes cools more rapidly than the air in the valleys. Then the cooler and denser mountain air sinks into the valley.

 **Explain** why valley breezes and mountain breezes are considered convection currents.

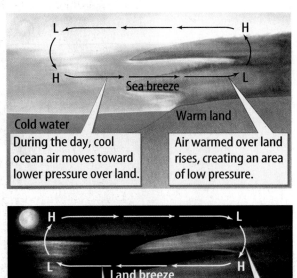

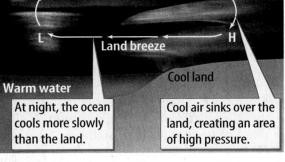

Cold water — During the day, cool ocean air moves toward lower pressure over land.

Warm land — Air warmed over land rises, creating an area of low pressure.

Warm water — At night, the ocean cools more slowly than the land.

Cool land — Cool air sinks over the land, creating an area of high pressure.

Figure 25 Sea breezes blow from the ocean to the land during the daytime. Land breezes are more common during nighttime.

Figure 26 Valley breezes and mountain breezes occur as air changes temperature.

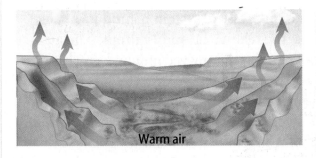

Warm air

Cool air

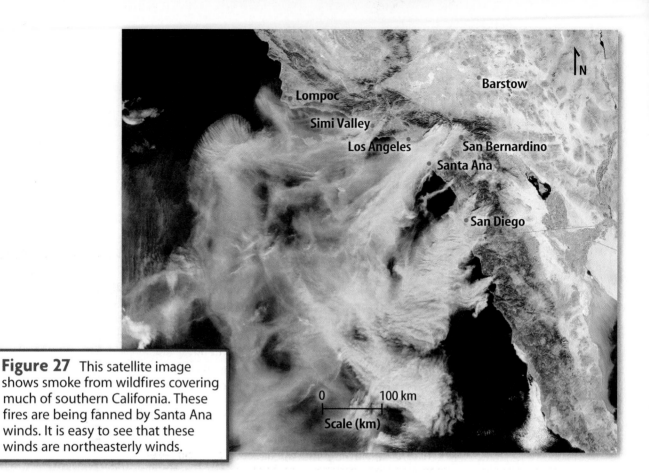

Figure 27 This satellite image shows smoke from wildfires covering much of southern California. These fires are being fanned by Santa Ana winds. It is easy to see that these winds are northeasterly winds.

Map labels: Lompoc, Barstow, Simi Valley, Los Angeles, San Bernardino, Santa Ana, San Diego, N

Scale: 0 — 100 km, Scale (km)

Santa Ana Winds

In southern California, hot, dry winds that blow from the east or northeast and continue toward the coast are often called **Santa Ana winds.** They usually blow from the mountain passes and canyons of southern California across the Los Angeles basin. Santa Ana winds, which are shown blowing smoke from wildfires in **Figure 27,** are common in the fall and winter months.

Although Santa Ana winds are sometimes fiercely hot and gusty winds, they do not start out that way. These winds develop from a cool, dense, and dry high-pressure air mass in the Great Basin. The air then flows toward the coastal areas through mountain passes and canyons. The dense air is compressed and warmed as it is forced downslope, through narrow canyons, by the high pressure system behind it. As the air temperature rises, the relative humidity drops. The air that started out cool and dry is now warmer and much drier.

The wind picks up speed and temperature as it is deflected and channeled through the canyons. One of theses canyons is the Santa Ana Canyon from which the wind gets its name. By the time the Santa Ana winds reach coastal areas, such as Los Angeles, they have high speeds and high temperatures.

Fires Santa Ana winds can be hazardous during fire season. Since the winds are hot, dry, and moving fast, they can dry out vegetation and increase the danger of fire. The potential for wildfires increases during drought years when winter rains are lower than normal. Other local winds influence the start and spread of fires during fire season. The names and descriptions of these winds are listed in **Table 1** and shown in the illustration below.

 Visual Check **Table 1** From which direction do the Mono winds blow?

Concepts In Motion
Interactive Table organize information about California winds at ca6.msscience.com.

Table 1 Significant Local California Winds During Fire Season	
Wind Name	**Characteristics and Facts**
1 Santa Ana wind	• blows from Great Basin into Los Angeles Basin and San Fernando Valley • named for blowing through Santa Ana Canyon
2 Mono wind	• blows from the Sierra Nevadas into the Great Central Valley • named for blowing from the direction of Mono Lake
3 California norther	• blows from the Siskiyou Mountains in northern California into the northern half of the Great Central Valley • brought temperature in Red Bluff to 119°F, August 8, 1978
4 Diablo wind	• blows from Mt. Diablo into San Francisco and Oakland • The costliest fire on record was in the Oakland hills, in October 1991.
5 Sundowner	• blows from Santa Ynez Mountains into Santa Barbara • named for blowing during late afternoon or early evening at about sundown • In the past, Goleta (near Santa Barbara) held the U.S. temperature record of 133°F from a Sundowner heat burst event on June 17, 1859.

California's Climate

California has two main climate regions—mediterranean and highland. Differences in precipitation and temperature define these two regions. In a Mediterranean climate, seasons are often referred to as *rainy* and *dry*. Some areas of California, particularly along the coast, experience fog many days of the year. Coastal areas may also experience sea breezes and land breezes during the day and at night. In the mountains, valley breezes and mountain breezes blow up and down the sides of mountains. Other areas of California, on the downwind slope of mountain ranges, are in rain shadows and experience low amounts of rainfall. Different winds that bring hot, dry air to an area can influence temperatures and increase the risk of fires throughout California.

LESSON 4 Review

Summarize

Create your own lesson summary as you write a script for a **television news report.**

1. **Review** the text after the **red** main headings and write one sentence about each. These are the headlines of your broadcast.

2. **Review** the text and write 2–3 sentences about each **blue** subheading. These sentences should tell *who, what, when, where,* and *why* information about each **red** heading.

3. **Include** descriptive details in your report, such as names of reporters and local places and events.

4. **Present** your news report to other classmates alone or with a team.

 **ELA6:** LS 1.4

Standards Check

Using Vocabulary

1. Distinguish between *rain shadow* and *mountain breeze.* **4.d**

2. In your own words, write the definition for *Santa Ana wind.* **4.e**

Understanding Main Ideas

3. Which best describes the difference between summer and winter seasons in California?

   A. hot and cold **4.e**
   B. foggy and wet
   C. dry and rainy
   D. fire and snow

4. **Outline** the main characteristics of California's climates. **4.e**

5. **Explain** how fog is formed in California's coastal regions. **4.e**

6. **Compare** the formation of sea breezes and land breezes. **4.d**

Applying Science

7. **Generate** a diagram that explains the difference between mountain breezes and valley breezes. **4.d**

8. **Sequence** Draw a graphic organizer like the one below. Fill in the boxes of the organizer with the correct order of steps of how a rain shadow forms. **4.e**

Science nline

For more practice, visit **Standards Check** at ca6.msscience.com.

MiniLab

How do Santa Ana winds move?

The Santa Ana winds in southwestern California are hot and dry. What can you learn about the path of the Santa Ana winds?

Procedure

1. Build a **clay** plateau in one corner of a **clear plastic tub.** This represents the Great Basin.

2. Make a clay dam around the perimeter of this high plateau. This represents the Sierra Nevada, San Gabriel, and other mountains.

3. Put some notches in the clay dam to represent Cajon Pass, the Santa Ana Canyon, and other mountain passes.

4. Fill the tub with **vegetable oil** so it covers the highest mountain peak. This represents low density air.

5. Pour **colored water** into the high plateau. This represents high density air from the Mojave Desert (Great Basin).

6. Pour the colored water until it spills over the mountain passes and into the basin below. This is the Los Angeles basin.

Analysis

1. **Evaluate** How does the model you made show the flow of Santa Ana winds from the high plateau into the Los Angeles basin?

2. **Infer** Santa Ana winds are described as hot, dry, dusty winds. What effect do these winds have on southern California?

 Science Content Standards

4.e Students know differences in pressure, heat, air movement, and humidity result in changes of weather.

Use the Internet:
How diverse is the natural landscape of California?

Materials

computer with internet access
references on California's bioregions

Science Content Standards

4.e Students know differences in pressure, heat, air movement, and humidity result in changes in weather.
7.d Communicate the steps and results from an investigation in written reports and oral presentations.
7.h Identify changes in natural phenomena over time without manipulating the phenomena (e.g., a tree limb, a grove of trees, a stream, a hillslope).

Problem

The state of California has within its borders deserts, redwood forests, prairies, wetlands, and many other types of natural landscapes. These result from the different kinds of climate that can be found within the state and encompass a variety of factors such as water, temperatures, and compositions of soil. The diversity of life that thrives in the state emerges from the varied weather patterns that can be tracked throughout California.

Hypothesis

You will be assigned a bioregion to research. **Write** a hypothesis that explains which factors cause the local weather patterns and climatic conditions that are considered normal for the region.

Collect Data

1. Locate your bioregion in the figure below. Use the Internet or library references to find information about the bioregion.

2. Research information about the weather factors that affect the area, including how much precipitation the area receives, and the annual temperature changes.

3. Research information about the climate conditions of the bioregion and whether they have changed over time.

4. Find out information about the type of habitats that are found in the bioregion and the type of plants and animals that live in those habitats.

5. Identify the soil type in the bioregion.

6. Choose one location in your bioregion and track any changes in the weather patterns for one week.

7. Construct a data table and post the results of your research at ca6.msscience.com.

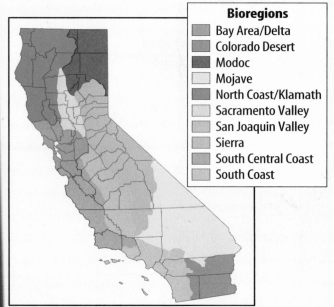

Bioregions
- Bay Area/Delta
- Colorado Desert
- Modoc
- Mojave
- North Coast/Klamath
- Sacramento Valley
- San Joaquin Valley
- Sierra
- South Central Coast
- South Coast

Analyze and Conclude

1. **Identify** What are the main characteristics of the climate of your bioregion?
2. **Describe** What plants and animals thrive because of these climatic factors?
3. **Identify** the weather factors that are part of the weather and climatic patterns in your region.
4. **Explain** How is the soil affected by the climate in your region?
5. **Evaluate** How do these local weather patterns fit into the big picture of climate for your bioregion?

Error Analysis

Each bioregion borders on another. Living things and different weather factors merge at those borders. When you researched, were you careful to focus on those factors and living things that were common especially to your region?

Coast Redwood trees, Redwood National Park, California

Communicate

WRITING in Science  ELA6: LS 2.2

Write a Report Present your findings to your class in the form of a report including graphs, tables, maps, photographs, and supporting documentation. Describe any changes that occured during your period of study.

Joshua trees at sunrise, Joshua Tree National Park, California

Real World Science

Chaparral

The climate was colder when Spanish explorers were spotted off the California coast in 1542. At that time, the Chumash house, or *'ap*, kept local natives warm. Although the Chumash relied heavily on the sea for food, they fished only seasonally and survived the winter months on sun-dried fish and acorn meal. The most important acorns were from the Coast Live Oak. The oaks reminded the Spanish of a similar scrub oak from home called *chaparro*. The word *chaparral* means thicket of chaparros, or place where chaparros grow.

Research the origin of the word *chaparral*. **Write** a brief history of the interactions between the Spanish and the native Chumash tribes.

ELA6: W 1.2

Mediterranean Climate

The mediterranean climate is often called a chaparral climate. Approximately 1,700 plant species make up the chaparral community in the Santa Monica Mountains National Recreation Area. This area represents one of the few mediterranean type ecosystems remaining in North America. Park rangers educate the public about the area by providing interpretive programs to approximately 500,000 visitors each year.

Visit **Careers** at ca6.msscience.com to learn more about the Santa Monica Mountains National Recreation Area. **Research** the area and **present** a program as though you were one of the park rangers.

SUPER SCOOPER

Sometimes, the summers in a chaparral climate are so hot that there are frequent fires and dry spells. The Super Scooper is able to skim the surface of a lake or an ocean, scoop up water, and then carry the water to the scene of a fire. The capability of landing on water, combined with the ability to reach remote locations, makes this aircraft ideally suited for fighting wildfires. This CL-214 Super Scooper, based in Van Nuys, California, was able to make round trips from a nearby lake to a fire in Simi Valley in approximately fifteen minutes.

Research to learn more about the Super Scooper. **Create** a table or a graph that compares the advantages and disadvantages of having a Super Scooper help fight fires in your area.

Science & Society

Fire's Role

Many chaparral plants are adapted to natural fires, which occur once every 10 to 30 or more years. Some studies suggest that fire actually benefits the chaparral community. However, recent studies suggest that the role of fire in these communities may be of recent origin. For example, during the 1700s, the Chumash were known to burn sage scrub and grasslands, but not chaparral. Today, the complete role of fires in Californian chaparral is still poorly understood. What is known is that the frequency of fires increases as human population continues to expand into the chaparral community. Without chaparral, many hillsides become vulnerable to erosion, mudslides, and landslides.

Visit Society at **ca6.msscience.com** to learn about some of the ways that fire affects chaparral. **List** some of the benefits and hazards of frequent fires in California chaparral.

The BIG Idea Many factors affect weather and climate.

Lesson 1 Weather
4.a, 4.e

Main Idea Weather describes the atmospheric conditions of a place at a certain time.

- Weather describes atmospheric conditions and the changes that take place over the short term.
- Condensation takes place when air reaches its dew point. This leads to the formation of clouds and fog.
- Water moves through the hydrosphere in a cycle driven by the energy of the Sun.

- **dew point** (p. 470)
- **humidity** (p. 469)
- **precipitation** (p. 472)
- **relative humidity** (p. 470)
- **water cycle** (p. 472)
- **weather** (p. 468)

Lesson 2 Weather Patterns
2.d, 4.e

Main Idea Several factors drive changes in weather.

- Several factors drive the changes in weather, such as differences in pressure, heat, air movement, and humidity.
- Short-term and long-term cycles can affect weather.
- Droughts and floods result from severe weather conditions.

- **air mass** (p. 475)
- **cold front** (p. 476)
- **drought** (p. 480)
- **flash flood** (p. 481)
- **season** (p. 478)
- **warm front** (p. 476)

Lesson 3 Climate
4.d, 4.e

Main Idea Climate is often defined by annual temperatures and precipitation amounts.

- Climate is the long-term average of the weather patterns of an area.
- Climate regions are often defined based on the average temperature and annual amounts of precipitation.

- **climate** (p. 483)
- **highland climate** (p. 484)
- **mediterranean climate** (p. 484)

Lesson 4 California Climate and Local Weather Patterns
4.d, 4.e, 7.d, 7.h

Main Idea California's climate is primarily mediterranean and highland.

- California's climate is mostly mediterranean and highland.
- Seasons in California are described best as rainy and dry.
- Rain shadows are responsible for low precipitation in some areas of California.

- **land breeze** (p. 493)
- **mountain breeze** (p. 493)
- **rain shadow** (p. 492)
- **Santa Ana wind** (p. 494)
- **sea breeze** (p. 493)
- **valley breeze** (p. 493)

STUDY TO GO Download quizzes, key terms, and flash cards from ca6.msscience.com.

Linking Vocabulary and Main Ideas

Use vocabulary terms from page 502 to complete this concept map.

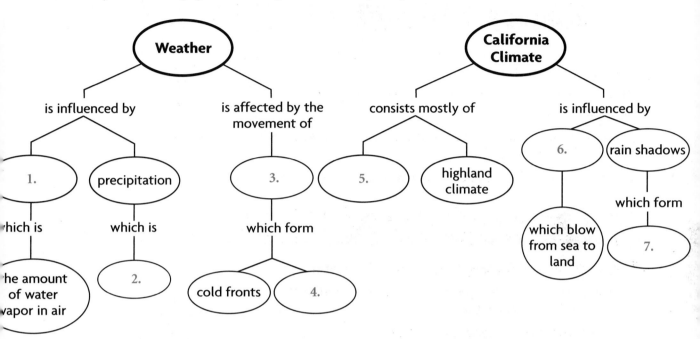

Using Vocabulary

Complete each statement using a word from the vocabulary list.

8. Water, in liquid or solid form, that falls from the atmosphere is _____ .

9. A(n) _____ is a body of air that has similar features.

10. When lighter, warmer air moves over heavier, colder air a(n) _____ is formed.

11. _____ is the long-term average of the weather patterns of an area.

12. The area of low rainfall on the downwind slope of a mountain is called a(n) _____.

13. A(n) _____ flows from the sea to land.

14. A(n) _____ flows from a mountain top downward.

Science Online

Visit ca6.msscience.com for:
▶ Vocabulary PuzzleMaker
▶ Vocabulary eFlashcards
▶ Multilingual Glossary

Understanding Main Ideas

Choose the word or phrase that best answers the question.

1. Which are winds that blow from the west to the east?
   A. easterlies   4.e
   B. westerlies
   C. north winds
   D. south winds

2. Which process occurs when air reaches its dew point?
   A. condensation   4.e
   B. evaporation
   C. A warm front is formed.
   D. A cold front is formed.

3. Which best describes the major climate regions in California?
   A. highland, tropical   4.e
   B. mediterranean, highland
   C. mediterranean, tropical
   D. temperate, mediterranean

4. The map below shows the formation of a weather system.

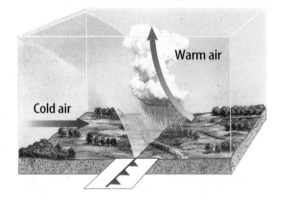

The formation of which is shown above?
A. a cold front   4.e
B. a warm front
C. a low-pressure system
D. a high-pressure system

5. Which statement is true?
   A. In a low-pressure system, air moves downward.   4.e
   B. In a high-pressure system, air moves downward.
   C. Low-pressure systems are associated with fair weather.
   D. High-pressure systems are associated with rainy weather.

6. Which does not influence the climate of a region?
   A. latitude   4.e
   B. altitude
   C. mountain barriers
   D. oxygen in the atmosphere

7. Which ocean current has the greatest influence on the climate of California?
   A. California Current   4.e
   B. Gulf Stream
   C. Equatorial Current
   D. Pacific Current

8. The figure below shows the formation of a breeze.

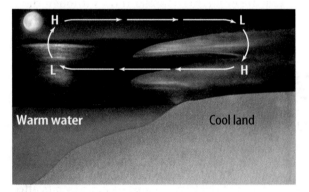

Which type of breeze is shown above?
A. sea breeze   4.e
B. mountain breeze
C. land breeze
D. valley breeze

Science Online Standards Review ca6.msscience.com

Applying Science

9. **Predict** if the relative humidity of Location A will be higher, lower, or equal to the relative humidity of Location B. In both locations, the amount of water vapor in air is the same, but Location A has a higher temperature. **4.e**

10. **Identify** the type of precipitation shown below. Explain how a cold front is involved in the formation of this type of precipitation. **4.e**

Cloud droplets

Warm

Warm

11. **Construct** your own diagram of the water cycle. Pick a body of water near you and a specific type of precipitation. Include arrows showing the direction of water flow and label the processes that are occurring as water moves through the hydrosphere. **Describe** how the Sun drives the water cycle. **4.a**

12. **Analyze** the factors that determine the differences in temperature observed with latitude, seasons, and altitude. **4.e**

13. **Examine** the reasons why the distribution of land and water influence the climate of a region. **4.d**

14. **Compare** how El Niño and La Niña events affect the weather. **4.e**

15. **Deduce** Suppose you are a surfer. Using your knowledge of sea breezes and land breezes, determine the best time to surf. **4.e**

16. **Analyze** the reasons why Santa Ana winds are dangerous for southern California. **4.e**

WRITING in Science

17. **Explain** Write a paragraph explaining whether you would expect to encounter higher or lower temperatures once you reach the summit of a mountain. **ELA6: W 1.2**

Cumulative Review

18. **Explain** how the Coriolis effect affects air and water as it moves around Earth's surface. **4.d**

19. **Describe** how ocean currents can affect the climate of a region. **4.d**

Applying Math

Use the table on page 489 to answer questions 20 and 21.

20. Find the average amount of precipitation in Orick, California, November 1–7, 2005. **MA6: SP 1.1**

21. Find the average amount of precipitation in Orick, California, November 15–21, 2005. **MA6: SP 1.1**

22. The table below shows the high and low temperatures for five days in April.

High and Low Temperatures		
Day	**High (°C)**	**Low (°C)**
Monday	22	14
Tuesday	20	12
Wednesday	23	15
Thursday	22	16
Friday	19	11

What was Wednesday's mean temperature? **MA6: SP 1.1**

Use the graph below to answer questions 1 and 2.

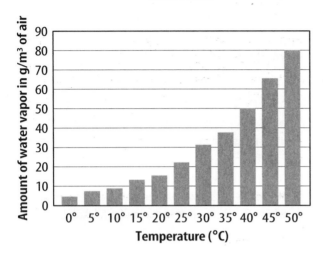

Dew Point

Amount of water vapor in g/m³ of air vs. **Temperature (°C)**

1 What amount of water vapor is in saturated air at a temperature of 45°C?

 A 55 g/m^3

 B 60 g/m^3

 C 65 g/m^3

 D 70 g/m^3 〔4.e〕

2 What conclusion about temperature and water vapor can be made based on data presented in the graph?

 A As the temperature decreases, the air can hold a greater amount of water vapor.

 B As the temperature increases, the air can hold a greater amount of water vapor.

 C As the temperature increases, the air holds a lesser amount of water vapor.

 D Temperature has no effect on the amount of water vapor the air can hold. 〔4.e〕

3 Which instrument is used to measure air pressure?

 A anemometer

 B barometer

 C rain gauge

 D thermometer 〔4.e〕

4 What process occurs when water vapor changes to a liquid?

 A condensation

 B conduction

 C evaporation

 D precipitation 〔4.e〕

5 The water cycle is illustrated in the image below.

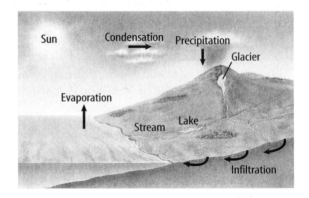

Which process transfers water from the atmosphere to Earth's surface?

 A condensation

 B evaporation

 C infiltration

 D precipitation 〔4.a〕

Science Online Standards Assessment ca6.msscience.com

6 Which causes sea and land breezes in California and other coastal states?

A conduction currents

B convection currents

C high-pressure systems

D low-pressure systems   **4.d**

Use the illustration below to answer questions 7 and 8.

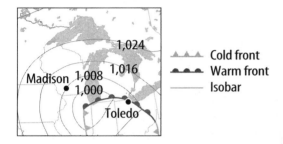

Cold front
Warm front
Isobar

7 What is the atmospheric pressure in the city of Madison, Wisconsin?

A 1,024 mb

B 1,016 mb

C 1,008 mb

D 1,000 mb   **4.e**

8 What type of front is near Toledo, Ohio?

A cold front

B occluded front

C stationary front

D warm front   **4.e**

9 Over a period of time, weather conditions were observed and recorded in a student's science journal. Some of the data from the journal is shown in the table below.

Weather Conditions on Monday			
Condition	**Morning**	**Afternoon**	**Evening**
Temperature	35	40	50
Humidity	low	high	high
Cloud cover	some	a lot	little
Rainfall	none	light rain	no rain

Which might have passed through the area while the student was recording data?

A a cold front

B a high-pressure system

C a low-pressure system

D a warm front   **4.e**

10 When building construction decreases vegetation that absorbs water runoff, it increases the effects of which event?

A earthquakes

B flash floods

C hurricanes

D tornados   **2.d**

11 Why is air temperature the warmest at tropical latitudes?

A Tropical latitudes receive the most solar radiation because there are no clouds.

B Tropical latitudes receive the most solar radiation because the Sun's angle is high.

C Tropical latitudes receive the least solar radiation because the Sun's angle is low.

D Tropical latitudes receive the least solar radiation because of heavy cloud cover.   **4.a**

Reading on Your Own...

Are you interested in learning more about the energy in Earth's atmosphere and oceans? If so, check out these great books.

Narrative Nonfiction

Nature's Fury: Eyewitness Reports of Natural Disasters, by Carole Vogel, uses eyewitness descriptions and newspaper excerpts to summarize 13 natural disasters that took place between 1871 and 1980. Included are the San Francisco earthquake, the Mount St. Helen's volcano eruption, and extreme weather. ***The content of this book is related to*** *Science Standard 6.2.*

Nonfiction

Global Warming: The Threat of Earth's Changing Climate, by Laurence Pringle, provides details on global warming, climatic changes, and the greenhouse effect. This book explains the potential future damage and offers solutions. The text is illustrated with numerous color photographs, diagrams, and charts. ***The content of this book is related to*** *Science Standard 6.3.*

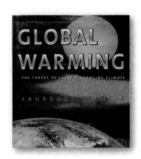

Nonfiction

Storms, by Seymour Simon, describes the atmospheric conditions that create thunderstorms, hailstorms, lightning, tornadoes, and hurricanes. This book tells how violent weather affects the environment and people. ***The content of this book is related to*** *Science Standard 6.4.*

Fiction

A Blizzard Year: Timmy's Almanac of the Seasons, by Gretel Ehrlich, features a year in Timmy's life while she works on her family's cattle ranch. She makes notes of the changes in the weather and how the family copes with the blizzards that could ruin their ranch. ***The content of this book is related to*** *Science Standard 6.4.*

Choose the word or phrase that best answers the question.

1. Which is the most abundant gas in the atmosphere?
   A. argon
   B. carbon dioxide
   C. oxygen
   D. nitrogen   **4.b**

2. Which are formed along subduction zones?
   A. continental shelves
   B. density currents
   C. mid-ocean ridges
   D. trenches   **1.a**

3. Which is the flattest feature of the ocean floor?
   A. abyssal plain
   B. continental rise
   C. continental slope
   D. mid-ocean ridge   **1.a**

4. The figure below shows the boundary between two air masses.

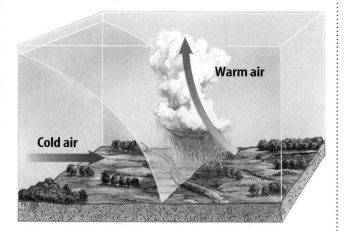

Cold air
Warm air

What forms when cold air moves toward warm air?
   A. cold front
   B. warm front
   C. high-pressure system
   D. low-pressure system   **4.e**

Write your responses on a sheet of paper.

5. **Analyze** why it is cooler at higher altitudes in the desert.   **4.e**

6. **Design an experiment** to find out how different surfaces, such as asphalt, soil, sand, and grass, absorb and reflect solar radiation.   **4.b**

7. **Discuss** why the inside of a parked car is hotter than the outdoor temperature on a sunny summer day.   **4.b**

8. **Analyze** At point A, a sound wave took 2 s to bounce off the ocean floor and reach the ship. At point B, it took 2.4 s. Which point is deeper? Support your answer with a labeled sketch.   **7.f**

9. **Predict** what would happen to global climates if the Sun emitted more energy.   **4.a**

10. **Explain** why air pressure decreases as altitude increases.   **4.e**

11. **Summarize** why sea breezes occur during the day but not at night.   **4.e**

12. **Design and complete** a comparison chart similar to the one below for the causes and effects of El Niño and La Niña.   **4.d**

El Niño	Both	La Niña

Field of Flowers California poppies cover this meadow in Antelope Valley, at the poppy reserve. This is the most consistent poppy bearing land in California.

West-Coast EVENTS

40,000–12,000 Years Ago
Scientists have described ancient organisms from fossils found in La Brea Tar Pits (California) and organized them according to what they ate and how they related to each other.

1861
First oil well in California is drilled by hand in Humboldt County.

1874–1892
John Muir writes about the natural beauty of northern California, leading to the establishment of Yosemite National Park.

A.D. 1   **1770**   **1840**   **1860**   **1880**   **190**

347
Chinese drill oil wells as deep as 240 m using bits attached to bamboo poles.

1770s
Jan Ingenhousz discovers how plants react to sunlight; this leads to understanding of photosynthesis.

August 1859
Edwin L. Drake drills first oil well in the United States in Titusville, Pennsylvania.

WORLD EVENTS

Science Online

To learn more about ecologists and their work, visit ca6.msscience.com.

Concepts In Motion

Interactive Time Line To learn more about these events and others, visit ca6.msscience.com.

1932
Third-largest producing oil field in the United States is discovered in Wilmington, California.

1980–1990
Wind farms are built in three California mountain passes, including near Palm Springs here shown.

1996–2005
Scientists climbing redwoods to measure their height discovered hemlock trees, huckleberry bushes, and small sea animals living in the tops of the redwoods.

920   **1940**   **1960**   **1980**   **2000**   **2020**

1931
Construction begins outside Las Vegas to dam the Colorado River. Hoover Dam was the largest dam at the time.

1957
First full-scale nuclear power plant in the world goes into service at Shippingport, Pennsylvania.

March 2005
Solar power station in Morocco projected to meet energy demands by 2008.

Ecological Roles

The BIG Idea

Living things interact with each other and with nonliving factors in ecosystems.

LESSON 1 4.a, 5.e, 7.c

Abiotic and Biotic Factors

Main Idea Living things and nonliving factors, such as air, water, sunlight, and soil, interact in Earth's ecosystems.

LESSON 2

5.c, 5.d, 5.e, 7.a, 7.b, 7.d

Organisms and Ecosystems

Main Idea Climates and the types of life they support define biomes on Earth. Ecological roles are the same in different biomes but may be filled by different species.

What's *so* special about this place?

This environment is located in the Sierra Nevada mountain range in California. It is characterized by plants that are specially adapted to survive a drought season. It also supports native animals, like these gray foxes that were once common throughout California, but are now isolated to these areas.

Science *Journal* Write three questions you have about this photo that you might like to explore further.

Launch Lab

How tangled is the life web?

How many things affect an animal? The sunshine and the air affect it. It may eat plants and animals. Think of an animal. Where does it live? What is that area like?

Procedure

1. Write the name of an animal in a circle in the center of a **blank sheet of paper.**

2. Write things in the environment that affect the animal every day in circles. Connect them to your animal.

3. Include interactions of the living and non-living factors that affect your organism.

Think About This

- **Determine** if the animal lives in water, soil, or trees.

- **Imagine** that a factor has suddenly disappeared from the animal's life. How would your animal survive without it?

 5.c, 7.h

Science Online

Visit ca6.msscience.com to:

▶ view Concepts in Motion

▶ explore Virtual Labs

▶ access content-related Web links

▶ take the Standards Check

 **FOLDABLES™ Study Organizer**

Abiotic Factors Make the following Foldable to identify the abiotic factors in an ecosystem.

▷ **STEP 1 Collect** three sheets of paper and layer them about 2 cm apart vertically. Keep the left edges even.

▷ **STEP 2 Fold** up the bottom edges of the paper to form five equal tabs. Crease the fold to hold the tabs in place.

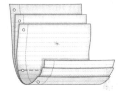

▷ **STEP 3 Staple** along the fold. **Label** as shown.

Abiotic Factors

 **Reading Skill** ELA6: R 2.4

Drawing Conclusions
As you read about the abiotic factors, write their names on the tabs. Include information about how each factor affects the number and types of organisms the environment can support.

Get Ready to Read

Make Inferences

①Learn It! When you make inferences, you draw conclusions that are not directly stated in the text. This means you "read between the lines." You interpret clues and draw upon prior knowledge. Authors rely on a reader's ability to infer because all the details are not always given.

②Practice It! Read the excerpt below and pay attention to highlighted words as you make inferences. Use this Think-Through chart to help you make inferences.

Human activities can impact **population size** and therefore **affect entire biomes.** For example, if people did not allow fires to burn in chaparrals, there would be no young pine trees. As older trees die, they would not be replaced. The animals that depend on the trees for food, shelter, and space would either die or move elsewhere.

—*from page 537*

Text	Question	Inferences
Human activities	What human activities?	Resource use?
Population size	How can human activities affect population size?	Death rates are high? Birth rates are low? Populations move?
affect entire biomes	How can it affect biomes?	Causes changes to abiotic factors? Biotic factors?

③Apply It! As you read this chapter, practice your skill at making inferences by making connections and asking questions.

Target Your Reading

Use this to focus on the main ideas as you read the chapter.

1 **Before you read** the chapter, respond to the statements below on your worksheet or on a numbered sheet of paper.

- Write an **A** if you **agree** with the statement.
- Write a **D** if you **disagree** with the statement.

2 **After you read** the chapter, look back to this page to see if you've changed your mind about any of the statements.

- If any of your answers changed, explain why.
- Change any false statements into true statements.
- Use your revised statements as a study guide.

Reading Tip

Sometimes you make inferences by using other reading skills, such as questioning and predicting.

Before You Read A or D	Statement	After You Read A or D
	1 A park in the city can be an ecosystem.	
	2 Animals, such as cats and dogs, use energy from the Sun to make food.	
	3 Earth is colder near the poles because it usually is cloudy.	
	4 Water moves through cycles.	
	5 Every organism in an ecosystem depends on other organisms.	
	6 Populations of organisms can grow without limits.	
	7 More species of organisms live in tropical rain forests because of the wet, warm climate.	
	8 There are six major zones on Earth that support different ecological communities.	
	9 There are many trees in the polar tundra.	
	10 Each species can fill many different roles in an ecosystem.	

Science Online

Print a worksheet of this page at ca6.msscience.com.

Science Content Standards

4.a Students know the sun is the major source of energy for phenomena on Earth's surface; it powers winds, ocean currents, and the water cycle.

5.e Students know the number and types of organisms an ecosystem can support depends on the resources available and on abiotic factors, such as quantities of light and water, a range of temperatures, and soil composition.

7.c Construct appropriate graphs from data and develop qualitative statements about the relationships between variables.

Reading Guide

What *You'll Learn*

▶ **Explain** how abiotic factors including light, temperature, air, water, and soil influence living things.

▶ **Describe** how systems depend on biotic and abiotic factors.

Why *It's Important*

Changes in living and nonliving factors can affect ecosystems.

Vocabulary

ecosystem
abiotic factor
humus
biotic factor
species
population
community
limiting factor

Review Vocabulary

climate: the long-term average of the weather patterns of an area (p. 483)

Abiotic and Biotic Factors

Main Idea Living things and nonliving factors, such as air, water, sunlight, and soil, interact in Earth's ecosystems.

Real-World Reading Connection What do you think of when you hear the word *ecosystem?* Perhaps you imagine a tropical rain forest with monkeys swinging from vine to vine, or California's Mojave Desert, with coyotes searching for food. These are examples of ecosystems, but ecosystems can be much smaller and closer to home.

What is an ecosystem?

An **ecosystem** consists of living things, called organisms, and the physical place they live. There are many types of ecosystems on Earth—coral reefs, woodlands, and ponds. The patch of weeds shown in **Figure 1** may not seem like an ecosystem. But if you look closely, you may find insects eating the plants or using them for shelter. Robins and other birds may be searching for earthworms and other food. Organisms in an ecosystem interact with each other as well as with nonliving parts of their environment.

Figure 1 A patch of weeds is an ecosystem.
Identify the interactions that may exist in this ecosystem

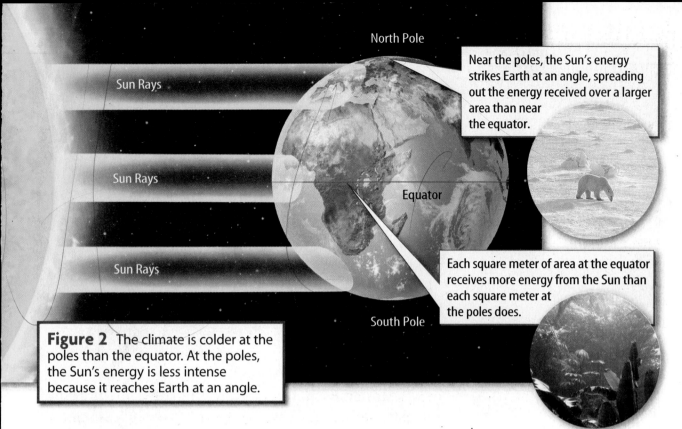

Near the poles, the Sun's energy strikes Earth at an angle, spreading out the energy received over a larger area than near the equator.

Each square meter of area at the equator receives more energy from the Sun than each square meter at the poles does.

Figure 2 The climate is colder at the poles than the equator. At the poles, the Sun's energy is less intense because it reaches Earth at an angle.

Abiotic Factors

Abiotic factors are the nonliving parts of an ecosystem. Suppose you want to grow a plant in your classroom. You could plant a seed in a pot of damp soil and place the pot on a sunny windowsill. What are the factors affecting this plant? Soil, water, and energy from the Sun are some of the abiotic factors that help control the growth of green plants.

The Sun

Most life on Earth depends on energy from the Sun, an important abiotic factor. Green plants use energy from the Sun to make food. Other animals eat those green plants. Still other animals eat those animals. Most organisms depend on green plants. The Sun's energy also controls many other abiotic factors in the environment.

Recall from Chapter 9 that areas of Earth receive different amounts of radiation from the Sun because Earth's surface is curved. **Figure 2** shows how rays of sunlight hit Earth at a low angle at the equator. This causes the land areas along the equator to be among the warmest places on Earth, and life is abundant. The poles are colder because the Sun's rays hit Earth at an angle, spreading the heat. There are fewer organisms at the poles due to the harsher climate.

 **Reading Check** Why are the poles colder than areas near the equator?

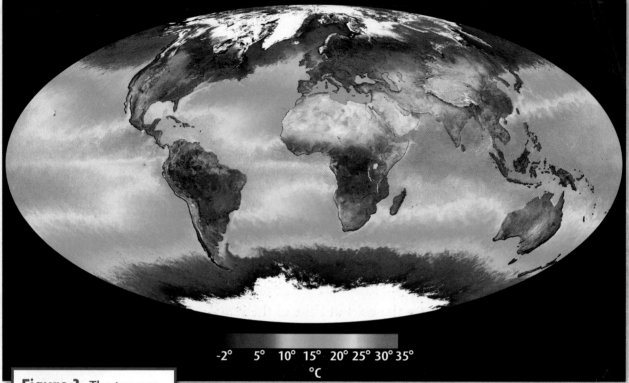

-2°   5° 10° 15° 20° 25° 30° 35°
°C

Figure 3 The temperature of the water at the surface of Earth's oceans ranges from 30°C near the equator to 0°C at the poles.

Temperature Without the Sun, Earth would be a frozen wasteland. The Sun's energy warms Earth and makes it possible for humans and other organisms to survive. The uneven warming of Earth's surface from the Sun's rays causes great temperature differences. The Sun influences temperature, an abiotic factor in ecosystems. **Figure 3** shows the temperatures of Earth's oceans.

Temperature is one of the abiotic factors that affects plant growth. Plants grow best in regions that are moderate—not too cold and not too hot. Plants also grow well when the temperature does not change greatly. This means that daily and seasonal temperatures are consistent. Because there are more plants, regions with constant warm temperatures support the greatest numbers of living things. Why do you think this is the case? Animal life is dependent on green plants. More types of plants and animals can live in areas where the temperatures are consistent. This explains why many more species live in tropical rain forests than in deserts or polar regions.

 Reading Check Why are there more plants and animals in the tropics than the desert?

Climate The Sun influences climate, another abiotic factor. When you need to decide what to wear each morning, perhaps you start by checking the weather forecast. In any area, the weather can change from day to day. Climate is the pattern of weather that occurs in an area over many years. You learned in Chapter 11 that climate does not change from day to day, so you would not check a climate forecast to decide whether to take your raincoat along when you head out for the day. Earth's climate patterns are controlled by the temperature differences between the equator and the poles. This difference creates winds and ocean currents that affect climate.

Scientists average temperature, precipitation, humidity, and the number of days of sunshine to determine an area's climate. For example, **Figure 4** shows zones of average temperature in the United States during the coldest time of the year. Farmers and gardeners use these zones to decide what kinds of plants to grow in their part of the country.

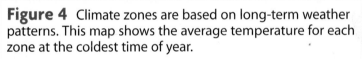

Figure 4 Climate zones are based on long-term weather patterns. This map shows the average temperature for each zone at the coldest time of year.

Identify which states contain the zone covering −6°C to −1°C?

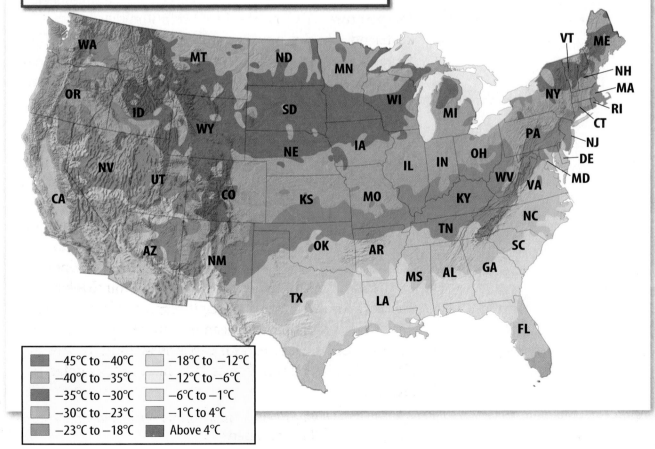

■ −45°C to −40°C	■ −18°C to −12°C
■ −40°C to −35°C	■ −12°C to −6°C
■ −35°C to −30°C	■ −6°C to −1°C
■ −30°C to −23°C	■ −1°C to 4°C
■ −23°C to −18°C	■ Above 4°C

The Water Cycle Water is an important abiotic factor to almost all life on Earth. Water helps organisms absorb nutrients and is important in ridding organisms of wastes. More organisms can survive in places with plenty of water.

What happens if you hang damp laundry outside on a sunny day? Heat energy from the Sun evaporates water from the clothes. Water also evaporates from rain-soaked highways, damp leaves in a forest, and the surfaces of lakes and oceans. Water that evaporates enters the atmosphere. **Figure 5** shows how water in the atmosphere condenses to form clouds. Water falls from clouds back to Earth as rain, snow, sleet, or hail. Some of this water flows over land into streams and rivers. Some soaks into the ground. Plants need this water to grow. They take up water through their roots and release it from their leaves. Animals play a part in the water cycle, too. They drink water and excrete it as waste. Extra water in the soil may drain downward and become groundwater.

 Figure 5 What happens to water that evaporates?

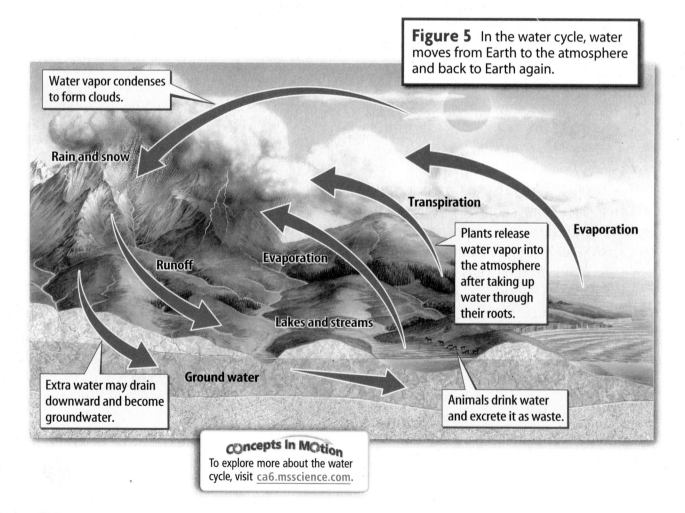

Figure 5 In the water cycle, water moves from Earth to the atmosphere and back to Earth again.

Water vapor condenses to form clouds.

Rain and snow

Runoff

Evaporation

Lakes and streams

Transpiration

Plants release water vapor into the atmosphere after taking up water through their roots.

Evaporation

Ground water

Extra water may drain downward and become groundwater.

Animals drink water and excrete it as waste.

Concepts In Motion
To explore more about the water cycle, visit ca6.msscience.com.

Soil

Earlier you read how to grow a plant. One of the things you would need is a pot filled with soil. This abiotic factor might seem simple, but soils are not all the same. For one thing, they contain different minerals, such as limestone or quartz. These minerals affect the chemistry of the soil, such as the acidity or alkalinity. Plants do not grow well if the soil is too acidic or too alkaline. Farmers and gardeners also measure the concentrations of important plant nutrients, including nitrogen, phosphorus, and potassium.

Dead plants and animals decay in soil, forming a dark-colored material that makes nutrients available to plants, called **humus** (HYEW mus). As you can see in **Figure 6,** humus lies in a thin layer at the soil surface. As it slowly breaks down, the organic matter in humus helps plants grow. Humus also supports plant growth by holding water in the soil. Like a sponge, it soaks up water and keeps the soil moist. In nature, the regions that support the most plant growth have soils that are rich in humus.

 Explain how humus helps plants grow.

Air

Like the Sun, air is also an important abiotic factor. Humans and many other organisms can only survive in places where the air contains enough oxygen. Many organisms use oxygen to help their cells release energy. Most organisms that use oxygen can only survive for five minutes without it before cells begin to die. At higher elevations, the mountain air contains less oxygen. Mountain climbers carry oxygen tanks to help them breathe. Some organisms can use less oxygen than can others. The air in a particular ecosystem determines the organisms that will live there.

Figure 6 • Humus, the organic matter in soil, helps plants to grow.

Humus layer

Oxygen

Carbon Dioxide

Figure 7 Oxygen and carbon dioxide flows between organisms.

SCIENCE USE V. COMMON USE

gas

Science Use a substance having neither definite shape nor volume that can expand indefinitely. *You exhale carbon dioxide gas.*

Common Use a fuel in internal-combustion engines. *We put gas in the car on our way home today.*

Exchanging Gases Some organisms, like humans, take oxygen from the air and release carbon dioxide. Other organisms do the opposite. Plants and some other organisms take in carbon dioxide and give off oxygen. **Figure 7** shows the exchange of oxygen in the environment. The grasses in this ecosystem take in carbon dioxide and release oxygen into the air. The lion takes in this oxygen and releases carbon dioxide for the grasses to take in. Green plants help keep these gases in balance.

 Figure 7 Contrast the gases given off by plants and animals.

Air Pollution Some human activities can pollute the air. Air pollution is the contamination of the air by harmful substances. Much of the pollution comes from vehicles and manufacturing. Organisms can have difficulty surviving in contaminated air. In humans, this can cause irritation in the respiratory system. It can also cause more serious problems, such as certain types of cancer. For wildlife, the pollution collects in the soil and water. This affects the place where the organism lives. It also affects the food supply. What can you do to help reduce pollution? Small changes, such as turning off electrical appliances when not in use, are helpful. Also, your family can walk, ride a bike, or take public transportation instead of driving. You can reduce the number of items you use, reuse items again, and recycle everyday items. The fewer new items you use means less items have to be manufactured.

yorba linda public library

UNPLUGGED

AMPED UP

MUSIC LENDING LIBRARY

Foster your love of music with the availability of instruments, sheet music, record players, and vinyl from the Yorba Linda Public Library.

ylpl.org/things

Biotic Factors

Biotic factors are the living parts of an ecosystem. Every organism in an ecosystem depends on other organisms. Consider the coral reef ecosystem in **Figure 8.** Corals are tiny marine animals. Corals contain microscopic algae in their tissues. While the algae provide food and oxygen, the corals provide a protected place for the algae to grow. This close interaction between two organisms is called symbiosis. Together, corals and algae build reef structures that provide homes to many organisms. Sea grasses and other plants provide shelter for fish looking for a place to hide from bigger fish looking for something to eat.

As in all types of ecosystems, plants and animals in coral reefs compete with others for food or living space. Some benefit each other by providing food, a place to live, or a way to hide from enemies. A **species** is a group of organisms that share similar characteristics and can reproduce among themselves producing fertile offspring. Each species plays a different role. Together they are the biotic factors of the ecosystem.

 Reading Check What are some biotic factors in a coral reef ecosystem?

Populations

How many people live in your town? The number of individuals of one species that occupy the same area is called a **population.** In ecosystems, the population of an organism refers to all the individuals of that species that live in a given location. For example, a scientist might study the population of condors in California or great white sharks in the Pacific Ocean. In a smaller ecosystem, a scientist could study the population of frogs in a pond.

WORD ORIGIN

population
from Latin *populus;* means
inhabitants

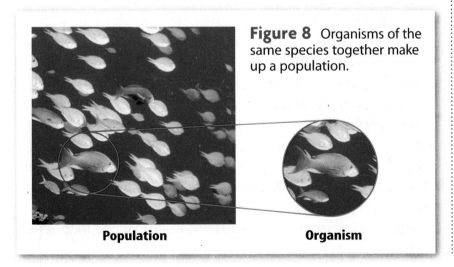

Figure 8 Organisms of the same species together make up a population.

Population   **Organism**

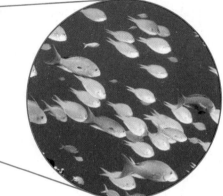

Figure 9 An ecosystem includes populations of organisms that interact with each other to form ecological communities.

Identify which abiotic factors affect this community.

Community   **Population**

Communities

A **community** is all the populations of species that occupy an area. You can see in **Figure 9** that a community includes more than one species. In the Pacific Ocean, you would find different communities near the shore than in the deep sea. Communities also change with distance from the equator. Remember that temperature is one of the abiotic factors that affects living things. As you go north or south from the equator, the temperature drops. Different ecological communities adapt to living in regions with different temperatures. **Figure 10** shows how biotic factors can interact in a Californian coastal scrub community.

 What makes up a community?

Limiting Factors

If you have ever kept a fish as a pet, you know there is a limit to how many fish you can add to the aquarium. If you add too many fish to an aquarium, oxygen levels can drop too low, and the fish can die. In nature, populations expand until the biotic or abiotic factors become limiting. A **limiting factor** is an environmental factor that limits how large a population can grow. If populations get too large, competition for resources, such as food and water, will cause some individuals to die.

WORD ORIGIN

community
from Latin *communitatem;*
means *fellowship*

524   **Chapter 12** • Ecological Roles

Visualizing Biotic Factors

Figure 10

Hawks, snakes, and many other organisms make up California's coastal scrub community. These communities have many shrubs and few trees. They are accustomed to hot, dry summers and mild, wet winters. During summer, many shrubs lose their leaves to conserve water. Animals living in this community must adjust to great differences between summer and winter water supplies.

Changing Factors

Limiting factors can change over time. For example, in ecosystems that normally have plenty of rain, water can become a limiting factor during a drought. Different factors limit different species. Sun-loving plants do not grow well in the shade or in dense forests. But mushrooms and forest wildflowers will grow well in the shade.

At the beginning of this chapter, you learned what would be needed to provide for a green plant to survive in your classroom. What do humans need to survive? All organisms, including humans, need food, water, shelter, and space.

Food

All organisms need food for energy. In some ecosystems, there is plenty of food. In others, food is scarce. What do you think causes this difference? Recall how abiotic factors affect plant growth. Temperature, types of soil, and amounts of sunlight and water are different from one ecosystem to another. These abiotic factors affect plant growth. The more plants that grow, the more food there is for other living things.

Compare the abiotic and biotic factors of two ecosystems shown in **Figure 11.** Which looks like it would support more organisms? In a tropical rain forest, constant warmth and plentiful water create perfect conditions for many plant species. The lush plant growth provides food to a variety of insects, birds, and other organisms. In deserts, plant life is limited, so less food is available. Less food means fewer organisms can live in deserts than in tropical rain forests.

Figure 11 Ecosystems that support more plant growth provide food to a greater number and variety of animals.

Identify the limiting factors in each of these environments.

Rain Forest

Desert

Water

All living things need water, but some need more water than others. Organisms that live in sandy deserts are **adapted** to a life that has little water during much of the year. An adaptation is any physical or behavioral characteristic that allows an organism to be better suited to the environment. Cactus have waxy coverings to prevent water loss from evaporation. Another way organisms survive in deserts is by staying inactive during the hot, dry summers. During the brief wet periods, wildflowers bloom, and animals gather food before the dry weather forces them to become inactive again. Some animals have the ability to cool themselves.

 How do organisms survive in the desert?

Other than deserts, where else on Earth do you think water could be a limiting factor? On snow-covered peaks, water may stay frozen all year. Even on peaks that do not stay frozen, lack of water can limit plant growth. If you drive or hike up a mountain, you will notice the tall trees you will see in the valley do not grow at higher elevations. The higher you go, the shorter the trees and bushes become. Finally, you reach an area like the one shown in **Figure 12.** At this high elevation, no trees grow, and the vegetation is limited to wildflowers, grasses, lichens, and mosses. What factors do you think might cause these differences in plant communities between the valley and the peak? One of the limiting factors is water. High in the mountains, the soils have a thin humus layer. They dry out quickly rather than remaining moist between storms.

ACADEMIC VOCABULARY
adapt (uh DAPT)
(*verb*) to make fit, often by modification
Juan adapted quickly to his new school by making friends and joining the drama club.

Treeline

Figure 12 High in the mountains, water and other abiotic factors limit plant growth.

Figure 13 In the oceans and on land, animals seek shelter from their enemies.

Identify the places for shelter in these images.

Shelter

If you lived in the Arctic how would the design of your home be different from your home if you lived in a tropical region? Wherever you live, you want your home to keep you dry when it rains or snows. You also might lock the doors for protection. Similarly, organisms in all types of ecosystems need shelter from both abiotic and biotic factors.

Animals living in harsh climates use shelter to protect themselves from abiotic factors. In deserts and polar regions, for example, frogs dig burrows to avoid the heat or cold.

As shown in **Figure 13,** shelter also protects organisms from their enemies. How do you think fish in the Pacific Ocean escape from sharks? One way is by hiding behind a rock or a clump of seaweed. On land, shrubs and piles of brush provide shelter for rabbits and mice, helping them to hide from eagles and owls.

 Why is shelter important?

Space

All organisms need space in which to live and grow. In a garden or a forest, plants will grow best if they do not have to compete with other plants. Animals also need space. Some animals defend their territories in order to protect enough land to meet their needs for food, water, and shelter.

Changes in Population

The life of sea otters, like the one in **Figure 14,** can show how abiotic and biotic factors affect ecosystems. Sea otters live in kelp beds—giant algae that grow more than 30 m. Sea urchins chew off the kelp where they are attached to the ocean floor. Otters eat urchins. When the otter population declines, the sea urchin population increases and the kelp beds float away. In populations of other organisms, including clams, snails, octopuses, and many different kinds of fish that depend on kelp for shelter, space and food also decline.

The sea otter population is not increasing as fast as scientists hoped. One reason may be that killer whales are eating more otters. The whales' usual food sources, seals and sea lions, have declined in recent years. Some scientists hypothesize that warmer ocean temperatures, an abiotic factor, and over-fishing may be responsible for the decline in seals and sea lions.

Figure 14 This sea otter is an important species in California. The ocean ecosystem depends on this animal.

LESSON 1 Review

Summarize

Create your own lesson summary as you organize an **outline.**

1. **Scan** the lesson. Find and list the first **red** main heading.

2. **Review** the text after the heading and list 2–3 details about the heading.

3. **Find** and list each **blue** subheading that follows the **red** main heading.

4. **List** 2–3 details, key terms, and definitions under each **blue** subheading.

5. **Review** additional **red** main headings and their supporting **blue** subheadings. List 2–3 details about each.

ELA6: R 2.4

Standards Check

Using Vocabulary

1. Use each term in a sentence: *ecosystem* and *biotic factor*. `5.e`

2. All the living organisms and the place they live is called a(n) _____. `5.e`

3. A(n) _____ is an environmental factor that limits how large a population can grow. `5.e`

Understanding Main Ideas

4. **Compare** two ecosystems near your home or school, and list some of the interactions you expect to occur in them. `5.e`

5. **List** several abiotic factors that control the numbers and types of organisms living in a desert. `5.e`

6. **Which** describes all of the sea urchins that live off the San Diego coast? `5.e`

   A. species
   B. population
   C. community
   D. ecosystem

Applying Science

7. **Summarizing** Complete this chart by adding links to examples of abiotic and biotic factors. `5.e`

Ecosystem

Abiotic factors

Biotic factors

For more practice, visit **Standards Check** at ca6.msscience.com.

Graphing Monthly Abiotic Factors

Data Collection

1. Go to ca6.msscience.com to find local weather information for a U.S. city.
2. Find the average monthly rainfall and temperature for a U.S. city for each month of an entire year.
3. Copy and fill in the data table and then graph your results.

Monthly Weather Reports			
Month	Average Rainfall	Average Temperature	Average Sunlight Hours
January			
February			
March			
April			
May			

Analysis

1. **Examine** your graphs of rainfall, temperature, and sunlight. How do these factors affect your life?
2. **Explain** how changing monthly factors might affect nature. What change might there be in plants over the year? In insects?

Science Content Standards

5.e Students know the number and types of organisms an ecosystem can support depends on the resources available and on abiotic factors, such as quantities of light and water, a range of temperatures, and soil composition.
7.c Construct appropriate graphs from data and develop qualitative statements about the relationships between variables.

Applying Math

Changing Soil pH

5.e

MA6: MR 2.5

The types and numbers of organisms found in soil depend on the pH of the soil. Aluminum sulfate and sulfur are two compounds often used to lower the pH of soil. The following table shows the kilograms of aluminum sulfate needed to lower the soil pH to a given level.

Example

How much more aluminum sulfate would be needed to lower the pH from 8.0 to 5.5 than to lower the pH from 8.0 to 6.0?

Kg of Aluminum Sulfate Needed to Lower Soil pH					
Present Soil pH	**Kg needed to Reach Desired pH**				
	pH 6.5	**pH 6.0**	**pH 5.5**	**pH 5.0**	**pH 4.5**
8.0	1.8	2.4	3.3	4.2	4.8
7.5	1.2	2.1	2.7	3.6	4.2
7.0	0.6	1.2	2.1	3.0	3.6
6.5	—	0.6	1.5	2.4	2.7
6.0	—	—	0.6	1.5	2.1

What you know:
- Amount of aluminum sulfate needed to lower the pH from 8.0 to 5.5: 3.3 kg
- Amount of aluminum sulfate needed to lower the pH from 8.0 to 6.0: 2.4 kg

What you need to find:
The difference in the amount of aluminum sulfate needed

Subtract:
   3.3 − 2.4 = 0.9 kg

Answer: You need 0.9 kg more of aluminum sulfate.

Practice Problems

1. How much more aluminum sulfate would be needed to lower the pH from 7.0 to 6.0 than to lower the pH from 7.0 to 6.5?

2. Which two pH changes could be made using 1.2 kg of aluminum sulfate?

Science Online
For help, visit
ca6.msscience.com.

Reading Guide

What *You'll Learn*

▶ **Explain** why different numbers and varieties of organisms live in different biomes.

▶ **Describe** how different organisms play similar roles in different ecosystems.

Why *It's Important*

Learning how species fill ecological roles in different climates will help you understand your environment.

Vocabulary
biome   habitat   niche

Review Vocabulary
latitude: the distance in degrees north or south of the equator (p. 49)

Organisms and Ecosystems

Main Idea Climates and the types of life that they support define biomes on Earth. Ecological roles are the same in different biomes but may be filled by different species.

Real-World Reading Connection Would you expect to find a penguin in the jungle? How about a lion at the north pole? Earth's communities of plants and animals live in a range of climates. Some organisms can survive only in tropical areas. Others are adapted to life in regions with ice and snow. What organisms live in your region?

Biomes

What does a desert in Arizona, like the one in **Figure 15,** have in common with a desert in Africa? Both are hot, have little rain, and have sandy soils. Organisms include water-conserving plants, lizards, snakes, and a few birds. Even widely separated regions of Earth can have similar organisms because they have similar climates and soils.

Earth has six major land zones, called **biomes,** that support a different type of ecological community.

Figure 15 Even ecosystems that have the same characteristics contain different organisms.

Arizona Desert

African Desert

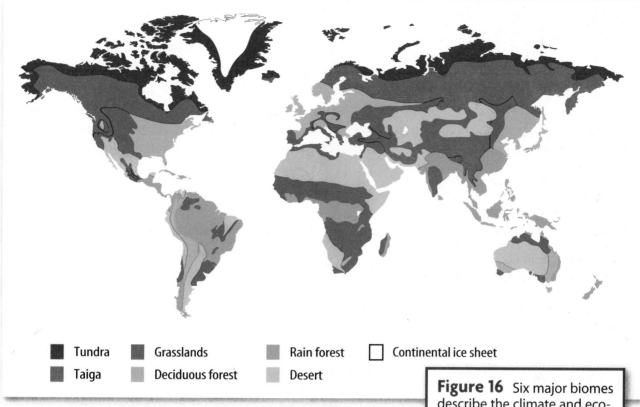

■ Tundra	■ Grasslands	■ Rain forest	□ Continental ice sheet
■ Taiga	■ Deciduous forest	■ Desert	

Figure 16 Six major biomes describe the climate and ecological communities of Earth.
Identify which biome you live in.

C*O*ncepts In M*O*tion

To see an animation about biomes, visit ca6.msscience.com.

Biome Types

Six major biomes, shown in **Figure 16,** describe the Earth's land regions. Each biome has a different climate and soil type.

Tundra The tundra biome lies near the north and south poles. During summer months, tundra biomes support many more organisms than during the harsh winters. Many species of shorebirds **migrate** to the tundra in the summer. There, they breed and raise their young. In the fall, they escape the harsh winter by flying south to warmer biomes where they can more easily find food. Tundra animals that do not migrate must hibernate or adapt to life in a cold, dark environment.

 Why do some tundra animals migrate south during the winter?

Taiga The largest biome on Earth, the taiga (TI guh), is a cold, forest region dominated by evergreen trees. Although the winter is long and cold, the taiga is warmer and wetter than the tundra. The taiga biome stretches across North America, northern Europe, and Asia.

ACADEMIC VOCABULARY
migrate (MI grayt)
(verb) to move from one place to another
Some Monarch butterflies migrate to California each fall.

How many organisms live here?

What lives in the desert? What lives in the forest? Where would you find more animals? Why do you think that is?

Procedure

1. Choose one biome to research.

2. List at least five common species of plants, animals, and other organisms important to the biome.

3. Create a fact poster to show your findings. Post your fact poster on the bulletin board beside posters of other biomes.

4. Make a table of characteristics in each biome.

Biome Characteristics			
Species	Biome 1	Biome 2	Biome 3

Analysis

1. **Explain** the characteristics the plants share in the biomes. How are they different?

2. **Identify** any special features the animals may have for living in each biome.

3. **Compare and contrast** the different biomes.

Rain Forests There are two types of rain forests. Because of abundant rainfall, both have lush, green plants throughout the year. Tropical rain forests, found near the equator, support more species than any other biome. Temperate rain forests are found in coastal regions, where moist air from the oceans drop between 150 to 500 cm of rain yearly.

 Name the two types of rain forest.

California Biomes

California is an area of diverse biome regions. The major biome types in California include the temperate deciduous forest, the desert, and grasslands.

Temperate Deciduous Forests Northern California has abundant rain, long summers, and cold winters. Plants in temperate deciduous forests, shown in **Table 1,** include many trees that lose their leaves in the winter. Some commonly found animals include insects, rodents, foxes, deer, and raccoons.

Desert The desert has very little rainfall—less than 2 cm per year. Life in the desert has adapted to conserve water. Desert plants, such as the cacti in **Table 1,** have tough, waxy coatings with thick leaves that store water. The animals include rodents, lizards, snakes, spiders, scorpions, and a few birds.

Grasslands Most grasslands have a dry season with little or no rain. This lack of moisture prevents forest development. Grasslands have different names. In Africa, grasslands are called savannahs. In the midwestern United States, they are called prairies. In southern California coastal areas, they are called chaparrals. Chaparral organisms are adapted to hot, dry summers and mild, rainy winters. Fires and droughts are common.

 Table 1 What differences do you notice between the biomes?

Table 1 Biomes of the World

Description	Biome
Tundra In the polar tundra, the ground may stay frozen all year. Little water or sunlight is available in winter. Tundra vegetation is limited to small plants that can survive the cold, dark winter.	
Taiga South of the tundra, most soils in the taiga thaw in the summer. Taigas are heavily forested and are home for moose, lynx, shrews, bears, and foxes.	
Temperate Deciduous Forest Temperate regions usually have four distinct seasons each year, with cold winters and hot summers. Between 75 and 150 cm of rain falls each year, but one season usually is not much wetter than another.	
Rain Forests Tropical rain forests (left) are located near the equator. More species of plants and animals are found here than anywhere else on Earth. Temperate rain forests (right), found near coasts, have huge evergreen trees.	
Grassland Grasslands usually have poor soil and uneven rainfall throughout the year. Animals found in grasslands include coyotes, eagles, bobcats, crickets, dung beetles, bison, and prairie dogs. A chaparral is a hot, dry grassland found in coastal regions. Winter is mild, usually about 10°C. Summers are so hot—up to 40°C—and dry that fires and droughts are common.	
Desert Deserts cover about one-fifth of Earth's land surface and are very dry. Some deserts are hot with very little rain throughout the year. Other deserts have cold winters with some snow.	

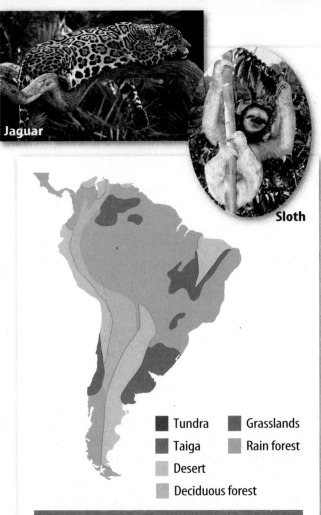

Jaguar

Sloth

Tundra **Grasslands**
Taiga **Rain forest**
Desert
Deciduous forest

South American Habitats

Figure 17 South America and Australia have many similar niches. Different species fill these niches.

Describe why these two places have similar niches.

Tundra **Grasslands**
Taiga **Rain forest**
Desert **Deciduous forest**

Australian Habitats

Habitat and Niche

While taking a walk in the forest you might notice a rotting log. If you look closely at the log, you might see termites eating the wood and ants eating the termites. Spiders might be hiding under the bark. Each organism in an ecosystem needs a place to live. The place an organism lives is called its **habitat.** The rotting log is the habitat for termites and spiders.

What is a niche?

One habitat might contain hundreds of species. You might think it impossible for so many species to occupy one habitat. However, each species has different requirements for its survival. They all eat different food, or use different shelter. Each species has its own **niche** (NICH), or role, in the environment—how it obtains food and shelter, cares for its young, and avoids danger.

Similar Niches

Every ecosystem contains similar niches. However, different species fill these niches in different ecosystems. The map in **Figure 17** shows that tropical rain forests exist on several continents at about the same latitude. Each of these rain forests has organisms that occupy similar niches. However, different species may fill these niches in different parts of the world. In South American rain forests, animals such as the harpy eagle and jaguars are hunters that catch and eat other organisms. In the rain forests of Australia, similar niches are filled by bandicoots and dingos.

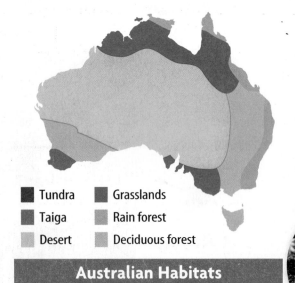

Bandicoot

Dingo

Human Impacts on Niches

What do you think would happen to an ecosystem if an organism no longer were able to fill its niche? Perhaps the species is hunted to extinction, or has had its habitat destroyed. Population decline affects the rest of the ecosystem because organisms no longer are able to fill their niche.

Human activities can impact population size and therefore affect entire biomes. For example, if people did not allow fires to burn in chaparrals, there would be no young pine trees. The animals would not have the trees for food, shelter, and space. In the next chapter, you will learn how organisms depend on each other in ecosystems.

LESSON 2 Review

Summarize

Create your own lesson summary as you write a **newsletter**.

1. **Write** this lesson title, number, and page numbers at the top of a sheet of paper.

2. **Review** the text after the **red** main headings and write one sentence about each. These will be the headlines of your newsletter.

3. **Review** the text and write 2–3 sentences about each **blue** subheading. These sentences should tell *who, what, when, where,* and *why* information about each headline.

4. **Illustrate** your newsletter with diagrams of important structures and processes next to each headline.

ELA6: W 1.2

Using Vocabulary

1. In your own words, write the definition of a *biome*. 5.d

2. What is the correct term for the place an organism lives?

   A. species
   B. niche
   C. habitat
   D. population   5.d

Understanding Main Ideas

3. **Identify** Copy and complete the concept map below. 5.d

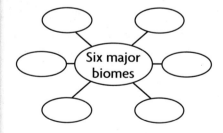

Six major biomes

Standards Check

4. **Explain** how a wolf and a lion can occupy the same niche in different environments. 5.e

5. **Compare and contrast** the number of organisms living in California's Mojave Desert and in a forest along the coast. 5.c

Applying Science

6. **Consider** living in a tundra biome. Give examples of how tundra climate would affect your life and how other organisms might cope. 5.e

7. **Predict** how living things in an ecosystem near your school might interact. 5.e

Science Online
For more practice, visit **Standards Check** at ca6.msscience.com.

Counting Species

Materials

wooden stakes (4)
survey ribbon or
  string (2 m)
pencil
colored pencils
ruler

Safety Precautions

Problem

When biologists study animals and plants, they often begin by finding out how many of each species are in an area. Since it would be impossible to count every single animal and plant, biologists take a sample plot and make a careful count of the animals and plants in that sample plot. Biologists use this sample plot to estimate how many species are in a larger area.

Form a Hypothesis

➤ **Review** the results from this chapter's laboratory investigations. Make a prediction about what would happen if one species disappeared. How would this change the ecosystem?

Collect Data and Make Observations

1. Read and complete the lab safety form.
2. In an undisturbed area, set up four wood stakes in a square. The stakes should be 30 cm apart on the sides of the square. Wrap 2 m of survey ribbon around the stakes to mark the square.
3. As accurately as you can, draw what you see inside your square.
4. List all the organisms you see.
5. Count each type of organism.
6. Design and construct a data table similar to the one below to organize your information from steps 4 and 5.

Plant or Animal Type	Number of Organisms

Science Content Standards

5.c Students know populations of organisms can be categorized by the functions they serve in an ecosystem.
5.e Students know the number and types of organisms an ecosystem can support depends on the resources available and on abiotic factors, such as quantities of light and water, a range of temperatures, and soil composition.
7.a Develop a hypothesis.
7.d Communicate the steps and results from an investigation in written reports and oral presentations.

Analyze and Conclude

1. **Predict** what larger animals might be part of your mini-square ecosystem, such as rabbits, cats, deer, cows, squirrels, or lizards.
2. **Identify** the climate zone for your plot.
3. **Describe** the area around your plot.
4. **Estimate** how many hours of sunlight your square gets per day. Describe any shading conditions.
5. **Describe** how wet your plot is. Is it near a stream, in a valley, in an upland area?
6. **Identify** evidence of erosion or other seasonal changes in the plot.
7. **Predict** what would happen if one species went extinct. How would this change the ecosystem?
8. **Error Analysis** Were there any parts to your procedure that you could have changed for more accurate results? What changes would you make to this activity for next time? Explain your ideas.
9. **Calculate** how many organisms you would find in an area 100 times larger than your study area.

Comunicate

WRITING in Science

Share Your Data How is your plot similar to other students' plots? How is it different? Compare your drawings and data tables with two classmates. Write a paragraph that explains how differences in climate and lighting might affect the types of organisms in each plot.

Real World Science

You can help save wildlife!

Conservation biology is a science that focuses on how to protect and restore the diversity of life on Earth. Conservation biologists work on a wide range of ecological problems, ranging from endangered species to regional conservation planning such as creation of wetland habitats and preservation of prairie dog ecosystems. This person is a wildlife biologist in West Africa.

Visit **Careers** at **ca6.msscience.com** to find out more about what a wildlife biologist does. Create a daily journal for a one-month research expedition to China to study pandas in the wild. List both the panda's and your daily activities, including travel experiences, hazards encountered, scenery, food, and supplies used.

Tracking Animals

Scientists attach tiny transmitters to wild animals, even animals as small as beetles. Radio tracking can help researchers determine a species' home range, population density, and key habitat elements. Radio collars transmit radio waves. Now scientists can track wildlife over much wider areas with receivers high above the ground in orbiting space satellites.

Visit **Technology** at **ca6.msscience.com** to find out more about radio tracking. See how animals are tracked in the Pacific Ocean using radio tags and satellites. Perform your own tracking study. For one week, track the occurrence of shoe color in your class. Use a table to track your observations then create a graph.

Prairie Dogs—Friend or Foe?

The prairie dog has played a central role in shaping the Great Plains. Prairie dog towns once covered much of the Plains. Ranchers believed they competed with cattle for grass. Elimination, including poisoning, began in the early 1900s and continues today, reducing prairie dog colonies. These measures have virtually eradicated the prairie dog and many of its predators, such as the black-footed ferret.

Visit **History** at ca6.msscience.com to find out more about prairie dogs and black-footed ferrets. Divide into small groups and research the pros and cons from the viewpoint of either a rancher and or a wildlife biologist. Debate the findings from the differing viewpoints.

Recovering Threatened Species

Populations of animals in the wild can be threatened by loss of viable habitat, industry, and environmental hazards such as pesticides and light pollution. Beginning in the 1960s and continuing today, society has begun to recognize the diminishing species and has started to repair the damage. Some good examples of recovering species are the California condor and the black-footed ferret. Through the work of conservation societies, zoos, and government wildlife agencies, these animals are slowly making a comeback.

Visit **Society** at ca6.msscience.com to learn more about these threatened species and related issues. Divide students into small groups and assign each group a particular man-made threat. Have each group compose a report to compare and contrast how the human impact has changed over the years.

The BIG Idea Living things interact with each other and with nonliving factors in ecosystems.

Lesson 1 Abiotic and Biotic Factors
4.a, 5.e, 7.c

Main Idea Living things and nonliving factors, such as air, water, sunlight, and soil, interact in Earth's ecosystem.

- Energy from the Sun influences abiotic factors such as climate and availability of water in different regions on Earth.

- Soil type is an abiotic factor that affects plant growth. Soils that are rich in humus support abundant plant growth.

- Air is an abiotic factor that affects and is affected by living things. Photosynthesis and respiration determine the concentrations of oxygen and carbon dioxide in the air.

- A population of organisms represents a single species. Populations of organisms in the same ecosystem interact as members of ecological communities.

- Population growth is limited by factors such as food, water, shelter, and space.

- Changes in populations can affect ecosystems.

- **abiotic factor** (p. 517)
- **biotic factor** (p. 523)
- **community** (p. 524)
- **ecosystem** (p. 516)
- **humus** (p. 521)
- **limiting factor** (p. 524)
- **population** (p. 523)
- **species** (p. 523)

Lesson 2 Organisms and Ecosystems
5.c, 5.d, 5.e, 7.a, 7.b, 7.d

Main Idea Climates and the types of life they support define biomes on Earth. Ecological roles are the same in different biomes but may be filled by different species.

- Ecosystems on Earth are categorized into biomes such as tundra, tropical rain forest, or grassland.

- Each biome has a different climate and supports different types of ecological communities.

- Biomes with mild climates support a greater variety of organisms than biomes with harsh climates.

- In every ecosystem, each species fills its own niche.

- A niche describes how a species interacts with other species and with abiotic factors in its environment.

- Different species may fill similar ecological roles in similar biomes.

- **biome** (p. 532)
- **habitat** (p. 536)
- **niche** (p. 536)

STUDY TO GO
Download quizzes, key terms, and flash cards from ca6.msscience.com.

Linking Vocabulary and Main Ideas

Use vocabulary terms from page 542 to complete this concept map.

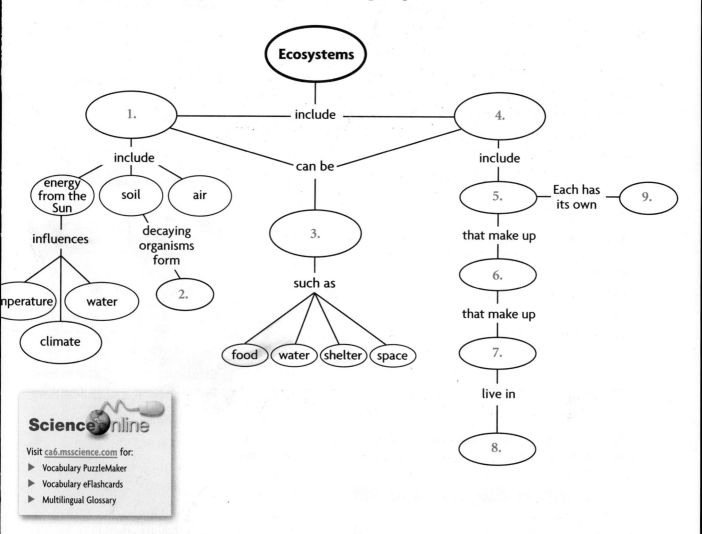

Using Vocabulary

Match a vocabulary term to each definition below.

10. the place an organism lives

11. the living parts of an ecosystem

12. Earth's major ecological communities

13. a factor that limits growth of a population

14. all the interacting species in an ecosystem

15. an organism's role in the environment—how it obtains food and shelter, cares for its young and avoids danger

Fill in each blank with the correct vocabulary term.

16. The nonliving parts of the environment are called _____ factors.

17. A(n) _____ is all the organisms in an area and the place they live.

18. A single species living in an area is a(n) _____.

Understanding Main Ideas

Choose the word or phrase that best answers the question.

1. Which would you use to describe a collection of sea anemones, sea stars, barnacles, and other organisms that live in a tide pool?
   A. a community   **5.c**
   B. a limiting factor
   C. an abiotic factor
   D. a population

2. Which measurement would you include in an assignment to measure biotic factors in a grassy area near your school?
   A. number of hours of sunlight   **5.e**
   B. number of species of insects
   C. average air temperature
   D. amount of rainfall

3. Because green plants need moisture, light, and soil, what do they depend on?
   A. biotic factors   **5.e**
   B. abiotic factors
   C. ecological communities
   D. shelter

4. The figure below shows how the Sun's rays hit Earth.

Equator

   Why are the poles colder than the equator?
   A. The Sun's rays hit Earth more dispersed at the poles.
   B. The Sun's rays hit Earth more dispersed at the equator.
   C. There are more green plants near the equator.
   D. Polar regions are usually cloudy.   **5.e**

5. Which group is a population?
   A. plants and animals in a meadow   **5.c**
   B. species of birds living in a rain forest
   C. Mojave rattlesnakes in the Mojave Desert
   D. fish and whales in the Pacific Ocean

6. In which type of biome would you find the greatest number of species of plants and animals?
   A. tundra   **5.d**
   B. desert
   C. chaparral
   D. rain forest

Use the figure below of a biome map to answer question 7.

   ■ Tundra   ■ Grasslands   ■ Rain forest
   ■ Taiga   ■ Deciduous forest ■ Desert

7. Which is the largest biome?
   A. desert   **5.d**
   B. tundra
   C. taiga
   D. rain forest

8. Plants and animals that are adapted to hot, dry summers and mild, rainy winters can be found in which biome?
   A. tundra   **5.d**
   B. chaparral
   C. desert
   D. tropical forest

9. What is the main reason different biomes support different plant and animal communities?
   A. They have different climates.   **5.d**
   B. They have different soil types.
   C. They have different niches.
   D. They have different biotic factors.

Applying Science

10. **Design an experiment** to test the effects of abiotic factors on the growth of bean seedlings. Describe your experiment and your predictions of the results. **5.e**

11. **Describe** how life in desert biomes depends on abiotic factors. **5.d**

12. **Infer** why you would expect to find more animals living in ecosystems with abundant plants than ecosystems with few plants. **5.e**

13. **Identify** three possible limiting factors for a biome of your choice. Evaluate how these factors might limit population growth in that biome. **5.d**

14. **Analyze** how polar bears adapt to their biome. **5.d**

15. **Analyze** In a California forest, deer eat plants. Mountain lions kill and eat some of the deer. What two ways do biotic factors affect the population of deer in this forest? **5.c**

16. **Compare and contrast** biotic and abiotic factors in an ecosystem of your choice. Include examples specific to that type of ecosystem. **5.e**

17. **Interpreting Graphs** The graph below shows yearly rainfall for four biomes.

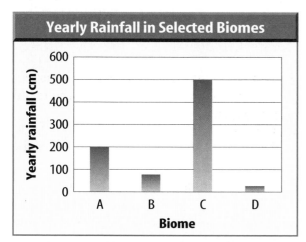

Yearly Rainfall in Selected Biomes

Which biome would you expect to have the greatest number of plant and animal species? Why? Explain how abiotic and biotic factors would influence life in the biome you selected. **5.d**

WRITING in Science

18. **Write** about a rotting log habitat. Explain how species and populations interact with biotic and abiotic factors. What limiting factors may affect this habitat? Describe each organism's niche. **ELA6: W 2.2**

Applying Math

Use the table below to answer questions 19–23.

Kg of Sulfur Needed to Lower Soil pH					
Present Soil pH	Desired pH				
	6.5	6.0	5.5	5.0	4.5
8.0	0.3	0.4	0.5	0.6	0.7
7.5	0.2	0.3	0.4	0.5	0.6
7.0	0.1	0.2	0.3	0.4	0.5
6.5		0.1	0.2	0.3	0.4
6.0			0.1	0.2	0.3

19. How much more aluminum sulfate would be needed to lower the pH from 7.0 to 5.5 than to lower the pH from 7.0 to 6.5? **MA6: MR 2.5**

20. Which pH changes could be made using 1.5 kg of aluminum sulfate? **MA6: MR 2.5**

21. How much more aluminum sulfate would be needed to lower the pH from 8.0 to 6.0 than to lower the pH from 8.0 to 6.5? **MA6: MR 2.5**

22. Which pH changes could be made using 0.6 kg of aluminum sulfate? **MA6: MR 2.5**

23. How much more aluminum sulfate would be needed to lower the pH from 7.0 to 5.5 than to lower the pH from 7.0 to 6.0? **MA6: MR 2.5**

Use the image below to answer questions 1–3.

1 Little light reaches the plants on the floor of this deciduous forest. Which season would let the bluebells pictured grow the best?

A spring

B summer

C fall

D winter   5.e

2 What process do these bluebells use to transform energy from the Sun into stored chemical energy for their life processes?

A desertification

B photosynthesis

C radiation

D respiration   5.e

3 The photo below shows a deciduous forest. What is the limiting factor for plants on the forest floor?

A food

B soil

C sunlight

D water   5.e

4 Which abiotic factor provides energy for nearly all life on Earth?

A air

B soil

C sunlight

D water   5.a

5 Which is characteristic of places at high elevations?

A fertile soil

B fewer molecules in the air

C tall trees

D warm temperatures   5.e

6 What two factors are most responsible for limiting life in a particular area?

A precipitation and sunlight

B soil conditions and precipitation

C sunlight and temperature

D temperature and precipitation   5.e

7 What helps determine the number of individuals that can survive in an area?

A biome

B ecosystem

C habitat

D limiting factors   5.c

8 Which is an example of a population?

A all the white-tailed deer in a forest

B all the trees, soil, and water in a forest

C all the plants and animals in a wetland

D the air, sunlight, and soil in a wetland   5.c

Use the image below of a grassland to answer questions 9 and 10.

9 **What are the biotic factors in this photo that are interacting?**

   A air and trees

   B antelopes and zebra

   C grass and sun

   D soil and water   5.c

10 **Which describes interactions between groups of antelope and zebras shown in the photo above?**

   A biome

   B community

   C ecosystem

   D population   5.c

11 **What term best describes populations living together in an area?**

   A niche

   B habitat

   C population density

   D community   5.c

Use the map below to answer questions 12–14.

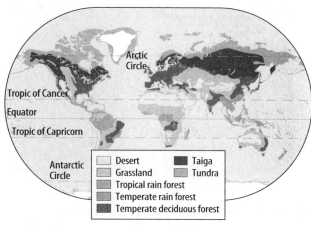

12 **What biome is located in the latitudes just south of the north pole?**

   A taiga

   B temperate deciduous rain forest

   C temperate rain forest

   D tundra   5.d

13 **Where is the tropical rainforest biome found?**

   A Arctic Circle

   B equator

   C tropic of cancer

   D tropic of capricorn   5.d

14 **What biome covers most of northern Africa?**

   A desert

   B temperate deciduous forest

   C temperate rain forest

   D tundra   5.d

Energy and Matter in Ecosystems

The BIG Idea

Matter cycles between organisms and the abiotic environment. Energy flows one way, from sunlight to producers to consumers and decomposers.

LESSON 1 5.a, 5.c, 7.a, 7.g
Producers and Consumers

Main Idea Producers make their own food, most using energy from the Sun. All other organisms depend on producers as their energy source.

LESSON 2
5.a, 5.b, 5.c, 7.b, 7.d, 7.e
Energy in Ecosystems

Main Idea Energy flows through ecosystems, from producers to consumers and decomposers.

LESSON 3
5.a, 5.b, 5.c, 7.a, 7.b, 7.d, 7.g
Matter in Ecosystems

Main Idea Matter cycles in ecosystems.

Why did she do *that?*

This lioness will gain energy from this porcupine if she can catch it. From where did the energy in the porcupine come? A porcupine eats plants for energy. From where do plants get their energy? Plants turn light energy from the Sun into energy they can use. So ultimately, from where does the energy for the lioness come?

Science *Journal* Write a paragraph describing what you know about energy and matter in ecosystems.

Launch Lab

Can you eat energy?

All living things on Earth need energy. Smiling uses energy, just as swimming laps in the pool takes energy. From where does the energy to work, play,  and study come? Make a healthful dinner menu for your family and see if you can trace the source of energy.

Procedure

1. Identify the food groups that make up a healthful diet.

2. Choose one or two from each group to make a dinner menu that you and your family would enjoy.

Think About This

- **Classify** each of the foods you chose for your menu as coming from a plant or animal.

- **Identify** where the animals get their energy.

- **Deduce** where you get your energy.

 5.a, 7.g

Science Online

Visit ca6.msscience.com to:

▶ view **Concepts in Motion**

▶ explore Virtual Labs

▶ access content-related Web links

▶ take the Standards Check

Energy Transfer Make the following Foldable to explain the transfer of energy in the environment.

▷ **STEP 1** **Collect** two sheets of paper and layer them about 2 cm apart vertically. Keep the left edges even.

▷ **STEP 2** **Fold** up the bottom edges of the paper to form 4 equal tabs. Crease the fold to hold the tabs in place.

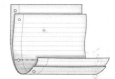

▷ **STEP 3** **Staple** along the fold. **Label** as shown.

| ↓ Energy from the Sun |
| ↓ Producers |
| ↓ Consumers |
| Decomposers |

Reading *Skill*  ELA6: R 2.4

Monitoring Your Comprehension
As you read this chapter, show how energy is transferred between organisms and the environment. Give examples at each level.

Get Ready to Read

Take Notes

ELA6: R.2.4

1 Learn It! The best way for you to remember information is to write it down, or take notes. Good note-taking is useful for studying and research. When you are taking notes, it is helpful to

- phrase the information in your own words;
- restate ideas in short, memorable phrases;
- stay focused on main ideas and only the most important supporting details.

2 Practice It! Make note-taking easier by using a chart to help you organize information clearly. Write the main ideas in the left column. Then write at least three supporting details in the right column. Read the text from Lesson 1 of this chapter under the heading *Consumers,* pages 556–558. Then take notes using a chart such as the one below.

Main Idea	Supporting Details
	1. 2. 3. 4. 5.
	1. 2. 3. 4. 5.

3 Apply It! As you read this chapter, make a chart of the main ideas. Next to each main idea, list at least two supporting details.

Target Your Reading

Use this to focus on the main ideas as you read the chapter.

1 **Before you read** the chapter, respond to the statements below on your worksheet or on a numbered sheet of paper.

- Write an **A** if you **agree** with the statement.
- Write a **D** if you **disagree** with the statement.

2 **After you read** the chapter, look back to this page to see if you've changed your mind about any of the statements.

- If any of your answers changed, explain why.
- Change any false statements into true statements.
- Use your revised statements as a study guide.

Reading Tip

Read one or two paragraphs first and take notes after you read. You are likely to take down too much information if you take notes as you read.

Science Online

Print a worksheet of this page at ca6.msscience.com.

Before You Read A or D	Statement	After You Read A or D
	1 Plants get their food from soil.	
	2 Plants are the only organisms that can make their own food.	
	3 The food you eat is used for energy and to help you grow.	
	4 Dead animals and plants do not need to be broken down to basic nutrients.	
	5 Energy flows only one way through ecosystems.	
	6 Many organisms can create their own energy.	
	7 Energy from the Sun is eventually captured by the top predators on Earth.	
	8 The amount of matter on Earth never changes.	
	9 When water evaporates, it leaves Earth's atmosphere, and more water is created when it rains.	
	10 Carbon is not very important for life on Earth.	

Science Content Standards

5.a Students know energy entering ecosystems as sunlight is transferred by producers into chemical energy through photosynthesis and then from organism to organism through food webs.
5.c Students know populations of organisms can be categorized by the functions they serve in an ecosystem.
7.a Develop a hypothesis.
7.g Interpret events by sequence and time from natural phenomena (e.g., the relative age of rocks and intrusions).

Reading Guide

What *You'll Learn*

▶ **Categorize** organisms into producers and consumers.

▶ **Classify** consumers into herbivores, carnivores, and omnivores.

Why *It's Important*

Learning about producers and consumers will help you understand the connection between all living things.

Vocabulary

ecology
producer
photosynthesis
consumer
protozoan
herbivore
carnivore
omnivore
decomposer
scavenger

Review Vocabulary

ecosystem: organisms and the physical place they live (p.516)

Producers and Consumers

Main Idea Producers make their own food, most using energy from the Sun. All other organisms depend on producers as their energy source.

Real-World Reading Connection When your body needs energy, you might eat a meal with your family or friends. If you were a green plant, you would soak up sunlight and make your own food. Ecosystems include organisms that make their own food and some that don't.

Ecosystems

Recall the discussion of ecosystems in the last chapter. Remember that each ecosystem includes biotic and abiotic factors. In the pond ecosystem shown in **Figure 1,** the biotic factors are the living things—fish, turtles, and plants. Abiotic factors, such as water, sunlight, and soil type, determine what sorts of organisms will be able to live in this ecosystem.

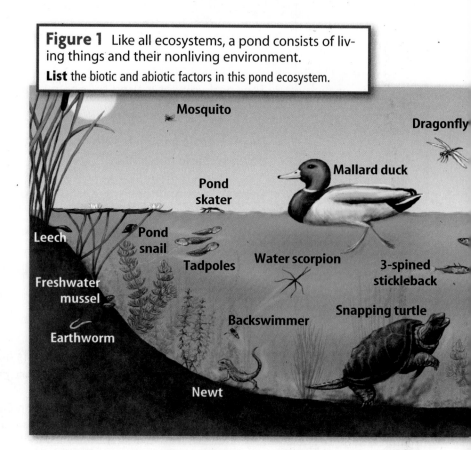

Figure 1 Like all ecosystems, a pond consists of living things and their nonliving environment.
List the biotic and abiotic factors in this pond ecosystem.

Mosquito

Dragonfly

Mallard duck

Pond skater

Leech

Pond snail

Tadpoles

Water scorpion

3-spined stickleback

Freshwater mussel

Backswimmer

Snapping turtle

Earthworm

Newt

How Organisms Relate

Ecology (ih KAH lu jee) is the study of the interactions between living things and their environment. It includes studying populations and communities and how energy and matter move through ecosystems.

WORD ORIGIN · · · · · · · · · · ·
ecology
from Greek *oikos* (means *house, dwelling place*) and *–logia* (means *study of*)

Producers

Matter is anything that has mass and takes up space. All organisms are made of matter. It takes energy to organize matter into food. **Producers** are organisms that use energy from the Sun or other chemical reactions to make their own food. Suppose you make a sandwich for lunch. Does this mean you are a producer? No. To be a producer, you would have to use energy from the Sun to make food. Most plants, algae, and some microorganisms are producers. Only a few types of producers on Earth make food without sunlight. Some bacteria in deep sea communities use energy from chemical reactions rather than from the Sun.

The Sun

Photosynthesis (foh toh SIHN thuh sus) is a process that producers use to make their own food using energy from sunlight. It is the main pathway by which energy and carbon enter the web of life. In **Figure 2,** you can see that producers use carbon dioxide and water to make chemical compounds, which they use as food.

WORD ORIGIN · · · · · · · · · · ·
photosynthesis
from Greek *photo* (means *light*) and *synthese* (means *synthesis*)

 **Reading Check** What process do producers use to make their own food using energy from sunlight?

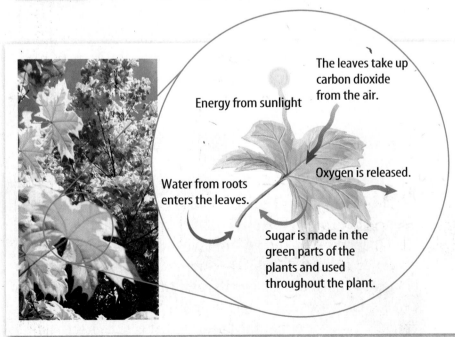

Energy from sunlight

The leaves take up carbon dioxide from the air.

Oxygen is released.

Water from roots enters the leaves.

Sugar is made in the green parts of the plants and used throughout the plant.

Figure 2 Sunlight to Food Through photosynthesis, producers use the Sun's energy to make their own food from carbon dioxide and water.

Plants

Most plants, like those in **Table 1,** are producers. Some people think that plants get food from the soil. This is not correct. Plants take up water from the soil and carbon dioxide from the air. Using these materials, producers make simple sugars.

When plants grow, they use the sugars produced during photosynthesis as energy and a source for carbon. The carbon combines with nitrogen and other nutrients. In this way, they create starches, proteins, oils, and other compounds. These compounds are the building blocks for the cells that make up the roots, stems, leaves, and seeds of each plant.

 Describe where plants get carbon and how they use it to grow.

Protists

You might think that producers have to be plants, but look at the protists in **Table 1.** Protists include algae, dinoflagellates, and euglenas. Euglenas do not have roots, stems, or leaves. They live in ponds and lakes. If you use a microscope to inspect a few drops of pond water, you might see euglena swimming. You may be surprised to learn that these single-celled swimmers are producers. Like all producers, they make their own food. All algae are protists that make their own food.

Bacteria

Bacteria are single-celled organisms found nearly everywhere on Earth. Some bacteria, called cyanobacteria, carry out photosynthesis. Cyanobacteria, like those in **Table 1,** have been on Earth for more than 3.5 billion years. Oxygen produced by ancient cyanobacteria helped create Earth's atmosphere as it exists today.

 Table 1 What do the organisms have in common?

Chemosynthesis

A few other types of bacteria are also producers. Instead of using energy from sunlight, however, these bacteria are able to make food using energy from chemical reactions in a process called chemosynthesis (kee moh SIHN thuh sus). Some chemosynthetic bacteria live deep in the ocean, where the Sun's rays never reach. Larger animals then eat the chemosynthetic bacteria or eat the animals that eat the bacteria.

Concepts In Motion
Interactive Table Organize information about different producers at ca6.msscience.com.

Table 1 Types of Producers

Organism	Characteristics
	**Plants** Most plants use energy from the Sun and take in water through their roots to make simple sugars. Plants use sugars as food to carry out their daily activities and to grow and reproduce. Sugars also help build plant structure.
	Protists Protists are single-celled or multicellular organisms that live in moist or wet surroundings. Some protists are plantlike. Some are animal-like. Some protists, including algae, have structures to make their own food.
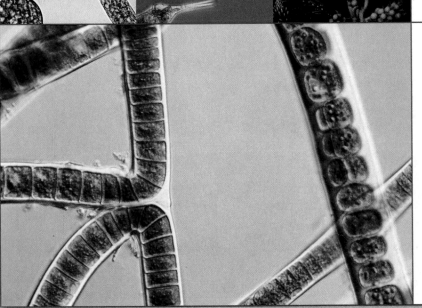	**Cyanobacteria** Cyanobacteria are single-celled organisms. They are an important source of food for some organisms in lakes, ponds, and oceans. Oxygen produced through photosynthesis is used by other aquatic organisms.

Consumers

Organisms that cannot make their own food are called **consumers.** All animals are consumers because they eat other organisms or their wastes. Some consumers eat producers, and some eat other consumers.

You are a consumer. You cannot carry out photosynthesis, so you depend on other organisms to make your food. In **Figure 3,** you can see where you get the parts of a familiar meal. If you eat lettuce or tomato, you are eating parts of producers. If you eat a chicken sandwich, the meat does not come directly from a producer. Instead, it comes from chicken, which is a consumer. Chickens get the energy they need by eating corn and other grains. If you drink milk or eat cheese, you too get some of the Sun's energy, passed from plants to the cow and then to you.

Some consumers are too small to be seen with the naked eye. Single-celled, animal-like protists, called **protozoans,** feed on living or dead organisms. These complex organisms have special **structures** to digest food and get rid of wastes. Protozoans are consumed by other, larger protozoans and by small, wormlike animals. These wormlike animals then become food for larger animals.

ACADEMIC VOCABULARY
structure (STRUHK chur)
(noun) the arrangement or formation of the tissues, organs, or other parts of an organism
Oak trees can be identified by the structure of their leaves.

Figure 3 **Are you a consumer?** Humans are consumers because we cannot make our own food. Most of our food comes from plants and animals.
Infer whether lettuce is a producer or consumer.

Types of Consumers

Think of the foods you eat. Foods such as fruits, nuts, rice, and vegetables come from plants. Meat, milk, and cheese come from animals. Different types of organisms get their energy from different types of food. Ecologists classify consumers into categories that describe the kinds of food they eat.

Herbivores Can you think of examples of animals that eat only plants? Elephants eat grasses. Caterpillars consume leaves. Squirrels eat nuts and seeds. Rabbits nibble garden plants. **Herbivores** are animals that eat only plants.

Carnivores Animals that only eat other animals are **carnivores.** Carnivores don't have to be big. Can you think of some smaller ones? How about a spider that traps insects in its web? Or a sea anemone that waits for creatures to swim into the reach of its sticky tentacles? Some animals, called predators, hunt and kill other organisms. The organisms they hunt and kill are called prey.

Can you imagine how a plant could be a carnivore? The Venus flytrap is an example of a carnivorous plant. Venus flytraps are producers because they get their energy through photosynthesis. They are also carnivores because they trap and digest insects. Venus flytraps grow in poor soils that are low in nitrogen. The insects they catch provide this needed nutrient. However, like all green plants, Venus flytraps get their energy from the Sun.

Omnivores Animals that feed on other animals and plants are **omnivores.** Grizzly bears are omnivores. They eat nuts, berries, seeds, and wildflowers. Grizzly bears also eat trout, elk, and insects. Unless you are a vegetarian, then you too are an omnivore.

 Reading Check What do omnivores eat?

Can you classify animals by diet?

Animals that eat only plants are herbivores. Carnivores are animals that eat only other animals. Animals that eat both plants and other animals are omnivores. Can you classify prairie animals by diet?

Data

1. Study the data in the table to determine what each animal eats.

2. Use the food the animal eats to classify it as an *herbivore,* a *carnivore* or an *omnivore.*

Classify Prairie Animals		
Animal	**Diet**	**Type**
Prairie dog	grass, roots, seeds, leaves	
Weasel	prairie dogs, voles	
Hawk	rabbits, squirrels, weasels, prairie dogs, grasshoppers	
Coyote	rabbits, mice, birds, deer, voles	
Vole	grasses, plants, seeds, birds' eggs	
Grasshopper	plants	

Data Analysis

1. **Evaluate** how plants and animals on the prairie are connected to each other.

2. **Hypothesize** what might happen if there were a drought and plants became scarce on the prairie.

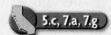

Figure 4 Dung beetles are scavengers. Young beetles, called larvae, feed on manure that the adults have rolled up.

Larvae | Adult

Decomposers and Scavengers Some organisms, called **decomposers,** break down dead organisms, animal droppings, leaves, and other wastes produced by living things. Decomposers make nitrogen and other nutrients available to support new life by breaking down dead organic matter. Many species of bacteria and fungi are decomposers as well as some insects, protists, and other invertebrates.

Reading Check How do decomposers make nutrients available to support new life?

Scavengers are organisms that feed on dead animals, like the crows or vultures that eat animals killed by traffic. Foxes and coyotes are predators, but can be scavengers too. When live prey is hard to find, these animals feed on dead animals.

Some scavengers eat wastes from other organisms. For example, the adult dung beetle in **Figure 4** rolls balls of manure from animal droppings. Then it lays its eggs inside these balls and buries them underground. When the eggs hatch, the dung provides food for the larvae as they grow and develop into adults.

Visual Check **Figure 4** Why are dung bettles classified as scavengers?

Think what would happen if decomposers and scavengers did not exist. Piles of dead plants and animals would cover Earth. Nitrogen, phosphorus, and other nutrients would limit new growth because these nutrients would remain in the bodies of dead organisms instead of returning to Earth. You will read in the next lesson about how they cycle through the ecosystem.

Organisms Depend on Each Other

You read in Chapter 12 that living things and nonliving factors interact in Earth's ecosystems. In this chapter, you read that producers, including most plants, some protists, and some bacteria, use energy from the Sun to make their own food. Consumers eat other organisms, including producers, and gain energy from them. Decomposers break down dead organic matter, making nutrients available for other organisms. Herbivores, such as cows and deer, eat only plants and plant materials. Carnivores, such as lions and eagles, eat other animals. Omnivores, such as humans and bears, eat both plants and animals. In the next two lessons, you will read about how energy and matter move through ecosystems.

LESSON 1 Review

Summarize

Create your own lesson summary as you design a **visual aid.**

1. **Write** the lesson title, number, and page numbers at the top of your poster.

2. **Scan** the lesson to find the **red** main headings. Organize these headings on your poster, leaving space between each.

3. **Design** an information box beneath each **red** heading. In the box, list 2–3 details, key terms, and definitions from each **blue** subheading.

4. **Illustrate** your poster with diagrams of important structures or processes next to each information box.

 ELA6: R 2.4

Using Vocabulary

1. In your own words, write the definition for *photosynthesis*. **5.a**

2. Consumers that feed on dead animals are called _____. **5.c**

Understanding Main Ideas

3. **Compare and contrast** producers and consumers. **5.c**

4. **Illustrate** how food moves through ecosystems. **5.a, 5.c**

5. **Identify** three types of organisms that are producers. **5.a**

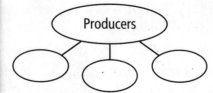

Producers

 ### Standards Check

6. Which of the following is a producer?

   A. spider   **5.a**

   B. oak tree

   C. coyote

   D. protozoan

7. **Determine** a good question to ask if you wanted to find out if an organism is a producer. **5.a**

Applying Science

8. **Predict** how life on Earth would be affected if there were no decomposers. **5.c**

9. **Evaluate** how a predator, such as a hawk, depends on producers for its survival. **5.c**

 **Science Online**

For more practice, visit **Standards Check** at ca6.msscience.com.

Reading Guide

What *You'll Learn*

▶ **Explain** how matter is transferred from one organism to another.

▶ **Draw** an energy pyramid showing loss of energy from one level to another.

Why *It's Important*

Learning how energy flows through ecosystems shows why organisms depend on each other.

Vocabulary

food chain   food web
primary consumer
secondary consumer
tertiary consumer

Energy in Ecosystems

Main Idea Energy flows through ecosystems, from producers to consumers and decomposers.

Real-World Reading Connection You might not think you need energy to read this page. However, you use energy all the time, even when you aren't active. All living things use energy to grow and carry out their daily lives.

Energy Through the Ecosystem

Think about all the ways in which the soccer players in **Figure 5** are using energy. They need energy to run and to kick the ball. They also need energy to walk, to talk, and even just to breathe. The spectators are less active than those out on the playing field, but they are still using energy.

Energy does not cycle through ecosystems. Instead, it moves in one direction—from the energy source to producers to consumers and decomposers. If producers stopped capturing energy from the Sun, all life on Earth would end because food supplies would run out.

Figure 5 Whether you are watching a game or running up and down the soccer field, your body is using energy.

Swainson's hawk

Mohave ground squirrel

Creosote bush

Organ pipe cactus

Mohave rattlesnake

Deer grass

Roadrunner

Barrel cacti

Kangaroo rat

 Tarantula

Creosote grasshopper

Figure 6 Energy flows one way through ecosystems.

Identify the producers and consumers.

Changing Energy

Organisms do not create energy. They also do not destroy it. Organisms change energy from one form to another. For example, producers change light energy into chemical energy through photosynthesis. When organisms use chemical energy in food, some of this energy is released as thermal energy.

Reading Check What happens to some of the chemical energy in food?

Food as Energy

The food you eat provides the sugars, starches, proteins, and fats your body needs to grow new cells. Your food also supplies chemical energy that your body uses as fuel. Energy passes through ecosystems as food. Producers, such as the desert grasses in **Figure 6,** capture energy from sunlight. When animals such as kangaroo rats eat desert plants, they gain energy from the plants. When hawks eat kangaroo rats, they too gain energy originally captured by producers.

ACADEMIC VOCABULARY
convert (kahn VURT)
(verb) to change something
into another form, substance,
state, or product
Boiling water converts to steam.

Food Chains

A **food chain** is an illustration of how energy moves through an ecosystem. Suppose a kangaroo rat nibbles on seeds from a bush in a California desert. The bush is a producer, so it **converts** sunlight energy into sugars. When a kangaroo rat eats seeds, it gains energy that has been stored by the bush. Now suppose a snake catches and eats the kangaroo rat. The snake gets energy from this food. Finally, suppose a hawk eats the snake. How is the hawk meeting its need for energy? The hawk gets energy from its food, the snake.

Reading Check Trace the path of energy from producer to predator.

Following the arrows in **Figure 7,** you'll notice it shows what each organism eats in this desert food chain. The arrows point in the direction of energy flow. Like all food chains, the one in **Figure 7** starts with the Sun. Then, in this case, a bush is the producer that brings the Sun's energy into the system. All organisms farther up the food chain depend on the bush to convert energy from sunlight into food.

Visual Check **Figure 7** Determine how energy flows through a food chain.

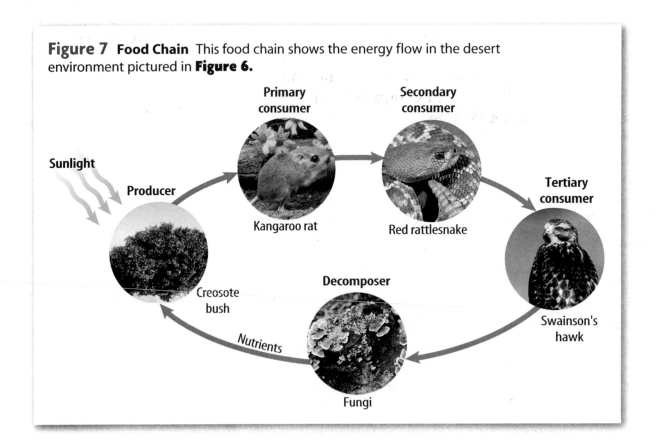

Figure 7 **Food Chain** This food chain shows the energy flow in the desert environment pictured in **Figure 6.**

Sunlight

Producer

Primary consumer
Kangaroo rat

Secondary consumer
Red rattlesnake

Tertiary consumer
Swainson's hawk

Creosote bush

Decomposer
Fungi

Nutrients

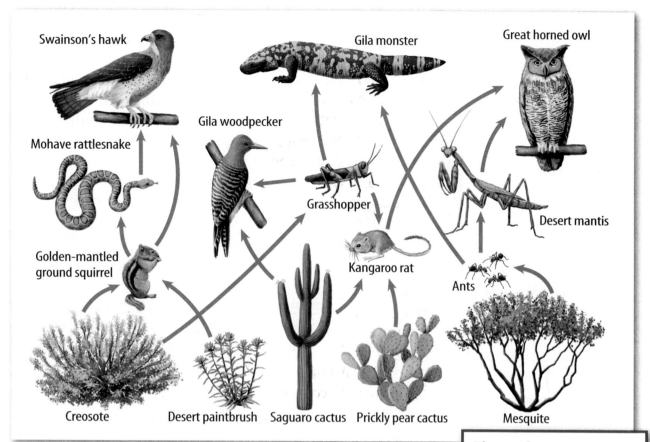

Swainson's hawk

Gila monster

Great horned owl

Gila woodpecker

Mohave rattlesnake

Grasshopper

Desert mantis

Golden-mantled ground squirrel

Kangaroo rat

Ants

Creosote   Desert paintbrush   Saguaro cactus   Prickly pear cactus   Mesquite

Figure 8 Food Web This food web shows the flow of energy from each organism in a desert environment.

Concepts In Motion

To see an animation of an antarctic food web, visit ca6.msscience.com.

Food Webs

A food chain is a simple model of energy flow, with each organism eating just one other type of organism. Actually, the picture is not so simple. An ecosystem contains more than one type of producer, and most organisms eat more than one type of food. A **food web,** shown in **Figure 8,** is a more complicated model of the flow of energy in an ecosystem.

You can see that as chemical energy passes through the desert ecosystem, it supports the life of many types of organisms. In this case, the producers are cacti, sagebrush, creosote bushes, and other desert plants. The consumers are the insects, lizards, snakes, foxes, and other organisms that eat these plants or other organisms. If you look carefully at **Figure 8,** can you find the food chain shown in **Figure 7?** It hasn't changed—you still should be able to follow arrows from the seeds to the kangaroo rat, the snake, and the hawk. What has changed? More organisms and arrows have been added to give a fuller picture of the variety of paths through which energy can flow.

 Figure 8 Starting with a producer, trace two ways a hawk can obtain energy.

MiniLab

00:15 minutes

What do they eat if they live in that biome?

Earth's biomes vary in climate, abiotic factors, and living organisms. Still, they all have plants and animals that need energy to live and grow. Energy is transferred from the Sun to plants to consumers. With research, you can become an expert on one food web that exists in one specific biome.

Procedure

1. Choose a biome according to your teacher's directions.

2. Use **science text** and **library materials** to research organisms included in your biome.

3. Draw a model food web for several of the plants and animals that live in the biome.

4. Use arrows to show the energy flow through the food web.

5. Label producers and primary, secondary, and tertiary consumers.

6. Discuss your web with the class.

Analysis

1. **Describe** the main producers in your biome.

2. **List** the animals in your biome that are also in the biomes of your classmates.

3. **Explain** the source of energy for all producers and consumers in your biome. How does this compare to the source for other biomes?

5.b, 7.b, 7.d

Try at Home

Energy Pyramids

Food webs show pathways of energy flow through ecosystems, from producers to consumers and decomposers. However, food webs do not show how much energy is available to each type of organism. For this, you need an energy pyramid, like the one in **Figure 9.**

The bottom layer is the largest and contains the producers. Herbivores are in the next level up. **Primary consumers,** such as insects, eat producers. Going up to the next level are **secondary consumers,** such as snakes, which eat herbivores. **Tertiary consumers** are at the top of the pyramid. These predators, such as hawks, prey on organisms in the levels below.

Releasing Thermal Energy Why do you think the energy pyramid gets smaller toward the top? Less energy is available to organisms in the upper levels because each organism releases some of the chemical energy in food to the air as thermal energy. All organisms, from single-celled algae to whales, release some food energy as thermal energy. This is why less total energy is available with each step up an energy pyramid.

Reading Check Why is less energy available to tertiary consumers?

Pyramid Size Compare the sizes of the energy pyramids in **Figure 9.** What can you conclude about the number of organisms supported in these biomes? Compare the bottom layer in the rainforest pyramid to the bottom layer in the desert pyramid. You can see that the producer layer is much larger for rain forests than deserts. This means that rain forests support a larger number of producers than you would find in a desert. The greater number of producers means that the upper layers of the rain forest pyramid can be larger too. This explains why more organisms live in rain forests than in desert biomes.

Figure 9 Energy Pyramids
The energy in an energy pyramid is dependent on the number of producers in an ecosystem.

Explain the shape of each of these energy pyramids.

3

2

1

Temperate Deciduous Forest

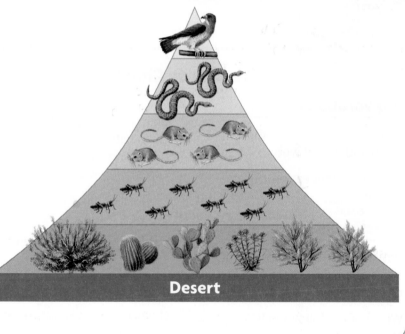

Desert

Tropical Rainforest

Concepts In Motion
To see an animation of an energy pyramid, visit ca6.msscience.com.

What do cars and organisms have in common?

The energy stored in gasoline burns in engines to power cars. Not all the energy in gasoline is used to make the car move. Much of the energy is released to the atmosphere as thermal energy. Once released, thermal energy cannot be recaptured.

As long as the Sun shines and producers are present, food will be made for life on Earth. Primary consumers eat producers for energy and nutrients. However, not all the energy stored in the producer is available for activities and growth of consumers. Like the car engine, much of the energy is released as thermal energy. Tertiary consumers have much less energy available to them than primary consumers. This is why energy pyramids are larger at the bottom than they are at the top.

LESSON 2 Review

Summarize

Create your own lesson summary as you design a **study web.**

1. **Write** the lesson title, number, and page numbers at the top of a sheet of paper.

2. **Scan** the lesson to find the **red** main headings.

3. **Organize** these headings clockwise on branches around the lesson title.

4. **Review** the information under each **red** heading to design a branch for each **blue** subheading.

5. **List** 2–3 details, key terms, and definitions from each **blue** subheading on branches extending from the main heading branches.

 **ELA6:** R 2.4

Standards Check

Using Vocabulary

1. Distinguish between a food chain and a food web. **5.a**

2. Organisms that eat producers are called _____. **5.c**

Understanding Main Ideas

3. **Construct** an energy pyramid and explain why the levels are different sizes. **5.a**

4. Which of the following is an example of a tertiary consumer? **5.a**

   A. mouse
   B. hawk
   C. creosote bush
   D. kangaroo rat

5. **Show** how a lion's life depends on producers. **5.a**

6. **Diagram** how energy from the Sun flows through an ecosystem of your choice. **5.a**

7. **Sequence** Recreate the diagram below. Fill in the food chain with desert organisms beginning with a producer and ending with a tertiary consumer. **5.b**

Applying Science

8. **Construct** your own food chain. Label organisms as producers or consumers and trace the path of energy. **5.c**

9. **Evaluate** why the bottom layer of the energy pyramid is larger in a tropical rain forest than one of an artic tundra. **5.c**

 Science nline

For more practice, visit **Standards Check** at <u>ca6.msscience.com</u>.

MiniLab

How much energy flows through an ecosystem?

Plants use only about 10 percent of the energy from the Sun to produce food. Each time an organism eats a plant or other animal, only 10 percent of the available food energy is retained by the consumer. The remaining 90 percent is released as thermal energy. Using orange juice will help you visualize how much food energy each level of consumer receives.

Procedure

1. Read and complete a lab safety form.

2. Draw a food chain beginning with a producer using the energy from the Sun. You could use the Sun, a plant, a grasshopper, and a frog.

3. Pour 1 L (1,000 mL) of **orange juice** into a **beaker**. This represents energy from the Sun.

4. Pour 100 mL (10 percent of 1,000 mL) of orange juice into a **graduated cylinder.** The plant only uses 10 percent of the original amount of energy available from the Sun.

5. From the 100 mL, pour 10 mL of orange juice into another **graduated cylinder.** This represents the energy that is captured by the grasshopper. The rest (90 mL) is released as thermal energy.

6. From the 10 mL, pour 1 mL into another **graduated cylinder.** The frog gains only this amount of energy that was originally supplied by the Sun.

Analysis

1. **Explain** what happens to most of the energy at each consumer level.

2. **Infer** why each biome has millions of insects, thousands of small animals, and only hundreds of large predators.

 **Science Content Standards**

5.a Students know energy entering ecosystems as sunlight is transferred by producers into chemical energy through photosynthesis and then from organism to organism through food webs.
7.b Select and use appropriate tools and technology to perform tests, collect data, and display data.
7.e Recognize whether evidence is consistent with a proposed explanation.

567

LESSON 3

Science Content Standards

5.b Students know matter is transferred over time from one organism to others in the food web and between organisms and the physical environment.
7.a Develop a hypothesis.
7.b Select and use appropriate tools and technology (including, calculators, computers, balances, spring scales, microscopes, and binoculars) to perform tests, collect data, and display data.
7.g Interpret events by sequence and time from natural phenomena (e.g., the relative ages of rocks and intrusions).
Also covers: 5.a, 5.c, 7.d

Reading Guide

What *You'll Learn*

▶ **Summarize** cycles of matter.

▶ **Explain** where matter comes from for plant growth.

Why *It's Important*

Matter needed for life on Earth is neither created nor destroyed, but is cycled through producers, consumers, and decomposers.

Vocabulary

nitrifying bacteria
nitrogen cycle
phosphorus cycle
carbon cycle

Review Vocabulary

water cycle: a model describing how water moves from Earth's surface to the atmosphere and back to the surface again (p. 472)

Matter in Ecosystems

Main Idea Matter cycles in ecosystems.

Real-World Reading Connection What do you think happens to leaves that fall from trees in a forest? Nobody rakes or sweeps them away, so you might think that they keep piling up year after year, building deeper and deeper piles. This doesn't happen. Dead organisms break down, making materials available for new growth.

Cycles of Matter

Leaves and other dead plant and animal materials, like the compost in **Figure 10,** gradually break down. Some of the chemicals they contain become part of the organic matter in soil. Others go into the air as gases. In this way, carbon, nitrogen, and other elements become available to support new life.

The amount of matter—anything that has mass and takes up space—on Earth never changes. Elements that make up matter cycle between living things and the nonliving environment.

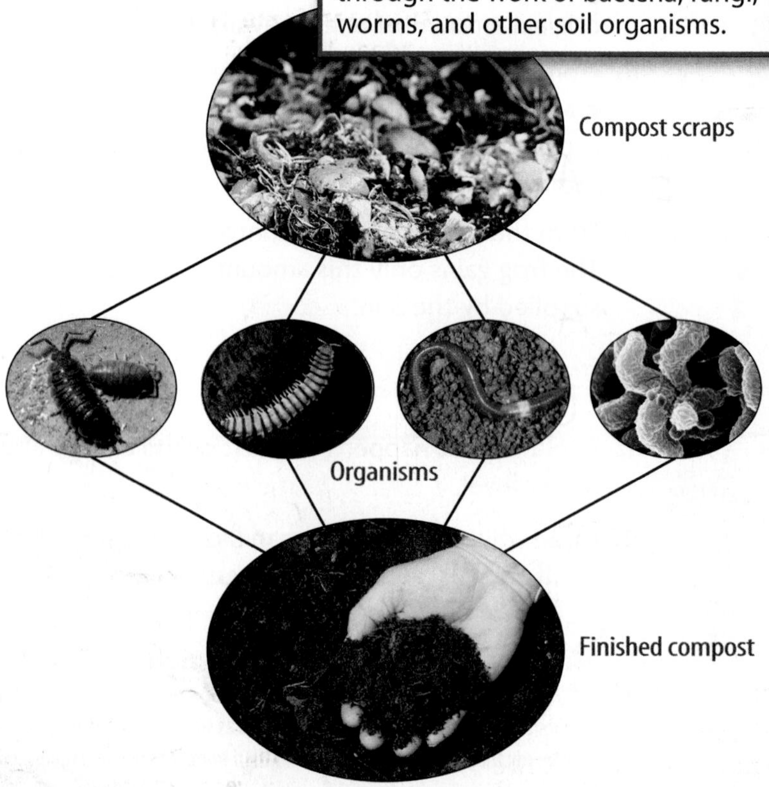

Figure 10 Leaves break down through the work of bacteria, fungi, worms, and other soil organisms.

Compost scraps

Organisms

Finished compost

Water Cycle

Think back to what you read about the water cycle in Chapter 12. Earth's supply of water is not growing or shrinking. Instead, water cycles from land to sea to air, then back to land. It is taken up by plants and animals and then released back into the environment.

Nitrogen Cycle

Like energy, elements are not created or destroyed. At times, these nutrients become part of the cells that make up organisms. At other times, these nutrients exist as abiotic factors in the environment.

 **Reading Check** What can limit plant growth?

Nitrogen makes up 78 percent of our air, but plants cannot use this nitrogen. Some soil bacteria, called **nitrifying bacteria,** change nitrogen into forms that plants can take up through their roots. Plants then build nitrogen into their tissues as they grow. Nitrogen continues up the food chain as one organism eats another. As dead organisms decay, the nitrogen goes back into the soil and air. The **nitrogen cycle,** shown in **Figure 11,** describes how nitrogen moves from the atmosphere to the soil, to living organisms, and then back to the atmosphere.

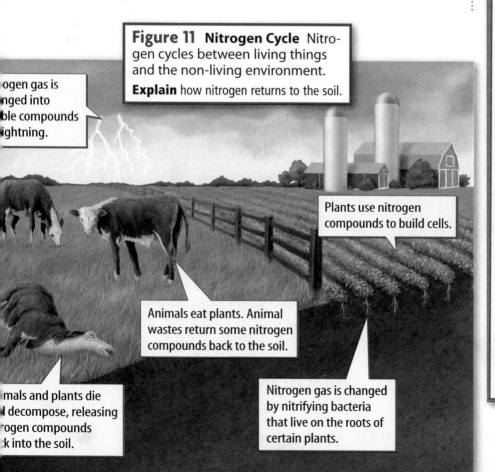

Figure 11 Nitrogen Cycle Nitrogen cycles between living things and the non-living environment.
Explain how nitrogen returns to the soil.

...ogen gas is ...nged into ...ble compounds ...ightning.

Plants use nitrogen compounds to build cells.

Animals eat plants. Animal wastes return some nitrogen compounds back to the soil.

...mals and plants die ...l decompose, releasing ...rogen compounds ...k into the soil.

Nitrogen gas is changed by nitrifying bacteria that live on the roots of certain plants.

00:15 minutes

Is your soil rich in nitrogen?

Plants need nitrogen to grow. Plants get nitrogen from the soil. Test your soil to see if it has a good percentage of nitrogen.

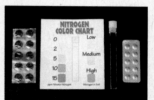

Procedure

1. Complete a lab safety form.
2. Follow directions carefully for using a **nitrogen soil test kit.**
3. Use the color chart to determine the quantity of nitrogen in your **soil sample.**
4. Compare with classmates to see if some soils have more nitrogen than others.

Analysis

1. **Determine** if the soil sample you brought doesn't have enough nitrogen.
2. **Hypothesize** why some of your classmates who live in the same region have soil that is rich or deficient in nitrogen.
3. **Deduce** how nitrogen got into your soil sample.

 5.b, 7.a, 7.b, 7.g

Phosphorus Cycle

Unlike nitrogen, phosphorus does not exist as a gas in Earth's atmosphere. Instead, phosphorus is in the soil from the weathering of rocks. The **phosphorus cycle,** shown in **Figure 12,** describes how phosphorus moves from soil to producers and consumers, and back to soil.

Like nitrogen, phosphorus moves from plants to animals when herbivores eat plants, and when carnivores eat herbivores. Phosphorus returns to the soil through animal wastes and when dead animals and plants decay.

 Reading Check How does phosphorus get into soils?

The Carbon Cycle

The **carbon cycle** describes how carbon moves between the living and nonliving environment. Life on Earth would not exist without carbon because carbon is the key element in the sugars, proteins, starches, and other compounds that make up living things.

Figure 13 shows how producers take carbon dioxide from the air during photosynthesis. Most organisms send carbon dioxide back into the air in a process called cellular respiration. In this way, carbon keeps cycling between the living and nonliving environment.

When a tree grows, where does its new matter come from? Using light energy, producers combine carbon dioxide from the air and water from the soil to make sugars and other compounds. The amounts of carbon, hydrogen, and oxygen in the ecosystem haven't changed. The matter simply has changed from air, water, and nutrients into living parts of the tree. As the tree grows, dies, and then decomposes, matter continues cycling between living and nonliving forms.

Visual Check **Figure 13** List three ways that carbon is released to the atmosphere.

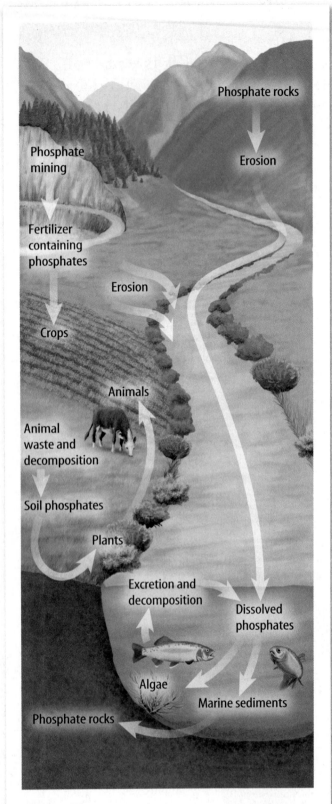

Figure 12 Cycling Phosphorus Phosphorus cycles between living things and the nonliving environment.

Visualizing the Carbon Cycle

Figure 13

Carbon—in the form of different kinds of carbon-containing molecules—moves through an endless cycle. The diagram below shows several stages of the carbon cycle. It begins when the plants and algae remove carbon from the environment during photosynthesis. This carbon returns to the environment through several carbon-cycle pathways.

A Air contains carbon dioxide in the form of carbon dioxide gas. Plants and algae use carbon dioxide to make sugars, which are energy-rich, carbon-containing compounds.

B Organisms break down sugar molecules made by plants and algae to obtain energy for life and growth. Carbon dioxide is released as a waste.

C Burning fossil fuels and wood releases carbon dioxide into the atmosphere.

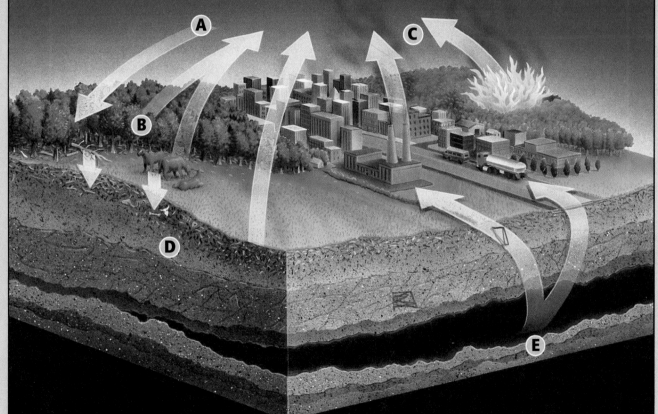

D When organisms die, their carbon-containing molecules become part of the soil. The molecules are broken down by fungi, bacteria, and other decomposers. During this decay process, carbon dioxide is released into the air.

E Under certain conditions, the remains of some dead organisms might be gradually changed into fossil fuels such as coal, gas, and oil. These carbon compounds are energy rich.

What have you learned?

Have you ever played with building blocks? They come in all shapes, colors, and sizes. Perhaps you've used the blocks to make buildings, or even a whole town. You can tear it down and use the same blocks to build a spaceship or a car. Like building blocks, matter on Earth is used to build organisms. When organisms die, decomposers tear them down to their basic building blocks.

It takes energy to organize the building blocks from individual pieces to a castle, a car, or a boat. Similarly, organisms, including people, need energy for growth and daily activities. In the next chapter, you will read about how people use Earth's energy and material resources. You'll also read about how the use of **resources** can affect ecosystems.

LESSON 3 Review

Summarize

Create your own lesson summary as you write a script for a **television news report.**

1. **Review** the text after the **red** main headings and write one sentence about each. These are the headlines of your broadcast.

2. **Review** the text and write 2–3 sentences about each **blue** subheading. These sentences should tell *who, what, when, where,* and *why* information about each **red** heading.

3. **Include** descriptive details in your report, such as names of reporters and local places and events.

4. **Present** your news report to other classmates alone or with a team.

 **ELA6:** LS 1.4

Standards Check

Using Vocabulary

Complete the sentences using the correct term.

phosphorus cycle
nitrifying bacteria

1. Nitrogen is changed to a form that plants can take up through their roots by _____. **5.b**

2. The _____ describes how phosphorus moves from the soil, to living organisms, and then back to the soil. **5.b**

Understanding Main Ideas

3. **Give examples** of how energy flows through ecosystems and matter cycles. **5.b**

4. **Explain** why carbon is needed for life on Earth. **5.b**

5. Where do trees get most of their matter for growth? **5.b**

 A. from air and water
 B. from soil
 C. from animals
 D. from other plants

6. **Summarize** Complete a chart like the one below demonstrating the carbon cycle. **5.b**

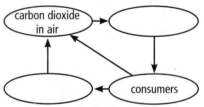

Applying Science

7. **Hypothesize** how your body gains matter as you grow. **5.b**

Science Online

For more practice, visit **Standards Check** at ca6.msscience.com.

Applying Math

Percent of Nitrogen in Soil

The amount of nitrogen in compost can vary in different samples. The amount of nitrogen available for plants to grow can be determined by the percentage of nitrogen in the compost and depth of compost.

Amount of Nitrogen in Compost at 1 Percent Concentration	
Depth (cm)	Mass (kg)
0.125 cm	0.5 kg
0.25 cm	1.1 kg
0.5 cm	2.2 kg
1 cm	4.4 kg
2 cm	8.8 kg

Amount of Nitrogen in Compost at 2 Percent Concentration	
Depth (cm)	Mass (kg)
0.125 cm	1.1 kg
0.25 cm	2.2 kg
0.5 cm	4.4 kg
1 cm	8.8 kg
2 cm	17.6 kg

Example

How much more nitrogen is available in a sample of compost that is 2 percent nitrogen if the depth is increased from 0.125 cm to 1 cm?

What you know:
- Amount of nitrogen available in 1 cm: 8.8 kg
- Amount of nitrogen available in 0.125 cm: 1.1 kg

What you need to find:

The difference in available nitrogen
$8.8 - 1.1 = 7.7$ kg

Answer: The difference in available nitrogen in a 2 percent nitrogen sample that increases in depth from $\frac{1}{8}$ cm to 1 cm is 7.7 kg.

Practice Problems

1. How much more nitrogen is available in a sample of compost that is 1 percent nitrogen if the depth is increased from 0.125 cm to 1 cm?

2. How much more nitrogen is available in a sample of compost that is 2 percent nitrogen if the depth is increased from 0.125 cm to 0.25 cm?

Science Online
For more math practice, visit **Math Practice** at ca6.msscience.com.

Is it primary, secondary, or tertiary?

Materials

a list of organisms
  from your teacher
photos of organisms
scissors
tape or glue

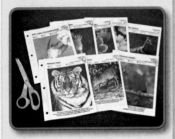

Safety Precautions:

WARNING: *Use caution with scissors.*

Science Content Standards

5.a Students know energy entering ecosystems as sunlight is transferred by producers into chemical energy through photosynthesis and then from organism to organism through food webs.
5.b Students know matter is transferred over time from one organism to others in the food web and between organisms and the physical environment.
5.c Students know populations of organisms can be categorized by the functions they serve in an ecosystem.
7.a Develop a hypothesis.
Also covers: 7.b, 7.d

Problem

Throughout this chapter, you have read many things about how plants and animals live in an ecosystem. You know which organisms are producers, consumers, decomposers, and scavengers. You can classify animals as herbivores, carnivores, and omnivores. You even know when animals are predators and when they are prey. Now you can organize your knowledge and demonstrate what you know.

Form a Hypothesis

➤ **Review** the results from this chapter's laboratory investigations.

➤ **Make a prediction** about how organisms interact in a biome.

Collect Data and Make Observations

1. Read and complete a lab safety form.
2. Look over the list from your teacher to determine which biome you and your partner are researching.
3. Research each organism on the list to classify it as a producer, a consumer, a predator, prey, an herbivore, a carnivore, an omnivore, a decomposer, or a scavenger.
4. Find photographs of each organism; you can also use copies and color them correctly.
5. Place photos in order of the food web and draw arrows for correct organization.
6. Classify organisms as producers and consumers.
7. Classify consumers as herbivores, carnivores and omnivores.
8. Include the abiotic factors that affect the ecosystem.
9. **Cleanup and Disposal:** Clean up paper cuttings, glue and scissors.

Analyze and Conclude

1. **Identify** which species are producers in the food web for your biome.
2. **Determine** which species are secondary consumers.
3. **Determine** which species are the top carnivores in the biome.
4. **Evaluate** whether there are any species without any predators in your food web.
5. **Hypothesize** what might help control the population of species that have no predators.
6. **Describe** how the climate, soil, and water availability in your biome determines which organisms survive there.
7. **Error Analysis:**
   - Check a source for your biome to be sure you have not included organisms that could not live there.
   - Check to be sure you have not included animals that do not live on the continent you are researching.
   - Check your energy flow; do the arrows move from the eaten to the eater?

Communicate

 **WRITING in Science**   ELA6: W 2.3

Write an oral report to present your research to the class explaining your choices and labels. Learn from the other partnerships.

Real World Science

You could be an oceanographer!

Oceanographers study the oceans. You might spend your time in the laboratory studying the water, microscopic creatures, or computers that model waves, wind currents, and the tides. Or, you could spend your time in the field, studying sea organisms and their habitat, or tracing the effects of humans on marine environments. To prepare for this career, you need to take courses in the sciences and mathematics.

Visit **Careers** at **ca6.msscience.com** to find out about oceanographers. Write a one-page journal entry of a day on the job. List your activities, who your employer is, and where you're working. Use proper grammar, spelling, punctuation, and capitalization.

TRACING MERCURY

Mercury is a liquid, dense, and toxic chemical not naturally found in the food we eat. Mercury in oceans can be taken in by fish. It can then spread within food chains as the fish are consumed by other organisms that humans eat. The technicians shown here are sampling for mercury.

Visit **Technology** at **ca6.msscience.com** to find out more about mercury in the food web. Create a poster showing the path of mercury in the food web from initial release into the environment to human consumption.

California's First Oceanographers

The Native Americans that flourished in the central coastal California region before the arrival of European settlers are now collectively known as the Chumash. Anthropologists have found evidence that suggests that the Chumash settlements out on the Channel Islands were among the first in North America. The Chumash were excellent shipbuilders. Their boats, called tomols, were built from redwood planks sealed with local natural substances such as tar.

Visit **History** at <u>ca6.msscience.com</u> to find out more about these early Californians. Divide the class into two groups and have each group reenact a day spent voyaging to new areas.

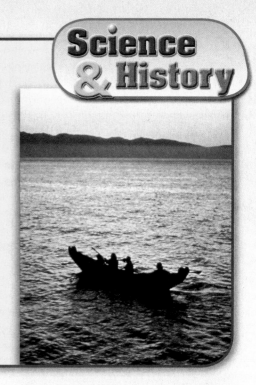

The California Sea Otter

The California sea otter is a keystone species. It plays a major role in the kelp forest community by controlling kelp eaters, such as sea urchins. In the 1700s they were hunted for their fur and nearly eliminated by the 1900s. By the 1930s, only a small group of 50 to 300 sea otters remained near Big Sur, California. There were therefore too many sea urchins, and kelp forests were almost destroyed. In 1977, they were added to the endangered species list and given some protection.

Visit **Society** at <u>ca6.msscience.com</u> to learn more about sea otters. Create a graph showing the world's approximate sea otter population around 1750, 1900, 1930, 1970, and today.

The BIG Idea Matter cycles between organisms and the abiotic environment. Energy flows one way, from sunlight to producers to consumers and decomposers.

Lesson 1 Producers and Consumers

5.a, 5.c, 7.a, 7.g

Main Idea Producers make their own food, most using energy from the Sun. All other organisms depend on producers as their energy source.

- Producers provide food energy to ecosystems.

- Through photosynthesis, producers capture energy from sunlight and use it to make food.

- Producers include green plants and some types of protists and bacteria.

- Consumers eat other organisms or their wastes.

- Consumers include carnivores, herbivores, omnivores, scavengers, and decomposers.

- **carnivore** (p. 557)
- **consumer** (p. 556)
- **decomposer** (p. 558)
- **ecology** (p. 553)
- **herbivore** (p. 557)
- **omnivore** (p. 557)
- **photosynthesis** (p. 553)
- **producer** (p. 553)
- **protozoan** (p.556)
- **scavenger** (p. 558)

Lesson 2 Energy in Ecosystems

5.a, 5.b, 5.c, 7.b, 7.d, 7.e

Main Idea Energy flows through ecosystems, from producers to consumers and decomposers.

- Energy flows one way from producers to consumers and decomposers.

- Ecosystems would run out of energy if producers stopped changing energy from sunlight into food.

- Each organism releases some energy to the environment as heat.

- Each step of a food chain contains less total energy than the step before.

- **food chain** (p. 562)
- **food web** (p. 563)
- **primary consumer** (p. 564)
- **secondary consumer** (p. 564)
- **tertiary consumer** (p. 564)

Lesson 3 Matter in Ecosystems

5.a, 5.b, 5.c, 7.a, 7.b, 7.d, 7.g

Main Idea Matter cycles in ecosystems.

- As organisms live, grow, die, and decompose, matter cycles between living and nonliving forms.

- Elements such as nitrogen and phosphorus are not created or destroyed.

- Nitrifying bacteria convert nitrogen in air into forms that plants can take up through their roots.

- Carbon becomes part of living things when green plants produce sugars through photosynthesis. These sugars are the basis for all food on Earth.

- Through respiration, organisms release carbon back into the air as carbon dioxide.

- **carbon cycle** (p. 570)
- **nitrifying bacteria** (p. 569)
- **nitrogen cycle** (p. 569)
- **phosphorus cycle** (p. 570)

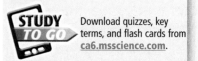 Download quizzes, key terms, and flash cards from ca6.msscience.com.

 Interactive Tutor ca6.msscience.com

Linking Vocabulary and Main Ideas

Use vocabulary terms from page 578 to complete this concept map.

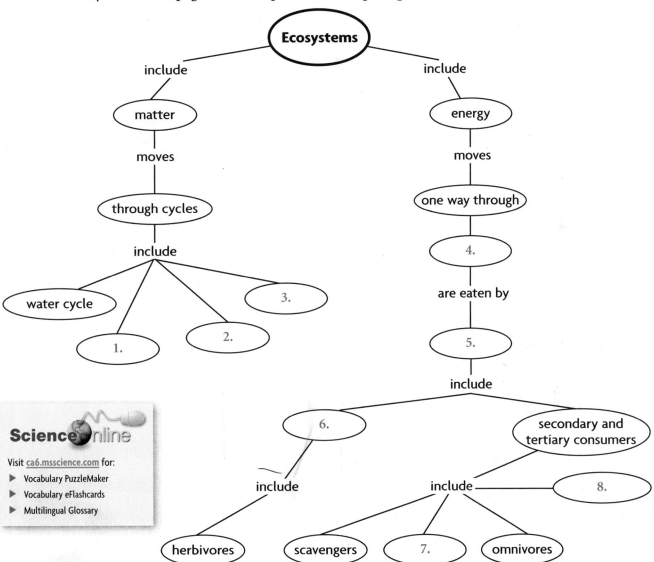

Using Vocabulary

Choose the word or phrase that best answers the question.

9. Which is a dung beetle?
  A. a producer   **C.** a scavenger
  B. an omnivore   **D.** an herbivore

10. Which is the process that producers use to make their own food using energy from the Sun?
  A. nitrogen cycle   **C.** photosynthesis
  B. chemosynthesis   **D.** energy pyramid

11. Which term describes a more complicated model of energy flow in an ecosystem?
  A. food web   **C.** phosphorus cycle
  B. food chain   **D.** carbon cycle

12. What are animals that eat herbivores called?
  A. omnivores   **C.** primary consumers
  B. producers   **D.** secondary consumers

Understanding Main Ideas

Choose the word or phrase that best answers the question.

1. Which describes a vegetarian?
   A. herbivore   **5.c**
   B. carnivore
   C. omnivore
   D. scavenger

2. Suppose you look through a microscope at a drop of pond water and see a protozoan eating another organism. From what you have seen, how would you classify the protozoan?
   A. scavenger   **5.c**
   B. producer
   C. omnivore
   D. carnivore

3. Why do all living things rely on producers?
   A. Producers recycle matter in ecosystems. **5.a**
   B. All living things eat producers.
   C. Producers change light energy into food.
   D. Producers eat other organisms.

Use the figure below to answer question 4.

The leaves take up carbon dioxide from the air.

Energy from sunlight

Water from roots enters the leaves.

Oxygen is released.

Sugar is made in the green parts of the plants and used throughout the plant.

4. What process is illustrated in the figure?
   A. water cycle   **5.a**
   B. nitrogen cycle
   C. photosynthesis
   D. food chain

5. What might happen if Earth had no decomposers?
   A. Scavengers would not be able to find food.
   B. Nitrogen and other nutrients would become limiting factors.
   C. Consumers would have nothing left to eat.
   D. Producers would run out of energy.   **5.b**

Use the figure below to answer questions 6–8.

6. What does the figure above show?
   A. a food plan   **5.b**
   B. a food chain
   C. an energy pyramid
   D. a food web

7. Why do the layers in the figure get smaller toward the top?
   A. Each layer has less energy than the layer below.   **5.b**
   B. There are more carnivores than the layer below.
   C. Each layer has less sunlight than the layer below.
   D. There are fewer herbivores than the layer below.

8. What are the organisms in the layer 1 called?
   A. herbivores   **5.c**
   B. producers
   C. carnivores
   D. consumers

Applying Science

9. **Explain** why you are not a producer. `5.a`

10. **Infer** Mushrooms cannot produce food through photosynthesis. Instead, they get their nutrition by digesting dead leaves and decaying organic matter. How would you classify mushrooms? Explain. `5.a`

11. **Illustrate** your own example of a food web. Label your organisms as *producers*, *primary consumers*, *secondary consumers*, *tertiary consumers*, *scavengers*, and *decomposers*. `5.a`

12. **Research** what elephants eat. In which level would you place an elephant on an energy pyramid? `5.a`

13. **Describe** the path of energy as it flows through a food chain. `5.a`

14. **Hypothesize** what happens to water when it evaporates. Design a way to demonstrate Earth's water cycle in a container in your classroom. Predict what processes will occur and form a hypothesis to explain why. `5.b`

Use the figure below to answer questions 15 and 16.

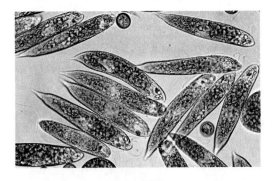

15. **Identify** the type of organism in the figure. `5.a`

16. **State** why this organism is classified as a producer. `5.a`

17. **Describe** the ecological roles of various organisms in a desert food web. `5.d`

WRITING in ▶Science

18. **Write** one paragraph explaining how carbon cycles through ecosystems. `ELA6: W 2.2`

Cumulative Review

19. **Explain** how limiting factors can affect food webs. `5.e`

20. **Infer** which type of biome you would expect to have the largest energy pyramid. `5.d`

Applying Math

Use the table below to answer questions 21 through 24.

Amount of Nitrogen in Compost at 2 Percent Concentration	
Depth (cm)	**Mass (kg)**
0.125 cm	1.3 kg
0.25 cm	2.7 kg
0.5 cm	5.4 kg
1 cm	10.9 kg
2 cm	x kg

21. How much nitrogen does compost contain at a depth of 0.125 cm? `MA6: NS 2.0`

22. As you dig deeper, would you expect the amount of nitrogen in compost to increase or decrease? `MA6: NS 2.0`

23. About how much nitrogen do you think would be contained in compost that is 2 cm deep? `MA6: NS 2.0`

24. How much more nitrogen is contained in compost that is 1 cm deep than in compost that is 0.5 cm deep? `MA6: NS 2.0`

1 Producers are different from the consumers because only the producers are able to do what?

A swim in deep water

B make their own food

C contribute to the marine food web

D digest the nutrients in other organisms 5.c

2 What would happen to consumers if all producers perished?

A Consumers would also die.

B Consumers would begin making their own food.

C Consumers would decompose organic matter.

D Consumers would move to a new environment. 5.c

3 Use the figure below to answer question 3.

What kinds of organisms are shown in the photo?

A producers and consumers

B consumers and decomposers

C predators and prey

D decomposers and producers 5.c

4 What is the flow of energy in an ecosystem?

A consumer→producer→Sun→decomposer

B decomposer→consumer→producer→Sun

C producer→consumer→Sun→decomposer

D Sun→producer→consumer→decomposer 5.a

Use the figure below to answer questions 5 and 6.

5 Which of the items shown in the diagram contribute to the nitrogen cycle by releasing AND absorbing nitrogen?

A the decaying organism only

B the trees only

C the trees and the grazing cows

D the lightning and the decaying organism 5.b

6 Which of the items shown in the diagram contribute to the nitrogen cycle by ONLY releasing nitrogen?

A the decaying organism only

B the trees only

C the trees and the grazing cows

D the lightning and the decaying organism 5.b

7 Which is an organism that can directly convert energy from the Sun into food?

A producers

B decomposers

C omnivores

D consumers 5.c

8 **Which organism might compete with the mouse for seeds?**

A hawk

B lion

C fox

D sparrow **5.b**

9 **Where is most of the energy found in an energy pyramid?**

A at the top level

B in the middle levels

C at the bottom level

D all levels are the same **5.a**

10 **Where is nitrogen-fixing bacteria found?**

A in deep sea vents

B in the air

C in the stomachs of certain animals, like cows and horses

D on or in the roots of certain plants such as peas, clover, beans and soybeans **5.b**

11 **How is carbon released into the atmosphere?**

A evaporation

B cellular respiration

C rocks

D specialized bacteria **5.b**

12 **What is the correct transfer of energy between a weasel, a rabbit, grasses, and a coyote?**

A coyote→grasses→rabbit→weasel

B grasses→rabbit→weasel→coyote

C rabbit→grasses→weasel→coyote

D weasel→grasses→rabbit→coyote **5.a**

13 **Use the figure below to answer question 13.**

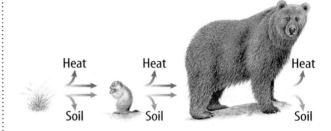

What term is used for the diagram above?

A energy ladder

B energy pyramid

C food chain

D food web **5.a**

14 **Which term describes organisms such as cows, horses, and goats?**

A carnivores

B detritivores

C herbivors

D omnivores **5.b**

15 **Why is the cycling of materials important for an ecosystem?**

A The amount of matter on Earth always changes, so the matter in an ecosystem must be conserved.

B The amount of matter on Earth changes at times, so the matter in an ecosystem must be kept in reserve.

C The amount of matter on Earth fluctuates, so the matter in an ecosystem must be balanced.

D The amount of matter on Earth never changes, so the matter in an ecosystem must be used over and over again. **5.b**

Resources

What can this field of gold be?

This is a field of crops that can eventually be used for energy. This energy source—called biodiesel—can be made from vegetable oils or animal fats. According to the National Biodiesel Board, as many as 250 school districts nationwide use biodiesel for their buses.

(Science *Journal*) Write down three things you already know that can save resources.

Launch Lab

Where did that come from?

Look around your classroom. Do you know what materials are used to make objects you use? Where did the materials come from?

Procedure

1. On a blank sheet of **paper,** draw a table with three columns—one for objects, one for materials, and one for original source.

2. Name five or six items in your classroom each made of a different material.

3. List the main material you think is used to make each item.

4. Guess the original source of the material for each item.

Think About This:

- **Determine** if it is obvious that some items are made from natural materials, such as wood.

- **List** the items you think might be made from oil or petroleum products.

6.c

Science Online

Visit ca6.msscience.com to:

▶ view **Concepts in Motion**

▶ explore Virtual Labs

▶ access content-related Web links

▶ take the Standards Check

Renewable and Nonrenewable Make the following Foldable to show the differences between renewable and nonrenewable resources.

▷ **STEP 1** **Fold** the bottom of a horizontal sheet of paper up about 4 cm.

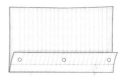

▷ **STEP 2** **Fold** in half.

▷ **STEP 3** **Unfold** once and dot with glue or staple to make two pockets. **Label** as shown.

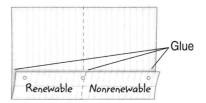

Glue

Renewable   Nonrenewable

Reading *Skill*  **ELA6:** R 2.4

Analyzing

As you read this chapter, use note cards to list natural and material resources. Explain why each resource is considered renewable or nonrenewable. Collect your note cards in the appropriate pocket.

Get Ready to Read

Questions and Answers

1 Learn It! Knowing how to find answers to questions will help you on reviews and tests. Some answers can be found in the textbook, while other answers require you to go beyond the textbook. These answers might be based on knowledge you already have or things you have experienced.

2 Practice It! Read the excerpt below. Answer the following questions and then discuss them with a partner.

> Most of us use natural energy resources each day without thinking about it. For example, you might awaken to an electric alarm clock, turn on the lights, and take a hot shower. You might listen to a radio as you eat pancakes cooked on a natural gas stove. You arrive at school on a bus and enter a classroom heated by oil or electricity. Before the school day even starts, you depended on many energy resources.
>
> —*from page 612*

- Can you name six ways you used energy resources today?
- What processes do you think were used to bring those resources to your home or school?
- What effect does using energy resources have on the environment?

3 Apply It! Look at some questions in the text. Which questions can be answered directly from the text? Which require you to go beyond the text?

Target Your Reading

Use this to focus on the main ideas as you read the chapter.

Reading Tip

As you read, keep track of questions you answer in the chapter. This will help you remember what you read.

1 Before you read the chapter, respond to the statements below on your worksheet or on a numbered sheet of paper.

- Write an **A** if you **agree** with the statement.
- Write a **D** if you **disagree** with the statement.

2 After you read the chapter, look back to this page to see if you've changed your mind about any of the statements.

- If any of your answers changed, explain why.
- Change any false statements into true statements.
- Use your revised statements as a study guide.

Before You Read A or D	Statement	After You Read A or D
	1 Items you use every day, such as clothes, books, and food, come from Earth's resources.	
	2 Nonrenewable resources are replaced faster than they can be used.	
	3 Water is a renewable resource because it moves through cycles on Earth.	
	4 Fossil fuels formed millions of years ago from decayed plants and animals.	
	5 Oil is the most abundant fossil fuel on Earth.	
	6 Most of our energy comes from solar power and wind because the energy can be readily stored for later use.	
	7 Energy sources other than fossil fuels have not been fully developed because they cause pollution.	
	8 Most plastic is made from oil.	

Science Online

Print a worksheet of this page at ca6.msscience.com.

Reading Guide

What *You'll Learn*

▶ **Identify** material resources used by people to meet their basic needs and to make their lives more comfortable.

▶ **Give examples** of material resources used to make common objects.

▶ **Classify** Earth's material resources as renewable or nonrenewable.

Why *It's Important*

Understanding Earth's natural resources will help you use them wisely.

Vocabulary

natural resource
renewable natural resource
estuary
nonrenewable natural resource

Review Vocabulary

magma: molten rock in the Earth's crust (p. 96)

Natural Resources

Main Idea People use a variety of materials from different parts of Earth to meet a diverse range of needs.

Real-World Reading Connection Almost all objects you use—from jewelry to baseball bats to textbooks—are formed, at least in part, from Earth's natural materials. You are probably holding or looking at something made from Earth's material resources right now.

Organic resources

Natural resources are materials and energy sources that are useful or necessary to meet the needs of Earth's organisms, including people. Plants and animals that are living, or were alive at sometime in the past, are organic material resources. Think about your food. Nearly all of it is either animal or plant material. How many items like those in **Figure 1** come from plants or animals? Some of your clothes, such as denim jeans, are made from cotton plants. Wool, silk, and leather come from animals. Many buildings are made of wood and most home furnishings are manufactured from plant materials. All of these are made from organic material resources.

Figure 1 Everything on this shelf is made of natural materials found on Earth.

Identify which products are from plant material. Which are from animals?

Inorganic resources

Not all natural resources are from plants or animals. Inorganic material resources, which include metals and minerals, do not come from organisms. Gold, boron, calcium carbonate, granite, and lanthanides are mined and exported to other countries from California. You are probably familiar with how gold is used. Boron is used in agriculture, detergents, ceramics, glass, fiberglass, wood treatments, and as a fire retardant. Calcium carbonate is used in paper, plastics, industry, agriculture, and food, and it has many other uses. Granite is used for making roads and building materials. Lanthanide products are used in cellular phones, televisions, computers, and transportation systems.

 Reading Check Name two examples of inorganic material resources.

Mining Costs A large amount of a mineral must be in one place to make it worth the cost of the mining operation. If the deposits are too small, it would cost more to separate the metal or minerals from the earth than what the materials are worth. The price of natural resources varies and so does the cost of the equipment to find and extract the resources. Throughout history, sometimes it was worthwhile to mine a particular resource and other times it was impractical to extract the same resource.

Building Structures Other inorganic resources include industrial and building materials, like those shown in **Figure 2.** Steel is made from iron that is extracted from Earth. Sand and gravel may not seem very valuable, but they are important resources for construction. Sand is used to make concrete that provides strength and stability to buildings, sidewalks, bridges, and other structures. Sand and gravel production in California is worth more than $1 billion each year. This is much more than the value of gold mined in the state.

Figure 2 Iron ore, sand, and gravel are important inorganic resources used for construction.

Determine which inorganic resource is used to make the steel structure in the bridge.

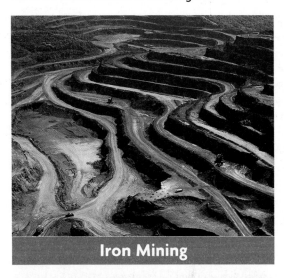

Iron Mining

Sand and Gravel

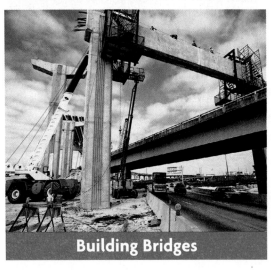

Building Bridges

How old is that tree?

Trees have adaptations that help them stay healthy and allow them to live a very long time. Do you think there is a correlation between the age and the height of a tree?

Data

1. Read the following facts about California trees and arrange them in a table.

   - Ponderosa pines grow to 50 m and can live 125 years.

   - Sequoias grow to 95 m and can live 3,200 years.

   - Redwoods grow to 112 m and can live 2,000 years.

   - White oaks grow to 21 m and can live 100 years.

2. Graph tree lifespan versus tree height.

Data Analysis

1. **Interpret** from your graph which trees live the longest.

2. **Give examples** of uses you think humans might have for these trees.

3. **Infer** which trees might not be considered a renewable resource.

Renewable Resources

Resources that can be replaced by nature as quickly as they are used are called **renewable natural resources.** Some renewable resources include trees, food resources, and animal resources.

Using up Renewable Resources

You might think that renewable resources will always be available for human use. But even renewable resources can be used up if we do not manage them carefully. For example, Douglas fir trees in California are used for lumber. These trees might be over 200 years old when they are cut down. It can be difficult to manage such slow-growing renewable resources if they can't be replaced as fast as they are cut. Some trees do grow faster than others, and the populations of these trees are easier to manage.

 Explain how trees can be a renewable resource.

Habitats as a Resource

Resources are more than the products that you use. For instance, a chicken needs food and water and protection from predators in order to become food for you. The chicken's habitat becomes a resource.

Forest Habitat Think about other resources that might be affected when cutting trees. Roads must be built to allow heavy machines into forested areas to cut trees. Road-building, cutting trees, and hauling them out of the forest can destroy forest habitat. Although young trees are usually planted to replace those that are lost, the forest habitat might not be completely replaced. In this case, the forest habitat is not a renewable resource. The trees are a renewable resource; they can replenish themselves. The habitat that a mature forest provides—food, shelter, and mates—is not a renewable resource. It would take more than a lifetime to replace the interactions.

Estuary Habitat When the rate of fish caught exceeds the rate of fish reproduction, this important renewable resource is affected. But fish populations can also decrease if areas the fish need for growth and reproduction are destroyed. An **estuary,** like the one in **Figure 3,** is a fertile area where a river meets an ocean. Estuaries contain a mixture of freshwater and salt water. Many species of fish and other organisms use estuaries to breed and raise their young because the calm waters and abundant food sources are ideal for young. But the ocean and its harbors are important resources for humans, too. Goods are transported by ships into harbors. The beautiful ocean view makes coastal areas ideal for marinas, houses, and hotels.

Habitat Loss In Colonial times, wetlands occupied more than 890,312 km^2 of the lower 48 states. By 1983, just 415,034 km^2 remained. This represents less than half the original wetlands. In California, 90 percent of wetlands have been developed. In fact, the San Francisco bay area alone has lost 85 percent of its wetlands. For fish species that depend on estuaries, the numbers of young fish that hatch each year and grow into adults decline as their nursery areas are destroyed. Many bird species depend on these coastal wetland habitats, too.

Effects of Habitat Loss Estuaries aren't the only areas in danger. Populations of many organisms in California and the United States are declining in numbers. California has nearly 300 threatened and endangered plant and animal species. Most of these declining numbers are due in some way to habitat loss. Not all of these species are resources used directly by humans. You learned in Chapter 13 that plants and animals are interconnected through food webs. A decrease in the population of one species whose habitat has been destroyed might have negative effects on species that serve as resources for people. For example, humans eat fish that are caught off the coast. Lost habitat means lost resources.

Estuary

Developed Harbor

Figure 3 Estuaries are important nesting and breeding grounds. They are also attractive places to live and work.

Describe how development in estuaries can affect populations of adult fish.

Nonrenewable Resources

Resources that are used more quickly than they can be replaced by natural processes are called **nonrenewable natural resources.** Some nonrenewable resources are used at rates far faster than their geologically slow formation rates. For example, gold, a nonrenewable resource, is deposited when hot water and molten rock, called magma, flows through spaces in underground rock. The hot magma heats water and gold travels with mineral solutions in the water. When the magma and solution cools, gold collects.

 How is gold deposited in rocks?

Gold is extracted from two types of mines in California, shown in **Figure 4.** Because gold is removed from Earth much faster than it can be deposited, it is a nonrenewable natural resource. Gold is worth a lot of money because it can easily can be formed into various shapes, it is pleasing to look at, and there is a limited amount of it on Earth.

California Gold Rush

In 1848, large veins of gold were discovered in California. Within months, thousands of people were traveling across the country in hopes of striking it rich. The gold seekers were called "49ers" because most left home in 1849. The California gold rush was a world attraction—gold seekers from all over the world flocked to California.

Although gold was easy to find at first, it quickly became difficult to make money because the mines yielded less and less gold. Those who did find gold often spent it on all the basic necessities of life. The people who made the most money were those who supplied the gold miners with food and other goods and services.

WORD ORIGIN

resource
from Latin *resurgere;* means *rise again*

Figure 4 Gold can be dislodged from rock through erosion. Placer mines are used to separate this gold from the rocks. Large veins are also taken from underground mines.

Placer Mine

Gold in Rock

Table 1 Average Water Use	
Daily Activity	**Water Used**
Flushing the toilet once	15 L
Taking a short shower	95 L
Taking a longer shower	190 L
Taking a bath	150 L
Washing clothes	190 L
Automatic dishwasher	38 L
Brushing your teeth while leaving the water running	7.5 L
Washing your hands while leaving the water running	30 L/min
Watering the lawn or plants with a hose	30 L/min

Water

Recall from Chapter 11 that water moves through cycles, evaporating from Earth's surface and condensing into clouds in the atmosphere. Water then returns to Earth's surface through precipitation. All the water on Earth is already here. Currently, there is no way to create new water. Freshwater is an important nonrenewable resource for California. Even though most of California is arid and dry, people use large amounts of water for irrigation, industry, and personal use. Most of the large soft-drink bottles hold two liters. Imagine how many liters of water you use for each activity in **Table 1.**

 Table 1 Estimate how much water your family uses every day.

California gets some of its freshwater from the Colorado River. Water from the lower Colorado River is divided between Arizona, California, Nevada, Native Americans, and Mexico. Many state and federal laws, as well as treaties with Mexico have been passed to try to **regulate** the use of this water resource. As states in the Colorado basin have increased their use of the river's water, there is concern about how long the Colorado River can be relied on as a water source. One way to solve this problem is to reuse water. Water that has been reclaimed from municipal wastewater, called sewage, can be treated and reused.

ACADEMIC VOCABULARY
regulate (REH gyuh layt)
(verb) to control or direct according to rule, principle, or law
Stoplights help regulate traffic flow.

What have you learned?

Earth has many natural resources that we have learned to use to manufacture materials. Renewable resources are those that can be replaced as quickly as they are used. Resources that are used faster than they can be replaced are nonrenewable. You will learn in the next section how some of Earth's natural resources are used for energy. As you read, think about how we can conserve natural resources without increasing our use of energy resources.

LESSON 1 Review

Summarize

Create your own lesson summary as you organize an **outline.**

1. **Scan** the lesson. Find and list the first **red** main heading.

2. **Review** the text after the heading and list 2–3 details about the heading.

3. **Find** and list each **blue** subheading that follows the **red** main heading.

4. **List** 2–3 details, key terms, and definitions under each **blue** subheading.

5. **Review** additional **red** main headings and their supporting **blue** subheadings. List 2–3 details about each.

 ELA6: R 2.4

Using Vocabulary

1. Materials and energy sources used to meet the needs of Earth's organisms are called _____. **6.b**

2. Which is a non-renewable resource?

   A. fish **6.b**

   B. water

   C. trees

   D. gold

Understanding Main Ideas

3. **Label** the following natural resources as either renewable or nonrenewable: silver, cotton, diamonds, fish, gravel, and lumber. **6.c**

4. **Explain** how the need for a particular resource can change its value. **6.b**

5. **Predict** how cutting down trees for development might change bird populations. **6.b**

 ## Standards Check

6. **Compare** Create a chart like the one below to compare renewable and nonrenewable resources. **6.b**

Renewable Resources	Nonrenewable Resources

7. **Deduce** how renewable resources could become depleted. **6.b**

Applying Science

8. **Recommend** a plan to save freshwater during the next week. Describe the steps and provide the amount of water you would save on the plan. **6.b**

9. **Explain** why sand and gravel are considered a natural material resource. **6.c**

Sciencenline

For more practice, visit **Standards Check** at ca6.msscience.com.

Reading Guide

What *You'll Learn*

▶ **Compare and contrast** the development and extraction of fossil fuels.

▶ **List** the energy resources that are used to create steam to turn turbines.

▶ **Assess** the advantages and disadvantages of each energy resource.

Why *It's Important*

People use vast amounts of energy resources daily.

Vocabulary

fossil fuel
geothermal energy
nuclear fission
nuclear fusion

Review Vocabulary

crust: the thin, rocky, outer layer of Earth (p. 103)

Energy Resources

Main Idea Some of Earth's natural resources can be used for energy, usually through conversion to electricity.

Real-World Reading Connection In 1962, Albert Szent-Gyorgyi, a scientist, wrote "discovery consists of seeing what everybody has seen and thinking what nobody has thought." How might this quote be useful for people searching for alternative energy sources?

Fossil Fuels

Fossil fuels are fuels formed in Earth's crust over hundreds of millions of years. Fossil fuels are nonrenewable resources because they cannot form as fast as they are used. Coal, oil, and natural gas are fossil fuels. Cars, buses, trains, and airplanes are powered by gasoline, diesel fuel, and jet fuel, which are made from oil. Coal is used in many power plants to produce electricity. Natural gas is used in manufacturing, for heating and cooking, and sometimes as a vehicle fuel. People on Earth are using more fossil fuels for energy today than they did in 1950, as shown in **Figure 5.**

Figure 5 Use of fossil fuels has increased since 1950.
Use Graphs How many more tons of fuel were used in 2000 than in 1970?

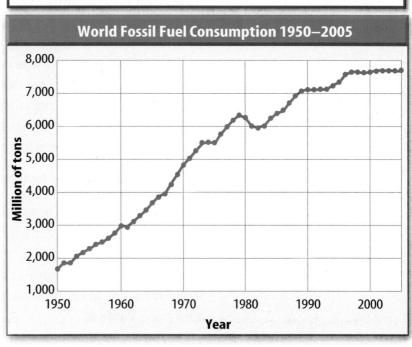

World Fossil Fuel Consumption 1950–2005

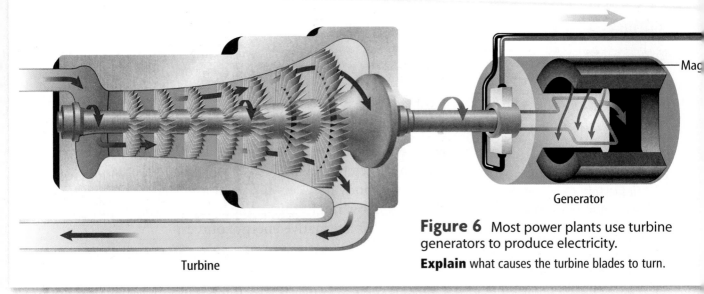

Mag

Generator

Figure 6 Most power plants use turbine generators to produce electricity.

Explain what causes the turbine blades to turn.

Turbine

ACADEMIC VOCABULARY

technology (tek NAH luh jee)

(noun) the application of science, especially for industrial or commercial use
Communication technology has help develop cellular telephones.

Energy from Fossil Fuels

The most useful energy sources can be readily converted to heat, electricity, and transportation. Today, people use fossil fuels as the primary source for energy because existing **technologies** readily convert energy from fossil fuels into forms we can use. Fossil fuels contain a lot of stored energy and burn easily to heat water and make steam. The steam then is used to turn large turbines that power generators, like the one in **Figure 6,** to create electricity.

Formation of Fossil Fuels

Fossil fuels are made from decayed plants and animals that have been preserved in Earth's crust by pressure, bacterial processes, and heat. This means that fossil fuels are organic natural resources. About 300 million years ago, before the age of the dinosaurs, Earth was covered with green, leafy plants. Earth's oceans and freshwater contained a large amount of algae and other small organisms. Over many hundreds of years, the decaying organisms were covered by sand and clay. The sand and clay layers formed into rock.

 What are fossil fuels made from?

As more rock was formed, the weight pressed down on the trapped remains of organisms. Eventually, over millions of years, heat and pressure from layers of rock pressing down the plant and animal remains turned them into the three main forms of fossil fuels—coal, oil, and natural gas—as shown in **Table 2.**

Table 2 Formation of Fossil Fuels

Fuel Formation	Description
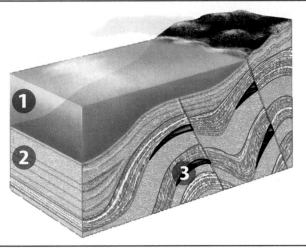	**Oil Formation** Microscopic plants and bacteria are the main source of oil. 1. Some of these organisms were producers, using energy from the Sun to make food for growth and reproduction. When they died, they fell to the seafloor. 2. The microscopic organisms were buried under clay. 3. Many layers of clay and mud increased the pressure and temperature, forming liquid oil.
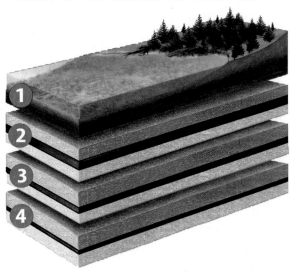	**Coal Formation** Coal is the most abundant of all the fossil fuels. 1. Coal formed from the incomplete decay of plants. 2. The partially decayed plant material, called peat, becomes sandwiched between layers of sediment. 3. Soft coal forms under moderate pressure and heat. 4. As more heat and pressure are applied, the soft coal becomes hard coal.
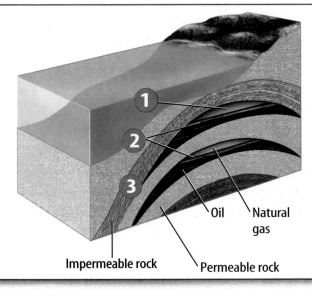 Impermeable rock Permeable rock Oil Natural gas	**Natural Gas Formation** Natural gas forms along with oil. 1. Because it is less dense than oil, natural gas is usually found above oil deposits. 2. At low temperatures, more oil is produced relative to natural gas. At higher temperatures, however, more natural gas is created, as opposed to oil. Natural gas is usually associated with oil in deposits that are 1–3 km below Earth's crust. 3. Natural gas is found in areas beneath layers of solid rock. The rock prevents the gas from escaping to the surface.

Figure 7 Oil is found deep within Earth's crust. Drill rigs can be located on land or in the ocean.

Oil

Oil, sometimes called petroleum, is used for heating and can be refined into gasoline, kerosene, or diesel fuel. As you will read later in this chapter, petroleum is also used for many other every day materials.

Oil formed mainly from ancient, microscopic plants and bacteria that lived in the ocean and saltwater seas. When these organisms died, they were quickly buried by clay and did not completely decay but formed muds rich in materials that form fossil fuels. The temperature and pressure increased as more clay and mud collected on top of the dead organisms. Further increases in temperature and pressure caused liquid oil to form. The oil flowed into spaces and pools in the rock.

To get the oil out of the rock spaces, geologists drill wells deep into Earth to the level of the reservoir of oil, as shown in **Figure 7.** Most of this oil is under pressure, so it flows to the surface, where it is collected. Oil collected out of the ground is sometimes called crude oil. Crude oil must be refined for use in cars, large trucks, airplanes, and trains.

 Figure 7 How is oil extracted from Earth?

Natural Gas

Natural gas forms along with oil. Because it is less dense, it most often is found on top of oil pools. Natural gas is valuable because it burns cleanly and can be transported easily in underground pipelines. Natural gas is used in many ways, such as for heating homes and cooking food.

Coal

Coal is the most abundant fossil fuel in the world. Coal forms where plant matter collects but is prevented from complete decay. The partially decayed plant material forms layers of spongy material called peat. Over time, the peat becomes buried and compressed. Increased pressure and temperature produce a solid, sedimentary rock known as coal.

Coal developed under moderate amounts of heat and pressure is called soft coal. If more heat and pressure is applied to soft coal, it can become hard coal. Soft coal retains some moisture and sulfur. When soft coal burns, it can release pollutants into the air. Hard coal is the cleanest-burning coal and has the greatest amount of carbon, so it provides more energy than soft coal. Coal is removed from Earth by strip mining and underground mining.

Figure 8 Coal near Earth's surface can be extracted by stripping away the land surface.

Strip Mines In strip mining, shown in **Figure 8,** large machinery is used to scrape the plants, soil, and rock layers off the ground above a vein of coal. Machines remove the surface of Earth, including the trees and other organic material, and place it beside the mine. Once the layer of coal is exposed, it can be broken apart and loaded into containers for transportation. Strip mining is cost-effective if the coal is close to the surface.

Underground Mines If the coal is deeper, underground mines, like the one in **Figure 9,** are used instead of strip mining. Underground mines are created by digging down into the earth at an angle to form tunnels. These tunnels go deeper and deeper until they reach the coal. Wooden beams and pillars are used to support the tunnel and make it safe for the miners. In some mines, tracks are laid to allow wagons to roll along the tunnels to move the coal toward the surface.

Mining coal produces a fine, black coal dust. If miners inhale coal dust, their lungs can be damaged. In the past, many miners suffered from a disease, called black lung. Today, miners wear protective clothing and masks to avoid inhaling coal dust.

 Why do underground coal miners need to wear masks?

Figure 9 Underground mines must be supported by beams and pillars. Miners must be protected from polluted air.

Alternatives to Fossil Fuels

Generating electricity doesn't have to come from burning fossil fuels. Alternative energy sources, including water, wind, ocean waves, and natural heat sources beneath Earth's surface can be used to produce electricity.

1 Hydroelectric Power

1 Hydroelectric Power

Hydroelectric power is a renewable energy source. Large dams block the flow of water from major rivers and create lakes behind the dams. As the water moves rapidly through the narrow openings in the dam, turbines generate electricity. More water can be released when needs are great.

2 Wind Energy

2 Wind Energy

Scientists are searching for cost-effective ways of harnessing energy from wind to generate enough energy to power local communities. Near Palm Springs, California, wind farms with long rows of wind towers connect to generators. This non-polluting, renewable method has proven successful on a small scale. Wind towers require a steady wind that is not too strong or too weak. The lack of consistent wind, however, prevents it from being reliable as a sole source of electricity.

3 Geothermal Energy

3 Geothermal Energy

Have you ever seen pictures of volcanoes? Perhaps you have visited a geyser spewing hot water. In some places on Earth, the magma is near Earth's surface. The heat energy in Earth's crust is called **geothermal energy.** The extreme heat found near geysers and volcanoes can be used to generate steam for electricity. Geothermal energy is clean, renewable, and safe, but there are only a few places where sufficient heat is near enough to the surface.

Can you explain why these alternative energy resources are renewable? Because water moves through cycles driven by solar energy, hydroelectric power is a renewable resource. Can wind be used up as it turns a turbine? Geothermal energy is renewable because magma supplies a continuous source of heat.

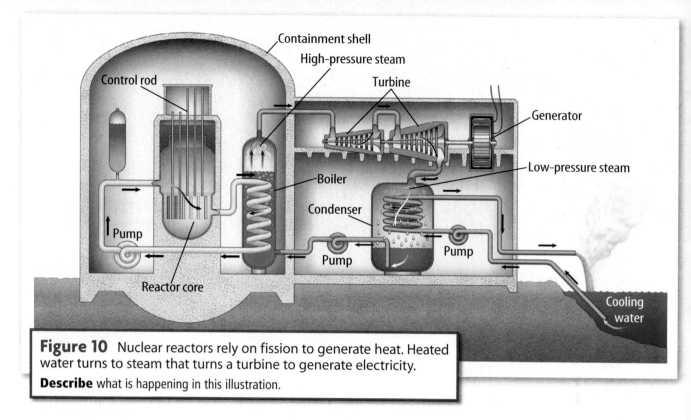

Figure 10 Nuclear reactors rely on fission to generate heat. Heated water turns to steam that turns a turbine to generate electricity.
Describe what is happening in this illustration.

Labels in figure: Containment shell, Control rod, High-pressure steam, Turbine, Generator, Low-pressure steam, Boiler, Condenser, Pump, Cooling water, Reactor core, Pump, Pump

Nuclear Energy

All matter is made of atoms—tiny particles that we cannot see. Even though atoms are very small, when they split, a large amount of energy is released. Splitting atoms to release energy is called **nuclear fission.** Atoms from uranium, an element that is mined from ore deposits in Earth, are split in a nuclear reactor, like the one in **Figure 10.** When atoms split, the energy that is released heats water in the reactor. Steam is produced and turns a turbine. The turbine runs a generator that creates electricity.

 Reading Check How is nuclear fission used to generate electricity?

Combining atoms also generates heat. In a **nuclear fusion** reaction, two atoms of hydrogen that are heavier than normal, called deuterium, join together to form one atom. It's the same type of reaction that powers the Sun. Like nuclear fission, nuclear fusion gives off large amounts of energy. Fusion reactions, however, are not easy to start. Have you ever tried to make the same ends of a magnet join together? Like the ends of a magnet, an atom's center has a positive charge. In order to get the atoms to fuse and start the fusion reaction, temperatures must be over 100,000,000°C.

Very little uranium is needed to supply a nuclear fission reactor; however, uranium is a nonrenewable resource. Deuterium can be made from water, which you learned is nonrenewable if not managed well.

WORD ORIGIN · · · · · · · · · · · ·
fission
from Latin *fissionem;* means *a breaking up, cleaving*

SCIENCE USE V. COMMON USE · ·
reaction
Science Use a change or transformation in which a substance decomposes or combines with other substances. *Sugars are made from carbon dioxide and water in a photosynthesis reaction.*
Common Use a response to a stimulus. *Poison ivy can cause a severe skin reaction.*

Figure 11 Science and engineering are exciting careers because there are so many new ideas to investigate. This is a new design to use the energy of the Sun for electricity.

Identify the part of the solar tower that generates electricity.

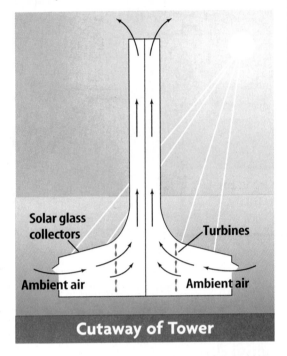

Solar glass collectors

Turbines

Ambient air   Ambient air

Cutaway of Tower

Outside of Tower

Solar Energy

Sunlight is a perpetual renewable resource because we can never use up all the energy from the Sun. Solar energy converts the light and heat energy from the Sun into electricity. Solar cells capture energy from the Sun and convert it to energy for calculators and other small appliances. Solar panels are made up of many solar cells and store the energy in a series of batteries for later use. For solar energy to be effective for larger-scale use, there must be a way to harness large amounts of energy from the Sun.

Solar Generated Power Have you ever noticed that it is hotter near the ceiling or on the upstairs level of a house? This is because warm air rises. Scientists in California are experimenting with panels constructed at the bottom of large towers, like the one in **Figure 11,** that collect sunlight and heat up. As the warm air rises, it turns turbines that generate electricity. Solar energy is a possibility for a clean, endless supply of energy in the future. We just need to develop reliable, cost-effective ways to get large amounts of energy from the Sun and store that energy for later use. **Figure 12** shows other ways solar energy is harnessed.

 Reading Check How does a solar tower generate electricity?

Drawbacks of Solar Power The Sun seems like the perfect source of clean energy. Currently though, solar energy is not cost-effective because the equipment is expensive, the batteries store a limited amount of energy, and solar panels do not collect energy at night or on very cloudy days. Solar energy is not a practical way to power cars and trucks. The weight of the batteries required to store the energy would use additional energy to transport. These added batteries would take up much needed space in a small car.

Visualizing Solar Energy

Figure 12

Sunlight is a renewable energy source that provides an alternative to fossil fuels. Solar technologies use the Sun's energy in many ways—from heating to electricity generation.

▼ **Electricity** Photovoltaic (PV) cells turn sunlight into electric current. They are commonly used to power small devices, such as calculators. Panels that combine many PV cells provide enough electricity for a home—or an orbiting satellite, such as the *International Space Station,*

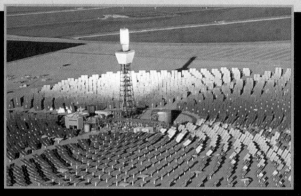

▲ **Power Plants** In California's Mojave Desert, an experimental solar power plant used hundreds of mirrors to focus sunlight on a water-filled tower. The steam produced by this system generates enough electricity to power 2,400 homes.

▼ **Cooking** In hot, sunny weather, a solar oven or panel cooker can be used to cook a pot of rice or heat water. The powerful solar cooker shown below reaches even higher temperatures. It is being used to fry food.

▼ **Indoor Heating** South-facing windows and heat-absorbing construction materials turn a room into a solar collector that can help heat an entire building, such as this Connecticut home.

▲ **Water Heating** Water is heated as it flows through small pipes in this roof-mounted solar heat collector. The hot water then flows into an insulated tank for storage.

Contributed by National Geographic

Biomass

Organic matter that makes up plants is known as biomass. Biomass can be used to produce fuels for electricity and transportation. Food crops, like corn and soybeans, grasses, trees, and even garbage are forms of biomass. However, most biomass must be converted, or refined, into usable energy forms. While there are many petroleum refineries today that convert oil into many different types of fuels and materials, there are few refineries for biomass.

 **Reading Check** What is biomass?

Wave Energy

Have you ever tried to stand up in the ocean when the waves were crashing over you? If so, you have experienced the energy of the waves. Harnessing the energy from the waves is challenging. It is difficult to find a location with regular, strong wave action, where a system of turbines would not be damaged on the rocks, and where it would work during low and high tides.

Total Energy Contributions

Figure 13 shows that in the United States, about 40 percent of energy use is from oil, 23 percent is from coal, 22 percent from natural gas, 8 percent from nuclear energy, and 3 percent from hydroelectric power. The rest is from a combination of geothermal, solar, and wind energy.

Figure 13 Both renewable and nonrenewable resources are used for energy in the United States.

Calculate how much energy comes from fossil fuels.

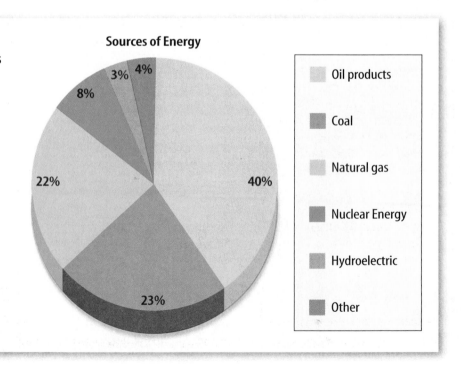

Sources of Energy

- Oil products 40%
- Coal 23%
- Natural gas 22%
- Nuclear Energy 8%
- Hydroelectric 3%
- Other 4%

What have you learned?

Billions of people all over the world use fossil fuels every day. Because fossil fuels are nonrenewable, Earth's supply of them is limited. In the future, they might become more expensive and difficult to obtain. You will learn in the next chapter how using fossil fuels can cause environmental problems. Developing alternative energy resources, such as wind and solar power, can help reduce pollution and conserve energy resources.

LESSON 2 Review

Summarize

Create your own lesson summary as you write a **newsletter.**

1. **Write** this lesson title, number, and page numbers at the top of a sheet of paper.

2. **Review** the text after the **red** main headings and write one sentence about each. These will be the headlines of your newsletter.

3. **Review** the text and write 2–3 sentences about each **blue** subheading. These sentences should tell *who, what, when, where,* and *why* information about each headline.

4. **Illustrate** your newsletter with diagrams of important structures and processes next to each headline.

 **ELA6: W 1.2**

Standards Check

Using Vocabulary

1. In your own words, write the definition of *fossil fuels*. **6.b**

2. The heat energy in Earth's crust is called _____. **6.a**

Understanding Main Ideas

3. **List** three types of fossil fuels. **6.b**

4. Complete the concept map about alternative fuels. **6.a**

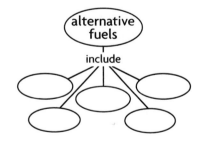

5. **Summarize** the effect on the environment from extracting fossil fuels. **6.a**

6. **Illustrate** how oil forms. **6.b**

7. Why is solar energy not cost-effective?

   A. The equipment is expensive.

   B. It creates too much pollution.

   C. Sunlight is a nonrenewable resource.

   D. It must be windy. **6.a**

Applying Science

8. **Explain** why hydroelectric power is not used everywhere in the United States. **6.a**

9. **Rank** the energy resources in order of risk to the natural environment. Explain your ranking. **6.a**

Science nline

For more practice, visit **Standards Check** at ca6.msscience.com.

MiniLab

How can you make the Sun work for you?

When you sit outside on a warm, sunny day, the Sun warms your skin. The Sun also helps the flowers bring color to your lawn and vegetables grow ripe and delicious. The Sun's rays bring energy, which you can use in many ways.

Procedure

1. Read and complete a lab safety form.
2. Paint one small **tin can** black and another same-sized **can** white.
3. Paint the **lids** the same color as the cans.
4. Use a **beaker** to measure equal amounts of water into each container.
5. Place a **thermometer** into each container. Be sure the bulb is covered by the water.
6. Place both cans on a sunny windowsill or outside in the sunlight.
7. Measure and record the temperature at the onset and at 15-minute intervals.

Analysis

1. **Compare** how the temperature changed at each 15-minute interval.
2. **Infer** Why did the temperature increase more in one color can than in the other?
3. **Design** a solar cooker based on your observations from this lab.
4. **Determine** if using a solar cooker would contribute to air pollution.

 Science Content Standards

6.a Students know the utility of energy sources is determined by factors that are involved in converting these sources to useful forms and the consequences of the conversion process.

7.b Select and use appropriate tools and technology (including calculators, computers, balances, spring scales, microscopes, and binoculars) to perform tests, collect data, and display data.

7.e Recognize whether evidence is consistent with a proposed explanation.

LESSON 3

Science Content Standards

6.a Students know the utility of energy sources is determined by factors that are involved in converting these sources to useful forms and the consequences of the conversion process.

6.b Students know different natural energy and material resources, including air, soil, rocks, minerals, petroleum, fresh water, wildlife, and forests, and know how to classify them as renewable or nonrenewable.

6.c Students know the natural origin of the materials used to make common objects.

7.a Develop a hypothesis.

7.d Communicate the steps and results from an investigation in written reports and oral presentations.

7.e Recognize whether evidence is consistent with a proposed explanation.

Reading Guide

What *You'll Learn*

▶ **Identify** common objects made from natural resources.

▶ **Suggest** strategies to conserve energy.

Why *It's Important*

Using Earth's material and energy resources wisely is important for all life on Earth.

Vocabulary

recycling
particulates
conservation

Review Vocabulary

global warming: an increase in Earth's average surface temperature (p. 399)

Using Energy Resources

Main Idea Conserving resources can help prevent shortages and reduce pollution.

Real-World Reading Connection Think about all the activities you do each day that depend on electricity. How would you change your schedule and activities if you only had electricity from 8 A.M. until 8 P.M. every day?

Location of Natural Resources

Different resources are found in different parts of the world. If the resource is rare or difficult to obtain, it usually is expensive. Buying and selling resources is an important part of the economy for different areas of the United States and other countries. Using **Table 3,** map the location of some resources in the United States. Location of natural resources is important to consider because it affects the amount of resources needed for transportation.

 Table 3 List the nonrenewable resources.

Table 3 Resources from Various States	
State	**Resources**
Alabama	Cement, limestone, cotton, lumber
Alaska	Oil, fish, lumber, zinc, gold, sand and gravel
Arizona	Cotton, copper, sand and gravel
California	Milk, grapes, flowers, sand and gravel, cement, boron
Delaware	Shellfish, soybeans, sand and gravel
Florida	Oranges and lemons, crushed stone, fish
Hawaii	Sugar, pineapple, nuts, fish
Minnesota	Lumber, iron, corn
Montana	Lumber, gold, coal, natural gas
Texas	Oil, natural gas, crushed stone
Washington	Lumber, sand and gravel, apples, fish

Is it made from plants or plastic?

Products are manufactured and designed in so many different ways because of modern technology and available processes. Sometimes it is difficult to tell if something you use is made from plant materials or from plastic.

Procedure

1. Examine **two shirts** and determine if they are cotton or synthetic.

2. Examine **floor samples** and determine if they are tile or hardwood.

3. Examine a **basket woven from reeds** and a **basket woven from plastic strips**.

4. Put your responses in a chart and explain what clue you used to determine the difference.

5. Find three items in the classroom that are probably made from plastic. List these on your chart.

Analysis

1. **List** the items you found that are not made of natural materials. From what do you think they are made from?

2. **Explain** Do you think oil is a key ingredient in any of the synthetic materials you found?

Manufacturing Common Objects

Fuel for energy is not the only product from fossil fuels. Oil is used to manufacture many common materials, including most plastics and synthetic clothing, such as nylon. Even lipstick has petroleum in it. Chemicals and other materials are manufactured from petroleum at petrochemical plants, like the one in **Figure 14.** When you use a plastic fork or open a plastic package, you are holding objects made from fossil fuels.

 What materials, other than fuels, are made from oil?

We use Earth's material and energy resources for making and selling most products, packaging the objects, transporting them, keeping the stores a comfortable temperature for working and shopping, and for advertising the products. For example, look around your classroom. Are the desks made of wood? The pencil you use contains wood and graphite, a soft carbon mineral mined from Earth. The eraser may be made from plants or oil. The metal holder for the eraser also is made from Earth's natural material resources. If any one of the steps can be removed—for instance, the product can be transported less distance or can have little or no packaging—the strain on our natural resources is greatly lessened.

Figure 14 Petrochemical plants use petroleum combined with chemicals to make a variety of materials you use every day, including most plastics.

Oil Spill

Cleaning the Beach

Caring for Wildlife

Figure 15 Oil must be transported from the drilling site to refineries. Unfortunately, accidents happen.

Explain how spilled crude oil can severely damage ocean and coastal ecosystems.

Making Resources Last

To counter the effects that manufacturing has on our natural resources, people can reduce the amount of products they purchase, reduce the amount of packaging of products they buy, reuse products, and recycle. **Recycling** is reprocessing an item or natural resource. Plastics and certain natural material resources, such as aluminum, can be recycled. Recycling uses less energy than extracting new natural resources and helps natural material and energy resources last longer.

Drawbacks of Using Fossil Fuels

There are many drawbacks to fossil fuel use. In order to refine oil into gasoline and other fuels, it must be transported from the drilling site to the refinery. Ships are the most economical way to transport crude oil. As illustrated in **Figure 15,** marine life can be greatly affected if crude oil spills from a ship.

Pollution

Burning fossil fuels releases many gases into the air. All fossil fuels give off carbon dioxide, a gas that contributes to global warming. Burning coal also gives off sulfur dioxide, a gas that causes acid rain. Gasoline exhaust contributes to urban smog. Burning fossil fuels also release tiny particles, called **particulates,** into the air. Particulates can damage lungs. Manufacturing products from fossil fuels can also pollute our air and water. There are national and state agencies that regulate how much pollution can be released into the environment. These regulations change and are influenced by local, state, and government policies and economics.

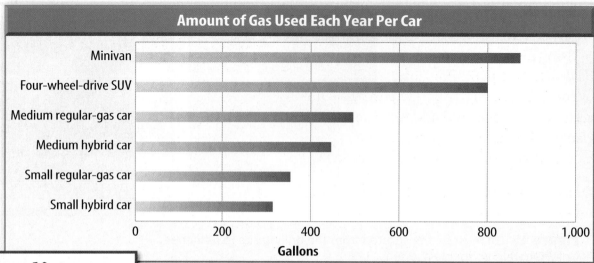

Amount of Gas Used Each Year Per Car

Car Type	Gallons
Minivan	
Four-wheel-drive SUV	
Medium regular-gas car	
Medium hybrid car	
Small regular-gas car	
Small hybird car	

Figure 16 Some cars get much better gas mileage than other cars. Hybrid cars use less gasoline because they have an electric motor that helps provide some of their energy.

Identify the cars that use the most and least amounts of gasoline.

ACADEMIC VOCABULARY

register (REH juh stuhr)
(verb) an official written record of names or events or transactions
The child was registered for kindergarten by early August.

Driving Cars Exhausts these Resources

Americans use almost one-quarter of the world's petroleum. Seventy-five percent of that is for vehicles. There are approximately 24 million vehicles **registered** in the state of California alone. The average vehicle can travel only 18 miles on one gallon of gas. In California, each driver travels about 36 miles every day. If everyone had cars with an average fuel economy of 30 miles per gallon, Californians would save more than 500 million gallons of gasoline each month, greatly reducing the strain to our resources. As **Figure 16** shows, driving smaller cars can save gasoline. Other ways to conserve gasoline include carpooling, bicycling, walking short distances, and using buses or trains instead of driving. Scientists are also developing electric hybrid technologies and researching ways to use fuels other than gasoline.

Using Coal

About 70 percent of the electricity produced in the United States is generated using fossil fuels, especially coal. Coal must be extracted from Earth, transported by rail or barge, processed, and burned to generate electricity.

Strip Mining Look back to **Figure 8.** Strip mining destroys the land surface over a large area, and the land that is left behind is unstable. Landslides, erosion, and polluted streams and lakes often result from this type of mining. Heavy machines used to strip the land contribute to air pollution.

Burning Coal When coal burns, gases are released. Some gases help to form acid rain, which can damage forests, lakes, and streams. Burning coal also releases particulates and some toxic metals, such as mercury, into the air.

Alternative Energy and the Environment

Using fossil fuels affects the environment by causing air pollution when they are burned. Extracting and transporting fossil fuels can damage land and water resources. Using alternative energy sources can also have negative effects on the environment.

Dams Although hydroelectric power is renewable, large lakes that are created behind dams destroy wildlife habitat that was along the sides of the river. Fish movement up and down the river is also affected. Fish ladders, like the one in **Figure 17,** are built to help fish swim over dams. Dams also increase the amount of sediment and erosion downstream when large volumes of water are suddenly released to generate more power when demand increases. Dam failure can cause devastating floods.

Nuclear Waste Nuclear fission power plants produce small volumes of dangerous radioactive waste. Great care must be taken to store the waste safely for 10,000 years. There are no polluting gases released into the air when a nuclear reactor generates electricity. However, a nuclear reactor generates a large amount of heat.

Like all engines, nuclear power plants must get rid of unusable heat energy. A car uses antifreeze pumped through a radiator. The radiator helps transfer the heat to the air. Nuclear power plants use water in the same way the car uses air. Heated water turns to steam and the excess heat is transferred to another water source. Heated water released into the environment can harm fish and other wildlife. However, this is a problem for most power plants, because most use steam to turn turbines.

 **Reading Check** How can a nuclear power plant harm fish and other wildlife?

Figure 17 This dam on the Columbia River in Washington has a fish ladder to help fish swim upstream to reproduce.

Table 4 The Average Amount of Electricity Used by Common Household Appliances	
Appliance	**Average Watts, or Units of Electricity, Used per Hour**
Lightbulb	100
Stereo	100
Television	230
Washing machine	250
Vacuum cleaner	750
Dishwasher	1,000
Toaster	1,200
Oven	1,300
Air conditioner	1,500
Hair dryer	1,500
Microwave	1,500
Clothes dryer	4,000
Freezer	5,100
Refrigerator/freezer	6,000

Wind Farms Wind towers don't require heating water and they don't pollute the air. Although wind farms do not destroy the environment permanently, they do stop birds from using the area for nesting or feeding habitats. The best place to locate wind farms might be near large bodies of water, where it usually is windy. This might be an obstruction of people's view of the ocean.

Using Energy Resources Wisely

Most of us use natural energy resources each day without thinking about it. For example, you might awaken to an electric alarm clock, turn on the lights, and take a hot shower. You might listen to a radio as you eat pancakes cooked on a natural gas stove. You arrive at school on a bus and enter a classroom heated by oil or electricity. Before the school day even starts, you depend on many energy resources.

 **Reading Check** How might you use energy resources before you get to school in the morning?

You might not think about energy very often because there is not an immediate shortage of resources. However, if people continue to use nonrenewable energy resources at current levels, there could be shortages in the future. Few people want to completely change their lives to avoid using nonrenewable energy resources, but there are many ways to reduce the amount of resources we use.

Conservation is the preservation and careful management of the environment, including natural resources. Conserving nonrenewable resources is one of the most effective ways to help prevent shortages. Examine **Table 4.** How much electricity would you estimate your family uses on an average day? How could you reduce this?

What have you learned?

Earth's abundant resources supply materials and energy for all of Earth's organisms. People use these resources for everyday living. But using resources can cause pollution and lead to shortages. Most of the world's energy is supplied by burning fossil fuels. Extracting, transporting, and burning fossil fuels can cause damage to the environment. Developing alternative energy sources will reduce demand for fossil fuels—conserving this nonrenewable resource and preventing pollution. Conservation and recycling can help conserve resources and reduce pollution.

LESSON 3 Review

Summarize

Create your own lesson summary as you design a **visual aid**.

1. **Write** the lesson title, number, and page numbers at the top of your poster.

2. **Scan** the lesson to find the **red** main headings. Organize these headings on your poster, leaving space between each.

3. **Design** an information box beneath each **red** heading. In the box, list 2–3 details, key terms, and definitions from each **blue** subheading.

4. **Illustrate** your poster with diagrams of important structures or processes next to each information box.

ELA6: R 2.4

Standards Check

Using Vocabulary

1. Use the term *conservation* in a sentence. **6.a**

2. Which describes reprocessing an item or natural resource? **6.a**

    A. particulate
    B. conservation
    C. recycling
    D. manufacturing

Understanding Main Ideas

3. **Explain** how people can conserve gasoline. **6.a**

4. **Categorize** the resources you have used today as natural material or energy resources. **6.b**

5. **Interview** your friends or family to learn which resources they think are used to make plastic and nylon. **6.c**

6. **Distinguish** between conserving and recycling. What are the benefits of each? **6.a**

7. **Compare** Create a chart like the one below and compare the effect on the environment from burning fossil fuels to using alternative sources for electricity. **6.a**

Fossil Fuels	Alternative Energy

Applying Science

8. **Design** a plan to conserve resources by recycling and reusing materials. **6.a**

9. **Compare** the natural resources in different states using **Table 3.** Are there any natural resources found in more than one state? Why or why not? **6.c**

Science Online

For more practice, visit **Standards Check** at <u>ca6.msscience.com</u>.

DataLab

 Try at Home

Do all vehicles require fuels from oil?

Most vehicles on the road today use either gasoline or diesel fuel only. Alternative fuels, either used alone or mixed with gas, can conserve nonrenewable resources and reduce harmful pollution. Which alternative fuel do you think is most practical?

Data Collection

Go to **ca6.msscience.com** to research alternative fuels for transportation. Find out the origin of the fuel. On the table below, list the advantages and disadvantages of each fuel type. An example is listed in the table to help you get started.

Comparing Alternative Fuels				
Alternative Fuel	**Source**	**Does it need to be blended with gas?**	**Advantages**	**Disadvantages**
Ethanol	corn or other crops	yes	renewable resource, reduces air pollution	Engines may need to be converted.

Data Analysis

1. **Identify** each alternative fuel as a renewable or a nonrenewable resource.

2. **Explain** how alternative fuels affect air pollution.

3. **Rank** your preferred choice of alternative fuel. Explain your ranking to the class.

 Science Content Standards    MA6: MR 1.1

6.a Students know the utility of energy sources is determined by factors that are involved in converting these sources to useful forms and the consequences of the conversion process.
7.d Communicate the steps and results from an investigation in written reports and oral presentations.

Applying Math

Energy Usage

The circle graph below shows national energy product data.

Example

Find the measure of the circle graph angle for fossil fuel usage in degrees.

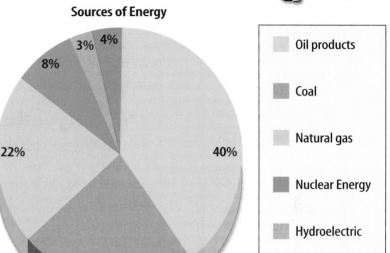

Sources of Energy

4%
3%
8%
22%
40%
23%

Oil products
Coal
Natural gas
Nuclear Energy
Hydroelectric
Other

MA6: MR 2.4

This is what you know:
In a circle graph, the sum of all the section percentages equals 100%.

Oil products: 40%,
coal: 23%,
natural gas: 22%

This is what you need to find:

- Percentage of non-fossil fuel usage
- Measure of the angle in degrees for non-fossil fuel usage

First, find the percentage of non-fossil fuel usage.

1 Add the percentages of the three fossil fuel sections together: $40\% + 23\% + 22\% = 85\%$

2 Subtract the total from 100%: $100\% - 85\% = 15\%$

Answer: The percentage of energy produced from 1999 was 15%.

Now find the measure of the angle in degrees for this section of the circle graph.

1 Write 15% as a fraction: $15\% = \frac{15}{100}$

2 Multiply $\frac{15}{100}$ by the number of degrees in the circle, 360°. $\frac{15}{100} \times 360° = 54°$

Answer: The measure of the angle for non-fossil fuel usage is 54°.

Practice Problems

1. Find the measure of the angle in degrees coal usage.
2. Find the measure of the angle in degrees renewable usage.

Science nline
For more math practice, visit **Math Practice** at ca6.msscience.com.

Design Your Own Lab

Become an Energy Expert

Problem

Think about all the ways you use energy every day. Transportation and manufacturing common items require energy resources. So does heating and cooling your home. Through research, you can become an expert on energy sources used for home heating.

Form a Hypothesis

Various energy sources can be used to heat a home. Some homes have electric furnaces, others use natural gas. Electricity can be generated from different resources. Decide which energy source you prefer.

Collect Data and Make Observations

1. Learn about a source of energy available to heat your home.
2. Choose from oil, natural gas, solar power, wood, and electricity.
3. Research the origin of the energy source and the necessary processes to make it usable.
4. Determine the resulting pollution and environmental consequences from any stage of energy production and usage.
5. Gather your information and organize it into a slideshow presentation or a series of posters to use as a basis for teaching your classmates.

Materials

documentaries
energy company brochures and materials
reference materials

Science Content Standards

6.a Students know the utility of energy sources is determined by factors that are involved in converting these sources to useful forms and the consequences of the conversion process.
7.a Develop a hypothesis.
7.d Communicate the steps and results from an investigation in written reports and oral presentations.

Energy Source Data			
Energy Source	Processes to Make Usable	Pollution	Other Environmental Consequences

Analyze and Conclude

1. **Calculate** which form of energy explained is the most expensive to acquire. Which is the least expensive?
2. **Describe** which form of energy explained is the most abundant. Least abundant?
3. **Determine** which form of energy is the cleanest for the environment.
4. **Rank** the forms of energy by how much pollution they cause.
5. **Research** ways in which pollution can be cleaned up.
6. **Evaluate** the cost of pollution cleanup from each energy source.
7. **List** three ways the general population can easily conserve energy.

Communicate

WRITING in Science  **ELA6: W 2.5**

Create a poster or computer presentation about the types of energy you researched, and a conclusion of your research. Are any energy types better than the others? Why or why not? Give your presentation to the class clearly and completely so they understand the pros and cons of the types of energy.

Real World Science

You could dig up coal, gold, or gravel!

Mining engineers find and remove resources to be used in many different industries. They might design mines, like the one shown here. They can design ways to move the materials from the mine or develop mining machinery. They may also focus on protection of the environment.

Visit **Careers** at **ca6.msscience.com** to find out more about mining. Create a help-wanted ad for an entry-level mining engineer. List the education required, the work environment, and the type of material. Be sure to use proper punctuation, spelling, grammar, and capitalization.

BURNING ICE AS FUEL?

Gas hydrates, shown here, are frozen deposits that look like ice but contain natural gas—usually methane. They are found in the polar and permafrost regions, on the outer continental shelf regions of the ocean, and in the deep sea because of the low temperatures and high pressures. It is believed the potential to use these as an energy resource exceeds the worldwide reserves of conventional oil and gas reservoirs and coal by a wide margin.

Visit **Technology** at **ca6.msscience.com** to find out more about gas hydrates. Write a short paper giving three challenges to using gas hydrates as a fuel source. Be sure to use proper punctuation, spelling, grammar, and capitalization.

ELA6: W 1.2

Burning Biomass for Fuel

Throughout history, man has burned biomass for fuel. Wood was once our main fuel source. Today, our most widely used fuel source is nonrenewable fossilized biomass, such as coal. A large amount of usable energy is available in the form of garbage, tree trimmings, sawdust, demolition and land-clearing waste, logs, low-grade waste paper and cardboard products, and all kinds of agricultural waste from corn cobs to rice hulls. These resources are locally available in various forms, often free or very cheap.

Visit **History** at **ca6.msscience.com** to find out more about different types of biomass. The average American produces more than 725 kg of waste each year. The United States burns 15 percent of its solid waste. What if more waste was sent to be burned for energy? Using this data, create a table showing the kilograms of waste burned and kept out of landfills at 15 percent, 25 percent, 33 percent, 50 percent, and 70 percent. What are some disadvantages?

Using Geothermal Energy

Geothermal energy could be the energy of the future. The steam from a geothermal vent is pictured here. The steam is pumped to the surface and used to drive turbines. These turbines can produce electricity. Currently, California is the leading state for producing and using geothermal energy. Its location at the tectonic plate junction makes it ideal for further development.

Visit **Society** at **ca6.msscience.com** to find out more about geothermal energy. Divide the class into two groups for a debate. Imagine there is a geothermal plant being proposed for your area. One group should debate the pros and one, the cons.

The BIG Idea Earth's resources provide materials and energy for everyday living.

Lesson 1 Natural Resources

6.b, 6.c, 7.c

Main Idea People use a variety of materials from different parts of Earth to meet a diverse range of needs.

- Natural resources are materials and energy sources that are useful or necessary to meet the needs of Earth's organisms.

- Plant and animal resources are used for food, clothing, shelter, and everyday objects.

- Minerals and metals are important resources for industry and construction.

- Renewable natural resources are constantly being replaced or recycled by nature.

- Nonrenewable natural resources are used more quickly than they can be replaced.

- **estuary** (p. 591)
- **natural resource** (p. 588)
- **nonrenewable natural resource** (p. 592)
- **renewable natural resource** (p. 590)

Lesson 2 Energy Resources

6.a, 6.b, 7.b, 7.e

Main Idea Some of Earth's natural resources can be used for energy, usually through conversion to electricity.

- Fossil fuels, such as oil, coal, and natural gas, are made from decayed plants and animals that have been preserved in Earth's crust for millions of years.

- Fossil fuels are burned, which heats water to make steam. Steam turns a turbine to generate electricity.

- Alternatives to fossil fuels include nuclear energy, geothermal energy, hydroelectric energy, wind, waves, and biomass.

- **fossil fuel** (p. 595)
- **geothermal energy** (p. 600)
- **nuclear fission** (p. 601)
- **nuclear fusion** (p. 601)

Lesson 3 Using Energy Resources

6.a, 6.b, 6.c, 7.a, 7.d, 7.e

Main Idea Conserving resources can help prevent shortages and reduce pollution.

- Resources you use every day come from all over the world.

- Chemicals, plastics, and other common materials are made from petroleum.

- Using energy resources can harm the environment.

- Conservation is an effective way to preserve nonrenewable resources.

- **conservation** (p. 612)
- **particulate** (p. 609)
- **recycling** (p. 609)

 **STUDY TO GO** Download quizzes, key terms, and flash cards from ca6.msscience.com.

Linking Vocabulary and Main Ideas

Use vocabulary terms from page 620 to complete this concept map.

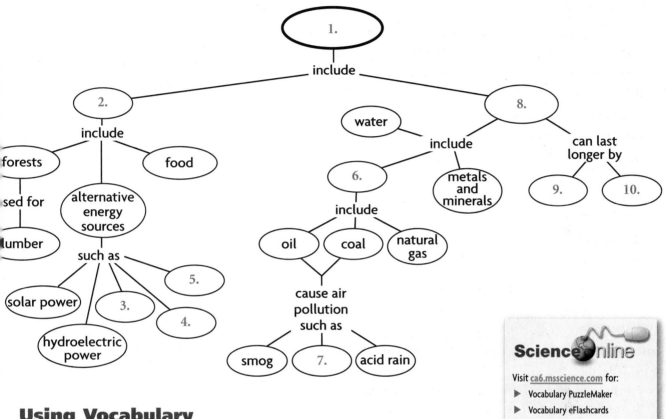

Using Vocabulary

Match the vocabulary terms with the correct phrase.

11. fuels formed in Earth's crust over hundreds of millions of years

12. splitting atoms to release energy

13. Earth's resources that are constantly being replaced or recycled by nature

14. a fertile area where a river meets an ocean

15. resources that are used more quickly than they can be replaced by natural processes

16. tiny particles in the air

17. the heat energy in Earth's crust

18. changing or reprocessing an item or natural resource for reuse

19. materials and energy sources that are useful or necessary to meet the needs of Earth's organisms

20. the preservation and careful management of the environment, including natural resources

Science online

Visit ca6.msscience.com for:
▶ Vocabulary PuzzleMaker
▶ Vocabulary eFlashcards
▶ Multilingual Glossary

Understanding Main Ideas

Choose the word or phrase that best answers the question.

1. Which is a fossil fuel?
   A. oil   **6.a**
   B. water behind a dam
   C. wood
   D. nuclear fission

2. Which is *not* an energy resource?
   A. oil   **6.b**
   B. coal
   C. silver
   D. geothermal energy

3. The chart below shows the sources for energy use in the United States.

Sources of Energy

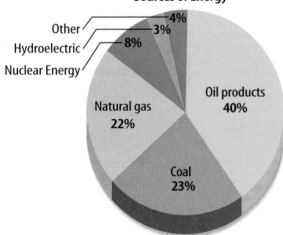

Other — 4%
— 3%
Hydroelectric — 8%
Nuclear Energy
Natural gas 22%
Oil products 40%
Coal 23%

What percentage of energy use is from fossil fuels?
   A. 40 percent   **6.a**
   B. 16 percent
   C. 85 percent
   D. 45 percent

4. Which is not a renewable resource, even if managed properly?
   A. coal   **6.b**
   B. fish
   C. cotton
   D. lumber

Use the figure below to answer questions 5 and 6.

5. How can this tower generate electricity?
   A. Solar heat collects, and hot air rises and turns a turbine.   **6.a**
   B. Steam is generated by burning fossil fuels.
   C. Atoms are split.
   D. Biomass is burned.

6. What is one disadvantage of this energy source?
   A. Energy can't be stored easily.   **6.a**
   B. It causes too much water and air pollution.
   C. It uses nonrenewable resources.
   D. The earth is damaged from mining.

7. In fission, what splits?
   A. turbines   **6.b**
   B. nucleus of an atom
   C. hard coal
   D. water

8. What is a major environmental problem with nuclear energy?
   A. It causes acid rain.   **6.a**
   B. A lot of uranium is used to generate electricity.
   C. The uranium is radioactive.
   D. Fish can't get up the ladder.

9. Which resource is used to make most plastics?
   A. forests   **6.c**
   B. uranium
   C. oil
   D. natural gas

Applying Science

10. Explain why certain resources are called fossil fuels. Why are they considered non-renewable? **6.a**

11. Analyze why all organic material resources aren't renewable. **6.b**

Use the table below to answer questions 12 and 13.

Average Water Use	
Daily Activity	**Water Used**
Flushing the toilet once	15 L
Taking a short shower	95 L
Taking a longer shower	190 L
Taking a bath	150 L
Washing clothes	190 L
Automatic dishwasher	38 L
Brushing your teeth while leaving the water running	7.5 L
Washing your hands while leaving the water running	30 L/min
Watering the lawn or plants with a hose	30 L/min

12. Calculate how much water you would save by turning off the water while brushing your teeth three times per day. **6.b**

13. Estimate how much less water you would use in one month if you turned off the water while brushing your teeth if you brushed three times per day. **6.b**

14. Diagram how oil is formed. **6.c**

WRITING in Science

15. Write a paragraph about alternatives to fossil fuels as energy resources. Explain the advantages and disadvantages of each type.

Cumulative Review

16. Analyze why composting is a form of recycling. Explain how matter is cycled in the process. **5.b**

17. Imagine you are teaching younger students about fossil fuels. Using words and diagrams, describe the biotic and abiotic factors involved in fossil fuel formation. Explain the energy flow from the Sun to using fossil fuels for energy. **5.e**

18. Hypothesize how an oil spill would damage a coastal ecosystem. What would happen if most of the top predators were killed? **5.c**

Applying Math

In California in 1999, the industrial use was 33%, the residential use was 17%, and the commercial use was 15%. Use these data for problems 19–22.

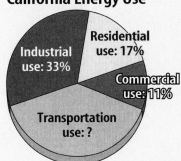

California Energy Use

Industrial use: 33%
Residential use: 17%
Commercial use: 11%
Transportation use: ?

19. Find the percentage of the circle graph that is transportation usage. **MA6: MR 2.4**

20. Find the measure of the angle in degrees for California's transportation usage. **MA6: MR 2.4**

21. Find the measure of the angle in degrees for California's industrial usage. **MA6: MR 2.4**

22. Find the measure of the angle in degrees for California's residential usage. **MA6: MR 2.4**

Use the illustration below to answer questions 1–3.

1 What is produced by the mechanism shown in the illustration?

A electricity   `6.a`

B coal

C petroleum

D plastic

2 In which section are the turbine blades found?

A A   `6.a`

B B

C C

D D

3 Which section represents the generator?

A A   `6.a`

B B

C C

D D

4 Which part of a CD player is made from crude oil?

A the screws that hold it together   `6.c`

B the plastic case

C the cardboard packaging

D the electrical parts that make it work

Use the image of an alternative energy source below to answer questions 5–6.

5 The illustration below shows a type of alternative energy. What type of alternative energy is being used?

A geothermal   `6.a`

B hydroelectric

C solar

D wind

6 What is one drawback to using the alternative energy shown above?

A available only when the Sun is shining   `6.a`

B available only where geysers or volcanoes are found

C available only when the wind is blowing

D available only where there is falling water

Use the figure below to answer questions 7–9.

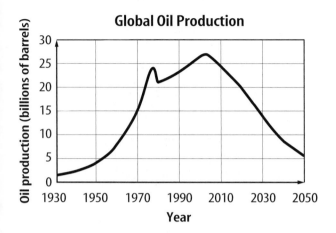

Global Oil Production

Use the circle graph below to answer questions 10–12.

Sources of Human-Caused Toxic Air Pollution

Mobile sources (50%) include cars, trucks, and planes.

Major sources (26%) include power plants and factories such as oil refineries and chemical manufacturers.

Area sources (24%) include businesses such as gas stations and dry cleaners.

7 According to the graph above, in which year will global oil production be at a maximum?

A 1974   `6.b`

B 2002

C 2010

D 2050

8 Approximately how many times greater was oil production in 1970 than oil production in 1950?

A 2 times   `6.b`

B 10 times

C 6 times

D 3 times

9 In which year will the production of oil be equal to the oil production in 1970?

A 2010   `6.b`

B 2015

C 2022

D 2028

10 Which are area sources of toxic air pollution?

A cars, trucks, and planes   `6.a`

B gas stations

C oil refineries

D power plants and factories

11 Which source emits the highest percentage of toxic air pollution?

A cars, trucks, and planes   `6.a`

B gas stations

C dry cleaners

D power plants and factories

12 Why are coal, petroleum and natural gas defined as fossil fuels?

A They are burned to release energy.   `6.b`

B They are nonrenewable and will run out.

C They cause air pollution when burned.

D They formed from remains of ancient life.

Are you interested in learning more about ecosystems and the organisms that inhabit them? If so, check out these great books.

Nonfiction

Compost Critters, by Bianca Lavies, describes what happens in a compost pile. Close-up color photographs show a variety of decomposers and other living things that help to produce humus. You may decide to make a compost pile after reading this book. ***The content of this book is related to*** *Science Standard 6.5.*

Biography

John Muir: My Life with Nature, by Joseph Cornell, is a short biography of John Muir, the "father of the national parks." The author writes using John Muir's own words. The book communicates Muir's deep respect for nature and animals. ***The content of this book is related to*** *Science Standard 6.5.*

Narrative Nonfiction

Once a Wolf: How Wildlife Biologists Fought to Bring Back the Grey Wolf, by Stephen Swinburne, surveys the history of the troubled relationship between wolves and humans. The conservation movement to restore this animal to the wild is described. Full-color photographs show the wolves in the wild. ***The content of this book is related to*** *Science Standard 6.5.*

Nonfiction

Pass the Energy, Please!, by Barbara McKinney, is a rhyming story about the exchange of energy and nutrients within the environment. Various food chains, including woodland, Arctic, meadow, and African plain, are presented. The role decomposers play is also discussed. ***The content of this book is related to*** *Science Standard 6.6.*

Choose the word or phrase that best answers the question.

1. Which are organisms that break down tissue and release nutrients and carbon dioxide back into the ecosystem?
   A. producers
   B. consumers
   C. decomposers
   D. photosynthetic **5.b**

2. The by-products of burning fossil fuels and wood are a primary part of which cycle of matter?
   A. the carbon cycle
   B. the hydrogen cycle
   C. the nitrogen cycle
   D. the water cycle **6.a**

3. How does harvesting crops affect the nitrogen cycle?
   A. Nitrogen increases in the soil.
   B. Plant decay cannot occur.
   C. The number of nitrogen-fixing bacteria increases in the soil.
   D. The plants will no longer release nitrogen. **5.e**

4. Which is a biotic factor within an ecosystem?
   A. air
   B. temperature
   C. water
   D. wood **5.b**

5. Which abiotic factor provides energy for nearly all life on Earth?
   A. air
   B. soil
   C. sunlight
   D. water **5.a**

Write your responses on a sheet of paper.

6. **Identify** the source of most air pollution. **6.a**

7. **Identify** how the terms *population*, *community*, and *ecosystem* are related. **5.c**

Use the figure below to answer questions 8 and 9.

8. **Describe** how the meadow ecosystem shown above is similar to a desert ecosystem. **5.e**

9. **Identify** which abiotic factors might affect organisms living in the meadow ecosystem. **5.e**

10. **Explain** how mosquitoes are a limiting factor for frogs if frogs feed on mosquitoes and other insects. What will happen to the frog population if the mosquito population rapidly increases or decreases? What will happen to the mosquitoes if the frogs disappear? **5.c**

11. **Explain** whether a food web or a food chain is a better choice to model the flow of energy in an ecosystem. **5.b**

12. **Evaluate** the advantages and disadvantage of nonrenewable and renewable resources. **6.b**

To Students and Their Families,

Welcome to sixth-grade Earth science. You will begin your journey by learning about the tools that Earth scientists use. Then you will continue with the structure of Earth and its resources.

Take a few moments each day to review what you have learned about Earth science. Test your knowledge of each Standard by answering the questions.

Remember, the knowledge and skills you will gain this year will be important beyond the classroom. They will help you to become environmentally aware and to better understand the planet on which you live.

Table of Contents

Standard Set 1: Plate Tectonics and Earth's Structure

Directions: Select the best answer for each of the following questions.

1 Earthquakes and volcanic eruptions often occur in the same regions of the world. What is the *best* explanation for this?
   A Earthquakes disturb magma below the surface and cause volcanic eruptions.
   B Volcanism weathers rocks, making them more likely to experience an earthquake.
   C Both earthquakes and volcanic eruptions are triggered by the gravitational pull of the Moon.
   D Both earthquakes and volcanic eruptions occur most frequently along plate boundaries.

2 The Marianas Islands are an arc of volcanic islands in the Pacific Ocean. Based on this fact only, which is *most* likely to be true?
   A Tectonic plates collide near the Marianas Islands.
   B The Marianas Islands are larger than most islands.
   C The Marianas Islands have a warm and wet climate.
   D There is very little seismic activity near the Marianas Islands.

3 The theory of continental drift and plate tectonics was first proposed by
   A Alfred Wegener.
   B Charles Darwin.
   C Christopher Columbus.
   D Isaac Newton.

4 The shape of Earth's mantle is *best* described as
   A a pie slice.
   B a solid sphere.
   C a hollow ball.
   D a rectangular prism.

5 Which of these physical processes causes the slow motion of lithospheric plates?
   A radiation
   B conduction
   C convection
   D evaporation

6 Which drawing should be used to model where Earth's mantle, crust, and core are found?
   A
   B
   C
   D

7 If a continental plate is moving at a rate of 3 cm/year, about how long would it take a continental plate to travel 1 m?
   A 6–8 months
   B 1–3 years
   C 20–50 years
   D 300–600 years

8 In which geologic layer does motion cause the movements of the continents?
   A crust
   B mantle
   C inner core
   D outer core

9 Earthquakes are caused by the motion of
   A tectonic plates.
   B the mantle.
   C hurricanes.
   D mountain ranges.

Science online Standards Practice   ca6.msscience.com

Standard Set 1: Plate Tectonics and Earth's Structure

Directions: Select the best answer for each of the following questions.

10 Once magma pushes through to Earth's surface, it is called
A gas.
B lava.
C ash.
D mud.

11 What is another name for a rupture in Earth's crust?
A earthquake
B volcano
C fault
D fissure

12 The process of one tectonic plate sliding under another is called
A faulting.
B spreading.
C subduction.
D folding.

13 Movements in Earth's mantle cause the
A regular patterns of the tides.
B flipping of the magnetic poles.
C slow weathering of mountains.
D changes in sizes of the ocean basins.

14

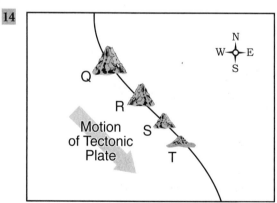

In the figure above, which is the order of the volcanoes, from youngest to oldest, that were formed by the motion of this tectonic plate?
A Q, R, S, T
B T, S, Q, R
C Q, T, S, R
D T, S, R, Q

15 *Most* of California sits on which continental plate?
A California Plate
B North American Plate
C Pacific Plate
D San Andreas Plate

16 The San Andreas fault is an example of which type of plate boundary?
A convergent
B divergent
C transform
D uniform

17 California includes diverse geologic features such as mountain ranges, lower valleys, and desert basins. Why don't most other U.S. states have the same diversity of landforms?
A Most other states do not border an ocean.
B The weather in California is unique.
C California is closer to the equator than other states.
D Most other states are not located on a major geologic fault.

18 Which type of seismic wave travels fastest?
A compression (P)
B shear (S)
C surface
D Rayleigh

19 How does an earthquake measuring 7.0 on the Richter scale compare to one measuring 6.0?
A A 7.0 earthquake is twice as powerful as a 6.0 earthquake.
B A 7.0 earthquake is 1/8th more powerful than a 6.0 earthquake.
C A 7.0 earthquake is ten times more powerful than a 6.0 earthquake.
D A 7.0 earthquake is half as powerful as a 6.0 earthquake.

20 Which type of seismic wave usually produces the *most* destruction?
A compression (P)
B shear (S)
C surface
D Love

Standard Set 2: Shaping Earth's Surface

Directions: Select the best answer for each of the following questions.

1 Which naturally occurring cause of landscape changes has *most* powerfully shaped Yosemite Valley?
A lightning
B sunlight
C running water
D strong winds

2 Which is an example of mass wasting?
A Construction crews dig out a silver mine.
B Rain loosens the soil, and it slides downhill.
C A large fissure is formed by a fault line.
D Grass growing through the sidewalk cracks the concrete.

3 Water erodes mountains and washes sediments to low areas where the soil is compacted over time and formed into new rock. Which is the name of this process?
A life cycle
B metamorphic cycle
C water cycle
D geologic cycle

4 Which is the *dominant* erosional process that has shaped California's landscape?
A blowing wind
B water running downhill
C fire
D earthquakes

5 Which is the primary force behind mass wasting?
A gravity
B electricity
C magnetism
D chemistry

6

Which is probably *most* responsible for the formation of the canyon in the figure above?
A wind
B water
C plants
D animals

7 Which two physical processes are at work in a flowing river?
A chemical weathering and electrolytic weathering
B mass wasting and biological weathering
C biological and mechanical weathering
D mechanical and chemical weathering

8 The energy of a flowing stream depends on which two factors?
A the temperature and the volume of the flowing water
B the slope of the land and the volume of the water flow
C the slope of the land and related geological faults
D the composition of the soil and the slope of the land

9 Which *best* explains why deltas usually form where rivers enter the ocean?
A A river's rate of flow decreases significantly, causing sediment to be deposited.
B A river's headward erosion removes sediment, causing the ocean to move onto land.
C The salty seawater mixes with the fresh river water, causing minerals to precipitate.
D The dense freshwater sinks beneath the salt water, causing sediment to be deposited.

Directions: Select the best answer for each of the following questions.

10 Rivers and streams carry sediments from one location to another. Which graph shows that the faster a river flows, the larger the particle the river can carry?

A
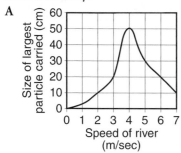

B
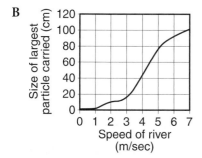

C
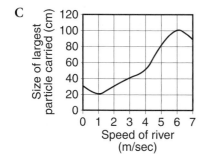

D
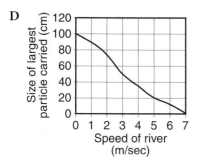

11 Which is the source for much of the sand present on a beach?

A rivers

B wind

C ocean waves

D volcanic eruptions

12 Which is the *primary* mechanism responsible for moving sediment along the beach?

A water currents

B evaporation

C wave action

D mass wasting

13 Which is *not* often found in the composition of sand on California beaches?

A quartz

B feldspar

C magnetite

D limestone

14 Beaches and coastlines are not static formations and will change naturally over time. Why?

A Wave action erodes and moves beach materials.

B Global warming influences water levels along the coastline.

C Human development rebuilds beaches for our own purposes.

D Periodic droughts pull more water away from the beach.

15 Silt and clay settle to the ocean bottom farther from the shore than rocks and pebbles because of their

A chemical composition.

B spherical shape.

C crystalline structure.

D low density.

16 An earthquake is *most* likely to affect a forest ecosystem by

A starting wildfires.

B causing landslides.

C decreasing rainfall.

D raising temperatures.

Standard Set 3: Heat (Thermal Energy) (Physical Science)

Directions: Select the best answer for each of the following questions.

1 Which phenomenon *cannot* transfer heat energy from one place to another?

A electromagnetic waves
B gravitational attraction
C infrared radiation
D movement of matter

2 Which is a unit of measure for heat?

A hertz
B volt
C degree
D calorie

3 An opera singer can shatter glass using only her voice. How is the energy transferred from the singer to the glass?

A air pressure
B electromagnetic waves
C heat flow
D sound waves

4 Which statement correctly describes the transfer of thermal energy?

A Thermal energy moves only between two hot objects.
B Thermal energy moves only between two cold objects.
C Thermal energy moves from a warmer object to a cooler object.
D Thermal energy moves from a cooler object to a warmer object.

5 Which is a description of heat?

A potential energy from atoms moving in many directions
B measure of kinetic energy of atoms in an object
C transfer of kinetic energy from one object to another
D when two objects cause each other to lose atoms

6 The energy from consumed fuel ultimately tends to transform into

A heat.
B work.
C light.
D waste.

7 When a moving automobile is stopped by the friction of applying its brakes, its kinetic energy is ultimately transformed into which form of energy?

A electricity
B gravity
C heat
D radioactivity

8 Which process is responsible for the heat flow in a solid?

A erosion
B conduction
C convection
D capacitance

9 Where is the energy stored in fuel before it is burned?

A in the surface layer of the material
B in the gravitational attraction of the material
C in the chemical bonds of the material
D in the electromagnetic properties of the material

10 What is the relationship between the waste product of a consumed fuel and the original fuel?

A The waste product has much less energy than the original fuel.
B The waste product has half the energy of the original fuel.
C The waste product has the same energy as the original fuel.
D The waste product has more energy than the original fuel.

11 Naomi placed one end of a metal spoon in a beaker of water and heated the beaker of water. Which hypothesis is *best* tested by Naomi's experiment?

A Fire is a chemical reaction.
B Hands can detect very cold temperatures.
C Metal does not burn.
D Metal conducts heat well.

Standard Set 3: Heat (Thermal Energy) (Physical Science)

Directions: Select the best answer for each of the following questions.

12

Heat Transfer

Type of Transfer	How It Transfers
Radiation	With rays or waves
Conduction	Contact of material
Convection	Flow of material

Heat can be transferred in several ways. The table above describes three types of heat transfer. Which is an example of conduction?

A the Sun shining on a metal chair

B a fire heating a room

C a metal pan burning a hand

D a hair dryer blowing hair

13 Which statement describes the flow of heat by conduction?

A Heat flows from a hot object to a cool object.

B Heated air rises while colder air descends.

C Heat flows from a cool object to a hot object.

D An object is heated when struck by light rays.

14 Which type of energy transfer occurs when the heat in beach sand transfers to a person's feet?

A convection

B conduction

C radiation

D kinetic

15 What is a medium through which energy can be transferred by radiation but not by convection or conduction?

A air

B water

C oxygen

D vacuum

16 Sarah fills a teapot with water and heats it using a kitchen stove. She uses a thermometer to record the temperature of the water 1 cm above the bottom of the teapot and 1 cm under the surface of the water. Which can Sarah conclude about the two temperatures?

A The water temperature at the bottom is hotter because it is closer to the heat source.

B The water temperature at the top is hotter because of convection currents.

C The two temperatures are the same due to conduction currents in the water.

D The two temperatures are the same due to convection currents in the water.

17 The diagram below shows how heat moves within a cup of hot cocoa.

Cold

Hot

This kind of heat transfer is called

A conduction.

B convection.

C radiation.

D reflection.

18 The energy transferred during radiation is carried by

A circulating currents.

B chemical compounds.

C ultrasound vibrations.

D electromagnetic waves.

Standard Set 4: Energy in the Earth System

Directions: Select the best answer for each of the following questions.

1 Energy from the Sun travels to Earth as
A seismic waves.
B electromagnetic waves.
C a uniform electric field.
D a uniform magnetic field.

2

Look at the figure above. Which thing depends *most* on solar radiation to supply energy to directly sustain it?
A bird
B boulder
C fire
D tree

3 Solar radiation is converted directly to stored energy in plants through
A the water cycle.
B photosynthesis.
C radiation.
D convection currents.

4 *Most* of the energy given off by the Sun arrives on Earth in what form?
A gamma rays
B infrared radiation
C visible light
D X rays

5 Energy from the Sun travels through space to Earths' surface by
A radiation.
B reflection.
C conduction.
D convection.

6

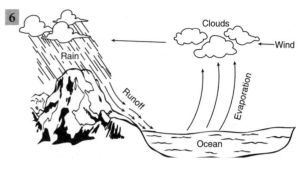

The diagram above shows
A the nitrogen cycle.
B the water cycle.
C the oxygen cycle.
D the soil cycle.

7 Why does the sky appear blue?
A Infrared light from the Sun appears blue to human eyes.
B Yellow light is filtered out by green plants, leaving only blue light.
C The atmosphere reflects the blue waters of the oceans.
D Blue wavelengths of visible light are scattered in Earth's atmosphere.

8 Light travels through space from the Sun to Earth by
A radiation.
B absorption.
C conduction.
D convection.

9 A volcano is an example of which type of heat transfer?
A convection
B radiation
C conduction
D erosion

10 Which of these natural materials is *not* a good conductor of heat?
A water
B rock
C air
D metal

Standard Set 4: Energy in the Earth System

Directions: Select the best answer for each of the following questions.

11 The movement of tectonic plates requires large amounts of energy. Where does this energy originate?

A convection currents in the mantle

B friction between continental plate boundaries

C solar radiation absorbed by Earth's crust

D weight of water in the oceans over the crust

12 Heat from Earth's interior moves in *which* direction?

A heat from the interior does not move

B toward the warmer core

C back and forth at the same elevation

D toward the cooler crust

13 When a large volcano erupts, its effects are seen in the atmosphere around the world. Which physical phenomenon causes this?

A convection currents in the atmosphere

B kinetic energy of the ejected contents

C magnetic force from Earth's poles

D radioactivity from the ejected contents

14 The trail of smoke from a burning match can *best* be used to observe what in the atmosphere?

A atmospheric pressure

B convection currents

C ambient temperature

D air quality

15 Which usually does *not* affect the movement of air currents in the atmosphere?

A Earth's rotation

B ocean temperatures

C magnetic field from Earth's poles

D warming of the atmosphere by the Sun

16 Dropping food coloring in a container of water that is being heated or cooled is a good way to observe

A convection currents in the water.

B the temperature of the water.

C contaminants in the water.

D electric currents in the water.

17 Students are interested in predicting whether or not a storm will arrive. Which datum will be *most* useful?

A time of sunrise

B time of high tide

C change in air pressure

D change in the Sun's position

18 Which is the *best* explanation why warmer air causes an increase in humidity and cooler air causes a decrease in humidity?

A Sunlight evaporates moisture.

B Saturated air cannot hold any more moisture.

C Cool air causes water molecules to join and condense.

D The dew point changes with the amount of moisture in the air.

19 Which usually does *not* result in a change in the weather?

A humidity variations

B changes in the Moon's phase

C differences in atmospheric pressure

D uneven heating of the land and the ocean

20

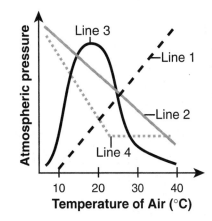

Which line above correctly represents the relationship of temperature to atmospheric pressure?

A 1

B 2

C 3

D 4

Standard Set 5: Ecology (Life Science)

Directions: Select the best answer for each of the following questions.

1 In order to survive, most plants need to make their own food through the process of photosynthesis. Which *best* describes the role of the Sun in photosynthesis?

A It warms the air.

B It carries away oxygen.

C It produces carbon dioxide.

D It provides a source of energy.

2

Energy Web

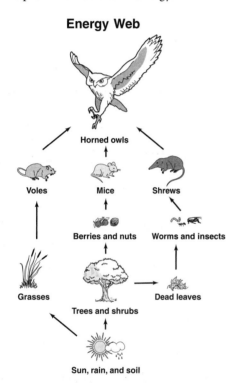

What would happen in the energy web above if the horned owl population were reduced by disease?

A The rate of photosynthesis would decrease.

B Berries and nuts would become more plentiful.

C The number of voles, mice, and shrews would increase.

D The amount of dead leaves would decrease.

3 Which of these is the ultimate source of energy in the energy pyramid?

A water

B air

C the Sun

D soil

4 Plants must take in carbon dioxide from the atmosphere in order to carry out photosynthesis. Which process contributes to the availability of carbon dioxide in the atmosphere?

A exhalation of air from an animal

B evaporation of water from a lake

C dissolution of nitrogen in the soil

D condensation of moisture in the air

5

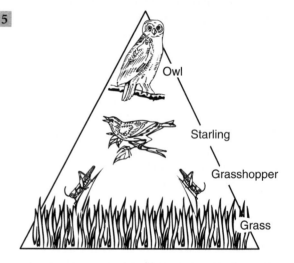

Examine the pyramid above. A food pyramid is one way of showing how much energy flows from one level to another in an ecosystem. Based on this pyramid, which organism has access to the *most* energy?

A grass

B grasshopper

C starling

D owl

Standard Set 5: Ecology (Life Science)

Directions: Select the best answer for each of the following questions.

6 Which shows the food chain in the correct order?
A zooplankton → bear → salmon
B salmon → zooplankton → bear
C zooplankton → salmon → bear
D bear → zooplankton → salmon

7 A local company dumped sewage into the wetlands. As a result, most of the fish in the wetlands died. This decrease in the number of fish will probably result in fewer
A mice.
B birds.
C spiders.
D insects.

8 Organisms in a population are categorized as either producers or consumers of
A gravitational potential energy.
B kinetic energy.
C electrical energy.
D chemical energy.

9 Which category of organisms does *not* include consumers?
A decomposers
B constructors
C predators
D scavengers

10 The location of a rabbit species stays the same, but its environment changes due to a drought that changes a swamp into a desert. Does the ecological role of this rabbit species stay the same?
A probably not
B most likely
C depends on the location
D depends on the species

11 Which defines the ecological role of a particular species of mouse found in the Andes Mountains of South America?
A environment
B geography
C food consumption
D rate of reproduction

12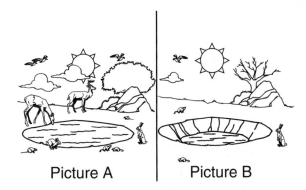

Picture A   Picture B

The ecosystem in Picture A is healthy, while the ecosystem in Picture B is not. What seems to be needed to help the organisms in picture B survive?
A more sunny days
B more water
C fewer predators
D more wind

13 Which is *not* an abiotic environmental factor?
A air
B animals
C soil
D water

Standard Set 6: Resources

Directions: Select the best answer for each of the following questions.

1 The evening news reports that modern technology is going to help reduce the problems of smog and the greenhouse effect. The new report is talking about
A finding replacements for CFCs.
B composting more garbage.
C using alternatives to fossil fuels.
D building containers for nuclear waste.

2 Although people once thought hydroelectric energy did not harm the environment, it is now known that building dams for hydroelectric energy plants
A increases global warming.
B interferes with the habitats of fish.
C causes the hole in the ozone layer.
D causes acid rain.

3 One of the disadvantages of nuclear energy is that it is difficult to dispose of the waste materials safely. This waste material is hazardous mainly because
A it causes acid rain.
B it releases greenhouse gases.
C it contains corrosive chemicals.
D it releases dangerous forms of energy.

4 Nuclear energy plants can have a negative effect on the environment in which way?
A They regularly release low-grade radioactive gases into the air.
B They regularly release low-grade radioactive wastes into rivers and lakes.
C They regularly release water that is at a higher temperature than the intake water.
D They regularly release carbon dioxide, sulfur dioxide, and nitrogen dioxide into the air.

5 A national environmental group has been monitoring the country's use of coal for generating energy. Their data indicate that the use of coal has increased over the past 5 years. What prediction could they make from this information?
A Lakes and rivers will be healthier.
B Many species of fish will thrive.
C The atmosphere might become more polluted.
D Levels of nitrogen in the atmosphere will drop.

6 Which is *not* a renewable energy source?
A solar energy
B wind energy
C geothermal energy
D fossil fuels

7 All are nonrenewable energy sources *except*
A coal.
B geothermal energy.
C natural gas.
D petroleum oil.

8
Some Natural Resources

Sun   Coal   Water

The figure above shows some natural resources for usable energy. Which is a valid statement about all natural resources?
A All natural resources are not renewable.
B All natural resources are renewable.
C Some natural resources are made in factories.
D Some natural resources are renewable, and some are not.

Standard Set 6: Resources

Directions: Select the best answer for each of the following questions.

9 Nonrenewable resources are natural resources that cannot be replaced quickly by nature when they are used. According to this definition, which is a nonrenewable resource?

A

B

C

PETROLEUM OIL

D

10 Which natural resource is renewable?
A coal
B natural gas
C oil
D water

11 Engineers sometimes put solar panels on the roofs of buildings. These solar panels can be used to
A grow plants on the roof.
B make stronger roofs.
C make heat and electrical energy.
D reduce the ozone layer hole.

12 Which is the *major* characteristic of nonrenewable energy resources?
A They will eventually run out.
B They are inexhaustible.
C They are made by humans.
D They are used only in cars.

13 What natural resource is used to make plastics?
A water
B sand
C oil
D wood

14 Which material is a renewable resource?
A oil
B wood
C natural gas
D coal

15 The cotton material that is used to make many of your shirts originates from
A a plant.
B a fish.
C oil.
D the soil.

16 Which is the natural origin of the gasoline used as energy in *most* vehicles?
A hydrogen
B oil
C water
D coal

17 Which is considered a renewable resource?
A coal
B water
C soda can
D petroleum

At-Home Standards Practice

Standard Set 7: Investigation and Experimentation

Directions: Select the best answer for each of the following questions.

1 Some Organisms' Embryonic Development

Organism	Embryo Week 2	Embryo Week 4	Embryo Week 6
1	gill slits	gill slits	
2	4 limbs	4 legs and a tail	
3	4 limbs	2 limbs and 2 wings	

The chart above is being filled in as data from an experiment become available. Which is the *most* likely hypothesis for this experiment?

A The three organisms share a common ancestor.

B Most embryos acquire their parents' characteristics.

C Each of these organisms requires six weeks to produce offspring.

D Gill slits are not essential to live.

2 Grass grows through a crack in a concrete sidewalk. Nearby, the root of an old tree is spreading across the top of the sidewalk. In this scenario, which material is the oldest?

A root of the tree

B concrete of the sidewalk

C grass growing in the crack

D soil beneath the sidewalk

3 Consumer / Producer Relationships

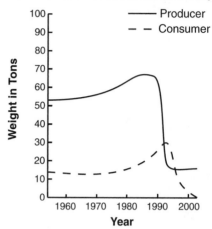

The graph above shows consumer/producer populations over several years. A reasonable hypothesis based on these data is that if the producer population decreases, then

A the predator population will increase.

B the producer population will become extinct.

C soil nutrients will decrease.

D the consumer population will decrease.

4 If you were going to examine the organisms living in a water sample from a nearby pond, which scientific tool would you *most* likely use?

A telescope

B binoculars

C microscope

D spring scale

5 A science class wants to measure the atmospheric pressure around the school. Which instrument would students use to record atmospheric pressure?

A thermometer

B barometer

C water gauge

D mass scale

Science Online Standards Practice ca6.msscience.com

Standard Set 7: Investigation and Experimentation

Directions: Select the best answer for each of the following questions.

6 The composition of the air we breathe is about 78 percent nitrogen, 21 percent oxygen, and 1 percent other gases. Which graph best illustrates this information?

A
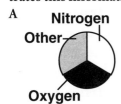

B
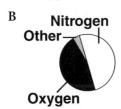

C
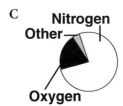

D
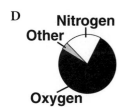

7 Information gathered from an investigation is called data and can be expressed

A as verbal, written, or numerical information.

B only in a graph or a table.

C as a hypothesis or a theory.

D as a peer-reviewed journal article.

8 Which is a reason for publishing the procedures and the results of scientific research?

A to make the research seem more scientific

B to allow the researchers to show their success to other scientists

C to allow other scientists to confirm that acceptable research methods were used

D to be able to obtain patents on the procedures so that the researchers can earn money from their research

9
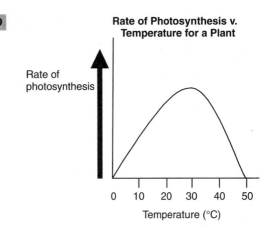

Which explanation of the graph above is *best* supported by the data?

A Photosynthesis does not occur at 0°C.

B Photosynthesis occurs at the fastest rate at 50°C.

C The rate of photosynthesis increases from 30°C to 40°C.

D The rate of photosynthesis at 45°C is greater than that at 25°C.

10 What type of map shows changes in elevation on Earth's surface?

A Mercator projection

B gnomonic projection

C contour map

D GPS

11
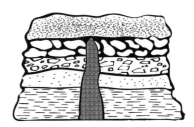

In the figure above, which layer of rock is *most* likely the youngest?

A top layer

B layer second from the top

C center layer

D bottom layer

At-Home Standards Practice

The answers for the At-Home Standards Practice presented on the previous pages are listed below. Use this answer key to check your understanding of the Standards. If you need help with a question, use the chapter and lesson reference to go back and review.

Standard Set 1 :
Pages 630–631

1 D Chp. 5, Lesson 1
2 A Chp. 2, Lesson 1
3 A Chp. 4, Lesson 1
4 C Chp. 2, Lesson 3
5 C Chp. 4, Lesson 3
6 C Chp. 2, Lesson 3
7 C Chp. 4, Lesson 3
8 B Chp. 4, Lesson 3
9 A Chp. 6, Lesson 1
10 B Chp. 7, Lesson 2
11 C Chp. 5, Lesson 1
12 C Chp. 5, Lesson 1
13 D Chp. 4, Lesson 2
14 D Chp. 7, Lesson 1
15 B Chp. 5, Lesson 2
16 C Chp. 5, Lesson 1
17 D Chp. 5, Lesson 2
18 A Chp. 6, Lesson 2
19 C Chp. 6, Lesson 3
20 C Chp. 6, Lesson 2

Standard Set 2:
Pages 632–633

1 C Chp. 8, Lesson 3
2 B Chp. 8, Lesson 2
3 D Chp. 8, Lesson 2
4 B Chp. 8, Lesson 3
5 A Chp. 8, Lesson 2
6 B Chp. 8, Lesson 2
7 D Chp. 8, Lesson 1
8 B Chp. 8, Lesson 2
9 A Chp. 8, Lesson 2
10 B Chp. 8, Lesson 2
11 A Chp. 8, Lesson 2
12 C Chp. 10, Lesson 3
13 D Chp. 10, Lesson 3
14 A Chp. 10, Lesson 3
15 D Chp. 8, Lesson 2
16 B Chp. 6, Lesson 4

Standard Set 3:
Pages 634–635

1 B Chp. 3, Lesson 2
2 D Chp. 3, Lesson 1
3 D Chp. 3, Lesson 2
4 C Chp. 3, Lesson 2
5 C Chp. 3, Lesson 3
6 A Chp. 3, Lesson 3
7 C Chp. 3, Lesson 2
8 B Chp. 3, Lesson 4
9 C Chp. 3, Lesson 1
10 A Chp. 3, Lesson 2
11 D Chp. 3, Lesson 4
12 C Chp. 3, Lesson 4
13 A Chp. 3, Lesson 4
14 B Chp. 3, Lesson 4
15 D Chp. 3, Lesson 4
16 A Chp. 3, Lesson 4
17 B Chp. 3, Lesson 4
18 D Chp. 3, Lesson 4

Standard Set 4:
Pages 636–637

1   B   Chp. 9, Lesson 1
2   D   Chp. 9, Lesson 1
3   B   Chp. 9, Lesson 1
4   C   Chp. 9, Lesson 1
5   A   Chp. 9, Lesson 1
6   B   Chp. 9, Lesson 1
7   D   Chp. 9, Lesson 1
8   A   Chp. 9, Lesson 1
9   A   Chp. 2, Lesson 3
10   B   Chp. 3, Lesson 4
11   A   Chp. 4, Lesson 3
12   D   Chp. 2, Lesson 3
13   A   Chp. 9, Lesson 3
14   B   Chp. 9, Lesson 3
15   C   Chp. 9, Lesson 3
16   A   Chp. 3, Lesson 4
17   C   Chp. 11, Lesson 2
18   C   Chp. 11, Lesson 1
19   B   Chp. 11, Lesson 2
20   B   Chp. 11, Lesson 1

Standard Set 5:
Pages 638–639

1   D   Chp. 13, Lesson 2
2   C   Chp. 13, Lesson 1
3   C   Chp. 13, Lesson 2
4   A   Chp. 13, Lesson 3
5   C   Chp. 13, Lesson 1
6   C   Chp. 13, Lesson 1
7   B   Chp. 13, Lesson 1
8   D   Chp. 13, Lesson 1
9   A   Chp. 13, Lesson 1
10   A   Chp. 12, Lesson 2
11   C   Chp. 12, Lesson 2
12   B   Chp. 12, Lesson 1
13   B   Chp. 2, Lesson 1

Standard Set 6:
Pages 640–641

1   A   Chp. 14, Lesson 3
2   B   Chp.14, Lesson 2
3   D   Chp.14, Lesson 2
4   C   Chp.14, Lesson 2
5   C   Chp.14, Lesson 3
6   D   Chp.14, Lesson 2
7   B   Chp.14, Lesson 3
8   D   Chp.14, Lesson 3
9   C   Chp.14, Lesson 3
10   D   Chp.14, Lesson 3
11   C   Chp.14, Lesson 2
12   A   Chp.14, Lesson 2
13   C   Chp.14, Lesson 3
14   B   Chp.14, Lesson 2
15   A   Chp.14, Lesson 3
16   B   Chp.14, Lesson 3
17   B   Chp.14, Lesson 2

Standard Set 7:
Page 642–643

1   D   Tools of the Scientist
2   D   Tools of the Scientist
3   D   Tools of the Scientist
4   C   Tools of the Scientist
5   B   Tools of the Scientist
6   C   Tools of the Scientist
7   A   Tools of the Scientist
8   C   Tools of the Scientist
9   C   Tools of the Scientist
10   C   Tools of the Scientist
11   A   Tools of the Scientist

Student Resources

For Students and Parents/Guardians

These resources are designed to help you achieve success in science. You will find useful information on laboratory safety, technology skills, and math skills. In addition, some Earth science reference materials are found in the Reference Handbook. You'll find the information you need to learn and sharpen your skills in these resources.

Student Resources Table of Contents

Science Safety Skill Handbook

These safety symbols are used in laboratory and field investigations in this book to indicate possible hazards. Learn the meaning of each symbol and refer to this page often. *Remember to wash your hands thoroughly after completing lab procedures.*

SAFETY SYMBOLS	HAZARD	EXAMPLES	PRECAUTION	REMEDY
DISPOSAL	Special disposal procedures need to be followed.	certain chemicals, living organisms	Do not dispose of these materials in the sink or trash can.	Dispose of wastes as directed by your teacher.
BIOLOGICAL	Organisms or other biological materials that might be harmful to humans	bacteria, fungi, blood, unpreserved tissues, plant materials	Avoid skin contact with these materials. Wear mask or gloves.	Notify your teacher if you suspect contact with material. Wash hands thoroughly.
EXTREME TEMPERATURE	Objects that can burn skin by being too cold or too hot	boiling liquids, hot plates, dry ice, liquid nitrogen	Use proper protection when handling.	Go to your teacher for first aid.
SHARP OBJECT	Use of tools or glassware that can easily puncture or slice skin	razor blades, pins, scalpels, pointed tools, dissecting probes, broken glass	Practice common-sense behavior and follow guidelines for use of the tool.	Go to your teacher for first aid.
FUME	Possible danger to respiratory tract from fumes	ammonia, acetone, nail polish remover, heated sulfur, moth balls	Make sure there is good ventilation. Never smell fumes directly. Wear a mask.	Leave foul area and notify your teacher immediately.
ELECTRICAL	Possible danger from electrical shock or burn	improper grounding, liquid spills, short circuits, exposed wires	Double-check setup with teacher. Check condition of wires and apparatus. Use GFI-protected outlets.	Do not attempt to fix electrical problems. Notify your teacher immediately.
IRRITANT	Substances that can irritate the skin or mucous membranes of the respiratory tract	pollen, moth balls, steel wool, fiberglass, potassium permanganate	Wear dust mask and gloves. Practice extra care when handling these materials.	Go to your teacher for first aid.
CHEMICAL	Chemicals that can react with and destroy tissue and other materials	bleaches such as hydrogen peroxide; acids such as sulfuric acid, hydrochloric acid; bases such as ammonia, sodium hydroxide	Wear goggles, gloves, and an apron.	Immediately flush the affected area with water and notify your teacher.
TOXIC	Substance may be poisonous if touched, inhaled, or swallowed.	mercury, many metal compounds, iodine, poinsettia plant parts	Follow your teacher's instructions.	Always wash hands thoroughly after use. Go to your teacher for first aid.
FLAMMABLE	Open flame may ignite flammable chemicals, loose clothing, or hair.	alcohol, kerosene, potassium permanganate, hair, clothing	Avoid open flames and heat when using flammable chemicals.	Notify your teacher immediately. Use fire safety equipment if applicable.
OPEN FLAME	Open flame in use, may cause fire.	hair, clothing, paper, synthetic materials	Tie back hair and loose clothing. Follow teacher's instructions on lighting and extinguishing flames.	Always wash hands thoroughly after use. Go to your teacher for first aid.

 Eye Safety Proper eye protection must be worn at all times by anyone performing or observing science activities.

 Clothing Protection This symbol appears when substances could stain or burn clothing.

 Animal Safety This symbol appears when safety of animals and students must be ensured.

 Handwashing After the lab, wash hands with soap and water before removing goggles

Safety in the Science Laboratory

Introduction to Science Safety

The science laboratory is a safe place to work if you follow standard safety procedures. Being responsible for your own safety helps to make the entire laboratory a safer place for everyone. When performing any lab, read and apply the caution statements and safety symbol listed at the beginning of the lab.

General Safety Rules

1. Complete the *Lab Safety Form* or other safety contract BEFORE starting any science lab.

2. Study the procedure. Ask your teacher any questions. Be sure you understand safety symbols shown on the page.

3. Notify your teacher about allergies or other health conditions which can affect your participation in a lab.

4. Learn and follow use and safety procedures for your equipment. If unsure, ask your teacher.

5. Never eat, drink, chew gum, apply cosmetics, or do any personal grooming in the lab. Never use lab glassware as food or drink containers. Keep your hands away from your face and mouth.

6. Know the location and proper use of the safety shower, eye wash, fire blanket, and fire alarm.

Prevent Accidents

1. Use the safety equipment provided to you. Goggles and a safety apron should be worn during investigations.

2. Do NOT use hair spray, mousse, or other flammable hair products. Tie back long hair and tie down loose clothing.

3. Do NOT wear sandals or other open-toed shoes in the lab.

4. Remove jewelry on hands and wrists. Loose jewelry, such as chains and long necklaces, should be removed to prevent them from getting caught in equipment.

5. Do not taste any substances or draw any material into a tube with your mouth.

6. Proper behavior is expected in the lab. Practical jokes and fooling around can lead to accidents and injury.

7. Keep your work area uncluttered.

Laboratory Work

1. Collect and carry all equipment and materials to your work area before beginning a lab.

2. Remain in your own work area unless given permission by your teacher to leave it.

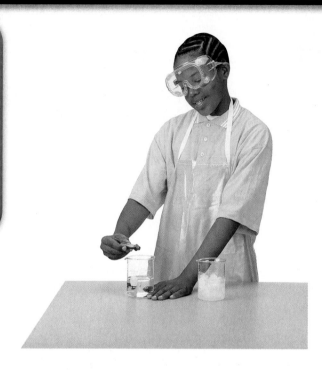

3. Always slant test tubes away from yourself and others when heating them, adding substances to them, or rinsing them.

4. If instructed to smell a substance in a container, hold the container a short distance away and fan vapors towards your nose.

5. Do NOT substitute other chemicals/substances for those in the materials list unless instructed to do so by your teacher.

6. Do NOT take any materials or chemicals outside of the laboratory.

7. Stay out of storage areas unless instructed to be there and supervised by your teacher.

Laboratory Cleanup

1. Turn off all burners, water, and gas, and disconnect all electrical devices.

2. Clean all pieces of equipment and return all materials to their proper places.

3. Dispose of chemicals and other materials as directed by your teacher. Place broken glass and solid substances in the proper containers. Never discard materials in the sink.

4. Clean your work area.

5. Wash your hands with soap and water thoroughly BEFORE removing your goggles.

Emergencies

1. Report any fire, electrical shock, glassware breakage, spill, or injury, no matter how small, to your teacher immediately. Follow his or her instructions.

2. If your clothing should catch fire, STOP, DROP, and ROLL. If possible, smother it with the fire blanket or get under a safety shower. NEVER RUN.

3. If a fire should occur, turn off all gas and leave the room according to established procedures.

4. In most instances, your teacher will clean up spills. Do NOT attempt to clean up spills unless you are given permission and instructions to do so.

5. If chemicals come into contact with your eyes or skin, notify your teacher immediately. Use the eyewash, or flush your skin or eyes with large quantities of water.

6. The fire extinguisher and first-aid kit should only be used by your teacher unless it is an extreme emergency and you have been given permission.

7. If someone is injured or becomes ill, only a professional medical provider or someone certified in first aid should perform first-aid procedures.

Computer Skills

People who study science rely on computer technology to do research, record experimental data, analyze results from investigations, and communicate with other scientists. Whether you work in a laboratory or just need to write a lab report, good computer skills are necessary.

Figure 1 Students and scientists rely on computers to gather data and communicate ideas.

Hardware Basics

Your personal computer is a system consisting of many components. The parts you can see and touch are called hardware.

Monitor   Screen   System unit

Speaker   Speaker

Keyboard   Mouse

Figure 2 Most desktop computers consist of the components shown above. Notebook computers have the same components in a compact unit.

Desktop systems, like the one shown in **Figure 2**, typically have most of these components. Notebook and tablet computers have most of the same components as a desktop computer, but the components are integrated into a single, book-sized portable unit.

Storing Your Data

When you save documents created on computers at your school, they probably are stored in a directory on your school's network. However, if you want to take the documents you have created home, you need to save them on something portable. Removable media, like those shown in **Figure 3**, are disks and drives that are designed to be moved from one computer to another.

Figure 3 Removable data storage is a convenient way to carry your documents from place to place.

Removable media vary from floppy disks and recordable CDs and DVDs to small solid-state storage. Tiny USB "keychain" drives have become popular because they can store large amounts of data and plug into any computer with a USB port. Each of these types of media stores different amounts of data. Be sure that you save your data to a medium that is compatible with your computer.

Getting Started with Word Processing Programs

A word processor is used for the composition, editing, and formatting of written material. Word processors vary from program to program, but most have the basic functions shown in **Figure 4**. Most word processors also can be used to make simple tables and graphics.

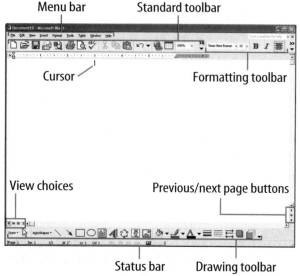

Figure 4 Word processors have functions that easily allow you to edit, format, view, and save text, tables, and images, making them useful for writing lab reports and research papers.

Word Processor Tips

- As you type, text will automatically wrap to the next line. Press *Enter* on your keyboard if you wish to start a new paragraph.

- You can move multiple lines of text around by using the *cut* and *paste* functions on the toolbar.

- If you make a typing or formatting error, use the *undo* function on the toolbar.

- Be sure to save your document early and often. This will prevent you from losing your work if your computer turns off unexpectedly.

- Use the *spell-check* function to check your spelling and grammar. Remember that *spell-check* will not catch words that are misspelled to look like other words, such as *cold* instead of *gold*. Reread your document to look for spelling and grammar mistakes.

- Graphics and spreadsheets can be added to your document by copying them from other programs and pasting them into your document.

- If you have questions about using your word processor, ask your teacher or use the program's *help* menu.

Getting Started with Spreadsheet Programs

A spreadsheet, like the one shown in **Figure 5**, helps you organize information into columns and rows. Spreadsheets are particularly useful for making data tables. Spreadsheets also can be used to perform mathematical calculations with your data. Then, you can use the spreadsheet to generate graphs and charts displaying your results.

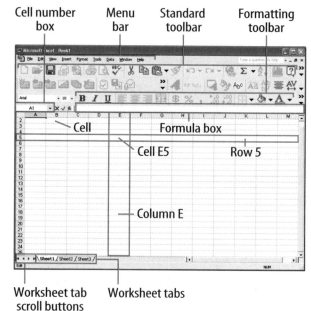

Figure 5 With formulas and graphs, spreadsheets help you organize and analyze your data.

Spreadsheet Tips

- Think about how to organize your data before you begin entering data.

- Each column (vertical) is assigned a letter and each row (horizontal) is assigned a number. Each point where a row and column intersect is called a cell, and is labeled according to where it is located. For example: column A, row 1 is cell A1.

- To edit the information in a cell, you must first activate the cell by clicking on it.

- When using a spreadsheet to generate a graph, make sure you use the type of graph that best represents the data. Review the *Science Skill Handbook* in this book for help with graphs.

- To learn more about using your spreadsheet program ask your teacher or use the program's Help menu.

Getting Started with Presentation Programs

There are many programs that help you orally communicate results of your research in an organized and interesting way. Many of these are slideshow programs, which allow you to organize text, graphs, digital photographs, sound, animations, and digital video into one multimedia presentation. Presentations can be printed onto paper or displayed on-screen. Slideshow programs are particularly effective when used with video projectors and interactive whiteboards, like the one shown in **Figure 6**. Although presentation programs are not the only way to communicate information publicly, they are an effective way to organize your presentation and remind your audience of major points.

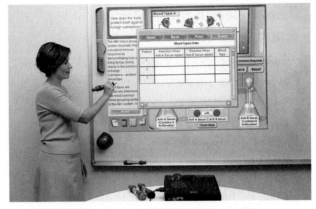

Figure 6 Video projectors and interactive whiteboards allow you to present information stored on a computer to an entire classroom. They are becoming increasingly common in the classrooms.

Presentation Program Tips

- Often, color and strong images will convey a point better than words alone. But, be sure to organize your presentation clearly. Don't let the graphics confuse the message.

- Most presentation programs will let you copy and paste text, spreadsheets, art and graphs from other programs.

- Most presentation programs have built-in templates that help you organize text and graphics.

- As with any kind of presentation, familiarize yourself with the equipment and practice your presentation before you present it to an audience.

- Most presentation programs will allow you to save your document in html format so that you can publish your document on a Web site.

- If you have questions about using your presentation software or hardware, ask your teacher or use the program's Help menu.

Doing Research with the World Wide Web

The Internet is a global network of computers where information can be stored and shared by anyone with an internet connection. One of the easiest ways to find information on the internet is by using the World Wide Web, a vast graphical system of documents written in the computer language, html (hypertext markup language). Web pages are arranged in collections of related material called "Web sites." The content on a Web site is viewed using a program called a Web browser. Web browsers, like the one shown in **Figure 7,** allow you to browse or surf the Web by clicking on highlighted hyperlinks, which move you from Web page to Web page. Web content can be searched by topic using a search engine. Search engines are located on Web sites which catalog key words on Web pages all over the World Wide Web.

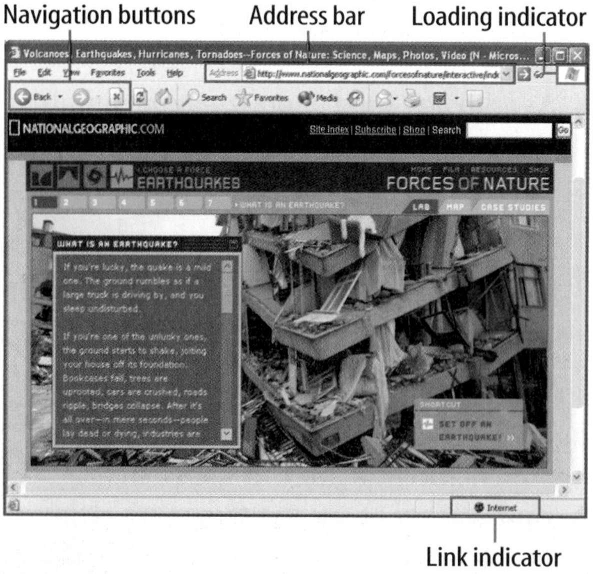

Navigation buttons   Address bar   Loading indicator

Link indicator

Figure 7 Web browsers have all the tools you need to navigate and view information on the Web.

World Wide Web Tips

- Search the Web using specific keywords. For example, if you want to research the element gold don't type *elements* into the search engine.

- When performing a Web search, enclose multiple keywords with quotes to narrow your results to the most relevant pages.

- The first hit your Web search results in is not always the best. Search results are arranged by popularity, not by relevance to your topic. Be patient and look at many links in your search results to find the best information.

- Think critically when you do science research on the Web. Compared to a traditional library, finding accurate information on the Web is not always easy because anyone can create a Web site. Some of the best places to start your research are websites for major newspapers and magazines, as well as U.S. government (*.gov*) and university (*.edu*) Web sites.

- Security is a major concern when browsing the Web. Your computer can be exposed to advertising software and computer viruses, which can hurt your computer's data and performance. *Do not download software at your school unless your teacher tells you to do so.*

- Cite information you find on the Web just as you would books and journals. An example of proper Web citation is the following:
  Menk, Amy J. (2004). *Urban Ecology.* Retrieved January 21, 2005, from McGraw-Hill Web site: http://www.mcgraw-hill.com/papers/urban.html

- The World Wide Web is a great resource for information, but don't forget to utilize local libraries, including your school library.

Math Review

Use Fractions (MA6: NS 1.0, NS 1.2, NS 2.0, NS 2.2, NS 2.4)

A fraction compares a part to a whole. In the fraction $\frac{2}{3}$, the 2 represents the part and is the numerator. The 3 represents the whole and is the denominator.

Reduce Fractions To reduce a fraction, you must find the largest factor that is common to both the numerator and the denominator, the greatest common factor (GCF). Divide both numbers by the GCF. The fraction has then been reduced, or it is in its simplest form.

Example Twelve of the 20 chemicals in the science lab are in powder form. What fraction of the chemicals used in the lab are in powder form?

Step 1 Write the fraction.

$$\frac{\text{part}}{\text{whole}} = \frac{12}{20}$$

Step 2 To find the GCF of the numerator and denominator, list all of the factors of each number.

Factors of 12: 1, 2, 3, 4, 6, 12 (the numbers that divide evenly into 12)

Factors of 20: 1, 2, 4, 5, 10, 20 (the numbers that divide evenly into 20)

Step 3 List the common factors.

1, 2, 4

Step 4 Choose the greatest factor in the list. The GCF of 12 and 20 is 4.

Step 5 Divide the numerator and denominator by the GCF.

$$\frac{12 \div 4}{20 \div 4} = \frac{3}{5}$$

In the lab, $\frac{3}{5}$ of the chemicals are in powder form.

Practice Problem At an amusement park, 66 of 90 rides have a height restriction. What fraction of the rides, in its simplest form, has a height restriction?

Add and Subtract Fractions with Like Denominators

To add or subtract fractions with the same denominator, add or subtract the numerators and write the sum or difference over the denominator. After finding the sum or difference, find the simplest form for your fraction.

Example 1 In the forest outside your house, $\frac{1}{8}$ of the animals are rabbits, $\frac{3}{8}$ are squirrels, and the remainder are birds and insects. How many are mammals?

Step 1 Add the numerators.

$$\frac{1}{8} + \frac{3}{8} = \frac{(1 + 3)}{8} = \frac{4}{8}$$

Step 2 Find the GCF.

$$\frac{4}{8} \text{ (GCF, 4)}$$

Step 3 Divide the numerator and denominator by the GCF.

$$\frac{4 \div 4}{8 \div 4} = \frac{1}{2}$$

$\frac{1}{2}$ of the animals are mammals.

Example 2 If $\frac{7}{16}$ of the Earth is covered by freshwater, and $\frac{1}{16}$ of that is in glaciers, how much freshwater is not frozen?

Step 1 Subtract the numerators.

$$\frac{7}{16} - \frac{1}{16} = \frac{(7 - 1)}{16} = \frac{6}{16}$$

Step 2 Find the GCF.

$$\frac{6}{16} \text{ (GCF, 2)}$$

Step 3 Divide the numerator and denominator by the GCF.

$$\frac{6 \div 2}{16 \div 2} = \frac{3}{8}$$

$\frac{3}{8}$ of the freshwater is not frozen.

Practice Problem A bicycle rider is riding at a rate of 15 km/h for $\frac{4}{9}$ of his ride, 10 km/h for $\frac{2}{9}$ of his ride, and 8 km/h for the remainder of the ride. How much of his ride is he riding at a rate greater than 8 km/h?

Math Skill Handbook

Add and Subtract Fractions with Unlike Denominators To add or subtract fractions with unlike denominators, first find the least common denominator (LCD). This is the smallest number that is a common multiple of both denominators. Rename each fraction with the LCD, and then add or subtract. Find the simplest form if necessary.

Example 1 A chemist makes a paste that is $\frac{1}{2}$ table salt (NaCl), $\frac{1}{3}$ sugar ($C_6H_{12}O_6$), and the remainder is water (H_2O). How much of the paste is a solid?

Step 1 Find the LCD of the fractions.

$$\frac{1}{2} + \frac{1}{3} \text{ (LCD, 6)}$$

Step 2 Rename each numerator and each denominator with the LCD.

Step 3 Add the numerators.

$$\frac{3}{6} + \frac{2}{6} = \frac{(3+2)}{6} = \frac{5}{6}$$

$\frac{5}{6}$ of the paste is a solid.

Example 2 The average precipitation in Grand Junction, CO, is $\frac{7}{10}$ inch in November, and $\frac{3}{5}$ inch in December. What is the total average precipitation?

Step 1 Find the LCD of the fractions.

$$\frac{7}{10} + \frac{3}{5} \text{ (LCD, 10)}$$

Step 2 Rename each numerator and each denominator with the LCD.

Step 3 Add the numerators.

$$\frac{7}{10} + \frac{6}{10} = \frac{(7+6)}{10} = \frac{13}{10}$$

$\frac{13}{10}$ inches total precipitation, or $1\frac{3}{10}$ inches.

Practice Problem On an electric bill, about $\frac{1}{8}$ of the energy is from solar energy and about $\frac{1}{10}$ is from wind power. How much of the total bill is from solar energy and wind power combined?

Example 3 In your body, $\frac{7}{10}$ of your muscle contractions are involuntary (cardiac and smooth muscle tissue). Smooth muscle makes $\frac{3}{15}$ of your muscle contractions. How many of your muscle contractions are made by cardiac muscle?

Step 1 Find the LCD of the fractions.

$$\frac{7}{10} - \frac{3}{15} \text{ (LCD, 30)}$$

Step 2 Rename each numerator and each denominator with the LCD.

$$\frac{7 \times 3}{10 \times 3} = \frac{21}{30}$$

$$\frac{3 \times 2}{15 \times 2} = \frac{6}{30}$$

Step 3 Subtract the numerators.

$$\frac{21}{30} - \frac{6}{30} = \frac{(21-6)}{30} = \frac{15}{30}$$

Step 4 Find the GCF.

$$\frac{15}{30} \text{ (GCF, 15)}$$

$$\frac{1}{2}$$

$\frac{1}{2}$ of all muscle contractions are cardiac muscle.

Example 4 Tony wants to make cookies that call for $\frac{3}{4}$ of a cup of flour, but he only has $\frac{1}{3}$ of a cup. How much more flour does he need?

Step 1 Find the LCD of the fractions.

$$\frac{3}{4} - \frac{1}{3} \text{ (LCD, 12)}$$

Step 2 Rename each numerator and each denominator with the LCD.

$$\frac{3 \times 3}{4 \times 3} = \frac{9}{12}$$

$$\frac{1 \times 4}{3 \times 4} = \frac{4}{12}$$

Step 3 Subtract the numerators.

$$\frac{9}{12} - \frac{4}{12} = \frac{(9-4)}{12} = \frac{5}{12}$$

$\frac{5}{12}$ of a cup of flour

Practice Problem Using the information provided to you in Example 3 above, determine how many muscle contractions are voluntary (skeletal muscle).

Multiply Fractions To multiply with fractions, multiply the numerators and multiply the denominators. Find the simplest form if necessary.

Example Multiply $\frac{3}{5}$ by $\frac{1}{3}$.

Step 1 Multiply the numerators and denominators.

$$\frac{3}{5} \times \frac{1}{3} = \frac{(3 \times 1)}{(5 \times 3)} = \frac{3}{15}$$

Step 2 Find the GCF.

$$\frac{3}{15} \text{ (GCF, 3)}$$

Step 3 Divide the numerator and denominator by the GCF.

$$\frac{3 \div 3}{15 \div 3} = \frac{1}{5}$$

$\frac{3}{5}$ multiplied by $\frac{1}{3}$ is $\frac{1}{5}$.

Practice Problem Multiply $\frac{3}{14}$ by $\frac{5}{16}$.

Find a Reciprocal Two numbers whose product is 1 are called multiplicative inverses, or reciprocals.

Example Find the reciprocal of $\frac{3}{8}$.

Step 1 Inverse the fraction by putting the denominator on top and the numerator on the bottom.

$$\frac{8}{3}$$

The reciprocal of $\frac{3}{8}$ is $\frac{8}{3}$.

Practice Problem Find the reciprocal of $\frac{4}{9}$.

Divide Fractions To divide one fraction by another fraction, multiply the dividend by the reciprocal of the divisor. Find the simplest form if necessary.

Example 1 Divide $\frac{1}{9}$ by $\frac{1}{3}$.

Step 1 Find the reciprocal of the divisor.

The reciprocal of $\frac{1}{3}$ is $\frac{3}{1}$.

Step 2 Multiply the dividend by the reciprocal of the divisor.

$$\frac{\frac{1}{9}}{\frac{1}{3}} = \frac{1}{9} \times \frac{3}{1} = \frac{(1 \times 3)}{(9 \times 1)} = \frac{3}{9}$$

Step 3 Find the GCF.

$$\frac{3}{9} \text{ (GCF, 3)}$$

Step 4 Divide the numerator and denominator by the GCF.

$$\frac{3 \div 3}{9 \div 3} = \frac{1}{3}$$

$\frac{1}{9}$ divided by $\frac{1}{3}$ is $\frac{1}{3}$.

Example 2 Divide $\frac{3}{5}$ by $\frac{1}{4}$.

Step 1 Find the reciprocal of the divisor.

The reciprocal of $\frac{1}{4}$ is $\frac{4}{1}$.

Step 2 Multiply the dividend by the reciprocal of the divisor.

$$\frac{\frac{3}{5}}{\frac{1}{4}} = \frac{3}{5} \times \frac{4}{1} = \frac{(3 \times 4)}{(5 \times 1)} = \frac{12}{5}$$

$\frac{3}{5}$ divided by $\frac{1}{4}$ is $\frac{12}{5}$ or $2\frac{2}{5}$.

Practice Problem Divide $\frac{3}{11}$ by $\frac{7}{10}$.

Math Skill Handbook

Use Ratios MA6: NS 1.2

When you compare two numbers by division, you are using a ratio. Ratios can be written 3 to 5, 3:5, or $\frac{3}{5}$. Ratios, like fractions, also can be written in simplest form.

Ratios can represent one type of probability, called odds. This is a ratio that compares the number of ways a certain outcome occurs to the number of possible outcomes. For example, if you flip a coin 100 times, what are the odds that it will come up heads? There are two possible outcomes, heads or tails, so the odds of coming up heads are 50:100. Another way to say this is that 50 out of 100 times the coin will come up heads. In its simplest form, the ratio is 1:2.

Example 1 A chemical solution contains 40 g of salt and 64 g of baking soda. What is the ratio of salt to baking soda as a fraction in simplest form?

Step 1   Write the ratio as a fraction.
$$\frac{salt}{baking\ soda} = \frac{40}{64}$$

Step 2   Express the fraction in simplest form. The GCF of 40 and 64 is 8.
$$\frac{40}{64} = \frac{40 \div 8}{64 \div 8} = \frac{5}{8}$$

The ratio of salt to baking soda in the sample is 5:8.

Example 2 Sean rolls a 6-sided die 6 times. What are the odds that the side with a 3 will show?

Step 1   Write the ratio as a fraction.
$$\frac{number\ of\ sides\ with\ a\ 3}{number\ of\ possible\ sides} = \frac{1}{6}$$

Step 2   Multiply by the number of attempts.
$$\frac{1}{6} \times 6\ attempts = \frac{6}{6}\ attempts = 1\ attempt$$

1 attempt out of 6 will show a 3.

Practice Problem Two metal rods measure 100 cm and 144 cm in length. What is the ratio of their lengths in simplest form?

Use Decimals MA6: NS 1.0

A fraction with a denominator that is a power of ten can be written as a decimal. For example, 0.27 means $\frac{27}{100}$. The decimal point separates the ones place from the tenths place.

Any fraction can be written as a decimal using division. For example, the fraction $\frac{5}{8}$ can be written as a decimal by dividing 5 by 8. Written as a decimal, it is 0.625.

Add or Subtract Decimals When adding and subtracting decimals, line up the decimal points before carrying out the operation.

Example 1 Find the sum of 47.68 and 7.80.

Step 1   Line up the decimal places when you write the numbers.
$$\begin{array}{r} 47.68 \\ + 7.80 \end{array}$$

Step 2   Add the decimals.
$$\begin{array}{r} {\scriptstyle 1\ 1} \\ 47.68 \\ + 7.80 \\ \hline 55.48 \end{array}$$

The sum of 47.68 and 7.80 is 55.48.

Example 2 Find the difference of 42.17 and 15.85.

Step 1   Line up the decimal places when you write the number.
$$\begin{array}{r} 42.17 \\ -15.85 \end{array}$$

Step 2   Subtract the decimals.
$$\begin{array}{r} {\scriptstyle 3\ 11} \\ 42.17 \\ -15.85 \\ \hline 26.32 \end{array}$$

The difference of 42.17 and 15.85 is 26.32.

Practice Problem Find the sum of 1.245 and 3.842.

Multiply Decimals To multiply decimals, multiply the numbers like numbers without decimal points. Count the decimal places in each factor. The product will have the same number of decimal places as the sum of the decimal places in the factors.

Example Multiply 2.4 by 5.9.

Step 1 Multiply the factors like two whole numbers.
$24 \times 59 = 1416$

Step 2 Find the sum of the number of decimal places in the factors. Each factor has one decimal place, for a sum of two decimal places.

Step 3 The product will have two decimal places.
14.16

The product of 2.4 and 5.9 is 14.16.

Practice Problem Multiply 4.6 by 2.2.

Divide Decimals When dividing decimals, change the divisor to a whole number. To do this, multiply both the divisor and the dividend by the same power of ten. Then place the decimal point in the quotient directly above the decimal point in the dividend. Then divide as you do with whole numbers.

Example Divide 8.84 by 3.4.

Step 1 Multiply both factors by 10.
$3.4 \times 10 = 34, 8.84 \times 10 = 88.4$

Step 2 Divide 88.4 by 34.

```
      2.6
34)88.4
   -68
    204
   -204
      0
```

8.84 divided by 3.4 is 2.6.

Practice Problem Divide 75.6 by 3.6.

Use Proportions   MA6: NS 1.3, AF 2.0

An equation that shows that two ratios are equivalent is a proportion. The ratios $\frac{2}{4}$ and $\frac{5}{10}$ are equivalent, so they can be written as $\frac{2}{4} = \frac{5}{10}$. This equation is a proportion.

When two ratios form a proportion, the cross products are equal. To find the cross products in the proportion $\frac{2}{4} = \frac{5}{10}$, multiply the 2 and the 10, and the 4 and the 5. Therefore $2 \times 10 = 4 \times 5$, or $20 = 20$.

Because you know that both ratios are equal, you can use cross products to find a missing term in a proportion. This is known as solving the proportion.

Example The heights of a tree and a pole are proportional to the lengths of their shadows. The tree casts a shadow of 24 m when a 6-m pole casts a shadow of 4 m. What is the height of the tree?

Step 1 Write a proportion.
$$\frac{\text{height of tree}}{\text{height of pole}} = \frac{\text{length of tree's shadow}}{\text{length of pole's shadow}}$$

Step 2 Substitute the known values into the proportion. Let h represent the unknown value, the height of the tree.
$$\frac{h}{6} = \frac{24}{4}$$

Step 3 Find the cross products.
$h \times 4 = 6 \times 24$

Step 4 Simplify the equation.
$4h = 144$

Step 5 Divide each side by 4.
$$\frac{4h}{4} = \frac{144}{4}$$
$$h = 36$$

The height of the tree is 36 m.

Practice Problem The ratios of the weights of two objects on the Moon and on Earth are in proportion. A rock weighing 3 N on the Moon weighs 18 N on Earth. How much would a rock that weighs 5 N on the Moon weigh on Earth?

Math Skill Handbook

Use Percentages MA6: NS 1.0

The word *percent* means "out of one hundred." It is a ratio that compares a number to 100. Suppose you read that 77 percent of the Earth's surface is covered by water. That is the same as reading that the fraction of the Earth's surface covered by water is $\frac{77}{100}$. To express a fraction as a percent, first find the equivalent decimal for the fraction. Then, multiply the decimal by 100 and add the percent symbol.

Example 1 Express $\frac{13}{20}$ as a percent.

Step 1 Find the equivalent decimal for the fraction.

$$
\begin{array}{r}
0.65 \\
20\overline{)13.00} \\
\underline{12\,0} \\
1\,00 \\
\underline{1\,00} \\
0
\end{array}
$$

Step 2 Rewrite the fraction $\frac{13}{20}$ as 0.65.

Step 3 Multiply 0.65 by 100 and add the % symbol.
$0.65 \times 100 = 65 = 65\%$

So, $\frac{13}{20} = 65\%$.

This also can be solved as a proportion.

Example 2 Express $\frac{13}{20}$ as a percent.

Step 1 Write a proportion.
$\frac{13}{20} = \frac{x}{100}$

Step 2 Find the cross products.
$1300 = 20x$

Step 3 Divide each side by 20.
$\frac{1300}{20} = \frac{20x}{20}$
$65\% = x$

Practice Problem In one year, 73 of 365 days were rainy in one city. What percent of the days in that city were rainy?

Solve One-Step Equations MA6: NS 1.0, MR 2.4, MR 2.5

A statement that two expressions are equal is an equation. For example, $A = B$ is an equation that states that A is equal to B.

An equation is solved when a variable is replaced with a value that makes both sides of the equation equal. To make both sides equal the inverse operation is used. Addition and subtraction are inverses, and multiplication and division are inverses.

Example 1 Solve the equation $x - 10 = 35$.

Step 1 Find the solution by adding 10 to each side of the equation.
$$x - 10 = 35$$
$$x - 10 + 10 = 35 + 10$$
$$x = 45$$

Step 2 Check the solution.
$$x - 10 = 35$$
$$45 - 10 = 35$$
$$35 = 35$$

Both sides of the equation are equal, so $x = 45$.

Example 2 In the formula $a = bc$, find the value of c if $a = 20$ and $b = 2$.

Step 1 Rearrange the formula so the unknown value is by itself on one side of the equation by dividing both sides by b.
$$a = bc$$
$$\frac{a}{b} = \frac{bc}{b}$$
$$\frac{a}{b} = c$$

Step 2 Replace the variables a and b with the values that are given.
$$\frac{a}{b} = c$$
$$\frac{20}{2} = c$$
$$10 = c$$

Step 3 Check the solution.
$$a = bc$$
$$20 = 2 \times 10$$
$$20 = 20$$

Both sides of the equation are equal, so $c = 10$ is the solution when $a = 20$ and $b = 2$.

Practice Problem In the formula $h = gd$, find the value of d if $g = 12.3$ and $h = 17.4$.

Use Statistics **MA6: SP 1.1, SP 1.4**

The branch of mathematics that deals with collecting, analyzing, and presenting data is statistics. In statistics, there are three common ways to summarize data with a single number—the mean, the median, and the mode.

The **mean** of a set of data is the arithmetic average. It is found by adding the numbers in the data set and dividing by the number of items in the set.

The **median** is the middle number in a set of data when the data are arranged in numerical order. If there were an even number of data points, the median would be the mean of the two middle numbers.

The **mode** of a set of data is the number or item that appears most often.

Another number that often is used to describe a set of data is the range. The **range** is the difference between the largest number and the smallest number in a set of data.

Example The speeds (in m/s) for a race car during five different time trials are 39, 37, 44, 36, and 44.

To find the mean:

Step 1 Find the sum of the numbers.
$39 + 37 + 44 + 36 + 44 = 200$

Step 2 Divide the sum by the number of items, which is 5.
$200 \div 5 = 40$

The mean is 40 m/s.

To find the median:

Step 1 Arrange the measures from least to greatest.
36, 37, 39, 44, 44

Step 2 Determine the middle measure.
36, 37, <u>39</u>, 44, 44

The median is 39 m/s.

To find the mode:

Step 1 Group the numbers that are the same together.
44, 44, 36, 37, 39

Step 2 Determine the number that occurs most in the set.
<u>44, 44</u>, 36, 37, 39

The mode is 44 m/s.

To find the range:

Step 1 Arrange the measures from greatest to least.
44, 44, 39, 37, 36

Step 2 Determine the greatest and least measures in the set.
<u>44</u>, 44, 39, 37, <u>36</u>

Step 3 Find the difference between the greatest and least measures.
$44 - 36 = 8$

The range is 8 m/s.

Practice Problem Find the mean, median, mode, and range for the data set 8, 4, 12, 8, 11, 14, 16.

A **frequency table** shows how many times each piece of data occurs, usually in a survey. **Table 1** below shows the results of a student survey on favorite color.

Table 1 Student Color Choice

Color	Tally	Frequency
red	IIII	4
blue	HHH	5
black	II	2
green	III	3
purple	HHH II	7
yellow	HHH I	6

Based on the frequency table data, which color is the favorite?

Math Skill Handbook

Use Geometry

MA6: MG 1.0, MG 1.1, MG 1.2, MG 1.3

The branch of mathematics that deals with the measurement, properties, and relationships of points, lines, angles, surfaces, and solids is called geometry.

Perimeter The **perimeter** (P) is the distance around a geometric figure. To find the perimeter of a rectangle, add the length and width and multiply that sum by two, or $2(l + w)$. To find perimeters of irregular figures, add the length of the sides.

Example 1 Find the perimeter of a rectangle that is 3 m long and 5 m wide.

Step 1 You know that the perimeter is 2 times the sum of the width and length.
$$P = 2(3\text{ m} + 5\text{ m})$$

Step 2 Find the sum of the width and length.
$$P = 2(8\text{ m})$$

Step 3 Multiply by 2.
$$P = 16\text{ m}$$

The perimeter is 16 m.

Example 2 Find the perimeter of a shape with sides measuring 2 cm, 5 cm, 6 cm, 3 cm.

Step 1 You know that the perimeter is the sum of all the sides.
$$P = 2 + 5 + 6 + 3$$

Step 2 Find the sum of the sides.
$$P = 2 + 5 + 6 + 3$$
$$P = 16$$

The perimeter is 16 cm.

Practice Problem Find the perimeter of a rectangle with a length of 18 m and a width of 7 m.

Practice Problem Find the perimeter of a triangle measuring 1.6 cm by 2.4 cm by 2.4 cm.

Area of a Rectangle The **area** (A) is the number of square units needed to cover a surface. To find the area of a rectangle, multiply the length times the width, or $l \times w$. When finding area, the units also are multiplied. Area is given in square units.

Example Find the area of a rectangle with a length of 1 cm and a width of 10 cm.

Step 1 You know that the area is the length multiplied by the width.
$$A = (1\text{ cm} \times 10\text{ cm})$$

Step 2 Multiply the length by the width. Also multiply the units.
$$A = 10\text{ cm}^2$$

The area is 10 cm².

Practice Problem Find the area of a square whose sides measure 4 m.

Area of a Triangle To find the area of a triangle, use the formula:
$$A = \tfrac{1}{2}(\text{base} \times \text{height})$$

The base of a triangle can be any of its sides. The height is the perpendicular distance from a base to the opposite endpoint, or vertex.

Example Find the area of a triangle with a base of 18 m and a height of 7 m.

Step 1 You know that the area is $\frac{1}{2}$ the base times the height.
$$A = \tfrac{1}{2}(18\text{ m} \times 7\text{ m})$$

Step 2 Multiply $\frac{1}{2}$ by the product of 18×7. Multiply the units.
$$A = \tfrac{1}{2}(126\text{ m}^2)$$
$$A = 63\text{ m}^2$$

The area is 63 m².

Practice Problem Find the area of a triangle with a base of 27 cm and a height of 17 cm.

Circumference of a Circle The **diameter** (*d*) of a circle is the distance across the circle through its center, and the **radius** (*r*) is the distance from the center to any point on the circle. The radius is half of the diameter. The distance around the circle is called the **circumference** (*C*). The formula for finding the circumference is:

$C = 2\pi r$ or $C = \pi d$

The circumference divided by the diameter is always equal to 3.1415926… This nonterminating and nonrepeating number is represented by the Greek letter π (pi). An approximation often used for π is 3.14.

Example 1 Find the circumference of a circle with a radius of 3 m.

Step 1 You know the formula for the circumference is 2 times the radius times π.

$C = 2\pi(3)$

Step 2 Multiply 2 times the radius.

$C = 6\pi$

Step 3 Multiply by π.

$C \approx 19$ m

The circumference is about 19 m.

Example 2 Find the circumference of a circle with a diameter of 24.0 cm.

Step 1 You know the formula for the circumference is the diameter times π.

$C = \pi(24.0)$

Step 2 Multiply the diameter by π.

$C \approx 75.4$ cm

The circumference is about 75.4 cm.

Practice Problem Find the circumference of a circle with a radius of 19 cm.

Area of a Circle The formula for the area of a circle is:

$A = \pi r^2$

Example 1 Find the area of a circle with a radius of 4.0 cm.

Step 1 $A = \pi(4.0)^2$

Step 2 Find the square of the radius.

$A = 16\pi$

Step 3 Multiply the square of the radius by π.

$A \approx 50$ cm^2

The area of the circle is about 50 cm^2.

Example 2 Find the area of a circle with a radius of 225 m.

Step 1 $A = \pi(225)^2$

Step 2 Find the square of the radius.

$A = 50625\pi$

Step 3 Multiply the square of the radius by π.

$A \approx 159043.1$

The area of the circle is about 159043.1 m^2.

Example 3 Find the area of a circle whose diameter is 20.0 mm.

Step 1 You know the formula for the area of a circle is the square of the radius times π, and that the radius is half of the diameter.

$A = \pi\left(\dfrac{20.0}{2}\right)^2$

Step 2 Find the radius.

$A = \pi(10.0)^2$

Step 3 Find the square of the radius.

$A = 100\pi$

Step 4 Multiply the square of the radius by π.

$A \approx 314$ mm^2

The area of is about 314 mm^2.

Practice Problem Find the area of a circle with a radius of 16 m.

Math Skill Handbook

Volume The measure of space occupied by a solid is the **volume** (V). To find the volume of a rectangular solid multiply the length times width times height, or $V = l \times w \times h$. It is measured in cubic units, such as cubic centimeters (cm^3).

Example Find the volume of a rectangular solid with a length of 2.0 m, a width of 4.0 m, and a height of 3.0 m.

Step 1 You know the formula for volume is the length times the width times the height.

$V = 2.0 \text{ m} \times 4.0 \text{ m} \times 3.0 \text{ m}$

Step 2 Multiply the length times the width times the height.

$V = 24 \text{ m}^3$

The volume is 24 m³.

Practice Problem Find the volume of a rectangular solid that is 8 m long, 4 m wide, and 4 m high.

To find the volume of other solids, multiply the area of the base times the height.

Example 1 Find the volume of a solid that has a triangular base with a length of 8.0 m and a height of 7.0 m. The height of the entire solid is 15.0 m.

Step 1 You know that the base is a triangle, and the area of a triangle is $\frac{1}{2}$ the base times the height, and the volume is the area of the base times the height.

$V = \left[\frac{1}{2}(b \times h)\right] \times 15$

Step 2 Find the area of the base.

$V = \left[\frac{1}{2}(8 \times 7)\right] \times 15$

$V = \left(\frac{1}{2} \times 56\right) \times 15$

Step 3 Multiply the area of the base by the height of the solid.

$V = 28 \times 15$

$V = 420 \text{ m}^3$

The volume is 420 m³.

Example 2 Find the volume of a cylinder that has a base with a radius of 12.0 cm, and a height of 21.0 cm.

Step 1 You know that the base is a circle, and the area of a circle is the square of the radius times π, and the volume is the area of the base times the height.

$V = (\pi r^2) \times 21$

$V = (\pi 12^2) \times 21$

Step 2 Find the area of the base.

$V = 144\pi \times 21$

$V = 452 \times 21$

Step 3 Multiply the area of the base by the height of the solid.

$V \approx 9,500 \text{ cm}^3$

The volume is about 9,500 cm³.

Example 3 Find the volume of a cylinder that has a diameter of 15 mm and a height of 4.8 mm.

Step 1 You know that the base is a circle with an area equal to the square of the radius times π. The radius is one-half the diameter. The volume is the area of the base times the height.

$V = (\pi r^2) \times 4.8$

$V = \left[\pi\left(\frac{1}{2} \times 15\right)^2\right] \times 4.8$

$V = (\pi 7.5^2) \times 4.8$

Step 2 Find the area of the base.

$V = 56.25\pi \times 4.8$

$V \approx 176.71 \times 4.8$

Step 3 Multiply the area of the base by the height of the solid.

$V \approx 848.2$

The volume is about 848.2 mm³.

Practice Problem Find the volume of a cylinder with a diameter of 7 cm in the base and a height of 16 cm.

Science Applications

Measure in SI

The metric system of measurement was developed in 1795. A modern form of the metric system, called the International System (SI), was adopted in 1960 and provides the standard measurements that all scientists around the world can understand.

The SI system is convenient because unit sizes vary by powers of 10. Prefixes are used to name units. Look at **Table 2** for some common SI prefixes and their meanings.

Table 2 Common SI Prefixes			
Prefix	**Symbol**	**Meaning**	
kilo-	k	1,000	thousandth
hecto-	h	100	hundred
deka-	da	10	ten
deci-	d	0.1	tenth
centi-	c	0.01	hundreth
milli-	m	0.001	thousandth

Example How many grams equal one kilogram?

Step 1 Find the prefix *kilo-* in **Table 2.**

Step 2 Using **Table 2,** determine the meaning of *kilo-*. According to the table, it means 1,000. When the prefix *kilo-* is added to a unit, it means that there are 1,000 of the units in a "kilounit."

Step 3 Apply the prefix to the units in the question. The units in the question are grams. There are 1,000 grams in a kilogram.

Practice Problem Is a milligram larger or smaller than a gram? How many of the smaller units equal one larger unit? What fraction of the larger unit does one smaller unit represent?

Dimensional Analysis `MA6: AF 2.1`

Convert SI Units In science, quantities such as length, mass, and time sometimes are measured using different units. A process called dimensional analysis can be used to change one unit of measure to another. This process involves multiplying your starting quantity and units by one or more conversion factors. A conversion factor is a ratio equal to one and can be made from any two equal quantities with different units. If 1,000 mL equal 1 L then two ratios can be made.

$$\frac{1,000 \text{ mL}}{1 \text{ L}} = \frac{1 \text{ L}}{1,000 \text{ mL}} = 1$$

One can convert between units in the SI system by using the equivalents in **Table 2** to make conversion factors.

Example How many cm are in 4 m?

Step 1 Write conversion factors for the units given. From **Table 2,** you know that 100 cm = 1 m. The conversion factors are $\frac{100 \text{ cm}}{1 \text{ m}}$ and $\frac{1 \text{ m}}{100 \text{ cm}}$

Step 2 Decide which conversion factor to use. Select the factor that has the units you are converting from (m) in the denominator and the units you are converting to (cm) in the numerator. $\frac{100 \text{ cm}}{1 \text{ m}}$

Step 3 Multiply the starting quantity and units by the conversion factor. Cancel the starting units with the units in the denominator. There are 400 cm in 4 m.

$$4 \text{ m} = \frac{100 \text{ cm}}{1 \text{ m}} = 400 \text{ cm}$$

Practice Problem How many milligrams are in one kilogram? (Hint: You will need to use two conversion factors from **Table 2.**)

Math Skill Handbook

Table 3 Unit System Equivalents	
Type of Measurement	**Equivalent**
Length	1 in = 2.54 cm 1 yd = 0.91 m 1 mi = 1.61 km
Mass and weight*	1 oz = 28.35 g 1 lb = 0.45 kg 1 ton (short) = 0.91 tonnes (metric tons) 1 lb = 4.45 N
Volume	$1 \text{ in}^3 = 16.39 \text{ cm}^3$ 1 qt = 0.95 L 1 gal = 3.78 L
Area	$1 \text{ in}^2 = 6.45 \text{ cm}^2$ $1 \text{ yd}^2 = 0.83 \text{ m}^2$ $1 \text{ mi}^2 = 2.59 \text{ km}^2$ 1 acre = 0.40 hectares
Temperature	$°C = \frac{(°F - 32)}{1.8}$ K = °C + 273

*Weight is measured in standard Earth gravity.

Convert Between Unit Systems **Table 3** gives a list of equivalents that can be used to convert between English and SI units.

Example If a meterstick has a length of 100 cm, how long is the meterstick in inches?

Step 1 Write the conversion factors for the units given. From **Table 3,** 1 in = 2.54 cm.

$$\frac{1 \text{ in}}{2.54 \text{ cm}} \quad and \quad \frac{2.54 \text{ cm}}{1 \text{ in}}$$

Step 2 Determine which conversion factor to use. You are converting from cm to in. Use the conversion factor with cm on the bottom.

$$\frac{1 \text{ in}}{2.54 \text{ cm}}$$

Step 3 Multiply the starting quantity and units by the conversion factor. Cancel the starting units with the units in the denominator. Round your answer to the nearest tenth.

$$100 \text{ cm} \times \frac{1 \text{ in}}{2.54 \text{ cm}} = 39.37 \text{ in}$$

The meterstick is about 39.4 in long.

Practice Problem 1 A book has a mass of 5 lb. What is the mass of the book in kg?

Practice Problem 2 Use the equivalent for in and cm (1 in = 2.54 cm) to show how $1 \text{ in}^3 \approx 16.39 \text{ cm}^3$.

Precision and Significant Digits MA6: MR 2.6

When you make a measurement, the value you record depends on the precision of the measuring instrument. This precision is represented by the number of significant digits recorded in the measurement. When counting the number of significant digits, all digits are counted except zeros at the end of a number with no decimal point such as 2,050, and zeros at the beginning of a decimal such as 0.03020. When adding or subtracting numbers with different precision, round the answer to the smallest number of decimal places of any number in the sum or difference. When multiplying or dividing, the answer is rounded to the smallest number of significant digits of any number being multiplied or divided.

Example The lengths 5.28 and 5.2 are measured in meters. Find the sum of these lengths and record your answer using the correct number of significant digits.

Step 1 Find the sum.

5.28 m	2 digits after the decimal
+ 5.2 m	1 digit after the decimal
10.48 m	

Step 2 Round to one digit after the decimal because the least number of digits after the decimal of the numbers being added is 1.

The sum is 10.5 m.

Practice Problem 1 How many significant digits are in the measurement 7,071,301 m? How many significant digits are in the measurement 0.003010 g?

Practice Problem 2 Multiply 5.28 and 5.2 using the rule for multiplying and dividing. Record the answer using the correct number of significant digits.

Scientific Notation

Many times numbers used in science are very small or very large. Because these numbers are difficult to work with scientists use scientific notation. To write numbers in scientific notation, move the decimal point until only one non-zero digit remains on the left. Then count the number of places you moved the decimal point and use that number as a power of ten. For example, the average distance from the Sun to Mars is 227,800,000,000 m. In scientific notation, this distance is 2.278×10^{11} m. Because you moved the decimal point to the left, the number is a positive power of ten.

The mass of an electron is about 0.000 000 000 000 000 000 000 000 000 000 911 kg. Expressed in scientific notation, this mass is 9.11×10^{-31} kg. Because the decimal point was moved to the right, the number is a negative power of ten.

Example Earth is 149,600,000 km from the Sun. Express this in scientific notation.

Step 1 Move the decimal point until one non-zero digit remains on the left.
1.496 000 00

Step 2 Count the number of decimal places you have moved. In this case, eight.

Step 2 Show that number as a power of ten, 10^8.

Earth is 1.496×10^8 km from the Sun.

Practice Problem 1 How many significant digits are in 149,600,000 km? How many significant digits are in 1.496×10^8 km?

Practice Problem 2 Parts used in a high performance car must be measured to 7×10^{-6} m. Express this number as a decimal.

Practice Problem 3 A CD is spinning at 539 revolutions per minute. Express this number in scientific notation.

Make and Use Graphs MA6: MR 2.4

Data in tables can be displayed in a graph—a visual representation of data. Common graph types include line graphs, bar graphs, and circle graphs.

Line Graph A line graph shows a relationship between two variables that change continuously. The independent variable is changed and is plotted on the *x*-axis. The dependent variable is observed, and is plotted on the *y*-axis.

Example Draw a line graph of the data below from a cyclist in a long-distance race.

Table 4 Bicycle Race Data	
Time (h)	Distance (km)
0	0
1	8
2	16
3	24
4	32
5	40

Step 1 Determine the *x*-axis and *y*-axis variables. Time varies independently of distance and is plotted on the *x*-axis. Distance is dependent on time and is plotted on the *y*-axis.

Step 2 Determine the scale of each axis. The *x*-axis data ranges from 0 to 5. The *y*-axis data ranges from 0 to 50.

Step 3 Using graph paper, draw and label the axes. Include units in the labels.

Step 4 Draw a point at the intersection of the time value on the *x*-axis and corresponding distance value on the *y*-axis. Connect the points and label the graph with a title, as shown in **Figure 8.**

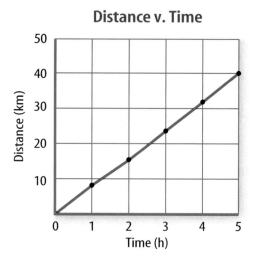

Distance v. Time

Figure 8 This line graph shows the relationship between distance and time during a bicycle ride.

Practice Problem A puppy's shoulder height is measured during the first year of her life. The following measurements were collected: (3 mo, 52 cm), (6 mo, 72 cm), (9 mo, 83 cm), (12 mo, 86 cm). Graph this data.

Find a Slope The slope of a straight line is the ratio of the vertical change, rise, to the horizontal change, run.

$$\text{Slope} = \frac{\text{vertical change (rise)}}{\text{horizontal change (run)}} = \frac{\text{change in } y}{\text{change in } x}$$

Example Find the slope of the graph in **Figure 8.**

Step 1 You know that the slope is the change in *y* divided by the change in *x*.

$$\text{Slope} = \frac{\text{change in } y}{\text{change in } x}$$

Step 2 Determine the data points you will be using. For a straight line, choose the two sets of points that are the farthest apart.

$$\text{Slope} = \frac{(40 - 0) \text{ km}}{(5 - 0) \text{ h}}$$

Step 3 Find the change in *y* and *x*.

$$\text{Slope} = \frac{40 \text{ km}}{5 \text{ h}}$$

Step 4 Divide the change in *y* by the change in *x*.

$$\text{Slope} = \frac{8 \text{ km}}{\text{h}}$$

The slope of the graph is 8 km/h.

Bar Graph To compare data that does not change continuously you might choose a bar graph. A bar graph uses bars to show the relationships between variables. The *x*-axis variable is divided into parts. The parts can be numbers such as years, or a category such as a type of animal. The *y*-axis is a number and increases continuously along the axis.

Example A recycling center collects 4.0 kg of aluminum on Monday, 1.0 kg on Wednesday, and 2.0 kg on Friday. Create a bar graph of this data.

Step 1 Select the *x*-axis and *y*-axis variables. The measured numbers (the masses of aluminum) should be placed on the *y*-axis. The variable divided into parts (collection days) is placed on the *x*-axis.

Step 2 Create a graph grid like you would for a line graph. Include labels and units.

Step 3 For each measured number, draw a vertical bar above the *x*-axis value up to the *y*-axis value. For the first data point, draw a vertical bar above Monday up to 4.0 kg.

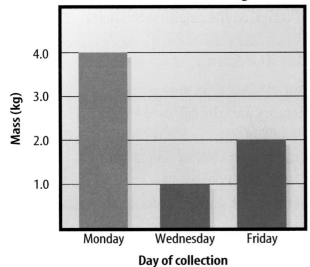

Aluminum Collected During Week

Practice Problem Draw a bar graph of the gases in air: 78% nitrogen, 21% oxygen, 1% other gases.

Circle Graph To display data as parts of a whole, you might use a circle graph. A circle graph is a circle divided into sections that represent the relative size of each piece of data. The entire circle represents 100%, half represents 50%, and so on.

Example Air is made up of 78% nitrogen, 21% oxygen, and 1% other gases. Display the composition of air in a circle graph.

Step 1 Multiply each percent by 360° and divide by 100 to find the angle of each section in the circle.

$$78\% \times \frac{360°}{100} = 280.8°$$

$$21\% \times \frac{360°}{100} = 75.6°$$

$$1\% \times \frac{360°}{100} = 3.6°$$

Step 2 Use a compass to draw a circle and to mark the center of the circle. Draw a straight line from the center to the edge of the circle.

Step 3 Use a protractor and the angles you calculated to divide the circle into parts. Place the center of the protractor over the center of the circle and line the base of the protractor over the straight line.

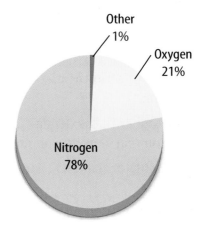

Practice Problem Draw a circle graph to represent the amount of aluminum collected during the week shown in the bar graph to the left.

Using a Calculator

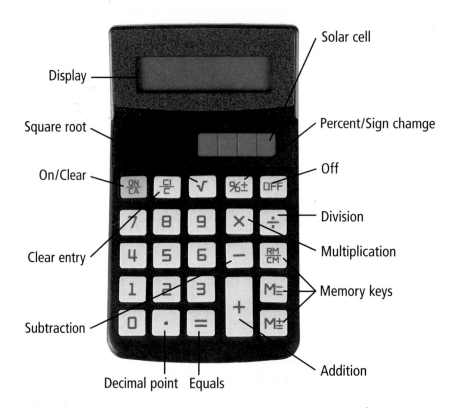

Display — Solar cell

Square root — Percent/Sign chamge

On/Clear — Off

Clear entry — Division

Multiplication

Subtraction — Memory keys

Decimal point   Equals — Addition

- Read the problem very carefully. Decide if you need the calculator to help you solve the problem.
- Clear the calculator by pressing the clear key when starting a new problem.
- If you see an E in the display, clear the error before you begin.
- If you see an M in the display, clear the memory and the calculator before you begin.
- If the number in the display is not one of the answer choices, check your work. You may have to round the number in the display.
- Your calculator will NOT automatically perform the correct order of operations.
- When working with calculators, use careful and deliberate keystrokes, and always remember to check your answer to make sure that it is reasonable. Calculators might display an incorrect answer if you press the keys too quickly.
- Check your answer to make sure that you have completed all of the necessary steps.

Understanding Scientific Terms

This list of prefixes, suffixes, and roots is provided to help you understand science terms used throughout this textbook. The list identifies whether the prefix, suffix, or root is of Greek *(G)* or Latin *(L)* origin. Also listed is the meaning of the prefix, suffix, or root and a science word in which it is used.

ORIGIN	MEANING	EXAMPLE
A		
ad (L)	to, toward	adaxial
aero (G)	air	aerobic
an (G)	without	anaerobic
ana (G)	up	anaphase
andro (G)	male	androecium
angio (G)	vessel	angiosperm
anth/o (G)	flower	anthophyte
anti (G)	against	antibody
aqu/a (L)	of water	aquatic
archae (G)	ancient	archaebacteria
arthro, artio (G)	jointed	arthropod
askos (G)	bag	ascospore
aster (G)	star	Asteroidea
autos (G)	self	autoimmune
B		
bi (L)	two	bipedal
bio (G)	life	biosphere
C		
carn (L)	flesh	carnivore
cephalo (G)	head	cephalopod
chlor (G)	light green	chlorophyll
chroma (G)	pigmented	chromosome
cide (L)	to kill	insecticide
circ (L)	circular	circadian
cocc/coccus (G)	small and round	streptococcus
con (L)	together	convergent
cyte (G)	cell	cytoplasm
D		
de (L)	remove	decompose
dendron (G)	tree	dendrite
dent (L)	tooth	edentate
derm (G)	skin	epidermis
di (G)	two	disaccharide

ORIGIN	MEANING	EXAMPLE
dia (G)	apart	diaphragm
dorm (L)	sleep	dormancy
E		
echino (G)	spiny	echinoderm
ec (G)	outer	ecosystem
endo (G)	within	endosperm
epi (G)	upon	epidermis
eu (G)	true	eukaryote
exo (G)	outside	exoskeleton
F		
fer (L)	to carry	conifer
G		
gastro (G)	stomach	gastropod
gen/(e)(o) (G)	kind	genotype
genesis (G)	to originate	oogenesis
gon (G)	reproductive	archegonium
gravi (L)	heavy	gravitropism
gymn/o (G)	naked	gymnosperm
gyn/e (G)	female	gynoecium
H		
hal(o) (G)	salt	halophyte
hapl(o) (G)	single	haploid
hemi (G)	half	hemisphere
hem(o) (G)	blood	hemoglobin
herb/a(i) (L)	vegetation	herbivore
heter/o (G)	different	heterotrophic
hom(e)/o (G)	same	homeostasis
hom (L)	human	hominid
hydr/o (G)	water	hydrolysis
I		
inter (L)	between	internode
intra (L)	within	intracellular
is/o (G)	equal	isotonic

ORIGIN	MEANING	EXAMPLE
K		
kary (G)	nucleus	eukaryote
kera (G)	hornlike	keratin
L		
leuc/o (G)	white	leukocyte
logy (G)	study of	biology
lymph/o (L)	water	lymphocyte
lysis (G)	break up	dialysis
M		
macr/o (G)	large	macromolecule
meg/a (G)	great	megaspore
meso (L)	in the middle	mesophyll
meta (G)	after	metaphase
micr/o (G)	small	microscope
mon/o (G)	only one	monocotyledon
morph/o (G)	form	morphology
N		
nema (G)	a thread	nematode
neuro (G)	nerve	neuron
nod (L)	knot	nodule
nomy(e) (G)	system of laws	taxonomy
O		
olig/o (G)	small, few	oligochaete
omni (L)	all	omnivore
orni(s) (G)	bird	ornithology
oste/o (G)	bone formation	osteocyte
ov (L)	an egg	oviduct
P		
pal(a)e/o (G)	ancient	paleontology
para (G)	beside	parathyroid
path/o (G)	suffering	pathogen
ped (L)	foot	centipede
per (L)	through	permeable
peri (G)	around, about	peristalsis
phag/o (G)	eating	phagocyte
phot/o (G)	light	photosynthesis
phyl (G)	race, class	phylogeny
phyll (G)	leaf	chlorophyll
phyte (G)	plant	epiphyte
pinna (L)	feather	pinnate

ORIGIN	MEANING	EXAMPLE
plasm/o (G)	to form	plasmodium
pod (G)	foot	gastropod
poly (G)	many	polymer
post (L)	after	posterior
pro (G) (L)	before	prokaryote
prot/o (G)	first	protocells
pseud/o (G)	false	pseudopodium
R		
re (L)	back to original	reproduce
rhiz/o (G)	root	rhizoid
S		
scope (G)	to look	microscope
some (G)	body	lysosome
sperm (G)	seed	gymnosperm
stasis (G)	remain constant	homeostasis
stom (G)	mouthlike opening	stomata
syn (G)	together	synapse
T		
tel/o (G)	end	telophase
terr (L)	of Earth	terrestrial
therm (G)	heat	endotherm
thylak (G)	sack	thylakoid
trans (L)	across	transpiration
trich (G)	hair	trichome
trop/o (G)	a change	gravitropism
trophic (G)	nourishment	heterotrophic
U		
uni (L)	one	unicellular
V		
vacc/a (L)	cow	vaccine
vore (L)	eat greedily	omnivore
X		
xer/o (G)	dry	xerophyte
Z		
zo/o (G)	living being	zoology
zygous (G)	two joined	homozygous

Topographic Map Symbols

Topographic Map Symbols

Primary highway, hard surface		Index contour
Secondary highway, hard surface		Supplementary contour
Light-duty road, hard or improved surface		Intermediate contour
Unimproved road		Depression contours
Railroad: single track		Boundaries: national
Railroad: multiple track		State
Railroads in juxtaposition		County, parish, municipal
		Civil township, precinct, town, barrio
Buildings		Incorporated city, village, town, hamlet
Schools, church, and cemetery		Reservation, national or state
Buildings (barn, warehouse, etc.)		Small park, cemetery, airport, etc.
Wells other than water (labeled as to type)		Land grant
Tanks: oil, water, etc. (labeled only if water)		Township or range line, U.S. land survey
Located or landmark object; windmill		Township or range line, approximate location
Open pit, mine, or quarry; prospect		
Marsh (swamp)		
Wooded marsh		Perennial streams
Woods or brushwood		Elevated aqueduct
Vineyard		Water well and spring
Land subject to controlled inundation		Small rapids
Submerged marsh		Large rapids
Mangrove		Intermittent lake
Orchard		Intermittent stream
Scrub		Aqueduct tunnel
Urban area		Glacier
		Small falls
x7369 Spot elevation		Large falls
670 Water elevation		Dry lake bed

Rocks

Rocks		
Rock Type	**Rock Name**	**Characteristics**
Igneous (intrusive)	Granite	Large mineral grains of quartz, feldspar, hornblende, and mica. Usually light in color.
	Diorite	Large mineral grains of feldspar, hornblende, and mica. Less quartz than granite. Intermediate in color.
	Gabbro	Large mineral grains of feldspar, augite, and olivine. No quartz. Dark in color.
Igneous (extrusive)	Rhyolite	Small mineral grains of quartz, feldspar, hornblende, and mica, or no visible grains. Light in color.
	Andesite	Small mineral grains of feldspar, hornblende, and mica or no visible grains. Intermediate in color.
	Basalt	Small mineral grains of feldspar, augite, and possibly olivine or no visible grains. No quartz. Dark in color.
	Obsidian	Glassy texture. No visible grains. Volcanic glass. Fracture looks like broken glass.
	Pumice	Frothy texture. Floats in water. Usually light in color.
Sedimentary (detrital)	Conglomerate	Coarse grained. Gravel or pebble-size grains.
	Sandstone	Sand-sized grains 1/16 to 2 mm.
	Siltstone	Grains are smaller than sand but larger than clay.
	Shale	Smallest grains. Often dark in color. Usually platy.
Sedimentary (chemical or organic)	Limestone	Major mineral is calcite. Usually forms in oceans and lakes. Often contains fossils.
	Coal	Forms in swampy areas. Compacted layers of organic material, mainly plant remains.
Sedimentary (chemical)	Rock Salt	Commonly forms by the evaporation of seawater.
Metamorphic (foliated)	Gneiss	Banding due to alternate layers of different minerals, of different colors. Parent rock often is granite.
	Schist	Parallel arrangement of sheetlike minerals, mainly micas. Forms from different parent rocks.
	Phyllite	Shiny or silky appearance. May look wrinkled. Common parent rocks are shale and slate.
	Slate	Harder, denser, and shinier than shale. Common parent rock is shale.
Metamorphic (nonfoliated)	Marble	Calcite or dolomite. Common parent rock is limestone.
	Soapstone	Mainly of talc. Soft with greasy feel.
	Quartzite	Hard with interlocking quartz crystals. Common parent rock is sandstone.

Minerals

Minerals					
Mineral (formula)	**Color**	**Streak**	**Hardness**	**Breakage Pattern**	**Uses and Other Properties**
Graphite (C)	black to gray	black to gray	1–1.5	basal cleavage (scales)	pencil lead, lubricants for locks, rods to control some small nuclear reactions, battery poles
Galena (PbS)	gray	gray to black	2.5	cubic cleavage perfect	source of lead, used for pipes, shields for X rays, fishing equipment sinkers
Hematite (Fe_2O_3)	black or reddish-brown	reddish-brown	5.5–6.5	irregular fracture	source of iron; converted to pig iron, made into steel
Magnetite (Fe_3O_4)	black	black	6	conchoidal fracture	source of iron, attracts a magnet
Pyrite (FeS_2)	light, brassy, yellow	greenish-black	6–6.5	uneven fracture	fool's gold
Talc ($Mg_3 Si_4O_{10} (OH)_2$)	white, greenish	white	1	cleavage in one direction	used for talcum powder, sculptures, paper, and tabletops
Gypsum ($CaSO_4 \cdot 2H_2O$)	colorless, gray, white, brown	white	2	basal cleavage	used in plaster of paris and dry wall for building construction
Sphalerite (ZnS)	brown, reddish-brown, greenish	light to dark brown	3.5–4	cleavage in six directions	main ore of zinc; used in paints, dyes, and medicine
Muscovite ($KAl_3Si_3 O_{10}(OH)_2$)	white, light gray, yellow, rose, green	colorless	2–2.5	basal cleavage	occurs in large, flexible plates; used as an insulator in electrical equipment, lubricant
Biotite ($K(Mg,Fe)_3 (AlSi_3O_{10}) (OH)_2$)	black to dark brown	colorless	2.5–3	basal cleavage	occurs in large, flexible plates
Halite (NaCl)	colorless, red, white, blue	colorless	2.5	cubic cleavage	salt; soluble in water; a preservative

Minerals

Minerals					
Mineral (formula)	Color	Streak	Hardness	Breakage Pattern	Uses and Other Properties
Calcite $(CaCO_3)$	colorless, white, pale blue	colorless, white	3	cleavage in three directions	fizzes when HCl is added; used in cements and other building materials
Dolomite $(CaMg (CO_3)_2)$	colorless, white, pink, green, gray, black	white	3.5—4	cleavage in three directions	concrete and cement; used as an ornamental building stone
Fluorite (CaF_2)	colorless, white, blue, green, red, yellow, purple	colorless	4	cleavage in four directions	used in the manufacture of optical equipment; glows under ultraviolet light
Hornblende $(CaNa)_{2-3}$ $(Mg,Al,$ $Fe)_5-(Al,Si)_2$ Si_6O_{22} $(OH)_2)$	green to black	gray to white	5—6	cleavage in two directions	will transmit light on thin edges; 6-sided cross section
Feldspar $(KAlSi_3O_8)$ $(NaAl$ $Si_3O_8),$ $(CaAl_2Si_2$ $O_8)$	colorless, white to gray, green	colorless	6	two cleavage planes meet at 90° angle	used in the manufacture of ceramics
Augite $((Ca,Na)$ (Mg,Fe,Al) $(Al,Si)_2 O_6)$	black	colorless	6	cleavage in two directions	square or 8-sided cross section
Olivine $((Mg,Fe)_2$ $SiO_4)$	olive, green	none	6.5—7	conchoidal fracture	gemstones, refractory sand
Quartz (SiO_2)	colorless, various colors	none	7	conchoidal fracture	used in glass manufacture, electronic equipment, radios, computers, watches, gemstones

Weather Map Symbols

Sample Station Model

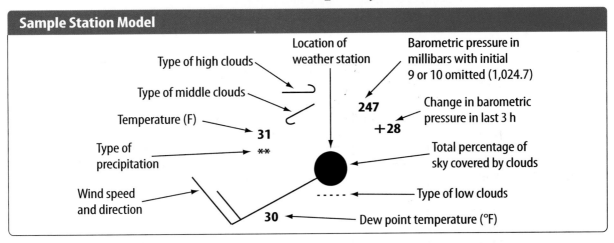

Sample Plotted Report at Each Station

Precipitation		Wind Speed and Direction		Sky Coverage		Some Types of High Clouds	
≡	Fog	○	0 calm	○	No cover	⌐	Scattered cirrus
★	Snow	╱	1–2 knots	◍	1/10 or less	⌐	Dense cirrus in patches
●	Rain	⌐	3–7 knots	◐	2/10 to 3/10	⌐	Veil of cirrus covering entire sky
⊼	Thunderstorm	⌐	8–12 knots	◑	4/10	⌐	Cirrus not covering entire sky
,	Drizzle	⌐	13–17 knots	◑	–		
▽	Showers	⌐	18–22 knots	◕	6/10		
		⌐	23–27 knots	◕	7/10		
		⌐	48–52 knots	◕	Overcast with openings		
		1 knot = 1.852 km/h		●	Completely overcast		

Some Types of Middle Clouds		Some Types of Low Clouds		Fronts and Pressure Systems	
∠	Thin altostratus layer	⌒	Cumulus of fair weather	Ⓗ or High Ⓛ or Low	Center of high- or low-pressure system
∠	Thick altostratus layer	⌣	Stratocumulus	▲▲▲	Cold front
⌐	Thin altostratus in patches	-----	Fractocumulus of bad weather	●●●	Warm front
⌐	Thin altostratus in bands	—	Stratus of fair weather	▲●▲	Occluded front
				▲▽	Stationary front

PERIODIC TABLE OF THE ELEMENTS

Columns of elements are called groups. Elements in the same group have similar chemical properties.

	Gas
	Liquid
	Solid
	Synthetic

Element — Hydrogen
Atomic number — 1
Symbol — H
Atomic mass — 1.008

State of matter

The first three symbols tell you the state of matter of the element at room temperature. The fourth symbol identifies elements that are not present in significant amounts on Earth. Useful amounts are made synthetically.

	1	2		3	4	5	6	7	8	9
1	Hydrogen 1 **H** 1.008									
2	Lithium 3 **Li** 6.941	Beryllium 4 **Be** 9.012								
3	Sodium 11 **Na** 22.990	Magnesium 12 **Mg** 24.305								
4	Potassium 19 **K** 39.098	Calcium 20 **Ca** 40.078	Scandium 21 **Sc** 44.956	Titanium 22 **Ti** 47.867	Vanadium 23 **V** 50.942	Chromium 24 **Cr** 51.996	Manganese 25 **Mn** 54.938	Iron 26 **Fe** 55.845	Cobalt 27 **Co** 58.933
5	Rubidium 37 **Rb** 85.468	Strontium 38 **Sr** 87.62	Yttrium 39 **Y** 88.906	Zirconium 40 **Zr** 91.224	Niobium 41 **Nb** 92.906	Molybdenum 42 **Mo** 95.94	Technetium 43 **Tc** (98)	Ruthenium 44 **Ru** 101.07	Rhodium 45 **Rh** 102.906
6	Cesium 55 **Cs** 132.905	Barium 56 **Ba** 137.327	Lanthanum 57 **La** 138.906	Hafnium 72 **Hf** 178.49	Tantalum 73 **Ta** 180.948	Tungsten 74 **W** 183.84	Rhenium 75 **Re** 186.207	Osmium 76 **Os** 190.23	Iridium 77 **Ir** 192.217
7	Francium 87 **Fr** (223)	Radium 88 **Ra** (226)	Actinium 89 **Ac** (227)	Rutherfordium 104 **Rf** (261)	Dubnium 105 **Db** (262)	Seaborgium 106 **Sg** (266)	Bohrium 107 **Bh** (264)	Hassium 108 **Hs** (277)	Meitnerium 109 **Mt** (268)

The number in parentheses is the mass number of the longest-lived isotope for that element.

Rows of elements are called periods. Atomic number increases across a period.

The arrow shows where these elements would fit into the periodic table. They are moved to the bottom of the table to save space.

Lanthanide series	Cerium 58 **Ce** 140.116	Praseodymium 59 **Pr** 140.908	Neodymium 60 **Nd** 144.24	Promethium 61 **Pm** (145)	Samarium 62 **Sm** 150.36
Actinide series	Thorium 90 **Th** 232.038	Protactinium 91 **Pa** 231.036	Uranium 92 **U** 238.029	Neptunium 93 **Np** (237)	Plutonium 94 **Pu** (244)

Science Online

Visit ca6.msscience.com for updates to the periodic table.

Metal

Metalloid

Nonmetal

The color of an element's block tells you if the element is a metal, nonmetal, or metalloid.

10	11	12	13	14	15	16	17	18
								Helium 2 **He** 4.003
			Boron 5 **B** 10.811	Carbon 6 **C** 12.011	Nitrogen 7 **N** 14.007	Oxygen 8 **O** 15.999	Fluorine 9 **F** 18.998	Neon 10 **Ne** 20.180
			Aluminum 13 **Al** 26.982	Silicon 14 **Si** 28.086	Phosphorus 15 **P** 30.974	Sulfur 16 **S** 32.065	Chlorine 17 **Cl** 35.453	Argon 18 **Ar** 39.948
Nickel 28 **Ni** 58.693	Copper 29 **Cu** 63.546	Zinc 30 **Zn** 65.409	Gallium 31 **Ga** 69.723	Germanium 32 **Ge** 72.64	Arsenic 33 **As** 74.922	Selenium 34 **Se** 78.96	Bromine 35 **Br** 79.904	Krypton 36 **Kr** 83.798
Palladium 46 **Pd** 106.42	Silver 47 **Ag** 107.868	Cadmium 48 **Cd** 112.411	Indium 49 **In** 114.818	Tin 50 **Sn** 118.710	Antimony 51 **Sb** 121.760	Tellurium 52 **Te** 127.60	Iodine 53 **I** 126.904	Xenon 54 **Xe** 131.293
Platinum 78 **Pt** 195.078	Gold 79 **Au** 196.967	Mercury 80 **Hg** 200.59	Thallium 81 **Tl** 204.383	Lead 82 **Pb** 207.2	Bismuth 83 **Bi** 208.980	Polonium 84 **Po** (209)	Astatine 85 **At** (210)	Radon 86 **Rn** (222)
Darmstadtium 110 **Ds** (281)	Roentgenium 111 **Rg** (272)	Ununbium * 112 **Uub** (285)		Ununquadium * 114 **Uuq** (289)				

* The names and symbols for elements 112–114 are temporary. Final names will be selected when the elements' discoveries are verified.

Europium 63 **Eu** 151.964	Gadolinium 64 **Gd** 157.25	Terbium 65 **Tb** 158.925	Dysprosium 66 **Dy** 162.500	Holmium 67 **Ho** 164.930	Erbium 68 **Er** 167.259	Thulium 69 **Tm** 168.934	Ytterbium 70 **Yb** 173.04	Lutetium 71 **Lu** 174.967
Americium 95 **Am** (243)	Curium 96 **Cm** (247)	Berkelium 97 **Bk** (247)	Californium 98 **Cf** (251)	Einsteinium 99 **Es** (252)	Fermium 100 **Fm** (257)	Mendelevium 101 **Md** (258)	Nobelium 102 **No** (259)	Lawrencium 103 **Lr** (262)

Glossary/Glosario

Cómo usar el glosario en español:

1. Busca el término en inglés que desees encontrar.
2. El término en español, junto con la definición, se encuentran en la columna de la derecha.

Pronunciation Key

Use the following key to help you sound out words in the glossary.

a	back (BAK)	ew	food (FEWD)
ay	day (DAY)	yoo	pure (PYOOR)
ah	father (FAH thur)	yew	few (FYEW)
ow	flower (FLOW ur)	uh	comma (CAH muh)
ar	car (CAR)	u (+ con)	rub (RUB)
e	less (LES)	sh	shelf (SHELF)
ee	leaf (LEEF)	ch	nature (NAY chur)
ih	trip (TRIHP)	g	gift (GIHFT)
i (i + com + e)	idea (i DEE uh)	j	gem (JEM)
oh	go (GOH)	ing	sing (SING)
aw	soft (SAWFT)	zh	vision (VIH zhun)
or	orbit (OR buht)	k	cake (KAYK)
oy	coin (COYN)	s	seed, cent (SEED, SENT)
oo	foot (FOOT)	z	zone, raise (ZOHN, RAYZ)

✳ Academic Vocabulary

English ─ A ─ Español

abiotic factor/atmosphere

factor abiótico/atmósfera

✳ **abiotic factor:** a nonliving part of an ecosystem (p. 517)

✳ **accumulate:** to increase slowly in quantity or number (p. 492)

✳ **adapt:** to make fit, often by modification (p. 527)

✳ **adjacent:** not distant, nearby (p. 224)

✳ **affect:** to have influence on or cause a change in (p. 486)

air mass: body of air that has consistent weather features (p. 475)

✳ **appreciate:** to grasp the nature, quality, worth, or significance of (p. 94)

arroyo: streambeds that are empty except during times of heavy rains or floods (p. 358)

asthenosphere: the solid, but plastic layer found in the upper mantle on which the crustal plates sit (p. 104)

atmosphere: mixture of gases that surrounds Earth (p. 382)

factor abiótico: pfactor abiótico: parte inerte de un ecosistema (p. 517)

acumular: aumentar lentamente la cantidad o número (p. 492)

adaptarse: volverse apto, a menudo mediante modificación (p. 527)

adyacente: no distante, cercano (p. 224)

afectar: influir en algo u ocasionar que algo cambie (p. 486)

masa de aire: cuerpo de aire que posee características meteorológicas uniformes (p. 475)

apreciar: comprender la naturaleza, calidad, valor o significado de algo (p. 94)

arroyo: lecho vacío de una corriente, excepto durante épocas de lluvias fuertes o inundaciones (p. 358)

astenosfera: capa sólida y plástica que se halla en el manto superior, donde se sientan las placas de la corteza (p. 104)

atmósfera: mezcla de gases que rodea a La Tierra (p. 382)

B

Basin and Range: a large area of the United States that consists of north-south trending mountain ranges and valleys (p. 358)

bathymetric (ba thuh MEH trihk) map: a map of the bottom of the ocean showing the contours of the ocean floor and its geologic features (p. 425)

beach: a landform consisting of loose sand and gravel found along a river, lake, sea, or ocean (p. 349)

biome: a zone of major ecological communities on Earth (p. 532)

biotic factor: a living part of an ecosystem (p. 523)

Cuenca y Cordillera: un área grande de los Estados Unidos formada por valles y cordilleras de montañas que se extienden de norte a sur (p. 358)

mapa batimétrico: mapa del fondo del océano que muestra su contorno y sus características geológicas (p. 425)

playa: accidente geográfico formado por arena y grava sueltas que se hallan a lo largo de los ríos, lagos, mares u océanos (p. 349)

bioma: área que alberga a las principales comunidades ecológicas de La Tierra (p. 532)

factor biótico: parte viviente de un ecosistema (p. 523)

C

California Current: a wide, slow-moving, cold-water current that flows southward along the California coast (p. 448)

carbon cycle: the process through which carbon moves between the living and nonliving environment (p. 570)

carnivore: animal that eats only other animals (p. 557)

chemical weathering: weathering that weakens and breaks down rocks using water and gases in the atmosphere (p. 335)

cinder cone volcano: small, loosely packed volcano formed when tephra falls to the ground (p. 304)

climate: long-term, average weather patterns of an area (p. 483)

cold front: the movement of a colder air mass toward a warmer air mass (p. 476)

community: all the species that occupy an area (p. 524)

composite volcano: volcano composed of alternating layers of tephra and lava that accumulate from explosive and quiet eruptions (p. 304)

conduction: the transfer of thermal energy by collisions between particles in matter (p. 145)

conductor: a material in which thermal energy moves quickly (p. 146)

consequence: something produced by a cause (p. 480)

conservation: the preservation and careful management of the environment, including natural resources (p. 612)

consumer: organism that cannot create its own food (p. 556)

constant: a factor in an experiment that remains the same (p. 29)

Corriente de California: Corriente de agua amplia, de movimiento lento, que fluye hacia el sur a lo largo de la costa californiana (p. 448)

ciclo del carbono: proceso mediante el cual el carbono se recicla entre los seres vivos y la parte inerte del medio ambiente (p. 570)

carnívoro: animal que se alimenta solamente de otros animales (p. 557)

desgaste químico: desgaste que debilita y fisiona las rocas con el agua y con los gases de la atmósfera (p. 335)

volcán de cono de cenizas: volcán pequeño, no muy sólido, que se forma al caer materia piroclástica al suelo (p. 304)

clima: patrón meteorológico prolongado promedio de una región (p. 483)

frente frío: movimiento de una masa de aire fría hacia una más cálida (p. 476)

comunidad: todas las especies que ocupan un área (p. 524)

volcán compuesto: volcán compuesto por capas alternantes de materia piroclástica y lava que se acumulan a partir de erupciones explosivas y no explosivas (p. 304)

conducción: transferencia de energía térmica producida por choques entre partículas de materia (p. 145)

conductor: material a través del cual la energía térmica se mueve rápidamente (p. 146)

consecuencia: algo producido por una causa (p. 480)

conservación: preservación y manejo cuidadoso del medio ambiente, incluyendo los recursos naturales (p. 612)

consumidor: organismo que no puede elaborar sus propios alimentos (p. 556)

constante: factor que permanece invariable en un experimento (p. 29)

Glossary/Glosario

contact: (1) the place where two rock formations occur next to each other (p. 59); (2) a union or junction of surfaces (p. 337)

continental drift: hypothesis stating that continents slowly drift parallel to Earth's surface (p. 167)

continental rifting: process that pulls a continent apart, usually in response to tension stress (p. 216)

continental shelf: an underwater portion of continental crust that extends from the continental shoreline and gently slopes toward the deeper parts of the ocean (p. 427)

control group: the part of a controlled experiment that contains all the same variables and constants as the experimental group but the independent variable is not changed (p. 29)

contour line: a line drawn on a map to join points of equal elevation (p. 55)

convection: transfer of thermal energy by the movement of matter from one place to another (p. 147)

convection current: the circular motion in a fluid that results when the fluid is heated from below (p. 149)

convergent plate boundary: boundary formed where two lithospheric plates are moving toward each other (p. 218)

convert: to change something into another form, substance, state, or product (p. 562)

core: the dense, metallic, central layer of Earth (p. 105)

Coriolis (kor ee OH lus) effect: a deflection in the movement of air and water caused by Earth's rotation (p. 405)

critical thinking: comparing what you already know with the explanation you are given in order to decide if you agree with it or not (p. 21)

crust: the thin, rocky, outer layer of Earth (p. 103)

cycle: a series of events that occur regularly and usually lead back to the starting point; a circular or spiral arrangement (p. 434)

contacto: (1) lugar donde dos formaciones rocosas ocurren una junto a la otra (p. 59); (2) unión o cruce de superficies (p. 337)

movimiento continental: hipótesis que afirma que los continentes se mueven lentamente en forma paralela a la superficie de La Tierra (p. 167)

fisura continental: proceso que separa a una masa continental, normalmente como respuesta al estrés producido por la tensión (p. 216)

plataforma continental: parte subterránea de la corteza continental que se extiende desde el litoral continental y se inclina ligeramente hacia las partes más profundas del océano (p. 427)

grupo de control: parte de un experimento controlado que contiene las mismas variables y constantes que el grupo experimental, pero en el que la variable independiente no se cambia (p. 29)

cota: línea que se traza en un mapa para unir los puntos que poseen la misma elevación (p. 55)

convección: transferencia de energía térmica mediante movimiento de material de un lugar a otro (p. 147)

corriente de convección: movimiento circular de un líquido que ocurre cuando éste se calienta por la parte inferior (p. 149)

límite de placa convergente: límite que se forma cuando dos placas litosféricas se mueven una hacia la otra (p. 218)

convertir: transformar algo cambiando una forma, sustancia, estado o producto (p. 562)

núcleo: capa densa, metálica y central de La Tierra (p. 105)

efecto Coriolis: desviación del movimiento del aire y del agua ocasionado por la rotación de La Tierra (p. 405)

pensamiento crítico: comparación de lo que sabemos con la explicación que recibimos para decidir si estamos de acuerdo o no con ella (p. 21)

corteza: capa delgada, rocosa y exterior de La Tierra (p. 103)

ciclo: serie de eventos que ocurren regularmente y que normalmente regresan al punto inicial; disposición circular o en espiral (p. 434)

data: factual information used as a basis for reading, discussion, or calculation (p. 168)

Davidson Current: a narrow, warm water, seasonal current that flows northward, close to the California coast, from lower latitudes to higher latitudes (p. 448)

datos: información factual que se usa como base para la lectura, discusión o cálculo (p. 168)

Corriente de Davidson: corriente de agua estrecha y estacional que fluye hacia el norte, cercana a la costa californiana, desde latitudes bajas hacia latitudes altas (p. 448)

decomposer: an organism that breaks down dead organisms, animal droppings, leaves, and other wastes produced by living things (p. 558)

✳ **define:** to fix or mark the limits of (p. 185)

density: the amount of matter an object has per unit volume (p. 91)

dependent variable: the factor you measure or observe during an experiment (p. 29)

deposition: the laying down of sediments in a new location by running water, wind, glaciers, or gravity (p. 342)

dew point: temperature at which air becomes fully saturated with water vapor and condensation forms (p. 470)

divergent plate boundary: boundary formed where two lithospheric plates are moving apart (p. 215)

downdraft: a sinking column of air (p. 404)

drought: period of time when precipitation is much lower than normal or absent (p. 480)

descomponedor: organismo que descompone organismos muertos, excremento de animales, hojas y otros desechos producidos por los seres vivos (p. 558)

definir: fijar o marcar los límites de algo (p. 185)

densidad: cantidad de material que posee un objeto por unidad de volumen (p. 91)

variable dependiente: factor que se mide u observa durante un experimento (p. 29)

deposición: depósito de sedimentos en un nuevo lugar a causa del agua corriente, el viento, los glaciares o la gravedad (p. 342)

punto de rocío: temperatura a la cual el aire se satura totalmente con vapor de agua y se produce condensación (p. 470)

límite de placa divergente: límite que se forma cuando dos placas litosféricas se distancian una de otra (p. 215)

corriente descendente: columna de aire que se mueve hacia abajo (p. 404)

sequía: período de tiempo durante el cual no hay precipitación, o ésta es mucho menor que lo normal (p. 480)

E

earthquake: rupture and sudden movement of rock along a fault (p. 246)

Earth science: the study of the changes that take place on and in Earth and the reasons for those changes (p. 2)

echo sounding: determination of the depth of water using sound waves (p. 426)

ecology: the study of the interactions between living things and their environment (p. 553)

ecosystem: all the plants and animals of an area and the physical place in which they live (p. 516)

elastic potential energy: energy stored when an object is stretched or compressed (p. 128)

elastic strain: energy stored as a change in shape (p. 247)

electromagnetic spectrum: the entire range of wavelengths or frequencies of electromagnetic radiation (p. 384)

✳ **emerge:** to rise from, to come out into view (p. 303)

energy: the ability to cause change (p. 124)

epicenter: point on Earth directly above the earthquake focus (p. 253)

erosion: the process of wearing away rock or soil (pp. 80, 342)

estuary: an extremely fertile area where a river meets ocean (p. 591)

experimental group: the part of a controlled experiment used to study relationships between variables you are interested in knowing more about (p. 29)

sismo: ruptura y movimiento repentino de las rocas a lo largo de una falla (p. 246)

ciencias de la Tierra: estudio de los cambios que tienen lugar en la Tierra y el origen de esos cambios (p. 2)

sonido de eco: determinación de la profundidad del agua mediante ondas de sonido (p. 426)

ecología: estudio de las interacciones entre los seres vivos y su medio ambiente (p. 553)

ecosistema: plantas y animales de una región y el lugar físico donde viven (p. 516)

energía potencial elástica: energía almacenada cuando un objeto se estira o se comprime (p. 128)

tensión elástica: energía almacenada como cambio de forma (p. 247)

espectro electromagnético: rango completo de longitudes de onda o frecuencias de la radiación electromagnética (p. 384)

emerger: surgir, hacerse visible (p. 303)

energía: capacidad de producir cambio (p. 124)

epicentro: punto de La Tierra que está directamente sobre el foco de un sismo (p. 253)

erosión: proceso de desgaste de las rocas o del suelo (pp. 80, 342)

estuario: área muy fértil donde un río se encuentra con el océano (p. 591)

grupo experimental: parte de un experimento controlado que se usa para estudiar las relaciones entre las variables que se desea investigar (p. 29)

Glossary/Glosario

F

fault: a fracture along which rocks on one side have moved relative to rocks on the other side (p. 211)

fissure eruption: eruption that occurs when magma escapes from narrow and elongated cracks in Earth's crust (p. 295)

flash flood: flood that takes place suddenly (p. 481)

flood: rising and overflowing the banks of a body of water (p. 348)

flood plain: the wide, flat valley that is located alongside some rivers and streams (p. 348)

fluid: a material made of particles that can easily change their locations (p. 147)

focus: location where energy is first released in an earthquake (p. 248)

food chain: an illustration of how food moves through an ecosystem (p. 562)

food web: a model more complex than a food chain that describes how food moves through an ecosystem (p. 563)

fossil fuel: fuel formed from organisms in Earth's crust over hundreds of millions of years (p. 595)

fracture: a break, or crack in rock (p. 211)

friction: force that opposes the sliding of two surfaces in contact (p. 136)

frost wedging: a process that uses expansion and contraction by freezing and thawing of water to break rocks (p. 337)

fuel: a material that can be burned to release energy (p. 135)

falla: fractura a lo largo de la cual las rocas de un lado se han movido con respecto a las del otro lado (p. 211)

erupción de fisura: erupción que ocurre cuando el magma escapa por las grietas estrechas y alargadas de la corteza terrestre (p. 295)

riada: inundación que ocurre repentinamente (p. 481)

inundación: crecida y desbordamiento de las riberas de una masa de agua (p. 348)

llanura de inundación: valle amplio y plano ubicado a lo largo de algunos ríos y corrientes (p. 348)

líquido: material formado por partículas que pueden cambiar fácilmente su ubicación (p. 147)

foco: lugar hacia donde se libera primero la energía de un sismo (p. 248)

cadena alimenticia: ilustración de la forma en que se mueven los alimentos a través de un ecosistema (p. 562)

red alimenticia: modelo más complejo que la cadena alimenticia y que describe la forma en que se mueven los alimentos a través de un ecosistema (p. 563)

combustible fósil: combustible formado en la corteza terrestre a partir de organismos durante cientos de millones de años (p. 595)

fractura: ruptura o grieta de una roca (p. 211)

fricción: fuerza que resiste el deslizamiento de dos superficies en contacto (p. 136)

presión de congelación: proceso que usa la expansión y la contracción producidas al congelarse y derretirse el agua para fisionar las rocas (p. 337)

combustible: material que puede quemarse para liberar energía (p. 135)

G

geologic formation: a three-dimensional body, or volume, of a certain kind of rock of a given age range (p. 58)

geologic map: a map that represents the geology of an area (p. 58)

geothermal energy: heat energy in Earth's crust (p. 600)

glacier: large masses of ice and snow that accumulate when the annual snowfall is greater than the melt off (p. 350)

Global Positioning System (GPS): a network of satellites used to determine locations on Earth (p. 191)

global warming: an increase in Earth's average surface temperature (p. 399)

formación geológica: cuerpo o volumen tridimensional de cierto tipo de rocas cuya edad está en un rango determinado (p. 58)

mapa geológico: mapa que representa la geología de una región (p. 58)

energía geotérmica: energía del calor de la corteza terrestre (p. 600)

glaciar: grandes masas de hielo y nieve que se acumulan cuando la tasa anual de nevadas supera a la tasa de derretimiento (p. 350)

Sistema de posición global (GPS, por sus siglas en inglés): red de satélites que se usa para determinar ubicaciones en La Tierra (p. 191)

calentamiento global: aumento de la temperatura promedio de la superficie terrestre (p. 399)

greenhouse gas: gas that strongly absorbs a portion of Earth's outgoing radiation (p. 399)

gyre (JI uhr): a cycle of ocean currents (p. 434)

gases del efecto invernadero: gases que absorben fuertemente una parte de la radiación que emite La Tierra (p. 399)

remolino: ciclo de corrientes oceánicas (p. 434)

habitat: the place an organism lives (pp. 450, 536)

heat: the movement of thermal energy from an object at a higher temperature to an object at a lower temperature (p. 142)

herbivore: animal that eats only plants (p. 557)

highland climate: a climate that is characterized by cool-to-cold temperatures and occurs in mountains and on high plateaus (p. 484)

hot spot: the result of an unusually hot area in Earth's interior that forms volcanoes when melted rock is forced upward and breaks through the crust (p. 297)

humidity: amount of water vapor per volume of air (p. 469)

humus (HYEW mus): a dark-colored layer of soil formed by decaying plants and animals (p. 521)

✳ **hypothesis:** a tentative explanation that can be tested with a scientific investigation (p. 176)

hábitat: lugar donde vive un organismo (pp. 450, 536)

calor: movimiento de la energía térmica de un objeto con una temperatura alta hacia un objeto con una temperatura más baja (p. 142)

herbívoro: animal que se alimenta solamente de plantas (p. 557)

clima de montaña: clima que se caracteriza por temperaturas que varían de frescas a frías y que ocurre en las montañas y en las mesetas altas (p. 484)

punto caliente: resultado de una región inusualmente cálida del interior de La Tierra que forma volcanes cuando la roca derretida es forzada a ascender e irrumpe a través de la corteza (p. 297)

humedad: cantidad de vapor de agua por volumen de aire (p. 469)

humus: capa de suelo oscura formada por plantas y animales en descomposición (p. 521)

hipótesis: explicación tentativa que puede ponerse a prueba mediante una investigación científica (p. 176)

✳ **inclined:** sloping, slanting, or leaning relative to the horizontal or vertical (p. 212)

independent variable: a factor in an experiment that is manipulated or changed by the investigator to observe how it affects a dependent variable (p. 29)

✳ **indicate:** to demonstrate or point out with precision (p. 262)

inference: a logical conclusion based on the information that is available to you (p. 20)

infrared wave: a portion of the electromagnetic spectrum with long wavelengths just beyond the range of visible light and sometimes felt as heat (p. 385)

✳ **interact:** to act on each other (p. 249)

✳ **internal:** existing or situated within the limits or surface of something (p. 257)

✳ **interval:** a space between objects, units, points, or states (p. 55)

inversion: an increase in air temperature as altitude increases (p. 396)

inclinado: pendiente o declive con respecto al plano horizontal o vertical (p. 212)

variable independiente: factor que el investigador manipula o cambia en un experimento para observar cómo afecta a una variable dependiente (p. 29)

indicar: demostrar o señalar con precisión (p. 262)

inferencia: conclusión lógica basada en la información que está disponible (p. 20)

onda infrarroja: parte del espectro electromagnético con longitudes de onda apenas más allá del rango de la luz visible que algunas veces se perciben como calor (p. 385)

interactuar: influir uno sobre el otro (p. 249)

interno: que existe o está situado dentro de los límites o de la superficie de algo (p. 257)

intervalo: espacio que hay entre objetos, unidades, puntos o estados (p. 55)

inversión: aumento de la temperatura del aire a medida que aumenta la altitud (p. 396)

Glossary/Glosario

J

jet stream: strong, continuous band of wind that moves at speeds of 200–250 km/h between the troposphere and the stratosphere (p. 408)

corriente de chorro: banda de viento fuerte y continua que se mueve a una velocidad de 250 a 250 kilómetros por hora entre la troposfera y la estratosfera (p. 408)

K

kinetic energy: the energy an object has because it is moving (p. 124)

energía cinética: energía que posee un objeto debido a su movimiento (p. 124)

L

lahar (LAH har): a rapidly flowing mixture of volcanic debris and water (p. 314)

land breeze: wind that blows from the land to the sea (p. 493)

landform: feature sculpted by processes on Earth's surface or resulting from forces within Earth (p. 79)

landslide: rapid, gravity-caused events that move soil, loose rock, and boulders (p. 343)

latitude: the distance in degrees north or south of the equator (p. 49)

lava: molten, liquid rock material found on Earth's surface (p. 96)

✳ **layer:** one thickness, course, or fold laid or lying over or under another (p. 105)

limiting factor: an environmental factor that limits the growth of a population of organisms in an ecosystem (p. 524)

liquefaction: process by which shaking during an earthquake causes loose sediment to act more like a liquid (p. 271)

lithosphere: the crust, and the solid uppermost layer of the mantle (p. 105)

lithospheric plate: large, brittle pieces of Earth's outer shell composed of crust and uppermost mantle (p. 183)

longitude: the distance in degrees east or west of the prime meridian (p. 49)

longshore current: a narrow current parallel to the shoreline (p. 440)

longshore drift: the combination of the movement of sand on the beach by breaking waves and the movement of sand in the longshore current (p. 440)

lahar: mezcla de desechos volcánicos y agua que fluye rápidamente (p. 314)

brisa terrestre: viento que sopla de la tierra hacia el mar (p. 493)

accidente geográfico: rasgo esculpido por los procesos que tienen lugar en la superficie terrestre o que resultan de las fuerzas en el interior de la Tierra (p. 79)

alud: evento rápido ocasionado por la gravedad que mueve al suelo, a las rocas sueltas y a los peñascos (p. 343)

latitud: distancia en grados al norte o al sur del ecuador (p. 49)

lava: material de roca derretida y líquida que se halla en la superficie terrestre (p. 96)

capa: espesor o estrato que yace sobre o debajo de otro (p. 105)

factor limitante: factor ambiental que limita el crecimiento de una población de organismos en un ecosistema (p. 524)

licuefacción: proceso mediante el cual el temblor que produce un sismo hace que los sedimentos sueltos actúen más como líquidos (p. 271)

litosfera: corteza, capa sólida exterior del manto (p. 105)

placa litosférica: pedazos grandes y quebradizos de la cubierta exterior de La Tierra formados por partes de la corteza y el manto exterior (p. 183)

longitud: distancia en grados al este o al oeste del meridiano principal (p. 49)

corriente a lo largo de la costa: corriente estrecha y paralela al litoral (p. 440)

desplazamiento a lo largo de la costa: combinación del movimiento de arena en la playa a causa de las olas que rompen y del movimiento de arena en la corriente a lo largo de la costa (p. 440)

magma: molten, liquid rock material found underground (p. 96)

mantle: the thick, middle layer of Earth (p. 104)

map legend: a list of all the symbols used on a map with an explanation of what each symbol means (p. 51)

map view: a view that is seen as if you were looking down on an area from above Earth's surface (p. 50)

marine: a reference to anything related to the ocean (p. 450)

mass wasting: the downhill movement of rocks or soil (p. 343)

meanders: the side to side curves in a river that occur when the land is relatively flat (p. 345)

Mediterranean climate: a climate that is characterized by mild, wet winters and hot, dry summers (p. 484)

✳ method: a way or process for doing something (p. 425)

mid-ocean ridge: long mountain range in the middle of the seafloor where new oceanic crust forms (p. 175)

✳ migrate: to move from one place to another (p. 533)

mineral: a substance that makes up rocks, is naturally occurring, is an inorganic solid, and has a definite chemical composition (p. 87)

mountain breeze: breeze that flows downward from mountains (p. 493)

magma: material de roca derretida líquida que se halla debajo del suelo (p. 96)

manto: capa gruesa central de La Tierra (p. 104)

leyenda de un mapa: lista de los símbolos usados en un mapa con una explicación de lo que significa cada uno de ellos (p. 51)

perspectiva de mapa: perspectiva que se ve como si estuviésemos mirando hacia abajo una región desde un área sobre la superficie terrestre (p. 50)

marino: referencia a cualquier aspecto que se relacione con el océano (p. 450)

derroche de masa: movimiento descendente de las rocas o del suelo (p. 343)

meandros: curvas de un río que van de lado a lado y que ocurren cuando la tierra es relativamente plana (p. 345)

clima mediterráneo: clima que se caracteriza por inviernos moderados y húmedos, y veranos cálidos y secos (p. 484)

método: manera o proceso para hacer algo (p. 425)

cordillera mesoceánica: larga cadena montañosa en medio del fondo del mar donde se forma una nueva corteza oceánica (p. 175)

migrar: trasladarse de un lugar a otro (p. 533)

mineral: sustancia que compone las rocas; ocurre en forma natural, es un sólido inorgánico y tiene una composición química definida (p. 87)

brisa de montaña: brisa que sopla hacia abajo desde las montañas (p. 493)

natural resource: a material or energy source that is useful or necessary to meet the needs of Earth's organisms, including people (p. 588)

niche: a unique ecological role of a species (p. 536)

nitrifying bacteria: a type of soil bacteria that can change nitrogen in the atmosphere to a form that plants can take up through their roots (p. 569)

nitrogen cycle: the process through which nitrogen moves from the atmosphere to the soil, to living organisms, and back to the atmosphere (p. 569)

nonrenewable natural resources: resources that are used more quickly than they can be replaced by natural processes (p. 592)

nuclear fission: energy given off when an atom splits and releases a neutron (p. 601)

nuclear fusion: combining two atoms to make one atom (p. 601)

recurso natural: material o fuente de energía útil o necesaria para satisfacer las necesidades de los organismos de La Tierra, incluyendo a las personas (p. 588)

nicho: función ecológica única de una especie (p. 536)

bacterias nitrificantes: tipo de bacterias del suelo que pueden transformar el nitrógeno de la atmósfera en una forma que las plantas pueden aprovechar a través de sus raíces (p. 569)

ciclo del nitrógeno: proceso mediante el cual el nitrógeno se recicla desde la atmósfera hacia el suelo y los seres vivos, y nuevamente hacia la atmósfera (p. 569)

recursos naturales no renovables: recursos que se usan con una rapidez mayor que aquella con la cual los procesos naturales pueden renovarlos (p. 592)

fisión nuclear: energía que se emite cuando un átomo se divide y libera un neutrón (p. 601)

fusión nuclear: combinación de dos átomos en uno solo (p. 601)

Glossary/Glosario

Glossary/Glosario

O

observation: act of watching something and taking note of what occurs (p. 18)

* **occur:** to happen (p. 126)

ocean current: ocean water that moves from place to place flowing like a river in the ocean (p. 430)

ocean floor: Earth's surface underneath the ocean water (p. 425)

ocean trench: long, narrow, deep parts of the seafloor formed where a subducting slab bends down into Earth (p. 185)

omnivore: animal that feeds on other animals and plants (p. 557)

observación: acto de contemplar algo atentamente y tomar nota de lo que ocurre (p. 18)

ocurrir: suceder (p. 126)

corriente oceánica: agua del océano que se mueve de un lugar a otro y que fluye como un río hacia el océano (p. 430)

fondo del mar: superficie terrestre por debajo del agua del océano (p. 425)

fosa marina: partes largas, angostas y profundas del fondo del mar que se forman cuando una plancha rocosa de subducción se flexiona hacia el interior de La Tierra (p. 185).

omnívoro: animal que se alimenta de otros animales y de plantas (p. 557)

P

Pangaea (pan JEE uh): ancient supercontinent that began breaking up about 200 million years ago (p. 167)

particulates: tiny particles released into the air (p. 609)

phosphorus cycle: the process through which phosphorus moves from the soil to producers and consumers, and back to soil (p. 570)

photosynthesis: a process that producers use to make their own food using energy from sunlight (p. 553)

physical weathering: weathering that breaks down rocks by contact without changing the mineral content of the rock (p. 337)

plate tectonics: theory that explains how lithospheric plates move and interact, causing major geologic features and events (p. 184)

population: the number of individuals of one species that occupy the same area (p. 523)

potential energy: stored energy (p. 127)

precipitation: water, in liquid or solid form, that falls from the atmosphere (p. 472)

prediction: to say in advance what will happen next in a sequence of events (pp. 19, 28)

primary consumer: an organism that eats producers (p. 564)

primary wave: compressional waves that vibrate rock particles back and forth parallel to the direction the wave travels (p. 254)

procedure: a sequence of instructions used to gather data in a scientific investigation (p. 31)

Pangaea: supercontinente de la antigüedad que comenzó a fisionarse hace alrededor de 200 millones de años (p. 167)

partículas: pequeños fragmentos que se liberan hacia el aire (p. 609)

ciclo del fósforo: proceso mediante el cual el fósforo se recicla desde el suelo hacia los productores y los consumidores, y nuevamente hacia el suelo (p. 570)

fotosíntesis: proceso que usan los productores para elaborar sus propios alimentos mediante el uso de la energía de la luz solar (p. 553)

desgaste físico: desgaste que fisiona las rocas por contacto sin cambiar su contenido mineral (p. 337)

placas tectónicas: teoría que explica cómo las placas litosféricas se mueven e interactúan para producir importantes rasgos y eventos geológicos (p. 184)

población: número de individuos de una especie que ocupan la misma área (p. 523)

energía potencial: energía almacenada (p. 127)

precipitación: agua en forma líquida o sólida que cae desde la atmósfera (p. 472)

predicción: declaración anticipada de lo que va a suceder próximamente en una secuencia de eventos (pp. 19, 28)

consumidor primario: organismo que se alimenta de los productores (p. 564)

ondas primarias: ondas compresivas que hacen vibrar a las partículas de las rocas de un lado a otro en dirección paralela a aquella en que viaja la onda (p. 254)

procedimiento: secuencia de instrucciones que se usa para recabar datos en una investigación científica (p. 31)

Glossary/Glosario

producer: organism that uses energy from the Sun or other chemical reactions to make its own food (p. 553)

profile view: a view that shows a vertical section of the ground (p. 50)

protozoan: single-celled, animal-like protist (p. 556)

pyroclastic flow: fast-moving, potentially deadly clouds of hot gases, ash, and other volcanic material produced by an explosive eruption (p. 315)

productor: organismo que usa la energía solar u otras reacciones químicas para elaborar sus propios alimentos (p. 553)

perspectiva de perfil: vista que muestra una sección vertical del suelo (p. 50)

protozoario: animal unicelular, protista animaloide (p. 556)

flujo piroclástico: nubes de gases calientes, cenizas y otro material volcánico, potencialmente mortales y de rápido movimiento, producidas por una erupción explosiva. (p. 315)

Q

qualitative data: descriptions of the natural world using words (pp. 18, 30)

quantitative data: descriptions of the natural world using numbers (pp. 18, 30)

datos cualitativos: descripción del mundo natural mediante palabras (pp. 18, 30)

datos cuantitativos: descripción del mundo natural mediante números (pp. 18, 30)

R

radiation: transfer of thermal energy by electromagnetic waves (p. 150)

rain shadow: area of low rainfall on the lee side, or downwind slope, of a mountain (p. 492)

✳ **ratio:** the relationship in quantity or size between two or more things (p. 51)

recycling: a form of reuse requires reprocessing or changing an item or natural resource (p. 609)

✳ **region:** a broad geographic area with similar characteristics (p. 449)

✳ **register:** an official written record of names, events, or transactions (p. 610)

✳ **regulate:** to control or direct according to rule, principle, or law (p. 593)

relative humidity: amount of water vapor in air relative to the maximum amount of water vapor the air can hold at that temperature before becoming saturated (p. 470)

✳ **release:** to set free from confinement (p. 315)

renewable natural resource: an Earth resource that is constantly being replaced or recycled by nature (p. 590)

✳ **resource:** a source of supply or support

rift valley: a long narrow valley that forms as blocks of rock slip down, mainly along normal faults (p. 216)

rip current: swift currents that flow away from the beach (p. 440)

rock: a natural, solid mixture of mineral crystal particles (p. 95)

radiación: transferencia de energía térmica por medio de ondas electromagnéticas (p. 150)

sombra lluviosa: área de poca precipitación en el costado de sotavento o pendiente de una montaña (p. 492)

proporción: relación de cantidad o tamaño entre dos o más cosas (p. 51)

reciclaje: forma de usar de nuevo algo que requiere el reprocesamiento o la transformación de un objeto o recurso natural (p. 609)

región: área geográfica vasta con características similares (p. 449)

registro: historial oficial escrito de nombres, eventos o transacciones (p. 610)

regular: controlar o dirigir de acuerdo con las reglas, principios o leyes (p. 593)

humedad relativa: cantidad de vapor de agua del aire en relación con la cantidad máxima de vapor de agua que el aire puede contener a la misma temperatura antes de saturarse (p. 470)

liberar: dejar libre de confinamiento (p. 315)

recurso natural renovable: recurso de La Tierra que la naturaleza repone o recicla constantemente (p. 590)

recurso: fuente de suministro o de apoyo

valle de fisura: valle largo y angosto que se forma cuando los bloques de rocas se deslizan, usualmente a lo largo de fallas normales (p. 216)

marea fuerte: corriente rápida que fluye y se aleja de la playa (p. 440)

roca: mezcla natural sólida de partículas de cristales minerales (p. 95)

Glossary/Glosario

rock cycle: a series of processes that change one rock into another (p. 100)

ciclo de la roca: serie de procesos que transforman una roca en otra (p. 100)

S

salinity: the amount of salt dissolved in a quantity of water (p. 433)

San Andreas Fault: fault zone that forms a transform plate boundary between the Pacific Plate and the North American Plate (p. 224)

sand: rocks that are between 0.0625 mm and 2.0 mm in diameter (p. 442)

Santa Ana winds: hot, dry wind that blows from the east or northeast across southern California (p. 494)

scavenger: organism that feeds on dead animals (p. 558)

science: the process of studying nature at all levels and the collection of information that is created through this process (p. 2)

scientific law: a rule that describes a pattern in nature (p. 6)

scientific theory: an explanation of things or events that is based on knowledge gained from many observations and investigations (p. 6)

sea breeze: wind that blows from the sea to the land (p. 493)

seafloor spreading: process by which new seafloor is made and then moves sideways away from mid-ocean ridges (p. 176)

sea level: the level of the ocean's surface halfway between high and low tides (p. 425)

season: regular changes in temperature and length of day that result from the tilt of the Earth's axis (p. 478)

secondary consumer: organism that eats herbivores (p. 564)

secondary wave: shearing wave that vibrate rock particles perpendicular to the direction of wave travels (p. 254)

securely: free from danger (p. 277)

sediment: rock that is broken down into smaller pieces or dissolved in water (p. 99)

seismic wave: wave of energy generated at the focus of an earthquake that travels through Earth (p. 252)

seismogram: tracing made by a seismograph, used to calculate the size of an earthquake and determine its location (p. 261)

seismograph: instrument used to measure and record ground movement caused by seismic waves (p. 261)

salinidad: cantidad de sal disuelta en un volumen de agua (p. 433)

Falla de San Andrés: zona de falla que forma un límite de placa transformante entre la Placa del Pacífico y la Placa de América del Norte (p. 224)

arena: rocas cuyo diámetro mide entre 0.0625 mm y 2.0 mm (p. 442)

vientos de Santa Ana: viento cálido y seco que sopla del este o el nordeste a través del sur de California (p. 494)

animal de carroña: organismo que se alimenta de animales muertos (p. 558)

ciencia: proceso mediante el cual se estudia la naturaleza a todos niveles, y la recopilación de información que se crea en este proceso (p. 2)

ley científica: regla que describe un patrón de la naturaleza (p. 6)

tèoria científica: explicación de circunstancias o eventos basada en el conocimiento obtenido a través de muchas observaciones e investigaciones (p. 6)

brisa marina: viento que sopla del mar hacia la tierra (p. 493)

dispersión del fondo del mar: proceso mediante el cual se forma un nuevo fondo oceánico y luego se mueve lateralmente alejándose de las cadenas montañosas mesoceánicas (p. 176)

nivel del mar: nivel de la superficie oceánica en el punto medio entre la marea alta y la marea baja (p. 425)

estación: cambios regulares de temperatura y de la duración del día que resultan de la inclinación del eje de La Tierra (p. 478)

consumidor secundario: organismo que se alimenta de herbívoros (p. 564)

onda secundaria: ondas de corte o cizalla que hacen vibrar a las partículas rocosas perpendicularmente a la dirección en que viaja la onda (p. 254)

de modo seguro: libre de peligro (p. 277)

sedimento: roca que se fisiona y forma rocas más pequeñas o que se disuelve en agua (p. 99)

onda sísmica: onda de energía que se genera en el foco de un sismo y que viaja a través de La Tierra (p. 252)

sismograma: trazado que hace el sismógrafo, el cual se usa para calcular la magnitud de un sismo y para determinar su ubicación (p. 261)

sismógrafo: instrumento que se usa para medir y registrar el movimiento del suelo ocasionado por las ondas sísmicas (p. 261)

shield volcano: broad, gently sloping volcano formed by quiet eruptions of basaltic lava (p. 304)

shore: area of land found between the lowest water level at low tide and the highest area of land that is affected by storm waves (p. 438)

shoreline: the place where the ocean meets the land (p. 438)

✹ **significant:** to have influence or effect (p. 361)

✹ **similar:** having characteristics in common (p. 399)

slab: portion of a lithospheric plate that sinks down into Earth's mantle during subduction (p. 189)

soil: a mixture of weathered rock, minerals, and organic matter on Earth's surface (p. 338)

✹ **source:** point of origin (p. 295)

species: a group of organisms that share similar characteristics and can reproduce among themselves producing fertile offspring (p. 523)

stratosphere: region of the atmosphere that extends from Earth's surface to a height of about 15–50 km (p. 383)

✹ **structure:** the arrangement or formation of the tissues, organs, or the parts of an organism (p. 556)

subduction: process by which one lithospheric plate is forced beneath another lithospheric plate, usually along a convergent plate boundary (p. 218)

✹ **summary:** a presentation of the content of a text in a condensed form or by reducing it to its main points (p. 150)

✹ **suspend:** to keep from falling or sinking by some invisible support; e.g. dust in the air (p. 440)

volcán de escudo: volcán amplio de pendiente leve que se forma por erupciones no explosivas de lava basáltica (p. 304)

costa: área de tierra que se halla entre el nivel más bajo del agua en la marea baja y el área más alta de tierra que es afectada por las olas de una tormenta (p. 438)

litoral: lugar donde se unen el mar y la tierra (p. 438)

significativo: que influye o tiene efecto en algo (p. 361)

similar: que tiene características en común (p. 399)

plancha rocosa: parte de una placa litosférica que desciende hacia el manto terrestre durante el proceso de subducción (p. 189)

suelo: mezcla de roca desgastada, minerales y materia orgánica en la superficie terrestre (p. 338)

fuente: punto de origen (p. 295)

especie: grupo de organismos que comparten características similares y pueden reproducirse entre sí para producir descendencia fértil (p. 523)

estratosfera: región de la atmósfera que se extiende desde la superficie terrestre hasta una altura aproximada de 15 a 50 km (p. 383)

estructura: disposición o formación de los tejidos u órganos o las partes de un organismo (p. 556)

subducción: proceso mediante el cual una placa litosférica es forzada a colocarse bajo otra placa litosférica, normalmente a lo largo de un límite de placa convergente (p. 218)

resumen: presentación del contenido de un texto en forma condensada o reducido a sus puntos principales (p. 150)

suspender: evitar que algo se caiga o se hunda mediante un apoyo invisible; por ejemplo, el polvo del aire (p. 440)

✹ **technology:** the application of science, especially for industrial or commercial use (p. 596)

temperature: measure of average kinetic energy of the particles in a material (p. 140)

✹ **temporarily:** lasting for a limited time (p. 403)

tephra (TEH fruh): fragment of rock or lava dropped from the air during an explosive volcanic eruption ranges in size from volcanic ash to huge boulders (p. 304)

tertiary consumer: organism, such as a predator, at the top of the energy pyramid (p. 564)

thermal energy: energy that moves from one place to another because of differences in temperature (p. 129)

thermal expansion: increase in the volume of a substance when the temperature increases (p. 140)

tecnología: aplicación de la ciencia, especialmente para uso industrial o comercial (p. 596)

temperatura: medida de la energía cinética promedio de las partículas de un material (p. 140)

temporalmente: duración por un tiempo limitado (p. 403)

materia piroclástica: fragmento de roca o lava que cae del aire durante una erupción volcánica explosiva, y cuyo tamaño oscila entre ceniza volcánica a inmensos peñascos (p. 304)

consumidor terciario: organismo, como un depredador, que está en la cima de la pirámide energética (p. 564)

energía térmica: energía que se mueve de un lugar a otro debido a diferencias de temperatura (p. 129)

expansión térmica: aumento del volumen de una sustancia cuando aumenta la temperatura (p. 140)

Glossary/Glosario

topographic map: a map that uses lines of equal elevation to show the shape of Earth's surface (p. 54)

✷ **traditionally:** an established or customary pattern or way of doing something (p. 469)

transfer: to move something from one place to another (p. 134)

transform plate boundary: boundary formed where two lithospheric plates scrape sideways past each another (p. 220)

✷ **transport:** to carry from one place to another (p. 80)

troposphere: region of the atmosphere that extends from Earth's surface to a height of about 8–15 km (p. 383)

tsunami: seismic sea wave caused by an earthquake and is highly destructive when it crashes on shore (p. 272)

mapa topográfico: mapa que usa líneas de igual elevación para mostrar la forma de la superficie terrestre (p. 54)

tradicionalmente: patrón o forma establecida o habitual de hacer algo (p. 469)

transferir: mover algo de un lugar a otro (p. 134)

límite de placa transformante: límite que se forma cuando dos placas litosféricas se rozan unas con otras (p. 220)

transportar: llevar de un lugar a otro (p. 80)

troposfera: región de la atmósfera que se extiende desde la superficie terrestre hasta una altura aproximada de 8 a 15 km (p. 383)

tsunami: onda sísmica marina ocasionada por un sismo, la cual es altamente destructiva cuando rompe en la costa (p. 272)

✷ **ultimate:** farthest, final, in the end (p. 344)

ultraviolet wave: a portion of the electromagnetic spectrum with short wavelengths just beyond the range of visible light (p. 385)

updraft: a rising column of air (p. 403)

uplift: any process that moves Earth's surface to a higher elevation (p. 79)

último: el más lejano, final, extremo (p. 344)

onda ultravioleta: parte del espectro electromagnético con longitudes de onda cortas, apenas más allá del rango de la luz visible, que algunas veces se perciben como calor (p. 385)

corriente ascendente: columna de aire ascendente (p. 403)

levantamiento: cualquier proceso que eleve la superficie terrestre (p. 79)

valley breeze: breeze that blows upward from the valley along the mountain slopes (p. 493)

variable: any factor in a scientific investigation that can have more than one value (p. 29)

vent: opening of a volcano where magma is forced up and flows out onto Earth's surface (p. 295)

viscosity (vihs KAH suh tee): a material's resistance to flow; the inability to flow easily (p. 302)

✷ **visible:** able to be seen (p. 384)

volcanic ash: fine-grained tephra or dust from a volcanic eruption; consisting of tiny, sharp mineral and glass-like particles (p. 313)

volcano: land or underwater feature that forms when magma reaches Earth's surface (p. 294)

✷ **volume:** the amount of space occupied by an object or a region of space (p. 140)

brisa de valle: brisa que sopla en dirección ascendente desde el valle a lo largo de las pendientes de las montañas (p. 493)

variable: factor de una investigación científica que puede tener más de un valor (p. 29)

respiradero: abertura de un volcán donde el magma es forzado a subir y fluye hacia la superficie terrestre (p. 295)

viscosidad: resistencia de un material a fluir; incapacidad para fluir con facilidad (p. 302)

visible: que puede verse (p. 384)

ceniza volcánica: material piroclástico de grano fino o polvo de una erupción volcánica formado por partículas minúsculas de mineral similares al vidrio (p. 313)

volcán: rasgo terrestre o subterráneo que se forma cuando el magma alcanza la superficie terrestre (p. 294)

volumen: cantidad de espacio que ocupa un objeto, o una región de un espacio (p. 140)

warm front: the movement of a lighter, warmer air mass over a heavier, colder air mass (p. 476)

water cycle: cycle in which water constantly moves between the hydrosphere and the atmosphere (p. 472)

wave: a disturbance in a material that transfers energy without transferring matter (p. 132)

weather: the atmospheric conditions, along with short term changes, of a certain place at a certain time (p. 468)

weathering: the process that breaks down and changes rocks on Earth's surface (p. 334)

wind: air that is in motion relative to Earth's surface (p. 401)

work: transfer of energy when a push or pull makes an object (p. 132)

frente cálido: movimiento de una masa de aire liviana y cálida hacia una más pesada y fría (p. 476)

ciclo hidrológico: ciclo mediante el cual el agua se mueve constantemente entre la hidrosfera y la atmósfera (p. 472)

onda: perturbación de un material que transfiere energía sin transferir materia (p. 132)

tiempo: las condiciones atmosféricas, así como los cambios a corto plazo de esas condiciones en un lugar y en un momento determinados (p. 468)

desgaste: proceso que fisiona y transforma las rocas de la superficie terrestre (p. 334)

viento: aire que está en movimiento con respecto a la superficie terrestre (p. 401)

trabajo: transferencia de energía al empujar un objeto o tirar de él (p. 132)

Index

Italic numbers = illustration/photo   **Bold numbers = vocabulary term**
lab = indicates a page on which the entry is used in a lab
act = indicates a page on which the entry is used in an activity

Index

Index

Index

Index

Index

Credits

Magnification Key: Magnifications listed are the magnifications at which images were originally photographed.

LM–Light Microscope

SEM–Scanning Electron Microscope

TEM–Transmission Electron Microscope

Acknowledgments: Glencoe would like to acknowledge the artists and agencies who participated in illustrating this program: Argosy Publishing; Articulate Graphics; Craig Attebery represented by Frank and Jeff Lavaty; Emily Damstra; Gary Hincks; Precision Graphics; Michael Rothman; Zoobotanica.

Photo Credits

List of Abbreviations:
AA=Animals Animals; CBS=Carolina Biological Supply; CB=Corbis-Bettmann; CP=Color-Pic; CMSP=Custom Medical Stock Photo; DRK=DRK Photo; ES=Earth Scenes; FPG=FPG International; GH=Grant Heilman Photography; LI=Liaison International; MP=Minden Pictures; OSF=Oxford Scientific Films; PA=Peter Arnold, Inc.; PR=Photo Researchers; PT=Phototake, NYC; SPL=Science Photo Library; SS=Science Source; TSM=The Stock Market; TSA=Tom Stack & Associates; TSI=Tony Stone Images; VU=Visuals Unlimited

Cover World Travel Images/Alamy Images; vii NASA/Goddard Space Flight Center; viii NASA/SPL/Photo Researchers; viiii Jim Corwin/Index Stock; x Jeff Greenberg/Omni-Photo Communications; xi FogStock LLC/Index Stock Imagery; xii CORBIS; xix CORBIS; xxv Chuck Place/Chuck Place Photography; **2** (bkgd t)Getty Images, (b)Cheryl Ravelo/Reuters/CORBIS; **4** (t)Science VU/USGS/Visuals Unlimited, (c)CORBIS, (b)Lawrence Migdale/SPL/Photo Researchers; **5** Paul Glendell/Alamy Images; **7** (t)Matt Meadows, (b)Amos Morgan/Getty Images; **8** (t)Steve Cole/Getty Images, (b)Andrew Lambert Photography/Photo Researchers; **9** (t)Matt Meadows, (b) Horizons Companies; **10 11** Matt Meadows; **12** (t)Comstock Images/Alamy Images, (b)Matt Meadows; **14** (t)Cordelia Molloy/SPL/Photo Researchers, (b)Matt Meadows; **15** (t)Robert Maust/Photo Agora, (b)NOAA; **16** (t)Breck P. Kent, (b)Jeff J. Daly/Visuals Unlimited; **18** (t)Martin Harvey/CORBIS, (b)Chie Nishio/Omni Photo Communications; **20** (tl tr)Icon Images, (b)Photographer's Choice/Getty Images; **22** (t) CORBIS, (r) Getty Images; **23** Michael Newman/PhotoEdit; **28** (l) Dennis MacDonald/Alamy Images, (r) Getty Images; **31** Brand X Pictures/Alamy Images; **34** John Russell/AFP/Getty Images; **37** Robert Butler; **38** (t)Barry Runk/Grant Heilman Photography, (b)Brian Atwater; **40** (t)Historical Picture Archive/CORBIS, (b)USGS; **41** (t)Doug Wilson/CORBIS, (b)Erik Fraser/The Eureka Times-Standard/AP/Wide World Photos; **42 43** Chris Collins/CORBIS; **43** NASA; **44** WorldSat International; **45** Horizons Companies; **54** Steve Smith/SuperStock/Alamy Images; **57** Chuck Place/Chuck Place Photography; **64 65** Horizons Companies; **66** (t)Roger Ressmeyer/CORBIS, (b)Detlev Van Ravenswaay/Photo Researchers; **67** MPI/Getty Images; **74** Martin Bond/Photo Researchers; **75** Matt Meadows; **78 79** Galen Rowell/CORBIS; **81** (t)Adam Jones/Visuals Unlimited, (b)Jim Corwin/Index Stock; **83** (t)Debra Ferguson, (b)Sam Camp/www.campphoto.com; **84** Krafft/Explorer/Science Source/Photo Researchers; **86** Aaron Haupt; **87** (l)Geolite/

http://www.geolite.com, (r)Roberto de Gugliemo/Photo Researchers; **88** (t)Charles D. Winters/Photo Researchers, (bl)Francois Gohier/Photo Researchers, (bcl)David Young-Wolff/PhotoEdit, (bcr)Chatham Tom Chatham/Created Gems, (br)Charles D. Winters/Photo Researchers; **89** (l)Paul Silverman/Fundamental Photographs NYC, (r)Dane A. Penland/Smithsonian Institute; **90** (t)Matt Meadows, (c)Paul Silverman/Fundamental Photographs NYC, (b)Mark A. Schneider/Visuals Unlimited; **91** (l)Matt Meadows, (r)R. Weller/Cochise College; **92** (l)Breck P. Kent, (r)Tim Courlas; **94** (l to r)Charles D. Winters/Photo Researchers, Traudel Sachs/PhotoTake NYC, Wayne Scherr/Photo Researchers, Thomas Hunn Co./Visuals Unlimited, Lawrence Lawry/Photo Researchers; **96** (l)George Bernard/Photo Researchers, (r)Carolina Biological Supply Company/PhotoTake NYC; **97** (l)Martin Miller, (tr)Steve Kaufman/CORBIS, (cr)Galen Rowell/Mountain Light, (br)David Muench/CORBIS; **98** (tl)Robert Folz/Visuals Unlimited, (tr)Brent Turner/BLT Productions, (bl)Larry Stepanowicz/Fundamental Photographs, (br)Breck P. Kent/Earth Scenes; **99** Bruce Burkhardt/CORBIS; **100** (l), George Bernard/Photo Researchers; (c) Keith Kent/Peter Arnold, Inc., (tr) Ralph Lee Hopkins/Photo Researchers, (cr) The Natural History Museum, London, (br) David Muench/CORBIS; **101** Paul Silverman/Fundamental Photographs; **102 104** NASA/Goddard Space Flight Center; **106** The Natural History Museum, London; **110** Horizons Companies; **112** (t) Bill Owen/California Department of Transportation, (b) NASA/EADS SPACE; **113** (t) John Cancalosi/Nature Picture Library, (b) Personnel of the NOAA Ship RAINIER; **116** Ken Lucas/Visuals Unlimited; **117** Tim Courlas; **120** Tom J. Ulrich/Visuals Unlimited; **121** Matt Meadows; **124** Runk/Schoenberger/Grant Heilman Photography; **125** (tl) The Extinction of the Dinosaurs by Eleanor Kish, reproduced by permission of the Canadian Museum of Nature, Ottawa, Canada, (tr) W. Cody/CORBIS, (cr) William Swartz/Index Stock/PictureQuest (bl) Duomo/CORBIS, (br) NASA/Goddard Space Flight Center; **127** (l) Jenni Carter/Biennale of Sydney, (c) Reuters/Will Burgess/Landov, (r) Matt Meadows; **129** CORBIS; **130** Jenni Carter/Biennale of Sydney; **131** (l) Odd Andersen/AFP/Getty Images, (r) Mark Cooper/CORBIS; **132** Matt Meadows; **133** Wolfgang Kaehler Photography; **134** (t) CORBIS, (bl) CORBIS, (bc) Blend Images/Getty Images, (br) Matt Meadows; **135** Mark Burnett; **136** (t) Matt Meadows, (b) Tony Freeman/PhotoEdit; **139** Tony Freeman/PhotoEdit; **140** Matt Meadows; **142** Condé Nast Archive/CORBIS; **143** Matt Meadows; **145** Theo Allofs/zefa/CORBIS; **146** Matt Meadows; **147** (l) Comstock Images/PictureQuest, (r) SPL/Photo Researchers; **148** (l) Matt Meadows, (r)CORBIS; **149** Matt Meadows; **150** (l) Jeff Greenberg/PhotoEdit, (tr) Theo Allofs/Zefa/CORBIS, (cr) Matt Meadows, (br) CORBIS; **152** Matt Meadows; **154** (t) CORBIS, (b) James Sugar/Black Star; **155** (t) Peter Walton/Index Stock, (b) Svenja-Foto/zefa/CORBIS; **158** Matt Meadows; **159** Wolfgang Kaehler Photography; **162** Simon Fraser/Photo Researchers; **163** Horizons Companies; **166** Christie's Images/CORBIS; **172** Horizons Companies; **174** USGS; **177** (l) Ralph White/CORBIS, (c) Davis Meltzer, (r) Ralph White/CORBIS;**178** NASA/Goddard Space Flight Center; **179** Astrium; **180** Lowell Georgia/CORBIS; **183** Gary Hincks/Photo Researchers; **187** (l) The Natural History Museum, London, (r) Marli Miller/Visuals Unlimited; **191** (t) NASA/Goddard Space Flight Center, (b) NASA/Photo Researchers; **192** (l), George Bernard/Photo Researchers; (c) Keith Kent/Peter Arnold, Inc., (tr) Ralph Lee Hopkins/Photo Researchers, (cr) The Natural History Museum, London, (br) David Muench/CORBIS; **194** Matt Meadows; **196** Horizons Companies; **197** Gary Hincks/Photo Researchers; **199** (t) Art Resource, NY, (b) ODP/TAMU Science Operator; **206** Roger Ressmeyer/CORBIS; **207** Matt Meadows; **210** NASA/JPL-Caltech; **212** Garry Hayes/Modesto Junior College; **213** (t) CORBIS, (b) Roger Ressmeyer/CORBIS; **214** Horizons Companies; **215** Dr Ken MacDonald/SPL/Photo Researchers; **216** Altitude/Peter Arnold, Inc.; **217** (inset) Dorling Kindersley, (bkgd) National Geographic Maps; **227 228** Horizons Companies; **230** (t) David Parker/SPL/Photo Researchers, (b) J.C. Tinsley/USGS; **231** (t) NASA/JPL/NIMA, (b) Gary Kazanjian/AP/Wide World Photo; **238** StudiOhio; **240** Ted Streshinsky/CORBIS; **240 241** (bkgd) Roger Ressmeyer/CORBIS; **241** (t) Roger Ressmeyer/CORBIS, (b) Peter Dejong/AP/Wide World Photos; **242** Pacific Press Service/Alamy Images; **243** Horizons Companies; **246** USGS; **250** Horizons Companies; **252** FireFly Productions/CORBIS; **256** Horizons Companies; **257** NASA/Goddard Space Flight Center; **260** NASA/JRC/IPSC/SES/A. Annunziato/C. Best; **264** (l) USGS, (r) Faisal Mahmood/Reuters/CORBIS; **265** AP/Wide World Photos; **266** Khalid Mosalam; **268** Matt Meadows; **270** USGS; **271** (t) George Hall/CORBIS, (c) Noburu Hashimoto/CORBIS, (b)David J. Cross/Peter Arnold, Inc.; **272** DigitalGlobe; **275** (l) David Sanger Photography/Alamy Images, (r) C.E. Meyer/USGS; **276** (t) Bettmann/CORBIS, (c) Craig Lovell/CORBIS, (b) James L. Stanfield/National Geographic Image Collection; **277** Roger Ressmeyer/CORBIS; **279** Matt Meadows; **280** Horizons Companies; **281** Tim Fuller; **282** (t) Alan Decker/Jacobs School of Engineering/UCSD Publications Office, (b) Strong Motion Instrumentation Program/California Geological Survey; **283** (t) USGS, (b) Wolfgang Langenstrassen/Epa/CORBIS; **289** David J. Cross/Peter Arnold, Inc.; **290** Roger Ressmeyer/CORBIS; **291** Horizons Companies; **294** Jeremy Bishop/SPL/Photo Researchers; **298** (t) Dr. Ken MacDonald/SPL/Photo Researchers, (b) NASA/SPL/Photo Researchers; **300** Horizons Companies; **301** Charles Rogers/Visuals Unlimited; **302** (t) USGS, (b) CORBIS; **303** (l) Stephen & Donna O'Meara/SPL/Photo Researchers, (r) Bettmann/CORBIS; **305** (t) Robert A. Jensen/USDA Forest Service, (c) Charles Rogers/Visuals Unlimited, (b) Steve Kaufman/Accent Alaska; **306** (l) Robert Hessler/Planet Earth Pictures, (tr) Krafft/HOA-QUI/Photo Researchers, (br) Paul Chesley, (bkgd) API/Explorer/Photo Researchers; **307** (tl) USGS, (tr) Christopher Woltemade, (bl) Anthony Dunn/Ambient Images/California Stock Photo, (br) Breck P. Kent/Earth Scenes; **308** (t) Rob & Ann Simpson/Visuals Unlimited, (b) Francois Gohier/Photo Researchers; **309** (t) Ernesto Burciaga/Omni-Photo Communications, (c) David McNew/Newsmakers/Getty Images, (b) Philip Rosenberg/Pacific Stock; **310** Randy Wells/Getty Images; **313** (l) Douglas Kirkland/CORBIS, (c) D.E. Wieprecht/USGS, (r) RMF/Visuals Unlimited; **314** (l) Bettmann/CORBIS, (r) Jacques Langevin/CORBIS; **315** (l) Jess Smith/Grant Heilman Photography, (c) Bernhard Edmaier/Photo Researchers, (r) Eric Kroll/Omni-Photo Communications; **316** (t) C. Neal/USGS, (b) USGS/Photo Researchers; **318 320** Horizons Companies; **322** (t) USGS/Cascades Volcano Observatory, (b) Mark Lennihan/AP/Wide World Photos; **323** (t) NASA/Photo Researchers, (b) Palmi Gudmundsson/Getty Images; **329** Kevin West/AP/Wide World Photos; **330** Paul A. Souders/CORBIS; **331** Horizons Companies; **334** (l) Jacob Halaska/Index Stock Imagery, (r) North Wind Picture Archives; **335** Jeff Greenberg/

Omni-Photo Communications; **336** (t) Karlene and Lowell Schwartz, (c) Marli Miller/Visuals Unlimited, (b) Jack Parsons/Omni-Photo Communications; **337** (l) Steve McCutcheon/Visuals Unlimited, (r) Zandria Muench Beraldo/CORBIS; **338** Matt Meadows; **339** (t) James D. Balog, (c) Martin Miller, (b) Kenneth H. Thomas/Photo Researchers, (bkgd) Stephen R. Wagner; **340** Deborah Kopp/Visuals Unlimited; **342** Bruce Chambers/Orange County Register/CORBIS; **343** (l) Jim Sugar/CORBIS, (c) Kevork Djansezian/AP/Wide World Photos, (r) Stouffer Productions/Animals Animals; **344** Dr. Marli Miller/Visuals Unlimited; **346** (t) JimWark@AirPhotoNA.com, (b) Jim Wark/Peter Arnold, Inc.; **347** (t) Norbert Rosing/Earth Scenes, (c) Bob Krist/CORBIS, (b) Brian Milne/Earth Scenes; **348** Justin Sullivan/Getty Images; **349** Marli Miller/University of Oregon/Visuals Unlimited; **350** Marli Miller/Visuals Unlimited; **352** (t) AP/Wide World Photos, (b) Grant Heilman/Grant Heilman Photography; **356** (t) Ethel Davies/Imagestate, (c) Fritz Polking/Visuals Unlimited, (b) Mark Gibson/Ambient Images/California Stock Photo; **357** (t) Jim Brandenburg/Minden Pictures, (b) Mark Gibson; **358** (t) Dr. Marli Miller/Visuals Unlimited, (b) Gary Crabbe/Alamy Images; **359** (t) Raymond Gehman/CORBIS, (b) NASA/CORBIS; **360** (tl tr) USGS, (b) Jim Goyjer/CORBIS; **366** (t) Novosti Press Agency/SPL/Photo Researchers, (b) Florida Images/Alamy Images; **367** (t) Cliff Hunt/Alamy Images, (b) Andrew Mcglenaghan/SPL/Photo Researchers; **370** (t) Dr. Marli Miller/Visuals Unlimited, (b) Deborah Kopp/Visuals Unlimited; **372** Kevork Djansezian/AP/Wide World Photos; **373** Andrew Mcglenaghan/SPL/Photo Researchers; **374** (tl bl br) StudiOhio, (tr)Doug Martin; **376** (t) Alaska Stock, (b) Jacob Halaska/Index Stock Imagery; **376/377** (bkgd) Dale O'Dell/CORBIS; **377** (t) Kevork Djansezian/AP/Wide World Photos, (b) Dave Martin/AP/Wide World Photos; **378** Shery Larson; **379** Matt Meadows; **385** (t) Photograph kindly provided by Dr. Glenn Tattersall/Brock University, (c) Leonard Lessin/Photo Researchers, (b) Leonard Lessin/Photo Researchers; **389** (t) Inga Spence/Visuals Unlimited, (b) Yann Arthus-Bertrand/CORBIS; **394** (t) Chuck Place Photography, (b) Barrie Fanton/Omni-Photo Communications; **395** Matt Meadows; **396** (t) CORBIS, (b) Ted Spiegel/CORBIS; **397** Matt Meadows; **403** (l) Yva Momatiuk/John Eastcott/Minden Pictures, (r)Dennis Macdonald/PhotoEdit; **407** (tl) Phil Schermeister/CORBIS, (tr) Gene Moore/PhotoTake NYC/PictureQuest, (c) Stephen R. Wagner, (bl) Joel W. Rogers, (br) Kevin Schafer/CORBIS; **410** Matt Meadows; **412** (t) University Corporation for Atmospheric Research, (b) Roger Ressmeyer/CORBIS; **413** (t) S.Feval/Le Matin/CORBIS, (b) NASA/GSFC/LaRC/JPL, MISR Team; **418** Yva Momatiuk/John Eastcott/Minden Pictures; **419** Jeremy Hoare/Getty Images; **420** Chuck Place/Chuck Place Photography; **421** Horizons Companies; **425** The Floor of the Oceans by Bruce C. Heezen and Marie Tharp, 1977 by Marie Tharp.; Reproduced by permission of Marie Tharp; **431** NASA/SPL/Photo Researchers; **432** (t) CORBIS, (b) Chuck Place/Chuck Place Photography; **433** Horizons Companies; **435** Raven/Explorer/Photo Researchers; **439** (t) Nik Wheeler/CORBIS, (c b) Joseph Melanson/Aero Photo Inc of www.skypic.com; **441** Sexto Sol/Getty Images; **442** (t) Darrell Gulin/CORBIS, (c) ADAM HART-DAVIS/SPL/Photo Researchers, (b) B. Runk/S. Schoenburger/Grant Heilman Photography; **443** Horizons Companies; **450** Brandon Cole Marine Photography/Alamy Images; **451** (l to r, t to b) Lloyd K. Townsend, Fred Whitehead/Animals Animals, Hal Beral/Visuals Unlimited, Andrew J.

Credits